Lecture Notes in Computer Scien

Edited by G. Goos, J. Hartmanis, and J. van L

Springer

Berlin
Heidelberg
New York
Barcelona
Hong Kong
London
Milan
Paris
Tokyo

Tiziana Margaria   Tom Melham (Eds.)

Correct Hardware Design and Verification Methods

11th IFIP WG 10.5 Advanced Research
Working Conference, CHARME 2001
Livingston, Scotland, UK, September 4-7, 2001
Proceedings

Springer

Series Editors

Gerhard Goos, Karlsruhe University, Germany
Juris Hartmanis, Cornell University, NY, USA
Jan van Leeuwen, Utrecht University, The Netherlands

Volume Editors

Tiziana Margaria
Universit t Dortmund, Lehrstuhl f r Programmiersysteme
Baroper Str. 301, 44221 Dortmund, Germany
E-mail: tiziana@ls5.cs.uni-dortmund.de

Tom Melham
University of Glasgow, Department of Computing Science
17 Lilybank Gardens, Glasgow G12 8QQ, UK
E-mail: tfm@dcs.gla.ac.uk

Cataloging-in-Publication Data applied for

Die Deutsche Bibliothek - CIP-Einheitsaufnahme

Correct hardware design and verification methods : 11th IFIP WG 10.5
advanced research working conference ; proceedings / CHARME 2001,
Livingston, Scotland, UK, September 4 - 7, 2001. Tiziana Margaria ; Tom
Melham (ed.). - Berlin ; Heidelberg ; New York ; Barcelona ; Hong Kong ;
London ; Milan ; Paris ; Tokyo : Springer, 2001
    (Lecture notes in computer science ; Vol. 2144)
    ISBN 3-540-42541-1

CR Subject Classification (1998): B, F.3.1, D.2.4, F.4.1, I.2.3, J.6

ISSN 0302-9743
ISBN 3-540-42541-1 Springer-Verlag Berlin Heidelberg New York

Springer-Verlag Berlin Heidelberg New York
a member of BertelsmannSpringer Science+Business Media GmbH

http://www.springer.de

' Springer-Verlag Berlin Heidelberg 2001
Printed in Germany

Typesetting: Camera-ready by author, data conversion by PTP-Berlin, Stefan Sossna
Printed on acid-free paper   SPIN: 10840208   06/3142   5 4 3 2 1 0

Preface

This volume contains the proceedings of CHARME 2001, the *Eleventh Advanced Research Working Conference on Correct Hardware Design and Verification Methods*. CHARME 2001 is the 11th in a series of working conferences devoted to the development and use of leading-edge formal techniques and tools for the design and verification of hardware and hardware-like systems.

Previous events in the 'CHARME' series were held in Bad Herrenalb (1999), Montreal (1997), Frankfurt (1995), Arles (1993), and Torino (1991). This series of meetings has been organized in cooperation with IFIP WG 10.5 and WG 10.2. Prior meetings, stretching back to the earliest days of formal hardware verification, were held under various names in Miami (1990), Leuven (1989), Glasgow (1988), Grenoble (1986), Edinburgh (1985), and Darmstadt (1984). The convention is now well-established whereby the European CHARME conference alternates with its biennial counterpart, the *International Conference on Formal Methods in Computer-Aided Design* (FMCAD), which is held on even-numbered years in the USA.

The conference took place during 4–7 September 2001 at the Institute for System Level Integration in Livingston, Scotland. It was co-hosted by the Institute and the Department of Computing Science of Glasgow University and co-sponsored by the IFIP TC10/WG10.5 Working Group on Design and Engineering of Electronic Systems. CHARME 2001 also included a scientific session and social program held jointly with the *14th International Conference on Theorem Proving in Higher Order Logics* (TPHOLs), which was co-located in nearby Edinburgh.

The CHARME 2001 scientific program comprised

- two **Invited Lectures**
    - by Steven D. Johnson (joint with TPHOLs) presenting *View from the Fringe of the Fringe,* and
    - by Alan Mycroft on *Hardware Synthesis Using SAFL and Application to Processor Design,*

as well as

- **Regular Sessions**, featuring 24 papers selected out of 56 submissions, ranging from foundational contributions to tool presentations,
- **Short Presentations**, featuring 8 short contributions accompanied by a poster presentation, and
- A morning of **Tutorials** aimed at industrial and academic interchange.

The conference, of course, also included informal tool demonstrations, not announced in the official program.

The paper *Applications of Hierarchical Verification in Model Checking*, by Robert Beers, Rajnish Ghughal, and Mark Aagaard, was presented at FMCAD

2000 but omitted inadvertently from the published proceedings of that conference. It is included in the present volume, since it contains appropriate content and the two conferences share a large common audience.

Warm thanks are due to the program committee and to all the referees for their assistance in selecting the papers, to Lorraine Fraser and to the ISLI Institute Director Steve Beaumont, to the TPHOLs organizers, and to the entire local organization team for their engaged and impressive effort on making the conference a smooth-running event.

The organizers are very grateful to Esterel Technologies, Intel, and Siemens ICN, whose sponsorship made a significant contribution to financing the event. Glasgow University provided financial support for publicity.

Warm recognition is due to the technical support team. Matthias Weiß at the University of Dortmund together with Ben Lindner and Martin Karusseit of METAFrame Technologies provided invaluable assistance to all the people involved in the online service during the crucial months. Finally, we are deeply indebted to Claudia Herbers for her first class support in the preparation of this volume.

September 2001   Tiziana Margaria and Tom Melham

Conference Chair

Tom Melham (University of Glasgow, Scotland)

Program Chair

Tiziana Margaria (University of Dortmund
and METAFrame Technologies GmbH, Germany)

Local Arrangements Chair

Andrew Ireland (Heriot-Watt University, Edinburgh, Scotland)

Publicity Chair

Simon Gay (University of Glasgow, Scotland)

Program Committee

Luca de Alfaro	(University of Berkeley, USA)
Dominique Borrione	(University of Grenoble, France)
Eduard Cerny	(Nortel Networks, USA)
Luc Claesen	(IMEC, Belgium)
Hans Eveking	(University of Darmstadt, Germany)
Kathi Fisler	(Worcester Polytechnic Institute, USA)
Daniel Geist	(IBM, Haifa, Israel)
Ganesh Gopalakrishnan	(University of Utah, USA)
Mark Greenstreet	(University of British Columbia, Canada, and SUN Microsystems, Palo Alto, USA)
Nicolas Halbwachs	(VERIMAG, Grenoble, France)
Hardi Hungar	(METAFrame Research, Dortmund, Germany)
Warren Hunt	(IBM and University of Texas, Austin, USA)
Steve Johnson	(Indiana University, USA)
Thomas Kropf	(Robert Bosch GmbH, Reutlingen, and University of Tübingen, Germany)
John Launchbury	(Oregon Graduate Institute, USA)
John O'Leary	(Intel, USA)
Laurence Pierre	(University of Marseille, France)
Mary Sheeran	(Chalmers University and Prover Technology, Göteborg, Sweden)

Local Organizers

Steve Beaumont (ISLI)
Lorriane Fraser (ISLI)
Tracy Gaffney   (ISLI)
Fiona Hunter   (ISLI)
Ken Baird   (University of Edinburgh)
Deirdre Burke   (University of Edinburgh)
Jennie Douglas (University of Edinburgh)

TPHOLs Organizing Committee

Richard Boulton (Conference Chair)
Louise Dennis   (Local Arrangements Co-chair)
Jacques Fleuriot (Local Arrangements Co-chair)
Simon Gay   (Publicity Chair)
Paul Jackson   (Programme Chair)

Reviewers

Will Adams	Cindy Eisner	Markus Müller-Olm
Evgeni Asarin	Martin Fränzle	Paritosh Pandya
Ritwik Bhattacharya	Mike Gordon	Marie-Laure Potet
Per Bjesse	Susanne Graf	Jun Sawada
Marius Bozga	Anna Gringauze	Ali Sezgin
Annette Bunker	Alan Hartman	Irit Shitsevalov
Prosenjit Chatterjee	Ravi Hosabettu	Ken Stevens
Koen Claessen	Damir Jamsek	Emmanuel Zarpas
Emil Dumitrescu	Michael Jones	
Niklas Een	Laurent Mounier	

Table of Contents

Algorithm Verification

Duration Calculus

View from the Fringe of the Fringe

(Extended Summary)

Steven D. Johnson*

Indiana University Computer Science Department, sjohnson@cs.indiana.edu

Abstract. Formal analysis remains outside the mainstream of system design practice. *Interactive* methods and tools are regarded by some to be on the margin of useful research in this area. Although it may seem relatively academic to some, it is vital that this the so-called "theorem proving approach" continue to be as vigorously explored as approaches favoring highly automated reasoning. *Design derivation*, a term for design formalisms based on transformations and equivalence, represents just a small twig on the theorem-proving branch of formal system analysis. A perspective on current trends is presented from this remote outpost, including a review of the author's work since the early 1980s.

In memory of Dexter Jerry Johnson, August 12, 1918 – June 23, 2001

1 On Behalf of Interactive Reasoning

"Formal methods for systems," the application of automated symbolic reasoning to system design and analysis, remains well outside the mainstream of engineering practice, even though it is philosophically central to the science of computing (for instance, eight (by my count) of the forty-one ACM Turing Award recipients are formal methods luminaries explicitly recognized for this aspect of their work.)

One could say, on the one hand, that formal methods research in software has significantly influenced practice. Type checking, structured and object-oriented programming languages, and other advances have been deeply assimilated. So, too, have equivalence checking and hardware description languages in hardware design practice. On the other hand, it is hard to tell whether, or to what extent, these tools have improved design methodology. Do more programmers use loop invariants? Does object oriented syntax improve the way engineers organize their thoughts, or does it merely impose structure, and if the latter, does that particular structure skew other aspects, such as communication and architecture?

Let us focus on verification. In the past twenty years, the question of *whether* formal analysis is meaningful or practical has turned into a question of *how and where* they can be most profitably deployed. The digital hardware industry is

---

* Work described in this paper was supported by the National Science Foundation under grants MIP8707067, MIP8921842, MIP9208745, and MIP9610358.

T. Margaria and T. Melham (Eds.): CHARME 2001, LNCS 2144, pp. 1–12, 2001.

the vanguard of this movement, the economic benefit having been established. These inroads are changing design processes, and consequently will change the perspective on and management of system design.

In many cases, perspectives are changing much faster than the rhetoric. In the early 1980s I was told by experts that equivalence checking is hopeless, that hardware description languages are unbearably inefficent, that programs are tantamount to proofs, that formal analysis (mine, in particular) bears no relationship to actual design, etc. One cannot refute these assertions; they remain true in certain contexts, both scientific and practical[1].

It is a sign of progress that one seldom hears these particular assertions at formal-methods meetings (as I did each of the examples above). The debate about practicality has become much more sophisticated. The old taunt, "Does it scale?" persists, but as often as not refers not to the capacity of the tools but to the population of engineers will use them. Even then, it is now more widely and credibly acknowledged that verification requires a fundamentally different skill set than designing.

I hope that these developments signal an end to the "algorithmic versus interactive" debate, or at least moves it to a higher level. It is certainly valid to say that any avenue into formal design analysis adopts a balance between high-level and incremental tools, and that more practical explorations will emphasize greater automation. It is equally true that successful *methodologies* will differ not in the degree of interaction, but in the nature of that interaction.

What constitutes a tolerable degree of interaction is principally a human factor, even if it is significantly influenced by the engineer's education, tool set, and design process. The nature of that interaction is another matter. This question is the crux of formalized design research. It must not be investigated from a single point of view. Speaking as an educator, it is crucial that feedback from industry distinguish between near-term practicability and far-term practicality.

In his usual clarion fashion, Hoare pinpoints the analogous distinction between top-down and bottom-up theories of programming [4]. The former supports the systematic development of correct programs but is of scant help in analyzing existing programs (i.e., debugging). The latter provides a basis for analyzing implementation properties but is not of much help in relating a program to its intended purpose. One perspective is specification oriented, and the other is implementation oriented. Until theory can rectify this dichotomy, it is the designer's task to resolve it in each instance. The same dialectic Hoare describes is present in applied research and exploratory practice.

Pnueli points out that, "... there exists no purely scientific solution [as yet] to the *system correctness problem*. We should concentrate on an engineering approach: assembling many tools and methods, ..." While acknowedging that interaction is fundamental, "One of the worst places to use ingenuity and creativity is an interactive deductive proof of verification conditions" [11].

---

[1] One of the statements was recently highlighted in a press release profiling the new CTO of a major technology company, who was reminiscing about how he had debunked the academic mythology of program verification.

I can think of places that can be at least as bad. Some examples: maneuvering a synthesizer by tweaking its input and output, performing ad hoc reductions to satisfy a model checker, reasoning abstractly about concrete data, and editing a **make** file for last-minute turnaround. To the extent that every design tool and environment dictates a style of expression or a mode of reasoning, it is a potential drain on ingenuity and creativity. Pnueli is arguing that the essence of engineering methodology lies in choosing the right tool for a given purpose. This is also precisely the goal of theorem proving. Of course, it is counterproductive to encumber this choice by an an overly restrictive tool set or abstruse notation, but so, too, is having an overwhelming, disorganized tool box.

What is called "theorem proving" in applied research today has much more to do with tool management than with proof editing. While it does require a proficiency in logic, gaining this proficiency is a minor task compared to assimilating the commands of an operating system or a CAD environment. Although theorem-proving remains far too restrictive (as I will argue in Sect. 3), interactive reasoning tools teach invaluable lessons about controling the interplay between tools and maintaining overall direction. All engineers would benefit from exploring this aspect, although not in the heat of a production schedule; that is a job for verification engineers.

This research community needs to become more expansive in its discussions of these questions, paying particular attention to avoid fading biases. With over twenty years of work, including accomplishments that could hardly have been conceived ten years ago, we can offer great deal of guidance to those who are only beginning explore practical formal methods. We might begin by neutralizing "automatic versus interactive" distinction by adopting the umbrella *automated interactive reasoning*. Since all approaches employ both, this might make it clearer that the real challenge is to effectively combine them.

2   Interlude: Are We Making Adequate Progress?

At a 1998 workshop on formal methods education [8], in the course of a discussion about mathematics in the CS/CE, Douglas Troeger made the remark, "Calculus is a formal method." I have been in a great many discussions concerning The Calculus in computer science, but I was struck for the first time by the literal truth of Troeger's statement. I thought that learning how The Calculus came into being might give me a fresh perspective on computing methodology. A colleague recommended Carl B. Boyer's *The History of the Calculus and its Conceptual Development* [2].

WARNING: Beware the musings of someone who has read just one history book.

I found it somewhat comforting that, according to Boyer's scholarly account, The Calculus was the product of no less than a millennium of active thought (elapsed time: around 2,500 years), with due acknowledgment to the Greeks, Babylonians, Egyptians, Hindus, Arabs, and Medieval scholars. However, it was with increasing dismay that, the more I read, Greek Age seemed more familiar than the Renaissance. I am fond of saying that computing has not yet seen its Newton (Liebnez if you prefer), but maybe I should be saying Archimedes (Eudoxus if you prefer). The Greeks also had, in Zeno, one of history's greatest nay-sayers. I'm not sure whether computing has seen its Zeno, although some candidates do come to mind.

The Greek Method of Exhaustion could solve integration problems, but was cumbersome and indirect; it lacked the logical foundation to deal even with primary ideas (e.g. measure and variable), much less key concepts (e.g. limit and convergence). Instead, The Method is based on the concept of approximation to to any desired degree of precision. If the Greeks had computers, that might have been the end of it.

It is fairer to say that Newton and Leibnez acquiesced to The Calculus than to say they discovered it; and they certainly didn't invent it. Invention cannot be attributed to any individual, but to a gradual evolution in thinking over several centuries. Newton and Liebnez, and perhaps others, discovered the surprising relationship between the integral and the derivative, and went on to use their discoveries. Two more centuries passed before a mathematical foundation was established to justify their methods. Throughout the history of The Calculus, formalism played both encouraging and inhibiting roles. In his conclusion, Boyer says,

> Perhaps the most manifest deterring force was the rigid insistence on the exclusion from mathematics of any idea not at the time allowing of strict logical interpretation. ... On the other hand, perhaps a more subtle, and therefore serious, hindrance to the development of the calculus was the failure, at various stages, to give to the concepts employed as concise and formal definition as was possible at the time.

While it is foolhardy to draw too many parallels between the history of the calculus of the physical world and the search for a calculus of system design (if, indeed, there is one), Boyer's comment surely applies to both. Formalism should describe, not prescribe, but it must be involved, and this applies not just to the mathematics, but also to its implementation in a computer.

I could not find in Boyer's account a clear sense of what was driving the persuit of The Calculus forward. Science and engineering, mathematics, metaphysics, theology, and politics all played a role, almost always in combination within each individual contributor. It is quite clear, of course, that The Calculus prevailed because its methods mad problems easier to solve. However, many of these problems were purely mathematical, predating any direct practical application. Once published, The Calculus became entrenched through scientific and engineering applications, despite rather strong criticism by formalists and mathematical theorists.

3   DDD – The Far Side of Theorem Proving

In Sect. 1, I spoke on behalf of interactive reasoning in automated formal verification. In this section I will describe some weaknesses in mainstream "theorem proving" research, from the standpoint of someone whose work investigates an alternate formalism.

If automated formal verification has established a beachhead in commercial integrated circuit design, interactive analysis has gained a foothold, at least. In the past few years, the successful infusion of formalized analysis in the design of commercial microprocessors includes a significant component of theorem proving. In research, automated and interactive verification are adapting to each other, and, in some cases, overlapping. The systems used in interactive reasoning, more properly called proof assistants than theorem provers, are quite advanced, having been developed over decades of applied research. Nearly all of them are based on predicate logic.

Just as most mathematical arguments involve both logic and algebra, interactive design analysis is improved by provisions for reasoning based on equivalence rather than implication. I use the term *derivation* to distinguish this mode of reasoning from *deductive* reasoning based on logical inference.

Before contrasting these two approaches, I should discuss what they have in common, for this is far more significant in the greater scheme of things. In all forms of interactive formal reasoning the object is to construct a *proof*, or sequence of commands invoking inference rules. A proof might be compared to a Unix `make` file; it is a command script that can be and is repeatedly executed. Specification and implementation expressions are byproducts of the proof building process. Derivational formalisms are sometimes described as "constructing" correct implementations, whereas deductive formalisms relate pre-existing design expressions. This is an invalid distinction for the most part. Regardless of the proof rules, proof construction is a creative process involving forward and backward reasoning, decomposition into subgoals, backtracking, specification revision, implementation adjustment, and so forth. The final form of the proof says almost nothing about how it was obtained.

In practice, proof construction is often systematic. All proof assistants automate some proof building tactics, ranging from embedded decision procedures to parameterized induction principles. Most proof assistants provide a metalanguage for composing basic patterns into proof-engineering strategies.

3.1   DDD and Its Origins

The acronym *DDD* refers either to "digital design derivation," the application of derivational reasoning to digital design, or the research tool that was developed to explore it. This line of investigation began in the early 1980s and continues to the present. Reference [6] contains a bibliography of DDD research for those interested in more details.

I originally described DDD as "functional programming applied to hardware," but eventually had to abandon this slogan. Even though any reasonable design formalism will encompass both hardware and software, equating hardware descriptions to software programs is still quite a leap for most people. Furthermore, the automation of reasoning based on functional model theory remains (so far as I can tell) out of step with the priority of language implementation over formal reasoning in the functional programming community. In any case, this work found a more receptive audience in formal methods the emerging formal verification area.

The work did begin with a language implementation, however. As a research assistant I explored the use of functional programming techniques in implementing operating systems (and later, distributed programming). I developed a programming language called *Daisy* to explore abstractions of communciation and concurrency, which were modeled as streams and represented by lazy lists.

Given the simplicity of digital devices at the time, and since I was more concerned with how processes communicated than with what they did, it was convenient to illustrate ideas about programming with with hardware models. This rather innocuous tactic immediately changed the direction of the work (although not its ultimate goal, which transcends hardware/software distinctions). For one thing, it induced a sharp distinction between description and realization, something I had trouble isolating in software studies. For another, designing hardware requires a more balanced regard for design aspects (principally architecture, behavior, coordination, and data).

A thesis emerged that each aspect of design has a distinct conceptual structure, and perhaps a distinct notation. A design formalism must be able to move among these structures, and cannot simply compose them without destroying one or more. Figure 1 gives a sense of this thesis while illustrating some of DDD's formal representations. Behavioral specification is represented by a recursive system of functions definitions. Architectureal specification is represented by a recursive system of streams. Data (abstract types and implementation relationships) and synchronization (e.g. as expressed in timing diagrams) aspects are not shown, but rather their infusion in architectural expressions.

As this is applied research, it is important to demonstrate that the formalism can be used to obtain real hardware of a reasonable complexity (given that the work is done at academic institution without engineering programs). Well over half of the research effort is spent doing case studies, the majority of that effort being to integrate DDD with a never-ending sequence of CAD tools and VLSI technologies. Two major demonstrations were the construction of a language-specific computer and comparative study of microprocessor verification:

- In a series of studies, DDD was used to derive working versions of Warren Hunt's FM8501 and FM9001 microprocessor, which Hunt verified using the *Nqthm* and *ACL2* theorem provers. The derived devices worked on first power-up, but were also extensively tested against Hunt's chip for instruction-level comformance. This study demonstrated that derivation alone is insufficient support for design; it must be integrated with a other formalisms to to deal with data and process abstractions, and ingenious design decisions.
- A language-specific computer for direct execution of compiled Scheme reflects the functional-programming roots of the research. DDD is implemented in Scheme and operates on Scheme expressions, so applying it to Scheme brings a kind of closure. DDD provides implementation verification of the

BEHAVIOR:

$$\mathsf{gcd}_b(x, y) \overset{\mathrm{df}}{=} \text{if } x = y \text{ then } x$$
$$\text{else if } x < y \text{ then } \mathsf{gcd}_b(x, y - x)$$
$$\text{else } \mathsf{gcd}_b(x - y, y)$$

ARCHITECTURE:

$$\mathsf{gcd}_a(\breve{x}, \breve{y}) \overset{\mathrm{df}}{=} x \text{ where}$$
$$x = \breve{x} : sel(x \geq y, x, y - x)$$
$$y = \breve{y} : sel(x > y, x - y, y)$$

COORDINATION: (request-acknowledge synchronization)

$$\mathsf{gcd}_c(x, y, req) \overset{\mathrm{df}}{=} (z, ack) \text{ where}$$
$$u = sel(req, x, sel(u \geq v, u - v, u))$$
$$v = sel(req, y sel(u > v, v, v - u))$$
$$z = u$$
$$ack = u = v$$

DATA: $(\langle Bit^2, +, \cdot, \ldots \rangle \text{ implementing } \langle Int, -, > \rangle)$

$$\mathsf{gcd}_d((\breve{x}_1, \breve{x}_0), (\breve{y}_1, \breve{y}_0)) \overset{\mathrm{df}}{=} (x_1, x_0) \text{ where}$$
$$x_0 = \breve{x}_0 : sel(ge?, x_0, d_0)$$
$$x_1 = \breve{x}_1 : sel(ge?, x_1, d_1)$$
$$y_0 = \breve{y}_0 : sel(gt?, d_0, y_0)$$
$$y_1 = \breve{y}_1 : sel(gt?, d_1, y_1)$$
$$gt? = x_1 \overline{y_0} + x_0 \overline{y_0}(x_1 + \overline{y_1})$$
$$ge? = (x_0 \odot y_0)(x_1 \odot y_1) + gt?$$
$$(d_1, d_0) = \mathsf{sub}(sel(ge?, (x_1, x_0), (y_1, y_0))$$
$$sel(ge?, (y_1, y_0), (x_1, x_0)))$$

Fig. 1. Four facets of system design represented as recursive equations. All but gcd_b are stream networks. The expression $\breve{v} : v$ denotes a stream whose head is the value $\breve{v}$ and whose tail is the stream v; sel is a two-way selector; the operations are boolean in gcd_d and arithmetic elsewhere. The tetrahedron symbolized the thesis that one expression cannot simultaneously reflect all design aspects, but must suppress one or more.

system components, a CPU, memory allocator, and garbage collector; but does not prove system-level noninterference properties; we will use a model checker for that. And DDD does not prove specifcation correctness; we will use a theorem prover for that. However, the CPU and garbage collector are extremely similar to the those of *VLISP* a Scheme implementation, whose compiler and run-time system were proved using rigorous but unmechanized mathematics [3]. In conjunction, these two studies represent a complete formal design whose scope of abstraction ranges from pure denotational semantics to hardware realization.

In the mid 1990s, students from the DDD group started a company to commercialize their research. They have produced two commercial products using formalized derivation, a synthesizable PCI bus interface and a configurable hardware core for Java byte code generation. The Java processor was one of the first to enter the marketplace. These accomplishments are evidence that formalized design derivation can be practical.

3.2   Toward Integrated Automated Interactive Reasoning

All of the studies just described involved the coordinated use of several reasoning and synthesis tools. I would now like to focus on some of the difficulties we have encountered doing these studies. They point to a need for improving the methodology, the tools, and also the attitudes, of interactive design analysis.

One of the more aggravating statements about DDD, appearing more that than once in published papers, is that it is not "formalized." This is an unwarrented detraction considered as a mathematical or logical statement. However, means merely that the soundness proofs of DDD's transformation rules have note been checked in some theorem prover or other. In that extremelyh parochial sense, the statement is correct but also somewhat hypocritcal. Not many theorem proving tools have proved their own inference rules; those that have are unlikley to have proven the implementation of those rules; and those that have done that, if any, have not greatly enhanced their utility by doing so.

The suggestion that all formalisms should be embedded in a single predicate logic is quaintly formalist, but the corollary that all interactive reasoning tools should be implemented in a particular theorem prover is simply wrong from both engineering and methodological standpoints.

3.3   Foundational Issues in Logic

Paul Miner explored the problems of interaction between DDD and the PVS theorem prover [9]. The problems are by no means unique to PVS; to the contrary, PVS was fairly adept for working around some of them. However, each work-around involves a semantic embedding with strategies tailored to make the embedding transparent. This may be possible on a case-by-case basis, but in combination the embeddings are certain to interfere with each other and break that transparency, leaving the user to untangle the consequences.

The most immediate problem is a lack of support for mutual recursions. This alone makes embedding DDD in PVS untenable, but that wasn't our goal anyway. We wanted instead to export verification conditions from DDD to PVS in order to verify ingenious design optimizations, especially those done on stream networks. However, streams domains are not well founded, raising a significant problem in the PVS logic. Miner was able to work around this problem, developing a *co-algebraic* theory of streams, based on their conventional representation as functions on the natural numbers, and a collection of higher order predicates and PVS strategies to mask that representation.

To illustrate the issue, consider the functions zip and map defined for streams:

$$\mathsf{zip}(a:A, b:B) \stackrel{\mathrm{df}}{=} a:b:\mathsf{zip}(A, B)$$
$$\mathsf{map}(f, a:A) \stackrel{\mathrm{df}}{=} f(a):\mathsf{map}(f, A)$$

Proving that map distributes over zip is intuitively straightforward.

$$
\begin{aligned}
&\mathsf{map}(f, \mathsf{zip}(a:A, b:B)) \\
&= \mathsf{map}(f, a:b:\mathsf{zip}(A,B)) && (\text{defn. zip}) \\
&= f(a):f(b):\mathsf{map}(f, \mathsf{zip}(A,B)) && (\text{defn. map}) \\
&\overset{\text{H}}{=} f(a):f(b):\mathsf{zip}(\mathsf{map}(f,A), \mathsf{map}(f,B)) && (\textit{induction}) \\
&= \mathsf{zip}(f(a):\mathsf{map}(f,A), f(b):\mathsf{map}(f,B)) && (\text{defn. zip})
\end{aligned}
$$

Although true, the '$\overset{\text{H}}{=}$' step is not a logically valid induction in PVS because streams are not well founded (There is no base case in zip's defintion, for example.) This inconvenience can be circumvented in a number of ways including (See [5] for one entry point into the literature on these matters).

(a) It is simply an abbreviated mathematical induction based on a representation of streams as functions over the natural numbers and extensionality.
(b) It is a recursion induction over the domain $S[\alpha] \overset{\text{df}}{=} \alpha \times S[\alpha]$, or (or a similar directed-complete partial order).
(c) It is an stylized proof that the relation:

$$
\sim = \{(\mathsf{map}(f, \mathsf{zip}(s, s')), \mathsf{zip}(\mathsf{map}(f,s), \mathsf{map}(f,s'))) \mid s \text{ and } s' \text{ streams}\}
$$

is a bisimulation. We should be using '$\sim$' rather than '$\overset{H}{=}$' in the third step.
(d) It is a co-induction.

Interpretation (a) is still fairly common, but it reminds me of the Greek Method of Exhaustion: cumbersome and unnecessarily indirect. Streams and other infinite objects are fundmental in computing theory, and should not be bound to a particular model [1]. Furthermore, the argument is a "forward" induction, not a standard one. Miner's PVS formalization is essentially (c) as justified by (a) and extensionality, but without the implicit assumption of equality. He codes strategies to perform bisimulation proofs to give the appearance of (d).

More direct mathematical foundations have recently appeared. Figure 2 shows a complete [10] logic for recursive equations with first-order substitution, due to Hurkens, McArthur, Moschovakis, Moss, and Whitney [5]. *First-order substitution* means that the recursive equations define non-parameterized objects, like streams. The generalization to first-order systems of recursive functions is in progress.

Figure 3 shows five transformation rules forming a basis for DDD architectural (stream) transformations. Introduction and identification are derived instances of the recursion rule in Fig. 2. The replacement rule extends equality to the underlying data type. Grouping is a structural rule allowing for the manipulation of multivalued functions. The collation rule, originally published as a multiplexing rule, is essentially a version of specialization.

Thus, the logic in Fig. 2 is much better suited to the DDD formalism than the PVS sequent calculus, suggesting that an "deep embedding" of DDD in PVS is not appropriate, even if it were practical to do so.

TAUTOLOGY: $\phi \vdash \phi$

EQUALITY: $\vdash A = A$, $A = B \vdash B = A$, and $A = B$, $B = C \vdash A = C$.

REPLACEMENT: $A = B \vdash E[A/x] = E[B/x]$, provided the substitutions are free.

SPECIALIZATION: $\forall x(\phi(x)) \vdash \phi(E)$, provided the substitution is free.

WEAKENING: If $\Gamma \vdash \phi$ then $\Gamma \cup \Delta \vdash \phi$.

CUT: If $\Gamma, \psi \vdash \phi$ and $\Gamma \vdash \psi$ then $\Gamma \vdash \phi$.

GENERALIZATION: If $\Gamma \vdash \phi(x)$ then $\Gamma \vdash \forall x(\phi(x))$ provided x is not free in Γ.

HEAD: $\vdash A(x_1, \ldots, x_n)$ where $\{\vec{x} = \vec{B}\}$
                  $= A(x_1$ where $\{\vec{x} = \vec{B}\}, \ldots, x_n$ where $\{\vec{x} = \vec{B}\})$

BEKIČ-SCOTT: $\vdash A$ where $\{\vec{y} = \vec{C}, \vec{x} = \vec{B}\}$
                  $= (A$ where $\{\vec{y} = \vec{C}\}$ where $\{\ldots, x_i = B_i$ where $\{\vec{y} = \vec{C}\}, \ldots\}$

FIXPOINT: $\vdash A$ where $\{x = A\} = x$ where $\{x = A\}$.

RECURSION: Given $\mathsf{A} \equiv A_0$ where $\{x_1 = A_1, \ldots, A_n = A_n\}$,
                  $\mathsf{B} \equiv B_0$ where $\{y_1 = B_1, \ldots, y_n = B_m\}$,
                  and a set, Σ of equations of the form $(x_i = y_j)$,
                  if $\Gamma, \Sigma \vdash A_0 = B_0$, and $\Gamma, \Sigma \vdash A_i = B_j$ for each $(x_i = y_j) \in \Sigma$
                  then $\Gamma \vdash \mathsf{A} = \mathsf{B}$.
                  provided No x_i/y_j occurs in B/A and no x_i or y_j occurs free in Γ.

Fig. 2. A complete logic for recursive systems with first-order substitution [5]

INTRODUCTION:
  A where $\{\vec{x} = \vec{B}\} \Leftrightarrow A$ where $\{\vec{x} = \vec{B}, y = C\}$
  Provided y is a new variable, and the r.h.s. is well formed.

IDENTIFICATION:
  A where $\{\ldots, x = B, y = C, \ldots\} \Leftrightarrow A$ where $\{\ldots, x = B[C/y], y = C, \ldots\}$.

REPLACEMENT:
  A where $\{\ldots, x = B[C/y], \ldots\} \Leftrightarrow A$ where $\{\ldots, x = B[D/y], \ldots\}$
  provided $\Delta \models C = D$.

SIGNAL GROUPING:
  A where $\{x = B, y = C, \ldots\} \Leftrightarrow A$ where $\{(x, y) = (B, C), \ldots\}$.
  with a suitable generalization of substitution.

COLLATION:
  $E[\natural/x] \Rightarrow E[B/x]$
  a one-way rewriting version of specialization, $\natural$ a generic "don't-care" constant.

Fig. 3. Adequate rules for DDD transformations on stream networks [7]. The Δ in the replacement rule stands for the underlying type theory.

Interactions between Interactive Formalisms. From the discussion above, the idea of a single logical framework serving as the umbrella environment for formalized analysis raises foundational problems. However, this is applied research. A derivation is the proof of a theorem about equivalence, and so can be stated in a logic whether proven or not. Conversely, I have already acknowledged that useful transformation system requires the support of a logic, for instance, to validate conditional transformations.

We have done a lot of work trying to integrate DDD with PVS (and some preliminary work to integrate with ACL2). We have found the software engineering hurdles to be more vexing than the semantic problems. To some extent, these problems mirror the conceit that all reasoning should be embedded in a one particular logic. Theorem provers, and DDD is no exception, are simply not designed to talk to other theorem provers as peers.

There are differences in look and feel between building a deductive proof and building a derivation. Taking PVS as an example, the two main proof windows contain the text of relevant theories (definitions and theorems), and the proof in the form of a sequent. Commands to the prover are mainly actions performed on the sequent, although the process typically induces revisions to the theories. In DDD, there are also two main windows, one the text of the expression being transformed, the other a history of the transformations that have been applied. The user focuses on the expression, not the derviation, so that interaction has the feel of a syntax directed expression editor.

Thus, DDD and PVS interactions are complementary. One can imagine a single environment supporting both, but such an environment tool would have two distinct modes of interaction. It being unlikely that a consolidated interface will appear any time soon, progress depends on the two systems talking to each other. Unfortunately, neither DDD nor PVS, nor any of the interactive tools we have used, has provisions for back-channel communication. These systems assume a single user and also impose strong restrictions on their file system.

Whether these problematic qualities are a consequence of a a formalist's pecking order, or lack of software engineering resources, they need to be repaired.

4   Conclusions

In formal methods research there is a moving boundary between what is practical and what is productive. Productivity is affected by both automation and expertise. The successful industrial users of formal verification are beginning to recognize "verification engineering" as a distinct skill set, to which design processes need to adapt. While it is important to focus on the transition from practical to productive, there is considerable danger in extrapolating from the past. It is certain that the best methods will eventually prevail and be reflected in automated interactive reasoning. Hastening that eventuality requires a healthy competition between "top-down" and "bottom-up" approaches, not only in theory, but in applied research and exploratory practice. What I have found, and have tried to illustrate in this paper, is that this dialectic does not arise in just one place but pervades every level of abstraction.

References

1. Jon Barwise and Lawrence Moss. *Vicious Circles*. CLSI Publications, Stanford, California, 1996.
2. Carl B. Boyer. *The History of the Calculus and its Conceptual Development*. Dover, New York, 1959. Republished 1949 edition.
3. Joshua D. Guttman, John D. Ramsdell, and Mitchell Wand. VLISP: a verified implementation of Scheme. *Lisp and Symbolic Computation*, 8:5–32, 1995.
4. C. A. R. Hoare. Theories of programming: Top-down and bottom-up meeting in the middle. In Jeannette M. Wing, Jim Woodcock, and Jim Davies, editors, *FM'99 - Formal Methods*. LNCS 1708.
5. A. J. C. Hurkens, Monica McArthur, Yiannis M. Moschovakis, Lawrence S. Moss, and Glen T. Whitney. The logic of recursive equations. *The Journal of Symbolic Logic*, 63(2):451–478, June 1998.
6. Steven D. Johnson. The Indiana University System Design Methods Laboratory home page. `http://www.cs.indiana.edu/hmg/hmg.html`.
7. Steven D. Johnson. Manipulating logical organization with system factorizations. In M. Leeser and G. Brown, editors, *Hardware Specification, Verification and Synthesis: Mathematical Aspects*, pages 260–281. LNCS 408, 1989.
8. Steven D. Johnson. A workshop on formal methods education: an aggregation of opinions. *International Journal on Software Tools for Technology Transfer*, 2(3):203–207, November 1999.
9. Steven D. Johnson and Paul S. Miner. Integrated reasoning support in system design: design derivation and theorem proving. In Hon F. Li and David K. Probst, editors, *Advances in Hardware Design and Verification (CHARME'97)*, pages 255–272. Chapman-Hall, 1997.
10. Lawrence S. Moss. Recursion and corecursion have the same equational logic. *Theoretical Computer Science*, to appear in 2002.
    `http://math.indiana.edu/home/moss/home.html`.
11. Amir Pnueli. These quotations are extracted from transparencies for invited talks at PODC'90, FM'99, and CAV'00. They can be found at
    `http://www.wisdom.weizmann.ac.il/~amir/invited-talks.html`.

Hardware Synthesis Using SAFL and Application to Processor Design

(Invited Talk)

Alan Mycroft[1,2] and Richard Sharp[1]

[1] Computer Laboratory, Cambridge University
New Museums Site, Pembroke Street, Cambridge CB2 3QG, UK
[2] AT&T Laboratories Cambridge
24a Trumpington Street, Cambridge CB2 1QA, UK
{am,rws26}@cl.cam.ac.uk

Abstract. We survey the work done so far in the FLaSH project (Functional Languages for Synthesising Hardware) in which the core ideas are (*i*) using a functional language SAFL to describe hardware computation; (*ii*) transforming SAFL programs using various meaning-preserving transformations to choose the area-time position (e.g. by resource duplication/sharing, specialisation, pipelining); and (*iii*) compiling the resultant program in a *resource-aware* manner (keeping the gross structure of the resulting program by a 1–1 mapping of function definitions to functional units while exploiting ease-of-analysis properties of SAFL to select an efficient mapping) into hierarchical RTL Verilog.

After this survey we consider how SAFL allows some of the design space concerning pipelining and superscalar techniques to be explored for a simple processor in the MIPS style. We also explore how ideas from partial evaluation (static and run-time data) can be used to unify the disparate approaches in Hydra/Lava/Hawk and SAFL and to allow processor specialisation.

1 Introduction

There are many formalisms which allow one to specify circuits. Commercially, the two most important are VHDL and Verilog. From the perspective of this work, they can be regarded as equivalent. They are rich languages which have multiple interpretations according to the intended use; different subsets are used according to the interpretation. It is worth summarising the interpretations/subsets here since we wish to use the terminology later:

Structural (netlist) subset: programs are seen as specifying the basic system components and their interconnecting wires.

RTL subset: extra primitives are provided to wait for events (e.g. a positive clock edge), compute expressions and transfer data between registers.

Behavioural subset: as well as the primitives available in the RTL subset, programs at the behavioural-level can exploit higher-level constructs such as sequencing, assignment and while-loops.

T. Margaria and T. Melham (Eds.): CHARME 2001, LNCS 2144, pp. 13–39, 2001.

Simulation interpretation: in this interpretation the whole language is acceptable; programs are seen as input to a discrete event simulator (useful for debugging the above forms).

One reason for the success of VHDL and Verilog has been their ability to represent circuits composed partly of structural components and partly of behavioural components (like embedding assembly code in a higher-level language). Another benefit is that the languages provide a powerful set of non-synthesisable constructs which are useful for simulation purposes (e.g. printing values to the screen).

Large investments have been made in algorithms and optimisations for synthesising hardware from Verilog/VHDL. However, the optimisations typically depend on the source program being standard in some sense (e.g. multiple components sharing a common clock event). Often one finds that less standard designs such as those forming asynchronous systems may not be synthesisable by a given tool, or worse still that a necessary asynchronous hardware cell required by a netlist is not in a vendor's standard cell library.

RTL-compilers, which have had a significant impact on the hardware design industry, translate RTL specifications to silicon (via a netlist representation). Tasks performed by an RTL-compiler include *Logic synthesis* – the translation of arithmetic and boolean expressions to efficient combinatorial logic and *Place and Route* – deciding where to position components on a chip and how to wire them up.

More recently *behavioural synthesisers* have become available. Behavioural synthesis (sometimes referred to as *high-level synthesis*) is the process of translating a program in a behavioural subset into a program in the RTL subset. Behavioural synthesis is often divided into three separate tasks [14]:

- *Allocation* is typically driven by user-supplied directives and involves choosing which resources will appear in the final circuit (e.g. three adders, two multipliers and an ALU).
- *Binding* is the process of assigning operations in the high-level specification to low-level resources – e.g. the + in line 4 of the source program will be computed by adder_1 whereas the + in line 10 will be computed by the ALU.
- *Scheduling* involves assigning start times to operations in the flow-graph such that no two operations will attempt to access a shared resource simultaneously. Mutually-exclusive access to shared resources is ensured by statically serialising operations during scheduling.

It is our belief that even the higher-level behavioural forms of VHDL and Verilog offer little in the way of abstraction mechanisms. We justify this claim by observing that, when writing in VHDL or Verilog, low-level implementation details and circuit structure tend to become fixed very early in the design process. Another example is that the protocol for communication between two functional units (often including exact timing relationships) is spread between their implementations and therefore is difficult to alter later. We develop this argument further in [25].

The problem we wish to identify is the conflict between (i) being able to make late, sweeping, changes to a system's implementation (though not to its logical specification); and (ii) the efficiency gained from low-level features (such as those found in VHDL and Verilog.) It is worth noting an analogy with programming languages here: efficiency can be gained by exploiting low-level features (e.g. machine code inserts or enthusiastic use of pointers) but the resulting system becomes harder for compilers or humans to analysis and optimise.

A principal aim of our work in the FLaSH project (Functional Languages for Synthesising Hardware), which started in 1999 at AT&T Research Laboratories Cambridge, was to adopt an aggressively high-level stance – we wanted to design a system in which (i) the programming language is clean (no 'implementation-defined subsets'); (ii) the *logical* structure can be specified but its realisation as *physical* structure can easily be modified even at late stages of design; and (iii) programs are susceptible to compiler analysis and optimisation facilitating the automatic synthesis of efficient circuits. We envisaged a single design framework in which we could, for example, select between synchronous and asynchronous design or rework the exact number of functional units, how they are shared and even the interconnection protocol (e.g. changing parallel-on-ready-signal to serial-starting-on-next-clock-tick) at any stage in the design process ('late-binding').

Of course the full aim as stated above has not yet been reached, but we have designed and implemented a language, SAFL 'Statically Allocated Parallel Function Language' embodying at least some of the principles above. One additional principle which we consider important about SAFL, or at least its intended implementation, is that it is *resource-aware*. By this we mean that its constructs map in a transparent way to hardware blocks, so that a program's approximate area-time consumption is clear from the SAFL source – compare the way that the space and time complexity of a C program is reasonably manifest in the program source because its primitives are associated with small machine code sequences.

The FLaSH project's optimising *compiler* translates SAFL into hierarchical RTL Verilog in a resource-aware manner. Each function definition compiles into a single hardware block; function calls are compiled into wiring between these functional blocks. In this framework multiple calls to the same source function corresponds to resource-sharing at the hardware level. Our compiler automatically deals with sharing issues by inserting multiplexers and static fixed-priority arbiters where required. Optimisations which minimise these costs have been implemented.

Accompanying the compiler is a *transformer*. This tool is intended to make semantics-preserving transformations on the SAFL source, typically with user-guidance, to facilitate architectural exploration. Although we have not yet built the transformer tool, we have published various transformation techniques which embody a wide range of implementation tradeoffs (e.g. functional unit duplication/sharing [19] and hardware-software co-design [20]). A key advantage of this approach is that it factors the current black-box user-view of synthesis tools ('this is what you get') into a source-to-source transformation tool which makes global changes visible to users followed by a more local-basis compilation tool.

Although a full comparison with other work is given in Sect. 1.1, we would like here to draw a distinction between this work and the framework used in Hydra [22], Lava [2] and Hawk [12]. Our aim in SAFL is to create a high-level *behavioural* language which is compiled into RTL Verilog whose compiler acts as an optimising back-end targeting silicon. SAFL can also run as an ML-style functional language for *simulation*. The Hydra/Lava/Hawk framework on the other hand uses the power of the functional language essentially for interconnecting lower-level components, and thus it is a *structural* language in our taxonomy. Note that the use of alternate interpretations of basis-functions (e.g. **and**, **or** etc.) means that Lava programs can be used for simulation as well as synthesis. We do not wish to claim either work is 'better', merely that they address different issues: we are looking for an "as high-level as possible" language from which to synthesise hardware and rely on others' compilers from RTL downwards; Hydra/Lava/Hawk concentrates on being able to describe hardware at a structural level (possibly down to geometric issues) in a flexible manner.

Although we have found SAFL a very useful vehicle to study issues of high-level behavioural synthesis, it is not the final word. In particular the function model used has a call-and-wait-for-result interface which does not readily interface with external state or multiple clock domains which are essential for a serious hardware design language. Recent work [25] has demonstrated that adding process-calculus features, even including restricted channel passing, can add to the expressive power of SAFL without losing resource-awareness; the resultant language is called SAFL+ and provides both I/O and interaction with components having state. However it is not so clear how much this increased expressive power costs in terms of the increased complexity of validating source-to-source transformations.

The following survey section of this paper draws on ideas from our other work: giving a general overview [18] at the start of the project, examining the theoretical basis of static allocation with parallelism in SAFL [19], describing the FLaSH compiler [23], examining hardware-software co-design [20] and studying the concept (so-called by analogy with soft typing) of soft scheduling [24].

After examining other related work in Sect. 1.1, the remainder of the paper is structured as follows: Section 2 introduces the SAFL language, how it might be compiled naïvely for synchronous hardware, the rôle of resource-awareness, how analyses aid compiler optimisations and concludes with a look at compiling to asynchronous or GALS (globally synchronous locally synchronous) hardware. Section 3 studies how source-to-source transformations at the SAFL level (together with the resource-awareness assumption) can reposition the hardware implementation on the area-time spectrum. These ideas are applied in Sect. 4 when a simple CPU is defined and SAFL transformations demonstrated which introduce pipeline and superscalar features. Section 5 considers how the ideas in Hydra/Lava/Hawk of using a functional language to express the *structural* composition of a system can be merged with the SAFL idea of *behavioural* definition;

this then leads on to a type system and to an investigation of partial evaluation in hardware. Finally, Sect. 6 concludes.

1.1   Comparison with Other Work

In the previous section we motivated our work by comparing SAFL to VHDL and Verilog. We also outlined the differences between our approach and that of Hydra/Lava/Hawk. This section continues by comparing SAFL with a number of other hardware design languages and methodologies.

We are not the first to observe that the mathematical properties of functional languages are desirable for hardware description and synthesis. A number of synchronous dataflow languages, the most notable being LUSTRE [5], have been used to synthesise hardware from declarative specifications. However, whereas LUSTRE is designed to specify reactive systems SAFL describes interactive systems (this taxonomy is introduced in [6]). Furthermore LUSTRE is inherently synchronous: specifications rely on the explicit definition of clock signals. This is in contrast to SAFL which could, for example, be compiled into either synchronous or asynchronous circuits.

The ELLA HDL is often described as functional. However, although constructs exist to define and use functions the language semantics forbid a resource-aware compilation strategy. This is illustrated by the following extract from the ELLA manual [17]: *"Once you have created a named function, you can use instances of it as required in other functions ... [each] instance of a function represents a distinct copy of the block of circuitry."* ELLA contains low-level constructs such as DELAY to create feedback loops, restricting high-level analysis. SAFL uses tail-recursion to represent loops at the semantic level; this strategy makes high-level analysis a more powerful technique.

Languages such as HardwareC [10] and Tangram [1] allow function definitions to be treated as shared resources. However, we feel that these projects have not gone as far as us in exploiting the potential benefits of resource-awareness. In particular:

- SAFL's dynamic scheduling policy (which relies on resource-awareness) leads to increased expressivity (and, in some cases, increased efficiency) over the more conventional static scheduling algorithms employed in HardwareC, Balsa and Tangram.
- We have developed a number of analyses and optimisations which are only made possible by structuring hardware as a series of function definitions [24, 23].
- We have investigated the impact of source-to-source transformations on SAFL and shown that it is a powerful tool for exploring the design-space. The functional properties of SAFL make it easier to apply transformations.

A number of languages have been developed which provide structural abstractions similar to those available in the Lava/Hawk framework. For example HML [11] is one such language based on Standard ML [15]; Jazz [9] combines a polymorphic type-system with object oriented features.

2   SAFL and Its Compiler

In this section we introduce the SAFL language and show how it can be mapped
to synchronous hardware. For space reasons, material described more fully elsewhere will only be summarised.

2.1   SAFL Language

SAFL has syntactic categories e (term) and p (program). First suppose that c
ranges over a set of constants, x over variables (occurring in let declarations
or as formal parameters), a over primitive functions (such as addition) and f
over user-defined functions. For typographical convenience we abbreviate formal
parameter lists $(x_1, \ldots, x_k)$ and actual parameter lists $(e_1, \ldots, e_k)$ to $\vec{x}$ and $\vec{e}$
respectively; the same abbreviations are used in let definitions. Then SAFL
programs p are given by recursion equations over expressions e; these are given
by:

$$e ::= c \mid x \mid \text{if } e_1 \text{ then } e_2 \text{ else } e_3 \mid \text{let } \vec{x} = \vec{e} \text{ in } e_0 \mid$$
$$a(e_1, \ldots, e_{arity(a)}) \mid f(e_1, \ldots, e_{arity(f)})$$
$$p ::= \text{fun } f_1(\vec{x}) = e_1; \ \ldots \ ; \text{fun } f_n(\vec{x}) = e_n;$$

It is sometimes convenient to extend this syntax slightly. In later examples
e1 ? e2:e3 is used as an abbreviated form of if-then-else; similarly we use a
case-expression instead of iterated tests; we also write e[n:m] to select a bit-field
[n..m] from the result of expression e (where n and m are integer constants).
    There is a syntactic restriction that whenever a call to function f_j from
function f_i is part of a cycle in the call graph of p then we require the call to be
a tail call.[1] Note that calls to a function not forming part of a cycle can occur
in an arbitrary expression context. This ensures that storage for the variables
and temporaries of p can be stored statically – in software terms the storage
is associated with the code of the compiled function; in hardware terms it is
associated with the logic to evaluate the function body.
    It is often convenient to assume that functions have been (re-)ordered so that
functions in a strongly connected component of the call graph are numbered
contiguously (we call such strongly connected components function *groups* [19]).
Apart from calls within groups, functions f_i can only call smaller numbered
$(j < i)$ functions f_j. It is also convenient to assume that the main entry point
to be called from the external environment, often written main(), is the last
function f_n.
    The other main feature of SAFL, apart from static allocatability, is that
its evaluation is limited only by data flow (and control flow at user-defined
call and conditional). Thus, in a let-expression let $\vec{x} = (e_1, \ldots, e_k)$ in e_0 or in
a call $f(e_1, \ldots, e_k)$ or $a(e_1, \ldots, e_k)$, all the $e_i(1 \leq i \leq k)$ are to be evaluated

---

[1] Tail calls consist of calls forming the whole of a function body, or nested solely within
the bodies of let-in expressions and consequents of if-then-else expressions.

concurrently. The body e_0 of a let is also evaluated concurrently subject only to data flow. In the conditional if e_1 then e_2 else e_3 we first evaluate (only) e_1; one of e_2 or e_3 is evaluated after its result is known.

This brings forth one significant interaction between static allocation and concurrency: in a call such as f(g(x),g(y)) the two calls to g must be serialised, since otherwise the same statically allocated storage (the space for g's argument) would be required simultaneously by competing processes. In hardware terms (see next section) a critical region, and arbiter to control entry to it, is synthesised around the implementation of g [24]. Section 2.3 shows how these arbiters can often be eliminated from the design while reference [19] discusses the interaction between static allocation and concurrency in more detail.

Note in the above example f(g(x),g(y)) there is little point in providing access to g by arbiter – simple compile-time choice of argument order is more effective. However in a more complicated case, such as f(g(h(x)),g(k(x))) where execution times for h and k are data-dependent, it can often make better sense for efficiency to keep run-time arbitration and to invoke first whichever call to g has its argument complete first, and then to invoke the other.

SAFL uses eager evaluation; we are sometimes asked why, especially given that Haskell-based tools (e.g. Lava and Hawk) are lazy. The resolution is that lazy-evaluation is problematic for our purposes. Firstly it is not so clear where to store the closures for suspended evaluations – simple tail recursion for n iterations in Haskell can often require $O(n)$ space; while this can sometimes be detected and optimised (strictness analysis) the formal problem is undecidable and so poses a difficulty for language design. Secondly lazy evaluation is inherently more sequential than eager evaluation – true laziness without strictness optimisation always has a single (i.e. sequential) idea of 'the next redex'.

Of course, a single tail-recursive function can be seen as a non-recursive function with a while-loop and assignment to parameters.[2] By repeatedly inline-expanding calls and making assignable local definitions to represent formal parameters, any SAFL program can be considered a nest of while-loops. However this translation has lost several aspects of the SAFL source: (*i*) the inlining can increase program size exponentially; (*ii*) not only is the association of meaningful names to loops lost, but the overall structure of the circuit (i.e. the way in which the program was partitioned into separate functional blocks) is also lost – thus resource-awareness is compromised; and (*iii*) concurrent function calls require a par construct to implement them as concurrent while-loops thus adding complexity.

---

[2] A function group can first be treated as a single function having the disjoint union of its component functions' formal parameters and having a body consisting of the component functions' bodies enclosed within a case-expression. The case-expression activates the relevant original function body based on the disjoint union actual parameter.

2.2   Naïve Translation to Synchronous Hardware

In this section we assume a global clock *clk* which progresses computation. All signals apart from *clk* are clocked by it, i.e. they become valid a small number of gate propagation delays after it and are therefore (with a suitable clock frequency) settled before the setup time for the next *clk*. We adopt the protocol/convention that signal wires (e.g. *request* and *ready*) are held high for exactly one pulse.

A simple translation of a SAFL function definition

$$\texttt{fun}\ f(\vec{x}) = e$$

is to a functional unit H_f which contains:

- an input register file, r, clocked by *clk*, of width given by $\vec{x}$;
- a *request* signal which causes data to be latched into r and computation of e to start;
- an output port P, here just specification of output wires to hold the value of e;
- a *ready* signal which is asserted when P is valid.

The SAFL primitives are compiled each having an output port and *request* and *ready* control signals; they may use the output ports of their subcomponents:

constants c: the constant value forms the output port and the *request* signal is merely copied to the *ready* signal.
variables x: since variables have been clocked into a register then they can be treated as constants, save that the output port is connected to the register.
if e_1 **then** e_2 **else** e_3: first e_1 is computed; when it is ready its boolean output routes a *request* signal either to e_2 or to e_3 and also routes the *ready* signal and output port of e_2 or e_3 to the result.
let $\vec{x} = \vec{e}$ **in** e_0: the $\vec{e}$ are computed concurrently (all their *requests* are activated); when all are complete e_0 is activated. Note that the results of the $\vec{e}$ are not latched.
built-in function calls $a(\vec{e})$: these are expanded in-line; the $\vec{e}$ are computed concurrently and their outputs placed as input to the logic for a; when all are complete this logic is activated;
user function calls $g(\vec{e})$: in the case that the called function only has one call (being non-recursive) the effect is similar to a built-in call – the $\vec{e}$ are computed concurrently and their outputs connected as to the input register file for g; when all are complete the *request* for g is activated; its output port and *ready* signal form those for $g(\vec{e})$. More complex function calls require more sophisticated treatment – see below.

The above explanation of calls to user-defined functions was incomplete in that it did not explain (i) how to implement a recursive call (SAFL restricts these to tail calls) nor (ii) how to control access to *shared* function blocks.

In the former case, the tail call $g(\vec{e})$ merely represents a loop (see below for mutual recursion); hence hardware is generated to route the values of $\vec{e}$ back to the input register for g and to re-invoke the *request* signal for the body of g when all the $\vec{e}$ have completed. In case (*ii*) we need to arbitrate between the possibly concurrent activation of g by this call and by other calls. The solution is to build a (synchronous) arbiter which: accepts *request* lines from the k call sites and keeps a record of which (at most one) of these have a call active. When one or more *requests* are present and no call is active, an implementation-defined *request* is selected and its values routed to the input register for g. On completion of g the *ready* signal of its body is routed back to the appropriate caller; the result value port can simply be connected to all callers.

One final subtlety in this compilation scheme concerns the values returned by functions invoked at multiple call sites (here not counting internal recursive calls). Consider a SAFL definition such as

```
f(x,y) = g(h(x+1),k(x+1),k(y))
```

where there are no other calls to h and k in the program. Assuming h clocks x+1 into its input register, then the output port P_h of h will continue to hold its value until it is clocked into the input register for g. However, assuming we compute the value of k(y) first, its result value produced on P_k will be lost on the subsequent computation of k(x+1). Therefore we insert a clocked *permanisor* register within the hardware functional unit, H_f corresponding to f(), which holds the value of k(y) (this should be thought of as a temporary used during expression evaluation in high-level languages). In the naïve compilation scheme we have discussed so far, we would insert a permanisor register for every function which can be called from multiple sites; the next section shows how static analysis can avoid this.

Adding permanisor registers at the output of resources, like k above, which may become valid and then invalid during a single call (to f above) explains our ability to avoid latching variables defined by let – permanisors have already ensured that signals of let-bound variables remain valid as long as is needed. The full details of the compilation process are described in [23].

2.3   Optimised Translation to Synchronous Hardware

Our compiler performs a number of optimisations based on whole-program analysis which improve the efficiency of the generated circuits (both in terms of time and area). This section briefly outlines some of these optimisations and refers the reader to papers which describe them in detail.

Removing Arbiters: Recall that our compiler generates (synchronous) arbiters to control access to shared function-blocks. In some cases we can infer that, even if a function-block is shared, calls to it will not occur simultaneously. For example, when evaluating f(f(x)) we know that the two calls to f must always occur sequentially since the outermost call cannot commence until the innermost call has been completed.

Whereas conventional high-level synthesis packages schedule access to shared resources by statically serialising conflicting operations, SAFL takes a contrasting approach: arbiters are automatically generated to resolve contention for all shared resources dynamically; static analysis techniques remove redundant scheduling logic. We call the SAFL approach *soft scheduling* to highlight the analogy with Soft Typing [4]: the aim is to retain the flexibility of dynamic scheduling whilst using static analysis to remove as many dynamic checks as possible. In [24] we compare and contrast soft scheduling to conventional static scheduling techniques and demonstrate that it can improve both the expressivity and efficiency of the language.

One of the key points of soft scheduling is that provides a convenient compromise between static and dynamic scheduling, allowing the programmer to choose which to adopt. For example, compiling `f(4)+f(5)` will generate an arbiter to serialise access to the shared resource H_f *dynamically*. Alternatively we can use a `let`-declaration to specify an ordering *statically*. The circuit corresponding to `let x=f(4) in x+f(5)` does not require dynamic arbitration; we have specified a static order of access to H_f. Note that program transformation can be used to explore static vs. dynamic scheduling tradeoffs.

Register Placement: In our naïve translation to hardware (previous section) we noted that a caller latches the result of a call into a register. We call such registers *permanising registers* since they are required to keep the result of a call to H_f permanent even if the value on H_f's output port subsequently changes (e.g. due to another caller accessing shared resource H_f). However, in many cases we can eliminate permanising registers: if we can infer that the result of a call to function f is guaranteed to remain valid (i.e. if no-one else can invoke f whilst the result of the call is required) then the register can be removed [23].

Cycle Counting: Consider translating the following SAFL program into synchronous hardware:

```
fun f(x) = g(h(x+1), h(k(x+2)))
```

Note that we can remove the arbiter for `h` if we can infer that the execution of `k` always requires more cycles than the execution of `h`.

Zero Cycle Functions: In the previous section we stated that it is the duty of a function to latch its arguments (this corresponds to callee-save in software terms). However, latching arguments necessarily takes time and area which, in some cases, may be considered unacceptable. For example, if we have a function representing a shared combinatorial multiplier (which takes a single cycle to compute its result), the overhead of latching the arguments (another cycle) doubles the latency.

The current implementation of the SAFL compiler [23] allows a user to specify, via pragma, certain function definitions as *caller-save* – i.e. it is then the duty of the caller to keep the arguments valid throughout the duration of the call. An extended register-placement analysis ensures that this obligation is kept, by adding (where necessary) permanising registers for such arguments at the call site. In some circumstances[3], this allows us to eliminate a resource's argument registers completely facilitating fine-grained, low-latency sharing of resources such as multipliers, adders etc.

There are a number of subtleties here. For example, consider a function f which adopts a caller-save convention and does not contain registers to latch its arguments. Note that f may return its result in the same cycle as it was called. Let us now define a function g as follows:

```
fun g(x) = f(f(x))
```

We have to be careful that the translation of g does not create a combinatorial loop by connecting f's output directly back into its input. In cases such as this *barriers* are inserted to ensure that circuit-level loops always pass through synchronous delay elements (i.e. registers or flip-flops).

2.4   Translation to Asynchronous Hardware

Although this has not been implemented yet, note that the design philosophy outlined in Sect. 2.2 made extensive use of request/acknowledge signals. Our current synchronous compiler models control events as 1-cycle pulses. With the change to edge events (either 2-phase or 4-phase signalling) and the removal of the global clock the design becomes asynchronous. The implementation of an asynchronous SAFL compiler is the topic of future work.

Note that the first two optimisations presented in the previous section (removal of arbiters and permanising registers) remain applicable in the asynchronous case since they are based on the causal-dependencies inherent in a program itself (e.g. when evaluating f(g(x)), the call to f cannot be executed until that to g has terminated). Although we cannot use the "cycle counting" optimisation as it stands, detailed feedback from model simulations incorporating layout delays may be enough to enable a similar type of optimisation in the asynchronous case.

2.5   Globally Asynchronous Locally Synchronous (GALS) Hardware

One recent development has been that of Globally Asynchronous Locally Synchronous (GALS) techniques where a number of separately clocked synchronous subsystems are connected via an asynchronous communication architecture. The GALS methodology is attractive as it offers a potential compromise between (*i*) the difficulty of distributing a fast clock in large synchronous systems; and (*ii*) the seeming area-time overhead of fully-asynchronous circuits. In a GALS circuit, various functional units are associated with different *clock domains*. Hardware to interface separate clock-domains is inserted at domain boundaries.

---

[3] Note that if the function is tail-recursive we cannot eliminate its argument registers since they are used as workspace during evaluation.

Our initial investigations of using SAFL for this approach have been very promising: clock domain information can be an annotation to a function definition; the SAFL compiler can then synthesise change-of-clock-domain interfaces exactly where needed.

3   Transformations in SAFL

As a result of our initial investigations, we believe that source-to-source transformation of SAFL is a powerful technique for exploring the design space. In particular:

- The functional properties of SAFL allow equational reasoning and hence make a wide range of transformations applicable (as we do not have to worry about side effects).
- The resource-aware properties of SAFL give fold/unfold transformations precise meaning at the design-level (e.g. we know that duplicating a function definition in the source is guaranteed to duplicate the corresponding resource in the generated circuit).

Although we have not yet built the transformer tool, we envisage it being driven with a GUI interface. There is also scope for semi-automatic exploration (cf. theorem proving), including perhaps hill-climbing.

In this section we give examples of a few of the transformations we have experimented with. We start with a very simple example, using *fold/unfold* transformations [3] to express resource duplication/sharing and unrolling of recursive definitions. Then a more complicated transformation is presented which allows one to collapse a number of function definitions into a single function providing their combined functionality. Finally we briefly outline a much larger, global transformation which allows one to investigate hardware/software co-design.

We have observed that the fold/unfold transformation is useful for trading area against time. As an example of this consider:

```
fun f x = ...
fun main(x,y) = g(f(x),f(y))
```

The two calls to f are serialised by mutual exclusion before g is called. Now use fold/unfold to duplicate f as f', replacing the second call to f with one to f'. This can be done using an unfold, a definition rule and a fold yielding

```
fun f  x = ...
fun f' x = ...
fun main(x,y) = g(f(x),f'(y))
```

The second program has more area than the original (by the size of f) but runs more quickly because the calls to f(x) and f'(y) execute in parallel.

Note that fold/unfold allows us to do more than resource/duplication sharing tradeoffs; folding/unfolding recursive function calls before compiling to synchronous hardware corresponds to trading the amount of work done per clock cycle against clock speed. For example, consider the following specification of a shift-add multiplier:

```
fun mult(x, y, acc) =
  if (x=0 or y=0) then acc
      else mult(x<<1, y>>1, if y[0:0] then acc+x else acc)
```

These 3 lines of SAFL produce over 150 lines of RTL Verilog. Synthesising a 16-bit version of `mult`, using Mentor Graphics' *Leonardo* tool, yields 1146 2-input equivalent gates. We can mechanically transform `mult` into:

```
fun mult(x, y, acc) =
  if (x=0 or y=0) then acc
  else let (x',y',acc') = (x<<1, y>>1,
                           if y[0:0] then acc+x else acc) in
      if (x'=0 or y'=0) then acc'
      else mult(x'<<1, y'>>1, if y'[0:0] then acc'+x' else acc')
```

which uses almost twice as much area and takes half as many clock cycles.

Another transformation we have found useful in practice is a form of loop collapsing. Consider a recursive main loop (for example a CPU) which may invoke one or more loops in certain cases (for example a multi-step division operation):

```
fun f(x,y) = if x=0 then y else f(x-1, y')
fun main(a,b,c) = if a=0 then b
                  else if ... then main(a', f(k,b), c')
                  else main(a',b',c')
```

Loop collapsing converts the two nested `while`-loops into a single loop which tests a flag each iteration to determine whether the outer, or an inner, loop body is to be executed. This test is almost free in hardware terms:

```
main(inner,x,a,b,c) =
  if inner then                         (* an f() iteration   *)
    if x=0 then main(0,x,a,b,c)         (* exit f()           *)
          else main(1,x-1,a,y',c)       (* another iteration  *)
  else
      if a=0 then b
      else if ... then main(1,k,a',b, c')  (* start f()       *)
      else main(0,x,a',b',c')              (* normal main step *)
```

This optimisation is useful because it allows us to collapse a number of function definitions, $(f_1, \ldots, f_n)$, into a single definition, F. At the hardware-level each function-block has a certain overhead associated with it (logic to remember who called it, latches to store arguments etc. – see Sect. 2). Hence this transformation allows us to save area by reducing the number of function-blocks in the final circuit. Note that the reduction in area comes at the cost of an increase in time: whereas previously $f_1, \ldots, f_n$ could be invoked in parallel, now only one invocation of F can be active at once.

Now consider applying this transformation to definitions $(f_1, \ldots, f_n)$ which enjoy some degree of commonality. Once we have combined these definitions into a single definition, F, we can save area further by applying a form of *Common Sub-expression Elimination* within the body of F. This amounts to exploiting let declarations to compute an expression once and read it many times (e.g. f(x)+f(x) would be transformed into let y=f(x) in y+y.)

Despite having a significant impact on the generated hardware, the transformations presented so far have been relatively simple. We have investigated more complicated transformations for exploring hardware/software co-design. Our method takes a SAFL specification, and a user-specified partition into hardware and software parts, and generates a specialised architecture (consisting of a network of heterogenous processors) to execute the software part. The basic idea involves representing processors and instruction memories (containing software) in SAFL itself and using a software compiler from SAFL to generate the code contained in the instruction memories. The details are beyond the scope of this paper; we refer the reader to [20] for more information.

4   Pipelines and Superscalar Expression in SAFL

```
fun cpu(pc, regs) =
  (let I = imem(pc)
   let (op,m,rd,ra) = (I[31:27], I[26], I[21:25], I[16:20])
   let (rb,imm) = (I[0:4], sext32(I[0:15]))
   let A = regs[ra]
   let B = m==0 ? imm : regs[rb]
   let C = alu(op, A, B)
   let D = case op of OPld => dmem(C,0,0)
                    | OPst => dmem(C,regs[rd],1)
                    | _    => C
   let regs' = case op of OPld => regs[D @ rd]
                        | OPadd => regs[D @ rd]
                        | OPxor => regs[D @ rd]
                        | _     => regs
   let pc' = case op of OPj => B
                      | OPbz => pc+4 + (A==0 ? imm : 0)
                      | _ => pc+4
   in (op==OPhalt ? regs'[0] : cpu(pc', regs')));
```

Fig. 1. Simple processor

As an example of how transformations work, consider the simple processor, resembling DLX [7] or MIPS, given in Fig. 1 (we have taken the liberty of removing most type/width information to concentrate on essentials and also omitted 'in' when it occurs before another 'let'). The processor has seven representative instructions, defined by enumeration

```
enum { OPhalt, OPj, OPbz, OPst, OPld, OPadd, OPxor };
```

and has two instruction formats (reg-reg-imm) and (reg-reg-reg) determined by a mode bit m. The processor is externally invoked by a call to cpu providing initial values of pc and registers; it returns the value in register zero when the OPhalt instruction is executed. There are two memories: imem which could be specified by a simple SAFL function expressing instruction ROM and dmem representing data RAM. The function dmem cannot be expressed directly in SAFL (although it can in SAFL+, our extended version [25]). It is declared by a native language interface and defined directly in Verilog: calls to dmem are serialised (just like calls to user-functions); if they could be concurrent a warning is generated. In this case it is clear that at most one call to dmem occurs per cycle of cpu. The intended behaviour of dmem(a,d,w) is to read from location a if w=0 and to write value d to location a if w=1. In the latter case the value of d is also returned. The use of functional arrays for regs and regs' is also to be noted: the definition let regs' = regs[v @ i] yields another array such that

$$regs'[i] = v$$
$$regs'[j] = regs[j] \quad \text{if } j \neq i$$

This can be seen as shorthand: the array regs corresponds to a tuple of simple variables, say (r0, r1, r2, r3)) and the value regs[i] is shorthand for the expression

$$(i==0 ? r0 : (i==1 ? r1 : (i==2 ? r2 : r3)))$$

and the array value regs[v @ i] is shorthand for the expression

$$((i==0 ? v : r0), (i==1 ? v : r1), (i==2 ? v : r2), (i==3 ? v : r3)).$$

Note that the SAFL restrictions mean that no dynamic storage is required, even when using array values as first-class objects. There is one advantage of the array notation in that it allows alternative implementation techniques; here we *may* infer that regs and regs' are never both live and share their storage as a single register file (when the conditionals above become multiplexors) but equally we may choose to use a rather less physically localised implementation, for example the rotary pipelines of Moore at al. [16].

Now let us turn to performance. We start by making three assumptions: first, that both imem and dmem take one clock tick; second, that the compiler also inserts a clocked register file at the head of cpu to handle the recursive loop; and that the alu() function is implemented without clocked registers which we will here count as just one delta[4] delay. We can now count clock and delta cycles,

---

[4] A delta cycle corresponds to a gate-propagation delay rather than a clock delay.

here just in terms of high-level SAFL data flow. Writing $n.m$ to mean n cycles and m delta cycles (relative to an external or recursive call to cpu being the 0.0 event), we can derive:

variable	cycle count
entry to body of cpu()	0.0
I	1.0
$(\mathrm{op}, \mathrm{m}, \mathrm{rd}, \mathrm{ra}), (\mathrm{rb}, \mathrm{imm})$	1.1
A, B	1.2
C	1.3
D	2.0 or 1.4[5]
regs'	2.1 or 1.5
pc'	1.3
recursive call to cpu()	2.2 or 1.6
next entry to body of cpu()	3.0 or 2.0

Note that we have counted SAFL-level data dependencies instead of true gate-delays; this is entirely analogous to counting the number of high-level statements in C to execute program speed instead of looking at the assembler output of a C compiler to count the exact number of instructions. The argument is that justifying many optimisations only needs this approximate count. A tool could easily annotate declarations with this information.

The result of this is that we have built a CPU which takes three clock cycles per memory reference instruction, two clock cycles for other instructions and with a critical path of length 6 delta cycles (which acts as a limit on the maximum clock rate). Actually, by a simple adjustment to the compiler, we could arrange that that cpu() is clocked at the same time as imem() and therefore achieve a one- or two-cycle instruction rate.

Now suppose we wish to make our simple CPU go faster; two textbook methods are adding a pipeline or some form of superscalar processing. We wish to reduce the number of clock cycles per instruction cycle and also to reduce the critical path length to increase the clock frequency.

The simplest form of pipelining occurs when we wish to enable the dmem and imem accesses to happen concurrently. The problem in the above design is that the memory address argument to dmem is only produced from the imem result. Hence we transform cpu() to cpu1a() as shown in Fig. 2; the suffix '1' on an identifier refers to a value which was logically produced one instruction ago. This transformation is always valid (as a transformation on a recursive program schema and thus computes the same SAFL function) but unfortunately the conditional test for OPhalt requires the calls to imem and dmem still to be serialised. To produce cpu2(), as shown in Fig. 3, we need to make a conscious adjustment to pre-fetch the instruction after the OPhalt by interchanging the if-then-else and the imem() call. This is now in contrast to cpu() and cpu1a() where instructions are only fetched when needed to execute. Now letting NOP stand for (OPbz<<27)+0 we see that the call cpu(pc,regs) is equivalent to the

---
[5] Depending on which path is taken.

```
fun cpu1a(pc, op1, regs1, C1, rd1) =
  (let D = case op1 of OPld => dmem(C1,0,0)
                     | OPst => dmem(C1,regs1[rd1],1)
                     | _    => C1
   let regs = case op1 of OPld => regs1[D @ rd1]
                        | OPadd => regs1[D @ rd1]
                        | OPxor => regs1[D @ rd1]
                        | _     => regs1
   in (op1==OPhalt ? regs[0] :          (* note this line *)
      let I = imem(pc)                   (* note this line *)
      let (op,m,rd,ra) = (I[31:27], I[26], I[21:25], I[16:20])
      let (rb,imm) = (I[0:4], sext32(I[0:15]))
      let A = regs[ra]
      let B = m==0 ? imm : regs[rb]
      let C = alu(op, A, B)
      let pc' = case op of OPj => B
                         | OPbz => pc+4 + (A==0 ? imm : 0)
                         | _ => pc+4
      in cpu1a(pc', op, regs, C, rd)));
```

Fig. 2. CPU after simple transformation

```
fun cpu2(pc, op1, regs1, C1, rd1) =
  (let D = case op1 of OPld => dmem(C1,0,0)
                     | OPst => dmem(C1,regs1[rd1],1)
                     | _    => C1
   let regs = case op1 of OPld => regs1[D @ rd1]
                        | OPadd => regs1[D @ rd1]
                        | OPxor => regs1[D @ rd1]
                        | _     => regs1
   let I = imem(pc)                       (* note this line *)
   in (op1==OPhalt ? regs[0] :           (* note this line *)
      let (op,m,rd,ra) = (I[31:27], I[26], I[21:25], I[16:20])
      let (rb,imm) = (I[0:4], sext32(I[0:15]))
      let A = regs[ra]
      let B = m==0 ? imm : regs[rb]
      let C = alu(op, A, B)
      let pc' = case op of OPj => B
                         | OPbz => pc+4 + (A==0 ? imm : 0)
                         | _ => pc+4
      in cpu2(pc', op, regs, C, rd)));
```

Fig. 3. CPU with pipelined memory access

call `cpu2(pc,NOP,regs,0,0)`, save that the latter requires only one clock for every instruction and that the instruction after an `OPhalt` instruction will now be fetched (but not executed). Note that the calls to `dmem` and `imem` in `cpu2()` are now concurrent and hence will happen on the same clock. It is pleasant to see such subtleties expressible in the high-level source instead of as hidden details.

We can now turn to exploring further the memory interface; in particular suppose we wish to retain separate `imem` (ROM) and `dmem` (RAM), each accessible in a single cycle, but wish both instruction- and data-fetches to occur from either source. This is form of a memory controller. In order to avoid concurrent access on every memory access instruction we wish it to be dual-ported, thus it will take two addresses and return two data values. When two concurrent accesses occur, either to `dmem` or to `imem` (e.g. because a `OPld` in one instruction refers to the memory bank which contains the following instruction), a *stall* will occur. A good memory controller will cause a stall only in this circumstance. Fig. 4 shows how this can be implemented in SAFL; `memctrl` is dual ported

```
fun memctrl(pc,a,d,r,w) =
  (let iv = is_dmem(pc) ? dmem(pc,0,0) : imem(pc)
   let dv = (r or w) ? (is_dmem(a) ? dmem(a,d,w) : imem(a)) : a
   in (iv,dv))
fun cpu3(pc, op0, regs0, C0, rd0) =
  (let (I,D) = memctrl(pc, C0, regs[rd0], op==OPld, op==OPst)
   in ...)
```

Fig. 4. CPU with memory controller

(two arguments and results) each memory access is directed (according to the, simple and presumably one delta cycle, function `is_dmem`) to the appropriate form of memory. The SAFL compiler detects the possible concurrent access to `imem` (and to `dmem`) and protects them both with an arbiter. The effect is as desired, a stall occurs only when the two accesses are to the same memory bank.

Another useful transformation is to reduce the length of the critical path in order to increase the clock rate. In `cpu2` this is likely to be the path through the `alu` function. Fig. 5 shows how the access to `alu` can be pipelined along with memory access to create a three-stage pipeline; here the the suffix '1' (resp. '2') on an identifier refers to a value which was logically produced one (resp. two) instructions ago. The processor `cpu4` works by concurrently fetching from `imem` the current instruction, doing the `alu` for the previous instruction op1 and doing memory access (and register write-back) for the second previous instruction op2. It is not quite equivalent to `cpu2` in that it exposes a *delay slot*; the result of an ALU or load instruction is not written back to `regs` until two instructions later, and thus the following instruction will still 'see' the old value. This is typically avoided by adding *forwarding* or *by-passing* hardware. In our terms this means

```
fun cpu4(pc, op1, A1, B1, rd1, op2, regs2, C2, rd2) =
  (let C = alu(op1, A1, B1)
   let D = case op2 of OPld => dmem(C2,0,0)
                      | OPst => dmem(C2,regs2[rd2],1)
                      | _    => C2
   let regs = case op2 of OPld => regs2[D @ rd2]
                        | OPadd => regs2[D @ rd2]
                        | OPxor => regs2[D @ rd2]
                        | _     => regs2
   let I = imem(pc)
   in (op2==OPhalt ? regs[0] :
      let (op,m,rd,ra) = (I[31:27], I[26], I[21:25], I[16:20])
      let (rb,imm) = (I[0:4], sext32(I[0:15]))
      (* forwarding (a.k.a. by-passing) would go here *)
      let A = regs[ra]
      let B = m==0 ? imm : regs[rb]
      let pc' = case op of OPj => B
                         | OPbz => pc+4 + (A==0 ? imm : 0)
                         | _ => pc+4
      in cpu4(pc', op, A, B, rd,  op1, regs, C, rd1)));
```

Fig. 5. CPU with pipelined ALU and memory access

comparing rd1 with ra and rb where indicated and using C instead of the value
from regs on equality.

Returning to the original cpu() form for simplicity of expression, we can
simply convert it to the superscalar processor shown in Fig. 6; since we have
dropped the pipeline we just use the '1' and '2' suffices for the 'left' and 'right'
instruction of a pair. As a processor this leaves quite a few things to be desired
– for example while the left (I1) and right (I2) instructions are serialised if I2
reads from a register written by I1, there is no such interlock on memory access
for concurrent writes. Similarly there is a *branch delay slot* in that I2 is carried
out even if I1 is a taken-branch. Further, to gain actual performance improvement one would need a single double-width imem64 function instead of the two
imem accesses; perhaps one can manage with a single-width dmem and require the
assembly code to be *scheduled* to pair memory access and non-memory-access
instructions. However, all structural hazards will be removed by the SAFL compiler by its insertion of arbiters around concurrently accessible resources. The
SAFL form explicitly lays out various options in the design space. For example,
as presented in Fig. 6 a single ALU is shared between the two separate instructions; duplicating this is clearly a good idea; however, less frequently used
components (perhaps a multiplier called by the ALU) could be provided in a
single form accessed by arbiter. A stall then only happens when the I1 and I2
instructions both use such a resource; we might choose to accept this point on
the speed/cost spectrum and again simply requiring compilers to schedule code
to avoid such stalls.

```
fun cpu5(pc, regs) =
  (let (I1,I2) = (imem(pc), imem(pc+4))
   let (op1,m1,rd1,ra1) = (I1[31:27], I1[26], I1[21:25], I1[16:20])
   let (rb1,imm1) = (I1[0:4], sext32(I1[0:15]))
   let (op2,m2,rd2,ra2) = (I2[31:27], I2[26], I2[21:25], I2[16:20])
   let (rb2,imm2) = (I2[0:4], sext32(I2[0:15]))
   if ((op1 == OPld or op1 == OPadd or op1 == OPxor) and
       (rd1 == ra2 or (m1==1 and rd1 == rb2)
                   or (op2 == OPst and rd1 == rd2))) then
     ...
     <I2 reads from a register written by I1 -- serialise>
     ...
   else
     let (A1,A2) = (regs[ra1], regs[ra2])
     let (B1,B2) = ((m1==0 ? imm1 : regs[rb1]),
                    (m2==0 ? imm2 : regs[rb2]))
     let (C1,C2) = (alu(op1, A1, B1), alu(op2, A2, B2))
     let D1 = case op1 of OPld => dmem(C1,0,0)
                        | OPst => dmem(C1,regs[rd1],1)
                        | _    => C1
     let D2 = case op2 of OPld => dmem(C2,0,0)
                        | OPst => dmem(C2,regs[rd2],1)
                        | _    => C2
     let regs' = case op1 of OPld => regs[D1 @ rd1]
                           | OPadd => regs[D1 @ rd1]
                           | OPxor => regs[D1 @ rd1]
                           | _     => regs
     let regs'' = case op2 of OPld => regs'[D2 @ rd2]
                            | OPadd => regs'[D2 @ rd2]
                            | OPxor => regs'[D2 @ rd2]
                            | _     => regs'
     let pc' = case op1 of OPj => B1
                         | OPbz => pc+8 + (A1==0 ? imm1 : 0)
                         | _ =>
                   case op of OPj => B2
                            | OPbz => pc+8 + (A2==0 ? imm2 : 0)
                            | _ => pc+8
     in (op1==OPhalt ? regs'[0]
         op2==OPhalt ? regs''[0] : cpu5(pc', regs'')));
```

Fig. 6. Simple superscalar processor

5 Compile-Time and Run-Time Types: Unifying SAFL and Lava

Based on the observation that Lava uses Haskell recursion to specify a circuit on the structural level (such as repetitive or nested circuits) whereas SAFL uses recursion to specify behavioural aspects, we now turn to a two-level language which can express both concepts in a single framework.

The idea is analogous to the distinction between static (compile-time) and (dynamic) run-time data in partial evaluation [8]; we consider partial evaluation for an extended SAFL in Sect. 5.1.

SAFL's type system (not explicitly spelt out previously, although always present in the implementation) is very simple, being of the form where values each have an associated size (n bits say) and therefore are ascribed type bit_n. Our implementation of SAFL currently requires that each constant, function argument and function result[6] is given an explicit width type in the style of the simply typed lambda-calculus. Function types, corresponding to hardware blocks, are then of type $bit_m \mapsto bit_n$. As in ML, functions can always be considered to have a single input and a single output as above; the addition of product forms to SAFL is then all that is required to model multiple arguments, results and `let` definitions. These can be seen as a family of bundling and unbundling operations:

$$join_{ij} : bit_i * bit_j \mapsto bit_{i+j} \tag{1}$$

$$split_{ij} : bit_{i+j} \mapsto bit_i * bit_j. \tag{2}$$

Similarly the family of sum forms $bit_i + bit_j$ can be represented as as type $bit_{\max(i,j)+1}$ in the usual manner.

SAFL functions are restricted to first order; allowing curried functions whose arguments and result are simple values poses few problems as the closure is of known size to the caller, but the gain in expressiveness does not seem worth the implementation effort. However allowing function values as arguments and results breaks the static allocatability requirement (counter-examples can be constructed based of the idea that any program can be expressed in continuation form using only tail recursion, see [19] for more details). Hence, given the syntactical separation between values and functions, the SAFL type system consists essentially of:

(values) bit_n
(functions) $bit_m \mapsto bit_n$.

Now let us consider the framework used in Lava. There, unwrapping the Haskell class treatment, circuits are essentially represented as graphs encoded as a datatype, say *circuit*. Thus, disregarding polymorphism at the moment and including $list(t)$ as representative of user-defined datatypes, the type system is essentially

$$\tau ::= int \mid circuit \mid \tau \to \tau \mid \tau \times \tau \mid list(\tau)$$

---

[6] Function results only need to be typed to ensure that all non-terminating recursive functions have a well-defined result type.

Circuit composition operations correspond to functions whose arguments or results are functions.

We can now combine these type systems as follows:

$$\sigma ::= bit_n \mid bit_m \mapsto bit_n$$
$$\tau ::= int \mid \sigma \mid \tau \to \tau \mid \tau \times \tau \mid list(\tau).$$

The interesting effect is that this type system has now become a two-level type
system (first studied by Nielson and Nielson [21]) which has two function constructors: '$\to$' representing compile-time, or structural, composition (cf. Verilog
module instantiation) and '$\mapsto$' representing run-time, or behavioural, computation (i.e. data movement in silicon).

Let us therefore invent a language, 2-SAFL, based on this type system, which:
allows use of general recursion to define compile-time functions (of types $\tau \to \tau$)
representing the structural design; and run-time functions (of types $bit_m \mapsto bit_n$
and respecting the SAFL static allocatability rules) representing the behavioural
core.[7] At its most primitive (using a $\lambda\vec{x}.e$ form to give function values instead
of having separate fun declarations) it has expressions, e, given by:

$$
\begin{array}{ll}
e ::= x \mid c \mid e\,e' & \\
\quad \mid \lambda^c \vec{x}.e & \mid \lambda^r \vec{x}.e \\
\quad \mid if^c\ e\ then\ e\ else\ e & \mid if^r\ e\ then\ e\ else\ e \\
\quad \mid let^c\ \vec{x} = e\ in\ e & \mid let^r\ \vec{x} = e\ in\ e
\end{array}
$$

Here the alternative left-hand-side forms λ^c, let^c etc., correspond to compile-time
constructions and those on the right λ^r, let^r etc., correspond to run-time, i.e.
SAFL, ones. We have constants (including fix to express recursion) and variables
of both forms and an overloaded application on both forms. Valid programs have
a set of well-formedness rules which express static allocatability restrictions on
the SAFL $\lambda^r\vec{x}.e$ form together with type and level constraints on the σ and τ as
in [21]; for example that compile-time computation (e.g. application of a value
defined by λ^c) cannot occur in the body of a $\lambda^r\vec{x}.e$ form. In examples below we
revert to the use of the $fun^c f(x) = e$ (and fun^r) form instead of the $\lambda^c\vec{x}.e$ (and
λ^r) used above.

This provides various interesting features, which we have not investigated in
detail. For example, it can be used to define SAFL functions which differ only
in type:

```
func multiplier(n) =
  local funr f(x:bit(n), y:bit(n), acc:bit(n)) : bit(n) =
    if y=0 then acc
    else f(x<<1, y>>1, if y[0:0] then acc+x else acc)
```

---

[7] There is a little surprise here: the structural level is outside the the behavioural level.
This is consonant with Lava in which compile-time, i.e. Haskell, functions allow one
to manipulate the primitive hardware cells which move values at run-time; all we
have done is to provide a richer, behavioural, run-time level in the form of SAFL.

```
      in f
      end;
  fun^r m1 = multiplier(16);
  fun^r m2 = m1;
  fun^r m3 = multiplier(16);
  fun^r m4 = multiplier(24);
  fun^r main(...) = ... m1 ... m2 ... m3 ... m4 ...
```

Here m1, m2 and m3 represent 16-bit multipliers and m4 a 24-bit multiplier. Note that resource-awareness is manifested here by m1 and m2 being synonyms for the same multiplier while m3 is a distinct multiplier (as is m4, but this is clear because it differs in width). The type of multiplier is then

$$(n \in int) \to (bit_n * bit_n * bit_n \mapsto bit_n).$$

Similarly, consider the way in which Lava circuits can be wired together by means of functions which operate on values of type *circuit*. We can capture this notion by means of a higher-order function taking arguments in the form of SAFL functions. In SAFL we can already define functional units f and g and then wire them together to make h as follows:

```
  fun^r f(x,y) = e_1;
  fun^r g(z) = e_2;
  fun^r h(x,y) = f(g(x+1),g(y+1));
```

The 2-SAFL language now allows, as in Lava, the particular combinator here used to define h to be abstracted and expressed as a higher-order value (suitable for re-use elsewhere in the system where similar combinators are required):

```
  fun^r f(x,y) = e_1;
  fun^r g(z) = e_2;
  fun^c combine(p,q,a) =
    local fun^r t(x,y) = p(g(x+1),q(y+a)) in t end;
  fun^r h = combine(f,g,1);
```

This example, although contrived, does show how compile-time functions such as combine can be used as in Lava. Supposing f has type $(bit_m * bit_n \mapsto bit_r)$ and g has type $(bit_i \mapsto bit_m)$, then the type of combine is:

$$(bit_m * bit_n \mapsto bit_r) \times (bit_j \mapsto bit_n) \times int \to (bit_i * bit_j \mapsto bit_r).$$

We could summarise the 2-SAFL extension as follows: in the Lava view, the only run-time computations arise from built-in primitives (AND, OR, flip-flops) whereas 2-SAFL has SAFL function definitions as primitive; iterative structure in Lava is represented via re-entrant graphs whereas in 2-SAFL it is represented via SAFL recursion.

We have not properly investigated type inference on this system – in many ways the compile-time types inherit from Lava, including widths, but the availability of SAFL functions at the run-time level instead of just hardware primitives may pose interesting questions of type inference. (For example, as above, run-time types, e.g. bit_n, can depend on compile-time values, e.g. n.) Instead of addressing this issue, we wish to turn attention to the possibility of exploring transformations which interchange the two forms of function (compile-time and run-time), i.e. partial evaluation.

5.1 Partial Evaluation

Although there is only space for a preliminary discussion of these ideas here, the 2-SAFL language provides a useful base for partial evaluation for hardware. In traditional partial evaluation [8], sometimes known as *program specialisation*, a generic program taking n inputs is specialised, with $k < n$ of its inputs being known values, to result in a specialised program taking $n-k$ inputs. The resultant program is generally more efficient that the original program. Few applications to hardware seem to exist; McKay and Singh [13] consider the problem of run-time specialisation of hardware implemented in FPGAs to increase speed.

Standard partial evaluation seems to fit in well with 2-SAFL; we show how a SAFL program of say three inputs can be reduced to a more efficient program of two arguments in the 2-SAFL framework. For example, consider a multi-cycle multiply-and-add defined by

```
fun mult(x, y, acc) =
    if y=0 then acc
    else mult(x<<1, y>>1, if y[0:0] then acc+x else acc)
```

We can trivially convert this to a multiply-by-13-and-add by defining

```
fun mult13(x,acc) = mult(x,13,acc).
```

Indeed, if `mult13` compiles into a zero-clock function (one which does not latch its inputs) then a call `mult13`(x,a) will compile into hardware identical to that of a call `mult`$(x,13,a)$.

However, the two-level type system can be used to create a specialised version of `mult13`. Writing `mult` first as

```
fun mult = λʳ(x, y, acc).
    if y=0 then acc
    else mult(x<<1, y>>1, if y[0:0] then acc+x else acc)
```

and then, curried on the two forms of λ, gives:

```
fun mult' = λᶜy. λʳ(x, acc).
    if y=0 then acc
    else mult'(y>>1)(x<<1, if y[0:0] then acc+x else acc).
```

This last form is not a valid 2-SAFL program since it contains a compile-time application `mult'`(y>>1) within a λ^r. However `mult'`(13) *can* be unfolded as:

```
fun mult13 = λʳ(x, acc). if 13=0 then acc
    else mult6(x<<1, if 13[0:0] then acc+x else acc)
```

```
fun mult6 = λʳ(x, acc). if 6=0 then acc
    else mult3(x<<1, if 6[0:0] then acc+x else acc)
fun mult3 = λʳ(x, acc). if 3=0 then acc
    else mult1(x<<1, if 3[0:0] then acc+x else acc)
fun mult1 = λʳ(x, acc). if 1=0 then acc
    else mult0(x<<1, if 1[0:0] then acc+x else acc)
fun mult0 = λʳ(x, acc). if 0=0 then acc
    else mult0(x<<1, if 0[0:0] then acc+x else acc).
```

which simplies to:

```
fun mult13 = λʳ(x, acc). mult6(x<<1, acc+x)
fun mult6 = λʳ(x, acc). mult3(x<<1, acc)
fun mult3 = λʳ(x, acc). mult1(x<<1, acc+x)
fun mult1 = λʳ(x, acc). mult0(x<<1, acc+x)
fun mult0 = λʳ(x, acc). acc
```

and hence (by unfolding, or just by compiling the used-once functions `mul6` to `mul0` into zero-clock functions) to:

```
fun mult13 = λʳ(x, acc).
    acc + x + ((x<<1)<<1) + (((x<<1)<<1)<<1)
```

or equivalently

```
funʳ mult13(x,acc) = acc + x + ((x<<1)<<1) + (((x<<1)<<1)<<1)
```

which now again adheres to the SAFL rules.

In this example at least, once the idea of using y as a static (λ^c) parameter had been mooted, the manipulation was forced by the type system.

Finally, let us observe that partial evaluation techniques applied to processors can produce interesting effects. A sequence of frequently occurring software instructions for a processor can be specialised into a single new instruction and the hardware necessary to execute this instruction (hopefully faster) automatically generated.

6   Conclusions and Further Work

We have found SAFL to be surprisingly powerful at describing hardware at a high-level in spite of its meagre features. In particular it seems to be very effective at describing processor design and transformations to adjust their area-time consumption as discussed in Sect. 4. An important aspect of this is resource-awareness. We can trust the SAFL compiler to optimise code without altering its gross structure; hence transformations on hardware structure can be seen as SAFL source-to-source transformations.

Another aspect of the same coin is that we would expect to be able to verify mechanically the transformations we make, and indeed to hope that semi-automatic tools can help the user to choose an area-time tradeoff.

At the moment SAFL is a self-contained language which compiles to Verilog. However we could also embed SAFL within VHDL or Verilog as a higher-level behavioural form – the main problem would be restricting access to lower-level details so that SAFL compiler optimisations and SAFL transformations which we have discussed remain valid.

The aspect about which we feel most exposed is that the pure functional call-and-wait-for-result interface provided by SAFL is sometimes too restrictive. Recent work [25] on SAFL+ suggests one possible way forward.

Acknowledgement. This work was supported by (UK) EPSRC grant GR/ N64256 "A Resource-Aware Functional Language for Hardware Synthesis"; the second author was also sponsored by AT&T Research Laboratories Cambridge. Phil Endecott, David Greaves and Paul Webster provided helpful comments.

References

1. Van Berkel, K. Handshake Circuits: an Asynchronous Architecture for VLSI Programming. International Series on Parallel Computation, vol. 5. Cambridge University Press, 1993.
2. Bjesse, P., Claessen, K., Sheeran, M. and Singh, S. Lava: Hardware Description in Haskell. Proc. 3rd ACM SIGPLAN International Conference on Functional Programming, 1998.
3. Burstall, R.M. and Darlington, J. A Transformation System for Developing Recursive Programs, JACM 24(1), 1979.
4. Cartwright, R. and Fagan, M. Soft Typing. Proc. ACM SIGPLAN Conference on Programming Language Design and Implementation, 1991.
5. Halbwachs, N., Caspi, P., Raymond, P. and Pilaud, D. The Synchronous Dataflow Programming Language LUSTRE. Proc. IEEE, vol. 79(9). September 1991.
6. Harel, D. and Pnueli, A. On the Development of Reactive Systems. Springer-Verlag NATO ASI Series, Series F, Computer and Systems Sciences, vol. 13, 1985.
7. Hennessy, J.L. and Patterson, D.A. Computer Architecture: A Quantitative Approach. Morgan Kaufmann, 1990.
8. Jones, N.D., Gomard, C.K. and Sestoft, P. Partial Evaluation and Automatic Program Generation. Prentice-Hall International Series in Computer Science, 1993.
9. The Jazz Synthesis System: http://www.exentis.com/jazz
10. Ku, D. and De Micheli, G. HardwareC – a Language for Hardware Design (version 2.0), Stanford University Technical Report: CSL-TR-90-419, 1990.
11. Li, Y. and Leeser, M. HML, a Novel Hardware Description Language and its Translation to VHDL. IEEE Transactions on VLSI Systems, vol. 8. no. 1. February 2000.
12. Matthews, J., Cook, B. and Launchbury, J. Microprocessor Specification in Hawk. Proc. IEEE International Conference on Computer Languages, 1998.
13. McKay, N and Singh, S. Dynamic Specialisation of XC6200 FPGAs by Parial Evaluation. Lecture Notes in Computer Science: Proc. FPL 1998, vol. 1482, Springer-Verlag, 1998.
14. De Micheli, G. Synthesis and Optimization of Digital Circuits. McGraw-Hill, 1994.
15. Milner, R., Tofte, M., Harper, R. and MacQueen, D. The Definition of Standard ML (Revised). MIT Press, 1997.

16. Moore, S.W., Robinson, P. and Wilcox, S.P. Rotary Pipeline Processors. IEE Part–E, Computers and Digital Techniques, Special Issue on Asynchronous Architectures, 143(5), September 1996.
17. Morison, J.D. and Clarke, A.S. ELLA 2000: A Language for Electronic System Design. Cambridge University Press 1994.
18. Mycroft, A. and Sharp, R.W. The FLaSH Project: Resource-aware Synthesis of Declarative Specifications. Proc. International Workshop on Logic Synthesis 2000.
19. Mycroft, A. and Sharp, R. A Statically Allocated Parallel Functional Language. Lecture Notes in Computer Science: Proc. 27th ICALP, vol. 1853, Springer-Verlag, 2000.
20. Mycroft, A. and Sharp, R. Hardware/Software Co-Design Using Functional Languages. Lecture Notes in Computer Science: Proc. TACAS'01, vol. 2031, Springer-Verlag, March 2001.
21. Nielson, F. and Nielson, H.R. Two Level Semantics and Code Generation. Theoretical Computer Science, 56(1):59-133, 1988.
22. O'Donnell, J.T. Hydra: Hardware Description in a Functional Language using Recursion Equations and High Order Combining Forms, The Fusion of Hardware Design and Verification, G. J. Milne (ed.), North-Holland, 1988.
23. Sharp, R. and Mycroft, A. The FLaSH Compiler: Efficient Circuits from Functional Specifications. AT&T Research Laboratories Cambridge Technical Report tr.2000.3, June 2000.
    Available from http://www.uk.research.att.com
24. Sharp, R. and Mycroft, A. Soft Scheduling for Hardware. Lecture Notes in Computer Science: Proc. SAS'01, vol. 2126, Springer-Verlag, July 2001.
25. Sharp, R. and Mycroft, A. A Higher-Level Language for Hardware Synthesis. Lecture Notes in Computer Science: Proc. CHARME'01, vol. 2144 (this volume), Springer-Verlag, September 2001.
26. Sheeran, M. muFP, a Language for VLSI Design. Proc. ACM Symp. on LISP and Functional Programming, 1984.

Applications of Hierarchical Verification in Model Checking

Robert Beers, Rajnish Ghughal, and Mark Aagaard

Performance Microprocessor Division, Intel Corporation,
RA2-401, 5200 NE Elam Young Parkway, Hillsboro, OR 97124, USA.

Abstract. The LTL model checker that we use provides sound decomposition mechanisms within a purely model checking environment. We have exploited these mechanisms to successfully verify a wide spectrum of large and complex circuits. This paper describes a variety of the decomposition techniques that we have used as part of a large industrial formal verification effort on the Intel Pentium® 4 (Willamette) processor.

1 Introduction

One of the characteristics that distinguishes industrial formal verification from that done in academia is that industry often works on large, complex circuits described at the register transfer level (RTL). In comparison, academic work usually verifies either small RTL circuits or high-level abstractions of complex circuits. Academia remains the primary driving force in the development of new verification algorithms, and many of these advances have been succesfuly transfered from academia to industry [2,4,6,7,9,12,15,16, 18]. However, improving the strategies for applying these algorithms in industry requires exposure to circuits of a size and complexity that are rarely available to academics.

An important task in continuing the spread of formal verification in industry is to document and promulgate techniques for applying formal verification tools. In this paper we describe a variety of decomposition strategies that we have used as part of the formal verification project for the Intel Pentium® 4 (Willamette) processor. This paper focuses on strategies taken from verifying two different implementations of queues and a floating-point adder.

The verification tool we use is an LTL model checker developed at Intel that supports a variety of abstraction and decomposition techniques [14,15]. Our strategies should be generally applicable to most model-checking-based verification tools. In theory, the techniques are also applicable for theorem-proving. But, because the capacity limitations of model checking and theorem proving are so very different, a strategy that is effective in reducing the size of a model checking task might not be the best way to reduce the size of the problem for theorem proving.

2 Overview

In this section we give a high-level view of LTL model checking and provide some intuition about how the model checker works. The focus of this paper is on the application

T. Margaria and T. Melham (Eds.): CHARME 2001, LNCS 2144, pp. 40–57, 2001.
© Springer-Verlag Berlin Heidelberg 2001

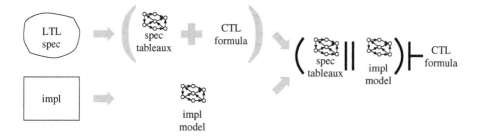

Fig. 1. Verification flow

of decomposition techniques, thus we concentrate on the usage of the model checker rather than the tool itself.

A high-level view of model checking for a linear temporal logic (LTL) is shown in Fig. 1. LTL model checking is done by converting the specification (which is an LTL formula) to a tableaux and a CTL formula. The tableaux is an automaton that recognizes traces that satisfy the LTL formula. The CTL formula checks that the tableaux never fails. The verification run computes the product machine of the tableaux and the implementation and checks that the product machine satisfies the CTL formula [8].

Because specifications are converted into automata, and model checking verifies automata against temporal formulas, specifications can themselves be verified against higher-level specifications. Figure 2 shows an implementation that is verified against a middle-level specification, which, in turn, is verified against a high-level specification.

We now step through a simple example of hierarchical decomposition, using the circuit `pri` shown in Fig. 3. The circuit `pri` takes three inputs (`i0`, `i1`, and `i2`) and outputs a `1` on the highest priority output line line whose input was a `1`. We model

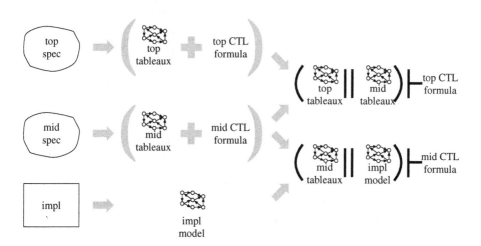

Fig. 2. Hierarchical verification flow

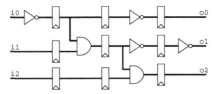

Fig. 3. Example circuit: `pri`

combinational logic as having zero delay and flip flops as being unit delay. We draw flip flops as rectangles with small triangles at the bottom.

Theorem 1 shows an example property that we will verify about `pri`. The property says that, for `pri`, if the output o0 is 1, then o2 will be 0. For the sake of illustration, we will pretend that verifying Theorem 1 is beyond the capacity of a model checker, so we decompose the problem. We carry out the verification in three steps by verifying Lemmas 1–3.

Theorem 1. $\mathtt{pri} \models (\mathtt{o0} \implies \neg\mathtt{o2})$

Lemma 1. $\mathtt{pri} \models (\mathtt{o0} \implies \neg\mathtt{o1})$

Lemma 2. $\mathtt{pri} \models (\neg\mathtt{o1} \implies \neg\mathtt{o2})$

Lemma 3. $(\mathtt{Lemma1}, \mathtt{Lemma2}) \models (\mathtt{o0} \implies \neg\mathtt{o2})$

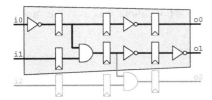

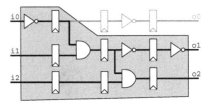

Fig. 4. Cone of influence reduction for Lemma 1

Fig. 5. Cone of influence reduction for Lemma 2

By decomposing the verification, we end up with multiple verification runs, but each one is smaller than Theorem 1, the original run (Table 1). The first two runs (Lemmas 1 and 2) are smaller than Theorem 1, because, using *cone of influence* reduction, Lemmas 1 and 2 do not need to look at the entire circuit. The last run, Lemma 3, glues together the earlier results to verify the top-level specification. When verifying Lemma 3, we do not need the circuit at all, and so its verification is very small and easy to run.

One advantage of this style of hierarchical verification is that it provides some theorem-proving-like capabilities in a purely model checking environment. Two disadvantages are that the approach is still subject to the expressiveness limitations and

Table 1. Summary of verification of `pri`

property	latches	inputs	total variables
Theorem 1	9	3	12
Lemma 1	6	2	8
Lemma 2	7	3	10
Lemma 3	0	3	3

some of the capacity limits of BDDs and that it does not support reasoning about the entire chain of reasoning. That is, we use Lemma 3 to verify the right-hand-side of Theorem 1 but we cannot explicitly prove Theorem 1, instead we use a database of specifications to track chains of dependencies.

3 Parallel Ready Queue

This section describes part of the verification of a "parallel ready queue" (`prqueue`) that operates in parallel on k channels (Fig. 6). Packets arrive, one at a time, at any of the `arrive` signals. A packet departs the queue if it is the oldest ready packet. Packet i becomes ready when `ready`$[i]$ is 1. Once a packet is ready, it remains ready until it has departed. Packets can become ready to leave the queue in any order. Hence, it is possible for a packet to leave the queue before an older packet if the older packet is not yet ready.

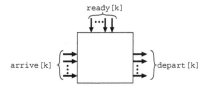

Fig. 6. Parallel queue circuit (`prqueue`)

There are a number of environmental constraints that are required for the circuit to work correctly:

1. A packet will not arrive on a channel if the queue currently contains a packet in that channel.
2. At most one packet will arrive at any time.
3. If a packet arrives on a channel, the corresponding ready signal will eventually be set to 1.
4. A channel will not be marked as ready if it does not contain a packet.

Note, the environment has the freedom to simultaneously mark multiple packets as ready.

The top level specification (Theorem 2) covers both liveness and safety.

Theorem 2. *PriQueue Specification*
  1. A packet that arrives on `arrive[i]` *will eventually depart on* `depart[i]`.
  2. If a packet departs, it is the oldest ready packet in the queue.

3.1 Implementation

We briefly describe the high-level algorithm for the implementation of `prqueue`. There are many sources of complexity, e.g., datapath power-saving control logic and various gated clock circuits, which we ignore for the sake of simplicity in the following description. The implementation of `prqueue` uses a two-dimensional $k \times k$ scoreboard (`scb`) to maintain the relative age of packets and an array of k `rdy` signals to remember if a channel is ready. Figure 7 shows the scoreboard and ready array for a queue with four channels.

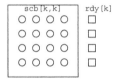

Fig. 7. Scoreboard in parallel ready queue

Initially, all of the elements in the scoreboard and ready array are set to 0. When a packet arrives on `arrive[i]`, all of the scoreboard elements in column i are set to 0 and all of the elements in row i are set to 1. The intuition behind this algorithm lies in two observations:

  - The scoreboard row for the youngest (most recently arrived) packet is the row with the most 1s, while the oldest packet has the row with the most 0s.
  - The packet for row i is older than the packet for row j if and only if $scb[i, j] = 1$.

The diagonal elements do not contain any semantically useful information. When `ready`$[i]$ is set to 1, packet i will be marked as ready to leave the queue. If at least one of the packets is marked as ready, the scoreboard values are examined to determine the relative age of all ready packets and the oldest among them departs the queue. The process of packets arriving, being marked as ready, and departing is illustrated in Fig. 8.

As is all too common when working with high-performance circuits, the specification and high-level description of the implementation appear deceivingly straightforward. This circuit has been optimized for area and performance and the actual implementation is much more complex than the high-level description. For example, the logic for `ready`$[i]$ signal involves computation of multiple conditions which all must be satisfied before the packet is ready to depart the scoreboard. We have ignored many such details of implementation for the purpose of the simplicity of the description. The actual circuit verified consists of hundreds of latches with significantly larger values for k.

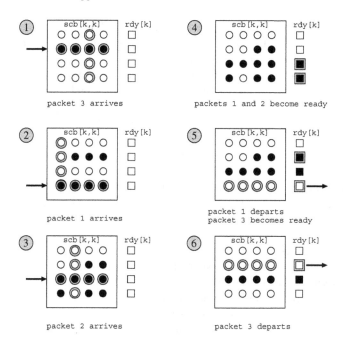

Fig. 8. Scoreboard in parallel ready queue

3.2 Verification

Theorem 3 is one of the major theorems used in verifying the circuit against the top-level specification 2.

Theorem 3 says: *if a packet is ready, then a packet will depart.* The signal depValid[i] is the "valid bit" for depart[i]; it says that the packet on row i is departing. There are some environmental assumptions such as issues with initialization that we do not encode in this specification for the sake of simplicity.

Theorem 3 (willDepart).
$(\exists i.\, \mathtt{rdy}[i]) \implies (\exists i.\, \mathtt{depValid}[i])$

Theorem 3 is a safety property of the implementation. The logic expression $\exists i.\, \mathtt{rdy}[i]$ stands for the boolean expression $\mathtt{rdy}[0] \lor \mathtt{rdy}[1] \lor \ldots \mathtt{rdy}[k]$. Theorem 3 is dependent upon the fanin cone for all k^2 entries in the scoreboard. The fanin cone of the scoreboard consists of hundreds of latches. This amount of circuitry and the temporal nature of information stored in the scoreboard array puts it well beyond the capacity of model checking for values of k found in realistic designs.

To carry out the verification, we decomposed the task into two parts. First, we verified that the scoreboard and ready array impart a transitive and anti-symmetric relationship that describes the relative ages of packets (Lemmas 4 and 5). Second, we verified that these properties imply Theorem 3.

Lemma 4 (Transitivity $i\,j\,k$).
$\mathtt{rdy}[i] \land \mathtt{rdy}[j] \land \mathtt{rdy}[k] \land \mathtt{scb}[i,j] \land \mathtt{scb}[j,k] \implies \mathtt{scb}[i,k]$

Lemma 4 says that if packet i is older than packet j and packet j is older than packet k, then packet i is older than packet k. (Recall that packet i is older than packet j if $\text{scb}[i, j] = 1$). The benefit of Lemma 4 is that we reduce the size of the circuit needed from k^2 scoreboard entries to just three entries. The cost of this reduction is that we had to verify the lemma for all possible combinations of scoreboard indices such that $i \neq j \neq k$, which resulted in $k \times (k - 1) \times (k - 2)$ verification runs

Lemma 5 (AntiSym i j).
$\text{rdy}[i] \wedge \text{rdy}[j] \implies \text{scb}[i, j] = \neg \text{scb}[j, i]$

Lemma 5 (anti-symmetry) says that if packet i is older than packet j, then packet j cannot be older than packet i (and vice versa.) It is easy to see why this lemma is true by observing that whenever a packet arrives the corresponding column values are set to 0 while the corresponding row values are set to 1. The anti-symmetry lemma relates only two values of the scoreboard entries. We verified Lemma 5 for all possible combinations of i and j (such that $i \neq j$), which resulted in $k \times (k - 1)$ verification runs.

After verifying Lemmas 4 and 5, we verified Theorem 3 by removing the fanin cone of the scoreboard array while assuming the two lemmas. Removing the fanin cone of the scoreboard array is essentially an abstraction which preserves its pertinent properties using transitivity and anti-symmetry lemmas. This particular abstraction preserves the truth of the safety property under consideration. The transitivity and anti-symmetry lemmas provide just enough restriction on the scoreboard values to establish a relative order of arrival among packets that are ready to exit. Note that the transitivity relationship between three scoreboard entries is sufficient to enforce proper values in the scoreboard array even when more than three packets are ready to depart.

The major sources of complexity in verifying Theorem 3 were:

- The large fanin cone of the scoreboard array, which builds up complex temporal information.
- The size of the scoreboard array: $k \times k$ entries.

We were able to verify Theorem 3 without this hierarchical decomposition for very small values of k. However, the verification task becomes increasingly more difficult for larger values of k. The hierarchical verification approach overcomes the model checking complexity for two reasons:

- The transitivity and anti-symmetry lemmas relate three and two values of the scoreboard entries. The verification complexity of these lemmas is relatively small and is independent of the size of the scoreboard.
- For the verification of Theorem 3, we were able to remove the fanin cone of the entire scoreboard. We enforce proper values in scoreboard entries by enforcing the transitivity and anti-symmetry safety properties which themselves are not of a temporal nature. However, these lemmas together provide enough information to build an implicit order of arrival of packets that allows the implementation to pick the eldest packet.

4 Floating-Point Adder

We briefly preview the basics of binary floating-point numbers [11] before diving into some of the decompositions used in verifying a floating point adder. A floating-point number is represented as a triple (s, e, m) where, s is a sign bit, e is a bit-vector representing the exponent, and m is a bit-vector representing the mantissa. We have verified the sign, mantisa and exponent for both *true addition* and *true subtraction* (the signs of the operands are the same or differ), but consider only the verification of the mantisa for true addition in this paper.

The real number represented by the triple (s, e, m) is:

$$(-1)^{\hat{s}} \times 2^{\hat{e} - bias} \times \hat{m} \times 2^{-n_m + 1}$$

where $\hat{x}$ is the unsigned integer encoding by the bit vector x and *bias* is a format-dependent *exponent bias*. The mantissa has an implicit leading 1, and so it represents the value $1.m$, which is in the range $[1, 2)$.

4.1 Implementation

A simple algorithm to calculate the result mantissa for floating-point addition is shown in Fig. 9.

input: two normal floating-point numbers $f_1 = (s_1, e_1, m_1)$ and $f_2 = (s_2, e_2, m_2)$

```
expdiff    := |ê₁ − ê₂|;                              absolute diff of exps
expcmp     := ê₁ > ê₂;                                f̂₁ is the larger number
bigman     := if expcmp then 1.m₁ else 1.m₂;          mant of larger number
smallman   := if expcmp then 1.m₂ else 1.m₁;          mant of smaller number
alignman   := shift_right(smallman, expdiff);         align small mant
addman     := bigman + alignman;                      add mantissas
resultman  := round(addman);                          round result
```

Fig. 9. Simple floating-point true addition algorithm

A simplified implementation for a floating-point true addition circuit `fpadder` is shown in Fig. 10. It follows the basic structure of the algorithm in Fig. 9. The actual implementation of the adder that we verified is much more complex than this simple depiction. For example, the actual implementation has more than just one adder and more than one parallel datapaths for the sake of performance.

4.2 Verification

We use a reference model similar to the algorithm depicted in Fig. 9 as the specification for floating-point addition. Theorem 4 is our top-level theorem for the mantissa.

Theorem 4 (Mantissa).
f_1 and f_2 are normal $\implies$ (`spec_resultman = impl_resultman`)

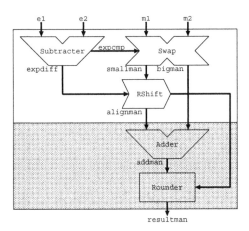

Fig. 10. Simplified circuit for floating-point true addition : `fpadder`

For the purpose of the verification, we essentially removed all state elements by *unlatching* the circuit thus arriving at a combinational model of the circuit. We used our model checker, instead of an equivalence checker, to verify the above property on the combinational model as only the model-checking environment provides support for hierarchical decomposition. Also, for the purpose of simplicity we do not include environmental assumptions in the description of Theorem 4.

The main challenge we faced in verifying Theorem 4 was BDD blow-up due to the combined presence of a shifter and adder. As has been reported by others [2,5], good variable orderings exist for fixed differences of exponents; however, the BDD size for the combination of variable shifting and addition is exponential with respect to the width of the datapath. Note that both the specification and implementation contain this combination, so we could not represent either the complete implementation or the complete specification with BDDs. Our strategy was to do structural decomposition on *both* the specification and implementation.

Lemma 6 (Shiftermatch).
f_1 and f_2 are normal $\implies$
  (`spec_alignman` = `impl_alignman`) $\wedge$
  (`spec_bigman` = `impl_bigman`) $\wedge$
  good_properties(`spec_alignman`, `spec_bigman`)

We decomposed the specification and implementation at the output of the shifter at the `alignman`, `bigman` and `sticky` signals. This decomposed the verification into two steps: from the inputs of the circuit through the shifter, and the adder and the rounder shown in Fig. 10. The two steps are shown in different shades. Through judicious use of cone of influence reductions and other optimizations to both the specification and implementation, we were able to verify the adder and the rounder without needing to include the shifter.

For the first half of the circuit, we verified that the output of the shifter in the specification is equivalent to that in the implementation. Our initial specification for the rounder was that for equal inputs, the specification and the implementation return equal outputs. However, we discovered that the implementation was optimized such that it returned correct results only for legal combinations of exponent differences and pre-rounded mantissas.

We then began a trial and error effort to derive a set of properties about the difference between the exponents and the pre-rounded mantissa. We eventually identified close to a dozen properties that, taken together, sufficiently constrained the inputs to the rounder such that the implementation matched the specification. Some of the properties are implementation dependent and some are of a more general nature. A few examples of implementation independent properties are:

- The leftmost exponent-difference number of bits of `spec_alignman` must be all $0s$.
- For exponent differences larger than a specific value, the `sticky` bit and the disjunction of a fixed implementation-dependent number of least significant bits in `spec_alignman` must be 1.
- The `expdiff+1`th leftmost bit of `spec_alignman` must be 1.

We modified the original lemma for the first half of the circuit to include these properties. We were able to verify Theorem 4 after removing both the shifters and enforcing the lemmas strengthened with good properties for the shifter and sticky bit.

The above approach illustrates how similar hierarchies present in the implementation and specification can be effectively used to decompose a verification task into smaller and more manageable sub-tasks. We have observed that this technique is particularly applicable when verifying an implementation against a reference model.

In [5], the authors present an approach for verification of floating-point adders based on word-level SMV and multiplicative power HDDs (*PHDDs). In their approach, the specifications of FP adders are divided into hundreds of implementation-independent sub-specifications based on case analysis using the exponent difference. They introduced a technique called short-circuiting to handle the complexity issues arising due to ordering problems.

Our approach uses a model-checker based on BDDs as opposed to word-level model-checker or *PHDDs. Our approach involves a small number of decompositions, as opposed to hundreds of cases. However in our approach, the decomposition specifications are implementation-dependent. We believe that for different implementations one can use the same basic approach towards decomposition. The *good_properties* will require the user to incorporate implementation-dependent details. Also, the hierarchical decomposition separates the shifter and the adder and hence does not require any special techniques such as short-circuiting to handle the verification complexity due to conflicting orders requirement.

5   Memory Arrays

Modern microprocessors contain dozens of first-in first-out (FIFO) queues. Certainly, the functional correctness of the processors depends upon the correct behavior of every FIFO in their designs. Due to the evolution of processors (further pipelining, wider datapaths, and so on), verifying just one typical FIFO can be a challenging task, especially for large FIFOs that are implemented as memory-arrays for area/performance optimizations. Verification of such FIFOs requires decomposing the problem into invariants about the control and datapath implementations. Our work does not refute this notion. Yet, although the proof structure of every FIFO we examined always required many invariants at various level of abstraction, we were able to leverage commonalities in the decomposition paths and methods, even for FIFOs being used in entirely different manners and being verified by different individuals.

5.1 Implementations of Memory Arrays

Memory arrays used as FIFOs are implemented as the array combined with a set of finite state machines that control a write (tail) pointer and a read (head) pointer (Fig. 11). Each pointer advances downward with its respective operation, incrementing by the number of elements written to or read from the array, and "wrapping" to the first (top) element when leaving the last (bottom) element. These pointers may take various forms depending upon the physical layout of the memory—linear arrays require only addresses whereas two-dimensional arrays may require an address and an offset into it, which may be implemented as slices of a single address.

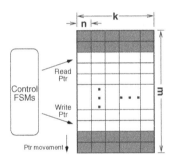

Fig. 11. Memory array circuit

Optimizations for area and performance usually lead to two-dimensional arrays rather than one-dimensional arrays. We use the term *slot* as the storage elements necessary for one queue element, and *row* as a group of slots associated with one address. Thus, an array consists of m rows with k slots per row containing n bits of data.

In one of the arrays we verified, all operations (writes and reads) access an entire row at a time. Other arrays allowed *spanning* writes and/or *partial* reads and writes (see Fig. 12). A spanning operation accesses slots in adjacent rows at the same time. A partial operation accesses fewer than k slots, and may or may not be spanning. In a partial access, the offset always changes but the row pointer may not, and vice versa for a spanning operation.

Optimizations in the control FSMs also lead to simplifications that tend to be problematic for verification and decomposition efforts. For example, the FSMs rarely, if ever,

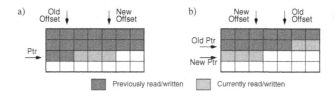

Fig. 12. a) Partial operation, b) Spanning (and partial) operation

take into consideration the position of both pointers. Thus, when examining only the read pointer logic there is nothing to prevent the read pointer from passing the write pointer (an unacceptable behavior). In the absence of constraint properties, only by including the logic of both pointers *and* the datapath could we prevent this from occurring. This detail alone nearly yields memory-array verification intractable for model checkers.

5.2 Overview of Memory Array Verification

Every FIFO in a processor has one or more correctness invariants that the design must satisfy. Verifying most top-level invariants requires reasoning about the behavior of the entire control and datapath. One array we worked with contained over 2100 variables in the necessary logic. Therefore, decomposition was a necessity.

Our technique for verifying memory-array invariants is summarized as follows:

1. Divide the memory array into symmetric, tool-manageable *slices* (e.g., slice equals one or more slots per row, or slice equals one or more rows.)
2. Define a set of properties applicable to every slice.
3. Define properties about the control FSMs that are as abstract as possible (for cone reduction) yet strong enough to guarantee the top-level invariant.
4. Verify the top-level invariants using the properties from 2 and 3.
5. For each slice, verify that if the properties about the control FSMs and the slice hold in a state, then the slice properties will hold in the next state.
6. For each property about the control FSMs, verify that if all properties about the control FSMS hold in a state, then the property will hold in the next state.
7. For any slice or control FSM property that is still intractable with the tools, repeat the decomposition procedure.

We identified a set of five invariants upon the read and write pointers that were used in nearly every memory array abstraction:

- The write pointer changes only when a write occurs.
- The read pointer changes only when a read occurs.
- If a row pointer changes, it increments by one (or more, depending upon the arrays input and output widths) and wraps at the maximum row.
- The read pointer will not pass the write pointer (begin reading bogus data)
- The write pointer will not pass the read pointer (overwrite existing data)

Arrays that allow partial and spanning operations contain offsets indicating which slots in a row are being accessed. Interactions between the offset circuitry and the memory array elements require additional invariants:

- If a read (write) occurs and the row pointer does not change, then the offset increments.
- If a read (write) occurs and the row pointer changes, then the offset decrements or stays the same.

With modern processor designs, reducing the complex nature of the circuitry controlling most FIFOs requires several hierarchical steps just to get to the abstractions above. These levels of the proof trees deal with implementation details for each particular FIFO.

The (deeply) pipelined nature of the array's datapath and control often requires temporally shifting the effects of these properties in order to substantially reduce the variable counts in "higher" proofs. In other words, using assumptions upon signals in different pipeline stages incurs a penalty if the signals must propagate to further pipestages. It is better to use existing assumptions, which cover the "wrong" pipestages, to prove similar specifications about the propagated signals, especially as the level of abstraction (and freedom given to the model checker) increases. FSM and datapath sizes and design complexities often required temporally shifting the same property to several different times to align with the various FIFO processing stages.

5.3 Array Examples

Here we present two examples of the memory-array verifications we carried out.

Our first example is a "write-once" invariant. This property requires that every slot is written to only once (to prevent destruction) until the read pointer passes it. The FIFO allowed for spanning and partial writes. Thus, every slot on every row has its own write enable line that is calculated from the write pointer, next offset, and global write enable.

For this invariant we sliced the array by row so that the final proofs involved m *predicate variables*. The predicate variables were defined as sticky bits that asserted if the write enable bits on the row behaved incorrectly. In this "write-once" context, incorrect behavior of the write enables includes the following:

1. The row is not being written to but one or more of the write enables asserted.
2. During a write to the row, a write enable for a slot behind the write pointer (and offset) and ahead of the read pointer (and offset) was asserted.

Note that there are other aberrant behaviors of the write enables but they fall outside the scope of the top-level invariant. They could be included in the predicate variable definitions without affecting the correctness of the final proof. However, needlessly extending the definitions runs the risk of undoing the necessary cone reductions.

The complexity in this verification comes from the inclusion of the read pointer in the predicate definitions. As mentioned before, the control FSMs generally do not take into account the behavior or values of the other pointers or FSMs. This FIFO's environmental requirement is that it will never receive more items than it can hold; otherwise, the write pointer could pass the read pointer. Even with this constraint we were forced to fully abstract the read and write pointers before attempting to prove that the predicate bits would never assert.

The last step was to sufficiently abstract the behavior of the predicate variables so they could all be used in the top-level proof. The following invariants were sufficient:

1. Initially, the predicate bit is not asserted.
2. If a row is not being written, its predicate bit does not change.
3. If a row is being written, its predicate bit does not transition from 0 to 1.

These invariants reduce the behavior of the predicate bits to a relation with the write pointer. Together they clearly satisfy the top-level specification (no predicate bit will ever become asserted, thus all write enables behaved as expected). While they may appear simple, the size of the memory array and the capacity limitations of model checking deemed them necessary.

Our second example was our first full-scale attempt at verifying an invariant upon an entire memory array. In this FIFO, control information needed by the read pointer's FSMs is written into the array by the write FSMs. The FIFO's correctness depends upon it being able to write valid and correct control data into the array and then safely and accurately read the data out.

Ideally we would have liked to have been able to prove the correctness of the read FSMs while allowing the control information to freely be any value of the model checker's choosing. However, it quickly became apparent that circuit optimizations required constraining the control data to a "*consistent*" set of legal values. Because rigorous verification work does not particularly enjoy the idea of correct data appearing for free, we were forced to show that control information read from the array always belonged to this consistent set. Fortunately, the idea of a valid set of bit values lends itself well to the notion of predicate variables.

The number of bits in the control data set required slicing the array twice: once by row, and then by slot per row. The slot-wise predicate bits indicated whether the slot's current control data was consistent. The row-wise bits, of which there were actually two sets, examined the predicate bits for the entire row while considering the write pointer and offset (or the read pointer and offset). Because the control information flows through the entire FIFO, we had to prove consistency in the following progression through the array's datapath:

- At the inputs to the memory array, for every slot that will be written, the control data is consistent.
- For each array row, if the write pointer moves onto or stays on the row, control data in each slot behind the (new) write offset is consistent.
- For each row, if the write pointer moves off the row, the entire row is consistent.
- For each row, if the write pointer neither moves onto nor moves off of the row, the consistency of each slot's control data does not change.
- For each row, if the read pointer is on the row, and if the write pointer is on the row, then all slots behind the write offset are consistent.
- For each row, if only the read pointer is on the row, all slots are consistent.
- At the memory array's output muxes, when a read occurs, the *active* output slots contain consistent control data, where *active* is defined by the write and read pointers

and offsets (i.e., if pointers on same row, active is between the offsets; otherwise active is above read offset.)

The last step establishes that we will always received consistent control data from the memory array.

As with the FIFO of the write-once invariant, we spent a significant effort reducing the read and write FSM logic to the minimum required behaviors and proving these behaviors at various stages in the pipeline (to align with the FIFO stages described above).

6   Conclusion

Related work in decomposition and model checking can be grouped into three large categories: sound combinations of model checking and theorem proving, ad-hoc combinations of different tools, and hard-coding inference rules into a model checker.

Sound and effective combinations of theorem proving and model checking has been a goal for almost ten years. Joyce and Seger experimented with using trajectory evaluation as a decision procedure for the HOL proof system [13]. They concluded that for hardware verification, most of the user's interaction is with the model checker, not the proof system. Consequently, using a model checker as a decision procedure in a proof system does not result in an effective hardware verification environment. The PVS proof system has included support for BDDs and mu-calculus model checking [17] and, over time, has received improved support for debugging. Aagaard et al have described extensions to the Voss verification system that includes a lightweight theorem proving tool tightly connected to the Voss model checking environment [3,1]. Gordon is experimenting with techniques to provide a low-level and tight connection between BDDs and theorem proving in the HOL proof system [10].

There are two advantages to combining theorem proving and model checking over a pure model checking based approach, such as we have used. First, general purpose logics provide greater expressability than specification languages tailored to model checking. Second, the sound integegration of model checking and theorem proving allows more rigorous results. The principal advantage of the approach outlined here was pragmatic: it enabled us to achieve a more significant result with less effort. Hierarchical model checking allowed an existing group of model checking users to extend the size and quality of their verifications with relatively minor costs in education.

Because finding an effective and sound combination of theorem proving and model checking has been so difficult, a variety of ad hoc combinations have been used to achieve effective solutions at the expense of mechanically guaranteed soundness. Representative of these efforts is separate work by Camilleri and Jang et al. Camilleri [4] has used both theorem-proving and model-checking tools in verifying properties of a cache-coherency algorithm. Jang et al [12] used CTL model checking to verify a collection of 76 properties about an embedded microcontroller and informal arguments to convince themselves that their collection of properties were sufficient to claim that they verified their high-level specification.

In an effort to gain some of the capacity improvements provided by a proof system without sacrificing the automation of model checking, McMillan has added inference rules for refinement and decomposition to the Cadence Berkely Labs SMV (CBL SMV) model checker [16]. Eiríksson has used CBL SMV to refine a high-level model of a protocol circuit to a pipelined implementation [9].

The model checker we use shares characteristics of ad hoc combinations of techniques and of adding theorem proving capabilities to a model checker. In essence, the model checker provides us with a decision procedure for propositional logic. This allows us to stay within a purely model checking environment and prove propositional formulae that can be solved by BDDs. However, because we cannot reason about the syntax of formulas, we cannot do Modus Ponens reasoning on arbitrary chains of formulas. Hence, the complexity of our decompositions is limited by the size of BDDs that we can build and we use a separate database facility to ensure the integrity of the overall decomposition.

Although we do not use a theorem prover, we have found that experience with theorem proving can be useful in finding effective decomposition strategies. The challenge is to find *sound* and *effective* decomposition techniques. Assuring the soundness of a decomposition technique benefits from a solid grounding in mathematical logic. Mathematical expertise can also be helpful by providing background knowledge of a menu of decomposition techniques to choose from (e.g., temporal induction, structural induction, data abstraction, etc.). In most situations, many sound decomposition techniques are applicable, but most will not be helpful in mitigating the capacity limitations of model checking. Picking an effective decomposition technique requires knowledge of both the circuit being verified and the model checker being used. We often found that slight modifications to specifications (e.g. the ordering of temporal or Boolean operators) dramatically affect the memory usage or runtime of a verification. Over time, our group developed heuristics for writing specifications so as to extract the maximum capacity from the model checker.

Decomposition techniques similar to the ones that we have described have become standard practice within our group. For this paper we selected illustrative examples to give some insights into how we use decomposition and why we have found it to be successful. Table 2 shows some statistics of a representative sample of the verifications that have used these techniques.

Table 2. Circuit sizes and verification results

Circuit	total latches	average number of latches per decomposition	number of verification runs	maximum memory usage	total run time	
t-mem	903	9	70	830MB	27h	
f-r-stall	210	160	10	150MB	23h	
rt-array	500	311	6	110MB	14h	
rq-array	2100	140	300	120MB	100h	Section 5
all-br	1500	175	200	600MB	170h	Section 5
m-buf-com	200	140	20	250MB	20h	
fpadd	2000	400	100	1200MB	2h	Section 4
fpsub	2500	500	100	1800MB	7h	

The formal verification effort for the Intel Pentium® 4 processor relied on both formal verification and conventional simulation-based validation. The goal, which in fact was achieved, was that simulation would catch most of the errors and that formal verification would locate pernicious, hard-to-find errors that had gone undetected by simulation.

Acknowledgments. We are indebted to Kent Smith for suggesting the decomposition strategy for the parallel ready queue and many other helpful discussions. We also would like to thank Bob Brennan for the opportunity to perform this work on circuits from Intel microprocessors and Ravi Bulusu for many discussions and helpful comments on the floating-point adder verification.

References

1. M. D. Aagaard, R. B. Jones, K. R. Kohatsu, R. Kaivola, and C.-J. H. Seger. Formal verification of iterative algorithms in microprocessors. In *DAC*, June 2000.
2. M. D. Aagaard, R. B. Jones, and C.-J. H. Seger. Formal verification using parametric representations of Boolean constraints. In *DAC*, July 1999. (Short paper).
3. M. D. Aagaard, R. B. Jones, and C.-J. H. Seger. Lifted-fl: A pragmatic implementation of combined model checking and theorem proving. In L. Thery, editor, *Theorem Proving in Higher Order Logics*, pages 323–340. Springer Verlag; New York, Sept. 1999.
4. A. Camilleri. A hybrid approach to verifying liveness in a symmetric multi-processor. In E. L. Gunter and A. Felty, editors, *Theorem Proving in Higher Order Logics*, pages 49–68. Springer Verlag; New York, Sept. 1997.
5. Y.-A. Chen and R. Bryant. Verification of floating-point adders. In A. J. Hu and M. Y. Vardi, editors, *CAV*, pages 488–499, July 1998.
6. E. M. Clarke, O. Grumberg, and D. E. Long. Model checking and abstraction. *ACM Trans. on Prog. Lang. and Systems*, 16(5):1512–1542, Sept. 1994.
7. E. M. Clarke, D. E. Long, and K. L. McMillan. Compositional model checking. In *LICS*, pages 353–362, 1989.
8. E. M. J. Clarke, O. Grumberg, and D. A. Peled. *Model Checking*. The MIT Press; Cambridge, MA, 1999.
9. A. P. Eiríkson. The formal design of 1m-gate ASICs. In P. Windley and G. Gopalakrishnan, editors, *Formal Methods in CAD*, pages 49–63. Springer Verlag; New York, Nov. 1998.
10. M. Gordon. Programming combinations of deduction and BDD-based symbolic calculation. Technical Report 480, Cambridge Comp. Lab, 1999.
11. IEEE. *IEEE Standard for binary floating-point arithmetic*. ANSI/IEEE Std 754-1985, 1985.
12. J.-Y. Jang, S. Qadeer, M. Kaufmann, and C. Pixley. Formal verification of FIRE: A case study. In *DAC*, pages 173–177, June 1997.
13. J. Joyce and C.-J. Seger. Linking BDD based symbolic evaluation to interactive theorem proving. In *DAC*, June 1993.
14. G. Kamhi, L. Fix, and O. Weissberg. Automatic datapath extraction for efficient usage of hdds. In O. Grumberg, editor, *CAV*, pages 95–106. Springer Verlag; New York, 1997.
15. S. Mador-Haim and L. Fix. Input elimination and abstraction in model checking. In P. Windley and G. Gopalakrishnan, editors, *Formal Methods in CAD*, pages 304–320. Springer Verlag; New York, Nov. 1998.
16. K. McMillan. Minimalist proof assistants: Interactions of technology and methodology in formal system level verification. In G. C. Gopalakrishnan and P. J. Windley, editors, *Formal Methods in CAD*, page 1. Springer Verlag; New York, Nov. 1998.

17. S. P. Miller and M. Srivas. Formal verification of the AAMP5 microprocessor: A case study in the industrial use of formal methods. In *Workshop on Industrial-Strength Formal Specification Techniques*, Apr. 1995.

18. Y. Xu, E. Cerny, A. Silburt, A. Coady, Y. Liu, and P. Pownall. Practical application of formal verification techniques on a frame mux/demux chip from Nortel Semiconductors. In L. Pierre and T. Kropf, editors, *CHARME*, pages 110–124. Springer Verlag; New York, Oct. 1999.

Pruning Techniques for the SAT-Based Bounded Model Checking Problem

Ofer Shtrichman

The Minerva Center for Verification of Reactive Systems, at the Dep. of Computer Science and Applied Mathematics, The Weizmann Institute of Science, Israel; and IBM Haifa Research Lab
ofers@summer.weizmann.ac.il

Abstract. Bounded Model Checking (BMC) is the problem of checking if a model satisfies a temporal property in paths with bounded length k. Propositional SAT-based BMC is conducted in a gradual manner, by solving a series of SAT instances corresponding to formulations of the problem with increasing k. We show how the gradual nature can be exploited for shortening the overall verification time. The concept is to reuse constraints on the search space which are deduced while checking a k instance, for speeding up the SAT checking of the consecutive $k+1$ instance. This technique can be seen as a generalization of 'pervasive clauses', a technique introduced by Silva and Sakallah in the context of Automatic Test Pattern Generation (ATPG). We define the general conditions for reusability of constraints, and define a simple procedure for evaluating them. This technique can theoretically be used in any solution that is based on solving a series of closely related SAT instances (instances with non-empty intersection between their set of clauses). We then continue by showing how a similar procedure can be used for restricting the search space of individual SAT instances corresponding to BMC invariant formulas. Experiments demonstrated that both techniques have consistent and significant positive effect.

1 Introduction

SAT-based verification of invariants ($\mathbf{AG}p$) has been practiced for quite some time (see, for example [6]) under different names and for various verification tasks. Biere et. al. recently introduced the notion of Bounded Model Checking (BMC) [1], which extends these methods to LTL and reduces the verification problem to a pure propositional satisfiability problem. By doing so, it enables to exploit the power of advanced standard CNF-SAT solvers.

The basic concept of BMC is to search for a counter example in executions whose length is bounded by some integer k. For every model M, there exists a finite bound D, called the *Diameter* of M, such that M satisfies a property p iff no trace shorter or equal to D contradicts p. Thus, for a large enough k, this method is complete.

The BMC problem can be efficiently reduced to a propositional satisfiability problem whose size, in terms of number of variables, is linear in k. Since SAT is

T. Margaria and T. Melham (Eds.): CHARME 2001, LNCS 2144, pp. 58–70, 2001.

worst case exponential in the number of variables, k has a crucial effect on the ability to efficiently solve the BMC instance. Verification with BMC is normally based on a gradual process, where k is increased until one of the following occurs: a bug is found, the diameter D is reached, or the problem becomes intractable. In fact, experiments with real designs have shown that it is seldom the case that unsatisfiable instances (corresponding to bug free designs) can be efficiently solved for $k = D^1$. Several methods were suggested recently to cope with this problem, including a procedure which can be seen as an extended version of the classic inductive proof: first, the property is proven correct up to cycle k. Then, the procedure checks whether this fact implies that the property is also true in cycle $k + 1$. If not, k is increased, with the hope that the process will stop before reaching $D[11]$. In any case, BMC seems to be far more successful in falsification than in verification.

The tool BMC that was developed as part of [1], which reduces SMV-compatible models to a corresponding CNF-SAT problem, made it possible to evaluate these methods in comparison with standard BDD-based model checkers. Several such comparisons [2,12], caused BMC to gain recognition as a technique that can frequently outperform classic BDD-based model checking.

In a previous research we demonstrated how the unique structure of BMC invariant formulas can be exploited for various optimizations in the SAT solver, including pre-computation of variable ordering and addition of constraints on the search space [12]. In this paper we continue to explore ways in which generic CNF SAT solvers can be tuned for BMC or for other domains with similar characteristics. In particular, we investigate the possibility of exploiting BMC's gradual nature for speeding up the overall verification time. We will show how it is possible to exploit information gathered while solving a k-instance, for solving faster the consecutive $k + 1$ instance. The basic idea is to reuse clauses that were deduced while solving previous instances. These clauses, called *conflict clauses* for reasons we will later explain, are naturally recorded in the standard SAT procedure with the aim of pruning parts of the search tree.

A similar idea was proposed by Silva and Sakallah [8] for the case of Automatic Test Pattern Generation (ATPG). They refer to the reused clauses as *pervasive clauses* and explain, in ATPG terms, under what conditions they are formed: if a circuit is tested with two fault models (i.e. the circuit formula is conjuncted with different formulas, each representing a different fault state. See the above reference for more details), the conflict clauses that were deduced from the circuit itself when checking the first instance are declared pervasive. These clauses can therefore be reused when checking the second instance. The authors define the more general question of 'when can clauses be declared pervasive' as an open problem. In this sense this paper addresses this challenge: we investigate the necessary conditions under which a conflict clause can be shared by two or more general SAT instances, and show a simple decision procedure for their evaluation. In Section 5 we will show how a similar procedure can be used as part

---

^{1} Finding D in of itself is a hard problem, which we will not discuss in this paper.

of a different technique, called *constraints replication*, to add more constraints to a *single* SAT instance.

Experiments with both techniques proved their effectiveness. In a significant number of the test cases the overall verification time was reduced by 50 percent or more. More important, these improvements are rather consistent. As far as our experiments can show (15 different designs), the new techniques consistently reduces or leaves almost unaffected the solving time. Consistency has a strong practical advantage: rather than implementing it as a one more user activated flag, it encourages a change in the default configuration of the relevant tools.

The rest of the paper is structured as follows. We begin by giving necessary background on BMC and SAT in the next two sections. In Section 4 we describe the technical details of the suggested decision mechanism, and prove its soundness. In Section 5 we show how the same technique can be used for restricting the search space within a given SAT instance, as long as this instance stems from an invariant formula. In Section 6 we describe another related work called *incremental satisfiability*, which we believe will be better understood after reading the suggested method. Experimental results from our benchmark are given in Section 7, and some conclusions and ideas for future research are presented in Section 8.

2   Bounded Model Checking of Invariants

We focus on bounded model checking of invariants ($\mathbf{AG}p$ formulas). The general structure of the corresponding BMC instance is the following:

$$\varphi_j^k : \quad I_0 \wedge \bigwedge_{i=0}^{k-1} \rho(i, i+1) \wedge (\bigvee_{i=j}^{k} \neg P_i) \tag{1}$$

where I_0 is the initial state, $\rho(i, i+1)$ is the transition between cycles i and $i+1$, and P_i is the property in cycle i. Thus, this formula can be satisfied iff there exists a reachable state in cycle i ($j \le i \le k$) which contradicts the property P_i. The values of j and k can vary according to the range in which we are looking for the bug. $j = 0$ and $j = k$ are the two extremes corresponding to a full and exact search, respectively.

Our experiments were made on top of the enhanced versions of BMC [1] and Grasp [13,14], as were described in [12]. BMC takes an SMV – compatible model and generates a propositional SAT instance according to Equation (1). It also uses various heuristics to generate a variable ordering file, which is later used by Grasp to accelerate its search.

3   SAT Checking and Resolution

In this section we briefly outline the principles adopted by modern propositional SAT-checkers, and in particular those which Grasp is based on. Our description follows closely the one in [13].

Most of the modern SAT-checkers are variations of the well known Davis-Putnam procedure [4]. The procedure is based on a backtracking search algorithm that, at each node in the search tree, first decides on an *assignment* (i.e. both a variable and a Boolean value, which determines the next sub-tree to be traversed) and then iteratively applies the unit clause rule. The procedure backtracks once the current partial assignment contradicts one of the clauses. Each time that such a conflict occurs, `Grasp` analyzes the cause of the conflict. This analysis produces two distinct pieces of information:

1. The decision level which the procedure should backtrack to. Unlike the original Davis-Putnam procedure, `Grasp` supports non-chronological backtracks, thus if the current decision level is d, it can jump back to $d-i$ where $1 \leq i \leq d$. The mechanism for computing the backtrack level is elaborated in [13].
2. New clauses, called *conflict clauses*, are resolved and added to the clause database, thereby avoiding future occurrences of the same conflict. For example, if the assignment $x = T, y = F, z = F$ inevitably leads to a conflict, the addition of the conflict clause $\pi = (\neg x \lor y \lor z)$ will cause the search procedure to backtrack immediately if the above assignment is repeated. In Section 4.1 we will further elaborate on the resolution process which `Grasp` uses for computing these new clauses.

Conflict clauses are of special interest to us, because they possess valuable information for restricting the search. They are a result of time-consuming reasoning process, which can potentially be shared between SAT instances. It should be noted here that adding clauses that are consistent with the SAT instance (without adding new variables) typically makes the instance easier to solve, because it prunes parts of the search tree. This is only an empirical observation, not a theoretical result. Additional clauses can also slow down the process. First, there is an overhead associated with more clauses. This overhead is significant especially when deciding dynamically the next variable. Typically the next variable is chosen by a procedure which loops over all literals, looking for e.g. the assignment which leads to the maximum number of satisfied clauses. More clauses, therefore, slows down this process[2]. Secondly, the added clauses are not equally effective. The addition of one clause can prevent the formation of another, more effective clause, by pruning the sub-tree in which the other clause would have been created. These potential overheads caused most modern SAT solvers to permit a user restriction on the size and number of added clauses.

4   Constraints Sharing

Sharing conflict clauses between SAT instances can be applied whenever solving two or more SAT instances with non-empty intersection between their clauses sets. Constraints sharing is thereby expected to be far more effective in cases

---

[2] In our case we used predetermined static variable ordering, which eliminates this particular overhead.

where the solution is based on solving a series of SAT instances which share a large number of clauses. BMC and AI Planning problems [9] are two such cases. Pervasive clauses, the restricted version of constraints sharing, was also used in the past for several EDA problems [8,5], as was previously mentioned.

We begin the description of this technique with several simple definitions. In the following discussion, we use the same variables to denote CNF formulas and their associated sets of clauses. The difference will be clear from the context.

Let S_1 and S_2 be two sets of clauses associated with two CNF SAT instances, and φ_0 represent the set of clauses that are common to S_1 and S_2, i.e. $\varphi_0 = S_1 \cap S_2$. We will also need $\varphi_i = S_i \setminus \varphi_0$ ($i \in \{1, 2\}$), the non-overlapping subsets of S_1 and S_2. Finally, let ψ be a set of clauses that is deducible from φ_0, denoted by $\varphi_0 \vdash \psi$. Based on the following claim, we will be able to reuse ψ (which is computed while checking S_1) when checking S_2, by checking $S_2 \wedge \psi$:

Claim. if $\varphi_0 \vdash \psi$ then S_2 is satisfiable iff $S_2 \wedge \psi$ is satisfiable.

The claim is easy to justify: since $\varphi_0 \vdash \psi$ then $S_2 \vdash \psi$, which implies that $S_2 \leftrightarrow S_2 \wedge \psi$. Thus, S_2 is satisfiable iff $S_2 \wedge \psi$ is satisfiable.

In the general case, it is not common that two SAT instances share a large number of clauses. There is also a difficulty in mapping the variables between the two instances. However, according to Equation (1) it is apparent that with the exception of the clause $c_p : (\bigvee_{i=j}^{k} \neg P_i)$, φ_j^k is a subset of φ_{k+1}^t for all $t > k$. Thus, φ_1 is comprised of the single clause c_p[3].

In order to compute ψ, we need to isolate it from the set of conflict clauses that are deduced while checking S_1. Only then we can reuse it while checking S_2. One solution to the isolation problem is to check φ_0 rather than S_1. In the BMC case this can be done by omitting c_p from S_1. However, there are two drawbacks to this solution. First, φ_0 represents the transition relation, which is assumed to be consistent and therefore satisfiable. Experiments with this option demonstrated that typically φ_0 is trivially satisfied and, as a result, only a small number of conflict clauses are computed. Second, unlike solving S_1, this is an extra computation task which we prefer to avoid.

Thus, we are looking for a method to isolate ψ while checking S_1. Before we suggest an isolation mechanism, in the next sub-section we describe in more detail the mechanism which `Grasp` uses for computing conflict clauses.

4.1   Derivation of Conflict Clauses

We explain the mechanism of deriving new conflict clauses by following a simplified version of an example first given by Silva and Sakallah in [13]. Assume the clause data base includes the clauses listed in Fig. 1(a), the current truth assignment is $\{x_5 = 0\}$, and the current decision assignment is $x_1 = 1$. Then the resulting *partial implication graph* depicted in Fig. 1 (b) describes the unit clause propagation process implied by this decision assignment.

---

[3] Note that this is true even if P_i is not an atomic proposition. In this case the equivalence $\bigvee_{i=0}^{n} P_i \equiv \bigwedge_{i=0}^{n} (p_i = P_i) \wedge \bigvee_{i=0}^{n} p_i$ is used, where p_i is a new propositional variable.

$c_1 = (\neg x_1 \vee x_2)$

$c_2 = (\neg x1 \vee x_3 \vee x_5)$

$c_3 = (\neg x_2 \vee x_4)$

$c_4 = (\neg x_3 \vee \neg x_4)$

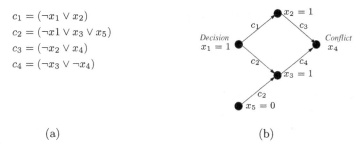

(a)   (b)

Fig. 1. A clause data base (a) and a partial implication graph (b) of the assignment $x_1 = 1$ shows how this assignment, together with assignments that were made in earlier decision levels, leads to a conflict.

Each node in this graph corresponds to a variable assignment. The incoming directed edges $(x_1, x_j)...(x_i, x_j)$ labeled by clause c represent the fact that $x_1...x_i, x_j$ are c's literals and that the current value of $x_1, ..., x_i$ implies the value of x_j according to the unit clause rule. Thus, vertices that have no incoming edges correspond to decision assignments. The partial implication graph in this case ends with a conflict vertex. Indeed the assignment $x_1 = 1$ leads to a conflict in the value of x_4, which implies that either c_3 or c_4 cannot be satisfied. When such a conflict is identified, **Grasp** determines those variable assignments that are directly responsible for the conflict. In the above example these are $\{x_1 = 1, x_5 = 0\}$. The conjunction of these assignments therefore represents a sufficient condition for the conflict to arise. Consequently, the negation of this conjunction must be satisfied if the SAT instance is satisfiable. We can thereby add the new conflict clause $\pi : (\neg x_1 \vee x_5)$ to the clause database, with the hope that it will speed up the search.

4.2   Isolating ψ

In Section 4 we argued that it is necessary to identify those conflict clauses which are deduced solely from φ_0. These will be the reusable pervasive clauses. The description of the derivation process in the previous subsection sheds light on how this can be achieved. Under the assumption that it is possible to identify in advance the partition of S_1 into φ_1 and φ_0, we suggest the following isolation procedure:

1. Mark φ_0 clauses.
2. For every conflict clause π, if all clauses leading to the conflict are marked, then mark π and add it to ψ.

In the BMC case, marking φ_0 clauses is easy, because we know that all clauses except c_p belong to φ_0.

We demonstrate the isolation procedure by considering a case in which c_1 from the example in Section 4.1 is in φ_1, and $c_2...c_4 \in \varphi_0$. According to step 1, $c_2...c_4$ are marked. While resolving the conflict in e.g. c_4, we observe that the unmarked clause c_1 is one of the clauses that lead to the conflict. We therefore do not mark the new conflicting clause π and do not add it to ψ.

Claim. By following the isolation procedure, we compute ψ such that $\varphi_0 \vdash \psi$.

Proof. The set of clauses in S_1 in any given time is comprised of three distinct subsets: φ_0, φ_1 and φ_π, where φ_π is the set of the dynamically added conflict clauses. To prove the claim we use induction on the size of φ_π. Initially φ_π and ψ are empty, thus obviously $\varphi_0 \vdash \psi$ is true in the base case. For the induction step, we assume $\varphi_0 \vdash \psi$ and add π to φ_π. We will focus on those cases where ψ is updated and denote the updated ψ by ψ', i.e. $\psi' = \psi \wedge \pi$. There are two such cases:

1. π is derived from φ_0 only. In this case $\varphi_0 \vdash \pi$. Together with the induction hypothesis we get $\varphi_0 \vdash \psi \wedge \pi$, which implies $\varphi_0 \vdash \psi'$.
2. Otherwise, π is derived from $\varphi_0 \cup \Pi$ (where Π is a subset of φ_π). We are interested in the case in which ψ is updated. According to step 2, this can only happen if all the clauses in Π are marked. In this case we have:

   (1) $\varphi_0 \vdash \psi$ (induction hypothesis)
   (2) $\varphi_0 \wedge \Pi \vdash \pi$ (assumption of case 2)
   (3) $\psi \vdash \Pi$ (Π's clauses are marked, therefore they were added to ψ.)
   (4) $\varphi_0 \vdash \Pi$ (from 1,3)
   (5) $\varphi_0 \vdash \pi$ (from 2,4)
   (6) $\varphi_0 \vdash \psi \wedge \pi$ (from 1, 5)

   and from (6), we have that $\varphi_0 \vdash \psi'$.

   $\square$

Once ψ is computed and saved to a file, we can simply merge it, after mapping the variable names, with S_2. $S_2 \wedge \psi$ should typically be solved faster than S_2 alone.

4.3   Implementation

While so far we referred to two SAT instances, the gradual nature of BMC allows to accumulate information from all previously checked instances with a lower k.

In the previous subsection we showed, given the list of φ_0 clauses, how to alter the SAT checker so it can generate ψ . In Fig. 2 we suggest a procedure which, based on this new feature, merges constraints sharing into the iterative BMC process. Constraints from previous runs are saved in a file called <model>.psi together with their corresponding k (line 2 in the procedure refers to this figure through the variable 'index'). These constraints are later merged into each new instance with a higher k. The procedure is bounded by the global variable D which holds the diameter of the design.

Bool **Solve** (model M, from j, to k, jump size jmp)
    1. Generate the file M.j-k.cnf where φ_0 clauses are marked.
    2. Add M.psi clauses with index $< k$ to M.j-k.cnf. Mark them as φ_0 clauses.
    3. SAT-solve M.j-k.cnf while adding ψ clauses to M.psi.
    4. If $result = unsatisfiable$
        if $(k < D)$ return **Solve** $(M, k + 1, k + jmp, jmp)$ else return True.
    else print trace, and return False.

Fig. 2. An iterative Bounded Model Checking procedure with constraints sharing.

5   Internal Constraints Replication

In [12] we suggested a technique called *constraints replication*, which adds constraints to φ_0^k based on the almost symmetric structure of this formula (without the initial state I_0, φ_0^k has k equal parts up to variable names). In order to describe this technique we will use two new notations: x_i denotes a variable x in cycle i, and $\pi^{(i)}$ denotes the clause obtained by shifting i cycles each variable in the clause π. For example, if $\pi = (x_3, y_6)$ then $\pi^{(2)} = (x_5, y_8)$. We will use a similar notation for set of clauses.

Let π be a conflict clause which is deduced from a set of clauses $S \subset \varphi_0^k$, i.e. $S \vdash \pi$. We claim that if $S^{(i)} \subset \varphi_0^k$, then $S^{(i)} \vdash \pi^{(i)}$. Consequently, the *replicated clause* $\pi^{(i)}$ is also a conflict clause which can be added to φ_0^k. The problem is that since φ_0^k is not completely symmetric (due to I_0), it is not always the case that $S^{(i)} \subset \varphi_0^k$ (ignoring, temporarily, the question of i's range). In [12] we suggested a two-step 'trial and error' approach to solve this problem. Given a conflict clause π, first generate all replicated clauses by simultaneously increasing or decreasing the variables indices in π, as long as they stay in the range $0..k$. In the second step, which we refer to as the simulation phase, check if the complement of each replicated clause indeed leads to a conflict (recall that by definition every assignment that satisfies the negation of a conflict clause must lead to a conflict). If yes, add $\pi^{(i)}$ to φ_0^k. If not – discard it.

5.1   An Alternative Solution

The problem with the simulation phase is that checking whether a given partial assignment leads to a conflict may require a large computational effort. If we choose to minimize the overhead by limiting the search time, we take the risk that some 'good' replicated clauses are discarded.

Based on a procedure similar to the one described in Section 4.2, we would now like to offer an alternative to the simulation phase. This method will always identify the good replicated clauses and will hardly require any overhead. For the sake of simplicity we will handle here φ_0^k rather than φ_j^k. Only minor adjustments are needed for handling the more general case.

Our goal is to check efficiently whether a given set of clauses S has a shifted set $S^{(i)}$, and compute the range of i. The following procedure utilizes φ_0^k's structure to achieve this goal:

1. While generating φ_0^k, mark each clause c if $c^{(i)} \in \varphi_0^k$ for $i = 0..k$ (all clauses except I_0 and c_p.
2. For every conflict clause π, if all clauses leading to the conflict (the S clauses) are marked, then mark π as 'replicable'. In addition, record l_S and h_S, the lowest and highest cycle index in S, respectively.
3. For each replicable clause π, add a replicated clause $\pi^{(i)}$ for i in the range $-l_S...(k - h_S)$.

Example 1. Consider the conflict clause $\pi : (\neg x_1 \vee z_2)$ which is deducible from the set of clauses S:

$$c_1' = (\neg x_1 \vee y_2) \qquad c_2' = (\neg x_1 \vee y_1 \vee z_2)$$
$$c_3' = (\neg y_2 \vee z_3) \qquad c_4' = (\neg y_1 \vee \neg z_3)$$

($c_1'..c_4'$ are structurally equivalent to $c_1..c_4$ of Fig. 1. Here we use the notation in which subscripts represent cycle numbers). If $c_1'..c_4'$ are marked in step 1, then π is replicable. We note that $l_S = 1$ and $h_S = 3$. Thus, if e.g. $k = 5$, we can add (in addition to π itself) the replicated clauses:

$$\pi^{(-1)} : (\neg x_0 \vee z_1) \qquad \pi^{(1)} : (\neg x_2 \vee z_3) \qquad \pi^{(2)} : (\neg x_3 \vee z_4).$$

5.2   When Do Replicated Clauses Become Pervasive?

After defining the conditions for adding clauses both outside (pervasive clauses) and inside (replicated clauses) the SAT instance, we now investigate the circumstances in which these two techniques can be combined, i.e. when do replicated clauses become pervasive. We once again use S to denote the set of clauses that imply the conflict clause π.

Claim. A replicated clause $\pi^{(i)}$ is pervasive if π is pervasive (where i is in the range as defined in step 2 of the procedure listed in Section 5.1).

Proof. Given that π is pervasive, it implies that $S \subset \varphi_0^{k+1}$. Since π is also replicable, S only includes clauses that were marked in step 1 of the procedure. Together this implies that $S^{(i)} \subset \varphi_0^{k+1}$ and therefore $\varphi_0^{k+1} \vdash \pi^{(i)}$. Thus, $\pi^{(i)}$ can be reused with φ_0^{k+1}. $\square$

6   More Related Work: Incremental Satisfiability

The idea of solving SAT instances incrementally can be attributed to Hooker [7]. He proposed an algorithm that given a satisfiable instance and an additional clause, it checks whether the satisfiability is preserved when the new clause is added to the formula. His experiments showed that solving large instances incrementally can be faster than solving them as one monolithic formula. It was later extended by Kim et al. [10] and used for path delay fault testing, a process in which the effect of faults on delays in certain paths is checked. The large number of paths typically requires the partition of the problem into a series of instances, each representing a subset of the tested paths. All the paths share the same prefix P, which empirically is far larger than the suffixes $s_1...s_i$. Incremental satisfiability is then used in the following way. A satisfying assignment for P is sought, and conflict clauses are added to P (those clauses that are deducible directly from P). If P is unsatisfiable, the process halts because the conjunction of P with S_i for all i is obviously unsatisfiable. Otherwise, the trace is used as an initial assignment when checking each of the instances $P \wedge S_i$ for all i. In case the initial trace does not lead to a satisfying assignment, the standard backtrack process is invoked.

The resemblance between their and our work is rather clear: the prefix P in their work is $\varphi_0 = S_1 \cap S_2$ in ours. The addition of conflict clauses that were computed while looking for a satisfying assignment to P is equivalent to the option of checking φ_0 directly. In Section 4 we argued that this is ineffective in the BMC case, because φ_0 is satisfied *too fast* for creating a substantial number of conflict clauses, which are essential for speeding up the search later.

7   Experimental Results

To experiment with the two suggested techniques, we randomly chose 15 different hardware designs from IBM's internal benchmark set. The results of this experiment show that constraints sharing has a consistent positive effect, or only marginal negative effect due to its overhead. However, as was explained in Section 3, this consistency can not be guaranteed for all future cases. Replicating clauses and sharing them, as described in Section 5, also had a very positive effect, although somewhat less consistent.

The results of the 15 cases can be divided into 3 groups: the first group includes 6 designs, which were solved at least 50 percent faster due to the suggested techniques. The second group includes 7 designs, which are solved very fast with or without the new techniques. The satisfying assignment is found in these cases before a significant number of conflict clauses are created, and therefore sharing them or replicating them has little effect, if any. In some of these cases the overhead is larger than the benefit, which results in a small negative effect. The last group includes 2 designs that timed-out with all methods.

In Fig. 3 we present results of five representative cases from the first two groups.

The last instances of designs 3, 8, 9 and 10 are satisfiable, while design 14 is unsatisfiable in all 5 instances. The *C-Sharing* strategy refers to constraints sharing, where the 'added clauses' line indicates the number of clauses that were added to each instance. The *Flip* strategy is a variation of the C-Sharing strategy: rather than using the same configuration for all instances (by 'configurations' we

Strategy		Design # 10					Design # 14				
	$k \rightarrow$	27	28	29	30	31	14	15	16	17	18
Normal	time(sec)	61	102	174	144	**14**	**10**	91	192	*	*
C-Sharing	time(sec)	63	77	80	47	16	**10**	58	**155**	1.6E4	*
	added clauses	0	973	1092	1208	1253	0	925	2117	3474	6116
Flip	time(sec)	-	50	-	62	-	-	**31**	-	**4219**	-
	added clauses	0	1112	1206	1361	1408	0	972	1827	3152	6057
C+rep	time	**48**	**21**	**19**	**44**	30	13	48	214	6211	*
	replicated	2094	1704	1216	1075	450	5932	5656	7778	1.7E4	*
	added clauses	0	482	1113	1536	2014	0	3374	5773	9806	1.6E4

Strategy		Design # 3					Design # 9				
	$k \rightarrow$	10	11	12	13	14	34	35	36	37	38
Normal	time (sec)	**3**	10	13	238	1	34	39	43	49	61
C-Sharing	time (sec)	**3**	12	**7**	75	1	35	38	44	47	58
	added clauses	0	207	571	955	1553	0	4	8	9	10
Flip	time (sec)	-	**9**	-	**8**	-	-	42	-	53	-
	added clauses	0	255	656	1126	1709	0	12	13	17	18
C + rep	time	4	14	11	23	2	**25**	**28**	**31**	**33**	**45**
	replicated	1229	1508	1954	2277	0	32	33	34	35	1.5E4
	added clauses	0	726	1877	3024	4380	0	33	67	102	138

strategy		Design # 8				
	$k \rightarrow$	31	32	33	34	35
Normal	time (sec)	**13**	**14**	**14**	**18**	38
C-Sharing	time (sec)	**13**	**14**	15	**18**	**29**
	added clauses	0	4	9	11	14
Flip	time (sec)	-	15	-	22	-
	added clauses	0	11	14	18	20
C + rep	time (sec)	15	16	18	19	**29**
	replicated	58	30	62	32	4638
	added clauses	0	60	91	155	188

Fig. 3. Representative results of four strategies show the advantage of constraints sharing and replicated clauses in reducing the overall verification time. *C-sharing* refers to the standard constraints sharing procedure, and the *Flip* strategy refers to a procedure where the search strategy alternates in each instance. *C + rep* is the same as C-Sharing, with the addition of internally replicated clauses. Best results are bold-faced, and asterisks (*) represent run times exceeding 20,000 sec.

refer to different ordering strategies, as were listed in [12]), we switched it every run. The instances in the odd columns were solved with an alternative configuration, and are therefore left empty to avoid confusion. The generally – better results are related to the different set of clauses that were added to each of these instances. This was a repeating phenomenon in the experiments we conducted, which indicates that adding clauses that were deduced by a different configuration can cause larger portions of the search space to be pruned. Obviously this can only be a good strategy if the alternative strategy similarly performs, on average, as the default one. The $C + rep$ strategy is the same as C-sharing, with the addition of replicated clauses. The 'replicated' line refers to the number of replicated clauses that were added. These clauses can become pervasive (see Section 5.2), which explains the increase in the number of 'added clauses' in the last line.

8 Summary

We introduced constraints sharing, a technique for sharing information between SAT instances whose clauses sets have a non empty intersection. This technique can be seen as a generalization of an older method called pervasive clauses, which was first introduced in the context of ATPG. We showed how this technique exploits the gradual nature of bounded model checking for shortening the overall verification time. We also showed how the same principle can be used, in the case of invariants checking, for adding constraints within a single SAT instance. Experimental results demonstrate the rather consistent positive effect that both of these methods have. Based on this observation, we implemented the two improvements as part of the default configuration of our versions of BMC and Grasp.

There are two experimental research directions that can be based on these techniques. First, using constraints sharing when checking two different properties of the same design. Although the percentage of shared clauses is expected to be smaller in this case (the property's clauses are different, and they impose a different *cone of influence*), it should nevertheless accelerate the overall verification time. Secondly, using the same techniques in other domains, such as AI Planning problems. SAT-based planning has been used in the past in a very similar way to BMC: a solution is found by solving a series of SAT instances, where each instance corresponds to a different number of allowed steps in the plan. See e.g. [3] and [9] for more details on this subject.

References

1. A. Biere, A. Cimatti, E. Clarke, and Y. Zhu. Symbolic model checking without BDDs. In *Proceedings of the Workshop on Tools and Algorithms for the Construction and Analysis of Systems (TACAS99)*, Lect. Notes in Comp. Sci. Springer-Verlag, 1999.
2. A. Biere, E. Clarke, R. Raimi, and Y. Zhu. Verifying safety properties of a power pc^{TM} microprocessor using symbolic model checking without bdds. In N. Halbwachs and D. Peled, editors, *Proc. 11th Intl. Conference on Computer Aided Verification (CAV'99)*, Lect. Notes in Comp. Sci. Springer-Verlag, 1999.
3. R. I. Brafman and H. H. Hoos. To encode or not to encode: linear planning. *IJCAI*, pages 988–993, 1999.
4. M. Davis and H. Putnam. A computing procedure for quantification theory. *J. ACM*, 7:201–215, 1960.
5. L.G.e. Silva, L.M. Silveira, and J.P.M. Silva. Algorithms for solving boolean satisfiability in combinational circuits. In *Proceedings of the IEEE/ACM Design, Automation and Test in Europe Conference (DATE)*, March 1999.

6. J.F. Groote, J.W.C. Koorn, and S.F.M. van Vlijmen. The safety guaranteeing system at station Hoorn-Kersenboogerd. Logic Group Preprint Series 121, Utrecht University, 1994.

7. J. N. Hooker. Solving the incremental satisfiability problem. *Journal of Logic Programming*, 15:177–186, 1993.

8. J.P.M. Silva and K. A. Sakallah. Robust search algorithms for test pattern generation. In *Proceedings of the IEEE Fault-Tolerant Computing Symposium*, June 1997.

9. H. Kautz and B. Selman. Planning as satisfiability. In *Proc. of the 10th European Conf. on AI*, pages 359–363, 1992.

10. J. Kim, J. Whittemore, J.P.M. Silva, and K. A. Sakallah. Incremental boolean satisfiability and its application to delay fault testing. In *IEEE/ACM International Workshop on Logic Synthesis (IWLS)*, June 1999.

11. M. Sheeran, S. Singh, and G. Stalmarck. Checking safety properties using induction and a sat-solver. In Hunt and Johnson, editors, *Proc. Int. Conf. on Formal Methods in Computer-Aided Design (FMCAD 2000)*, 2000.

12. O. Shtrichman. Tuning SAT checkers for bounded model checking. In E.A. Emerson and A.P. Sistla, editors, *Proc. 12th Intl. Conference on Computer Aided Verification (CAV'00)*, Lect. Notes in Comp. Sci. Springer-Verlag, 2000.

13. J.P.M. Silva and K. A. Sakallah. GRASP – a new search algorithm for satisfiability. Technical Report TR-CSE-292996, Univerisity of Michigen, 1996.

14. J.P.M. Silva and K. A. Sakallah. GRASP: A search algorithm for propositional satisfiability. *IEEE Transactions on Computers*, 48:506–516, 1999.

Heuristics for Hierarchical Partitioning with Application to Model Checking*

M. Oliver Möller[1] and Rajeev Alur[2]

[1] ≣BRICS †, Department of Computer Science, University of Aarhus,
Ny Munkegade, Building 540, DK – 8000 Århus C, Denmark
Fax.: (+45) 8942 3255
omoeller@brics.dk

[2] University of Pennsylvania, Computer & Information Science, Moore Building,
200 South 33rd Street, Philadelphia, PA 19104, USA
Fax.: (+1) 215 898-0587
alur@cis.upenn.edu

Abstract. Given a collection of connected components, it is often desired to cluster together parts of strong correspondence, yielding a hierarchical structure. We address the automation of this process and apply heuristics to battle the combinatorial and computational complexity.

We define a cost function that captures the quality of a structure relative to the connections and favors shallow structures with a low degree of branching. Finding a structure with minimal cost is *NP*-complete. We present a greedy polynomial-time algorithm that approximates good solutions incrementally by local evaluation of a heuristic function. We argue for a heuristic function based on four criteria: the number of enclosed connections, the number of components, the number of touched connections and the depth of the structure.

We report on an application in the context of formal verification, where our algorithm serves as a preprocessor for a temporal scaling technique, called *"Next" heuristic* [2]. The latter is applicable in reachability analysis and is included in a recent version of the MOCHA model checking tool. We demonstrate performance and benefits of our method and use an asynchronous parity computer and an opinion poll protocol as case studies.

1 Introduction

Imposing a hierarchical structure on a collection of components is helpful in many contexts for different reasons, such as better understanding and better analysis.

Consider four items, call them A, B, C, and D. They may be connected in some way, say by a mutual dependency. Let us assume that this gives rise to

---

* A longer version of this paper is available as technical report [12].

† Basic Research in Computer Science, Center of the Danish National Research Foundation.

T. Margaria and T. Melham (Eds.): CHARME 2001, LNCS 2144, pp. 71–85, 2001.

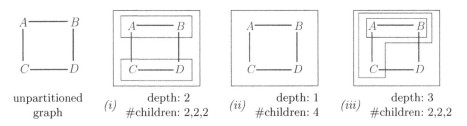

unpartitioned graph

(i) depth: 2 #children: 2,2,2

(ii) depth: 1 #children: 4

(iii) depth: 3 #children: 2,2,2

Fig. 1. Different ways to hierarchically partition a square

a ring structure, like in Fig. 1. Then, instead of viewing the system as a *set* with 4 elements, we can understand it structured as $\{\{A, B\}, \{C, D\}\}$, like in *(i)*. Here, only two connections A-C and B-D need to be visible (or understood) at the top level. With the new compounds $\{A, B\}$ and $\{C, D\}$ we found a more abstract description of the same data, that can be refined on demand. Another possible partition is $\{\{A, D\}, \{B, C\}\}$, but this requires all connections to be visible at the top-level, and thus should be rejected in favor of the first. We can partition in a hierarchic fashion: for example, $\{\{\{A, B\}, C\}, D\}$. In general, given a set, we partition it, and apply the process recursively to each set in the partition.

The set of distinguishable *hierarchical partitions* is adequately described as the set of rooted trees over leaf nodes $\{A, B, C, D\}$. In Fig. 1, we draw these trees as cascading boxes that may contain other boxes. Every box corresponds to an intermediate node and the outermost box to the root. As a rule, we favor trees that have a low degree of branching and are nevertheless shallow. The diagrams *(ii)* and *(iii)* depict both not very good trees, since they are either too broad or too deep. Moreover, it is desirable to minimize dependencies among remote tree parts, i.e., the number of links crossing box boundaries should be low.

We can regard this as a general design problem, where trees form an architectural hierarchy over atomic units. This *modular* description helps to see the same system on different levels of abstraction or detail. The emphasis on modularity and hierarchy is a central theme in software engineering, particularly in software design notations such as Statecharts [10] and UML [3]. While the most appropriate hierarchical structure can best be chosen by the designer, automatically constructing a hierarchical partition is required if no manually chosen structure is available, or if the original structure is lost during translations between models (e.g., during the process of abstraction).

Formal verification is a field where structure is particularly useful, since it is generally considered infeasible to deal with unorganized descriptions. Structure helps to spot design flaws, but it can also be exploited to make algorithmic treatment more efficient, or even possible at all. Well-known examples for this are model checking problems. *Model checking* [4, 11] is a powerful technique for discovering inconsistencies in high-level designs in hardware and communication protocols. Since it typically requires search in the global state-space, much research aims at providing heuristics to make this step less time- and space-consuming.

Consider once more the example with the four components. Let us interpret each atom as a process and each connection as the ability to synchronize on some action. We view the system hierarchically decomposed as in Fig. 1 *(i)*. Tools such as the concurrency workbench [5] can analyze it in the following way. First take the product of processes A and B. Now their synchronization can be viewed as internal to this composite process, and we can apply a reduction based on weak bisimulation minimizing the size. Analogously, compose C with D and minimize. The obtained description is still adequate, since it shows *the same behavior* (modulo the internal synchronization), but questions about this behavior are algorithmically easier to answer.

An alternative method that benefits from a hierarchical structure is implemented in a recent version of the model checking tool MOCHA, and will form the basis of the experiments in this paper. The technique, called *"Next" heuristic*, is a heuristic for on-the-fly search based on compressing unobservable transitions to a single meta-transition [2]. The basic idea is to describe the implementation P in a hierarchical manner, so that P is a tree whose leaves are atomic processes, and internal nodes compose their children and hide as many variables as possible. The basic reduction strategy, proposed by many researchers, is simple: while computing the successors of a state of a process, apply the transition relation repeatedly until a shared variable is accessed. This is applicable since changes to a private state are treated as stuttering steps. The benefit is greatly amplified by applying the reduction in a *recursive* manner exploiting the hierarchical structure, and has been shown to give significant reductions in space and time requirements, particularly for well-structured systems such as rings and trees.

As it turns out, the gain depends heavily on the hierarchical partition we impose. Ultimately, the run-time of the model checking algorithm can be seen as a measure on how to *compare* different choices. However, in practice we would look for qualifications that are faster to evaluate. And academically, run-time comparisons are too dependent on low-level implementation details to give clear analytic data. Thus we strive for a more abstract notion of comparison by means of a cost measure. For typical measures, the problem of finding the *best* hierarchical decomposition is likely to be *NP*-hard, and hence we must look for heuristics that are to be validated by experimentation.

In this paper, we strive to "discover" good hierarchies. We give a cost measure that allows us to compare hierarchical partitions, whenever the means of connection can be adequately described by a hypergraph. Determining the best structure for this measure is *NP*-complete. We present a greedy polynomial-time algorithm, that approximates good hierarchical partitions by local evaluation of a heuristic function. We corroborate applicability and usefulness via two case studies with our implementation of this algorithm in the MOCHA verification tool. When applied to a tree-shaped topology, this results in significant time- and memory-savings. In the second class of examples, the run-time performance is not drastically improved.

Plan. The next section gives a formal definition of the problem and classifies its complexity. In Sect. 3 we develop the algorithm to approximate good solutions.

Table 1. The combinatorial explosion in the number of distinguishable tree-indexings

n	1	2	3	4	5	6	7	8	9	10
$T(n)$	1	1	4	26	236	2'752	39'208	660'032	12'818'912	282'137'824

Section 4 reports on experiments with sample problems. Section 5 reflects on the limitation of our method, contrasts it to related work and lists open problems.

2   The Tree-Indexing Problem

In the following, we describe systems as hypergraphs, where the atomic units correspond to vertices and their connections are represented by hyperedges. E.g., in a reactive module description every hyperedge would correspond to a variable shared by the modules it connects to. Hierarchical partitions introduce an additional tree structure on top of this hypergraph and are augmented with a cost value. We briefly treat combinatorial and computational complexity of finding a tree of minimal cost.

A *hypergraph* $\mathcal{H} = (\mathcal{C}, \mathcal{E})$ is a finite set of vertices $\mathcal{C}$ together with a multi set $\mathcal{E}$, where every *hyperedge* $e \in \mathcal{E}$ is a subset of $\mathcal{C}$. We assume that every e corresponds to a unique label ℓ_e. Hyperedges of size 0 or 1 are disallowed. We draw hyperedges as branching lines. This coincides with common graph representation for the special case that every hyperedge is of size 2.

A *tree-indexing* $\mathcal{T}$ of a hypergraph $\mathcal{H} = (\mathcal{C}, \mathcal{E})$ is a rooted tree over leaf nodes $\mathcal{C}$, where every internal node has at least two children. We draw internal nodes as polygons, all contained polygons and vertices $v \in \mathcal{C}$ are children of this node. The outermost polygon corresponds to the root. Every tree-indexing is qualified by a cost value dependent on $\mathcal{E}$. For instance, in Fig. 1, the tree-indexing $\{\{A, B\}, \{C, D\}\}$ is better than $\{A, B, C, D\}$, and thus should be of lower cost.

Combinatorial Complexity. Given a hypergraph with n labeled vertices, we want to determine the number $T(n)$ of distinguishable tree-indexings. This is in an equivalent formulation recorded as Schröder's fourth problem [13]. It can be solved (for every fixed n) via a generating function method. Let $\varphi(z)$ be the ordinary generating function, where the n^{TH} coefficient corresponds to $T(n)$. Let $\hat{\varphi}(z)$ be its exponential transform. We can construct an equation, that $\hat{\varphi}(z)$ has to satisfy according to the theory of admissible constructions [7]. Every tree-indexing is either atomic, i.e. represented as z, or a set of at least two other tree-indexings, namely its children. This can be expressed using the admissible constructions 𝔘nion and 𝔖et.

$$\hat{\varphi}(z) \;=\; \text{𝔘nion}\,(z,\; \text{𝔖et}\,(\hat{\varphi}(z),\, \text{cardinality} \geq 2)) \tag{1}$$

Equation (1) can be transcribed as follows.

$$\hat{\varphi}(z) \quad = \quad z \ + \ \sum_{k \geq 2} \frac{1}{k!} \cdot (\hat{\varphi}(z))^k$$
$$\Longleftrightarrow \qquad \hat{\varphi}(z) \quad = \quad z \ + \ \exp{(\hat{\varphi}(z))} \ - \ \hat{\varphi}(z) \ - \ 1 \qquad (2)$$
$$\Longleftrightarrow \qquad \exp{(\hat{\varphi}(z))} \quad = \quad 2\,\hat{\varphi}(z) - z + 1$$

There is no closed form known for $\hat{\varphi}(z)$, $\varphi(z)$, or $T(n)$. However, for every fixed n we can extract the n^{TH} coefficient of $\varphi(z)$ with algebraic methods and thus approximate $T(n)$, as done in [6]. Table 1 gives an impression how fast this series grows. Thus we have only little hope to perform an exhaustive search on the domain of possible tree-indexings.

Computational Complexity. We can formulate the problem of finding a good tree-indexing as an optimization problem relative to a fixed *cost* function. This function should punish both deep structures and hyperedges that span over big subtrees. For every $e \in \mathcal{E}$ let $\mathcal{T}_e$ denote the smallest complete subtree of $\mathcal{T}$, such that every vertex $v \in e$ is a leaf of $\mathcal{T}_e$. With *leaves*$(\mathcal{T})$ we denote the set of leaf nodes in a tree $\mathcal{T}$. The *depth* of $\mathcal{T}$ is the length of the longest descending path from its root. The *depth cost* of a tree $\mathcal{T}$ is defined as a function

$$depth\_cost(\mathcal{T}) \quad := \quad \begin{cases} 2 & \text{if } depth(\mathcal{T}) = 1 \\ depth(\mathcal{T}) & \text{otherwise}. \end{cases} \qquad (3)$$

The cost of a tree-indexing $\mathcal{T}$ is then defined relative to $\mathcal{H} = (\mathcal{C}, \mathcal{E})$.

$$cost(\mathcal{T}) \quad := \quad \sum_{e \in \mathcal{E}} depth\_cost(\mathcal{T}_e) \cdot |leaves\,(\mathcal{T}_e)| \qquad (4)$$

For example, the tree-indexing *(i)* in Fig. 1 has cost $2 \cdot 2 \cdot 2 + 2 \cdot 2 \cdot 4 = 24$, which is preferable to tree-indexings *(ii)* and *(iii)* with costs $4 \cdot 2 \cdot 4 = 32$ and $2 \cdot 2 + 2 \cdot 3 + 2 \cdot 3 \cdot 4 = 34$ respectively.

EDGE-GUIDED TREE-INDEXING: Given a hypergraph $\mathcal{H} = (\mathcal{C}, \mathcal{E})$ and a number $K \in I\!N$. Decide whether there exists a tree-indexing of cost at most K.

The problem EDGE-GUIDED TREE-INDEXING is *NP*-complete, even if we restrict to the special case where $\mathcal{H}$ is a multi graph. Containment in *NP* holds, since we can guess any tree-indexing and compute its cost in (non-deterministic) polynomial time. A *NP*-hardness proof by reduction from MINIMUM CUT INTO EQUAL-SIZED SUBSETS is given in [12]. We expect EDGE-GUIDED TREE-INDEXING to remain *NP*-hard for other non-trivial definitions of a cost function like $\sum_e |leaves(\mathcal{T}_e)|$, though we do not have a proof for this. This precludes the possibility to determine an *optimal* tree-indexing in polynomial time[1] and suggests the application of heuristics in order to find a *reasonably good* tree-indexing efficiently.

---

[1] Unless *NP* turns out to be equal to *P*.

3   A Greedy Algorithm to Partition Hierarchically

In this section we develop a greedy-style algorithm that constructs a tree-indexing by successively grouping together sets with strong correspondence. The choice of these candidates relies on heuristics, which make use of a key observation: strong correspondences are likely to represented by a large number of connections.

A schematic description of our proposed algorithm is given in Fig. 2. The variable $\mathcal{F}$ is used to maintain a partial tree-indexing, i.e., a forest $\mathcal{F}$ with leaves $\mathcal{C}$. It is initialized as the forest with $|\mathcal{C}|$ trees, each consisting of a single node. The priority queue Q is ordered according to a rating function $\mathbf{r} : \wp^{\mathcal{D}} \times 2^{\mathcal{C}}_{\star} \to I\!R$. $\wp^{\mathcal{D}}$ is the set of forests over leaves $\mathcal{D} \subseteq \mathcal{C}$ and thus contains all possible sub-forests of $\mathcal{F}$. $2^{\mathcal{C}}_{\star}$ denotes a multi set of hyperedges and initially corresponds to $\mathcal{E}$. The top element of the queue is a subset of $\mathcal{F}$ with maximal $\mathbf{r}$-value.

The algorithm proceeds as follows. An initial set of candidates proposed for grouping together is inserted in the priority queue. Then a small number of executions of the while-loop follow. In each execution, the most promising candidate $\mathcal{A}$ is dequeued and the data is updated: in the forest, the trees in $\mathcal{A}$ are replaced by a tree with the fresh root $\mathcal{A}'$ and children $t \in \mathcal{A}$. Every set containing trees $t \in \mathcal{A}$ is removed from the priority queue and new candidates containing $\mathcal{A}'$ are inserted. Hyperedges $e \subseteq leaves(\mathcal{A}')$ are deleted, since they should not influence later selections.

Algorithm: *partition_incrementally*
    input: hypergraph $\mathcal{H} = (\mathcal{C}, \mathcal{E})$
    output: tree-indexing over leaves $\mathcal{C}$

    PriorityQueue Q := emptyQueue
    Forest $\mathcal{F}$:= $\mathcal{C}$

    FORALL considered candidates $\mathcal{A} \subseteq \mathcal{F}$
        insert($\mathcal{A}$, Q)
    WHILE *notempty(Q)*
        $\mathcal{A}$:= *top(Q)*   /⋆ pick best candidate ⋆/
        let $\mathcal{A}'$:= fresh root node with children $t \in \mathcal{A}$
        $\mathcal{F}$:= $(\mathcal{F} \setminus \mathcal{A}) \cup \{\mathcal{A}'\}$
        $\mathcal{E}$:= $\mathcal{E} \setminus \{e \mid e \subseteq leaves(\mathcal{A}')\}$ /⋆ remove covered hyperedges ⋆/
        update($\mathcal{E}$, $\mathcal{A}$, $\mathcal{A}'$)   /⋆ replace all $t \in \mathcal{A}$ by $\mathcal{A}'$ ⋆/
        FORALL $\mathcal{B} \in Q$ with $\mathcal{B} \cap \mathcal{A} \neq \emptyset$
            remove($\mathcal{B}$, Q)
        FORALL new candidates $\mathcal{D}$ containing $\mathcal{A}'$
            insert($\mathcal{D}$, Q)
    RETURN $\mathcal{F}$

Fig. 2. Incremental algorithm for constructing a tree-indexing

This description leaves open the questions, what should be used as a rating function and which candidates should be considered. We explain these aspects of the algorithm in the following.

Developing a Good rating function. The local choice of the best candidate could be performed by means of the *cost* function defined in (4), i.e., by picking the candidate with lowest cost after clustering. We chose not to do so for two reasons. First, the specific cost function was derived such that an *NP*-complete problem could be encoded into it; for small variations of this definition, the proof failed – thus, this particular definition is somehow artificial. Second, we would like to tune the choice by means of parameters in the rating function. Doing this with the cost function would almost certainly destroy the provable *NP*-hardness, and thus the justification for the choice.

Instead, we develop a rating function subsequently by taking into account the– – supposedly – crucial factors concerning the structure of the proposed candidate. Most importantly, we want to know the number of additional hyperedges that are completely covered by this set, and thus can be hidden from the outside without losing information.

Def 1 (cover number) *Let $\mathcal{H} = (\mathcal{C}, \mathcal{E})$ be a hypergraph, $\mathcal{F}$ a forest over leaves $\mathcal{C}$, $\mathcal{A} = \{\mathcal{T}_1, \ldots, \mathcal{T}_k\} \subseteq \mathcal{F}$. The* cover number *of $\mathcal{A}$, in symbols $\langle\!\langle \mathcal{A} \rangle\!\rangle$, is defined as the number of hyperedges covered by the trees in $\mathcal{A}$.*

$$\langle\!\langle \{\mathcal{T}_1, \ldots, \mathcal{T}_k\} \rangle\!\rangle \; := \; \left| \{ \ell_e \mid e \in \mathcal{E}, \; e \subseteq \text{leaves}(\mathcal{A}), \; \forall i. \; e \not\subseteq \text{leaves}(\mathcal{T}_i) \} \right|$$

Though this value tells a lot about a candidate, it is isolated not a good guideline. Recall that the set $\mathcal{C}$ has naturally always the highest possible cover number $|\{\ell_e \mid e \in \mathcal{E}\}|$.

We relate the cover number to the *size* n of a candidate, where size matters in terms of *possible* connections, which is $\binom{n}{2} = \mathcal{O}(n^2)$. We propose the following rating function.

$$\mathbf{r}_{pref}(\mathcal{A}) \quad := \quad \frac{\langle\!\langle \mathcal{A} \rangle\!\rangle}{|\mathcal{A}|^2} \tag{5}$$

In the following we refine $\mathbf{r}_{pref}$ by adding more structural information.

Def 2 (touch of a candidate) *Let $\mathcal{H} = (\mathcal{C}, \mathcal{E})$ be a hypergraph and $\mathcal{F}$ be a forest over leaves $\mathcal{C}$. Then the* touch *of $\mathcal{A} \subseteq \mathcal{F}$ is defined as the labels from hyperedges that connect $\mathcal{A}$ with the rest of $\mathcal{H}$.*

$$touch(\mathcal{A}) \quad := \quad \{ \ell_e \mid e \cap \text{leaves}(\mathcal{A}) \neq \emptyset \wedge e \not\subseteq \text{leaves}(\mathcal{A}) \}$$

Def 3 (depth of a candidate) *The* depth *of a tree with only one node equals 0. Let $\mathcal{F}$ be a forest, $\mathcal{A} = \{\mathcal{T}_1, \ldots, \mathcal{T}_k\} \subseteq \mathcal{F}$. The depth of $\mathcal{A}$ is defined as*

$$depth(\{\mathcal{T}_1, \ldots, \mathcal{T}_k\}) \quad := \quad 1 + \max_{1 \leq i \leq k} depth(\mathcal{T}_i)$$

We do not want to cut out subsystems, that are multiply connected to the rest, i.e., those who share many hyperedges with their complement. This is reflected by the number of labels in the touch: if it is small, the candidate is more attractive. Also, it is perceivable that preference should be given to candidates with small depth. Hence we propose the following improved rating function.

$$\mathbf{r}^{+}_{pref}(\mathcal{A}) := \frac{\langle\!\langle \mathcal{A}\rangle\!\rangle}{|\mathcal{A}|^2} + \frac{\varepsilon_1}{|touch(\mathcal{A})|} + \frac{\varepsilon_2}{depth(\mathcal{A})} \tag{6}$$

The parameters ε_1 and ε_2 are supposed to be chosen small and positive. For the experiments in Sect. 4.1, the assignments $\varepsilon_1 := 1/1000$, $\varepsilon_2 := 1/100000$ were used.

Restricting the Set of Considered Candidates. In our formulation of the algorithm *partition_incrementally* we remained unclear what the considered candidates are. We want to weed out hopeless candidates, e.g., those not sharing any labels, before adding them to our priority queue. In a positive formulation, consider only candidates, that are extensions of interesting pairs.

Def 4 (interesting pair) *Given a hypergraph* $\mathcal{H}(\mathcal{C}, \mathcal{E})$ *and a forest* $\mathcal{F}$ *over leaves* $\mathcal{C}$. *An* interesting pair $\{\mathcal{T}_1, \mathcal{T}_2\}$ *is a subset of* $\mathcal{F}$, *such that* $touch(\mathcal{T}_1) \cap touch(\mathcal{T}_2) \neq \emptyset$.

Clearly, every candidate that is *not* a superset of an interesting pair has cover number 0 and thus can be neglected. As it turns out in our implementation, the expensive part of the algorithm is the computation of the cover numbers. First computing interesting pairs and then extending them to candidates is an advantage.

   The number of candidates can still be excessive. Consider a hyperedge connecting all vertices, then all pairs are interesting pairs. Since the number of subsets of $\mathcal{C}$ is exponential in $|\mathcal{C}|$, an exhaustive enumeration is not feasible for large systems. If conservative techniques (like considering just extensions of interesting pairs) do not suffice, we have to apply a more rigorous pruning, even for the price of thereby ignoring good candidates. An obvious suggestion is to consider only candidates up to a certain size k, thus establishing an upper bound of $n^{k+1} - n - 1$ candidates. This k can be adjusted according to n, which provides a simple and reasonable method to prune the search.

   In the algorithm, the number of forests – initially n – decreases by one with each execution of the while loop. Operations like evaluating the rating function, testing $\mathcal{B} \cap \mathcal{A} \neq \emptyset$, and constructing new candidates containing $\mathcal{A}'$ can be assumed to be $\mathcal{O}(n)$, thus one execution of the while loop has the run-time bound $\mathcal{O}(n \cdot n^{k+1})$. The whole algorithm *partition_incrementally* has n executions of the while loop, which yields the polynomial bound $\mathcal{O}(n^{k+3})$ on its run-time.

4   Experimental Results

We implemented the algorithm from Sect. 3 in an experimental version of the MOCHA verification tool [1]. For symbolic (BDD-based) model checking, the

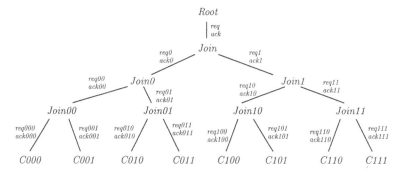

Fig. 3. Layout of an asynchronous parity computer with eight clients

Java implementation makes use of native libraries. However, our experiments do not make use of this option and perform the check in a purely enumerative manner. Therefore, given run-times and memory requirements are those of the Java Virtual Machine, executing on a Sun Enterprise 450 with UltraSPARC-II processors, 300 MHz. A contingent of 128 MB of memory was allocated, run-times are in milliseconds. Together with an optimization in the enumerative check called *"Next" heuristic* [2], we are able to corroborate effectiveness and usability of our algorithm in some simple examples. We consider an asynchronous parity computer and an opinion poll protocol. The MOCHA specifications are given in [12].

Note that in these experiments the checked property influences the obtained structure. In MOCHA every property relies on *variables* of the system. These variables can not be hidden and therefore are neglected in the partition algorithm, i.e., they are ignored in the evaluation of the rating function.

4.1 Asynchronous Parity Computer

This example models a parity computer, designed as a binary tree (Fig. 3). The leaf nodes are *Client* modules (abbreviated with *C*), communicating a binary value to the next higher *Join*. A simple hand-shake protocol is devised by the two variables *req* and *ack*. All components are supposed to move asynchronously. Thus the join nodes have to wait for both values to be present, before reporting their exclusive-or upwards. The *Root* component, after receiving the result of the computation, hands down an acknowledgment. When a client receives an acknowledgment, it is able to devise a fresh value.

We consider binary trees with N client nodes, where N varies from 3 to 8. The number of variables increases linearly with N, whereas the state-space grows exponentially. The sample question we pose is whether the module *Root* will ever output a value *zero* or *one*. We expect our model checking algorithm to falsify the claim, that it never will.

Reachability involves computing the successors of every state encountered, starting with the initial states. Consider the set S of all the processes. Then, successors of a state are computed by executing one step of one of the processes in S. Now suppose, we cluster the processes *Join00*, *C000*, and *C001* into one

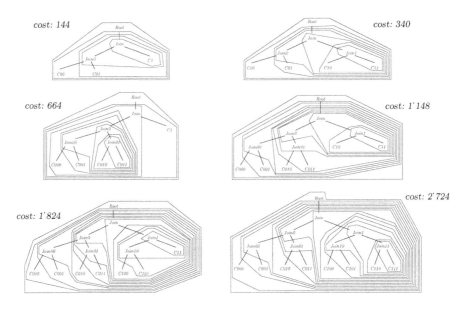

Fig. 4. Parity computers $N = 3, \ldots, 8$, partitioned with rating function $\mathbf{r}_{pref}$

composite process called P, and replace these three processes in S with P. It is clear that the communication between $J00$ with its children clients can be hidden from the rest of the system. Consequently, in reachability analysis of S, when we compute the successors due to execution of P, we can let the sub-processes in P repeatedly execute until $Join00$ communicates with its parent $Join0$. This is formalized in MOCHA by substituting P by a construct *next* Θ *for* P, where atomic transitions correspond to sequences of transitions of P until a variable shared with the remaining system is accessed. The modified search yields an improved performance as it cuts down on unnecessary interleavings.[2] This scheme can be applied repeatedly. It should be clear that the effectiveness of the scheme depends on the hierarchical partition.

An intuitively good choice for this hierarchical partition is grouping to-gether bottom up. Detecting this algorithmically is subtle. E.g., the difference between $\{Join0, Join00\}$ and $\{Client000, Join00\}$ is only minor, since both pairs cover exactly two variables. An incautious technique easily runs into er-rands, as to be seen in Fig. 4. Using $\mathbf{r}_{pref}$ as rating function in the algorithm *partition_incrementally* leads to uncomfortably deep hierarchies.

The more sophisticated rating function $\mathbf{r}_{pref}^{+}$ performs far better, as seen in Fig. 5. The parameters ε_1 and ε_2 were calibrated to $\varepsilon_1 := 1/1000$ and $\varepsilon_2 := 1/100000$, giving shallow structures a smaller bonus than those touching only few variables.

---

[2] A well-known method for reducing state-space in asynchronous systems is based on partial-order reductions [9]. The "Next" heuristic is incomparable to this method, see [2].

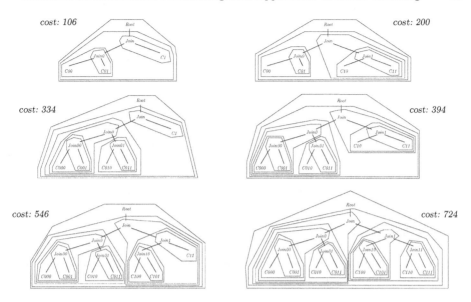

Fig. 5. r^+_{pref} yields shallower hierarchical partitions with lower cost values than r_{pref}

Table 2. Parity computer: Comparison of two heuristic functions

N	partition	|table|	check	N	partition	|table|	check	N	partition	|table|	check
3	162	95	1'121	3	105	51	973	3	57	51	404
4	853	645	4'921	4	148	117	1'707	4	75	117	1'097
5	740	1'943	17'086	5	627	139	1'780	5	127	139	1'726
6	2'811	16'045	161'394	6	2'097	205	3'000	6	516	205	2'929
7	9'928	58'351	694'834	7	10'592	271	4'395	7	247	271	4'364
8	47'239	410'901	5'442'315	8	50'664	469	8'322	8	342	469	8'184

Using r_{pref} as rating function r^+_{pref}, $\varepsilon_1 := 10^{-3}, \varepsilon_2 := 10^{-5}$ r^+_{pref} with $|\mathcal{A}| \leq 2$

The deep hierarchical structure obtained by using r_{pref} lead to excessive number of explored states, whereas with r^+_{pref} the growth of the explored state space with increasing N is only moderate. This gap is also reflected by the significantly higher cost values. Table 2 shows the run-time data in detail. With "partition" we denote the preprocessing time used by *partition_incrementally* and "check" corresponds to the run-time of the model checking algorithm. The number of explored states is recorded under "|table|", MOCHA keeps the states in a hash table. Note that the property we check does not hold, thus the model checking algorithm is able to abort without exploring all reachable states.

In the left and middle table, the time consumed for computing the hierarchical partition exceeds the model checking time for bigger examples. This is because we chose not to restrict the candidate size here, which yields a number of candidates increasing exponentially in N. The obtained hierarchical partitions

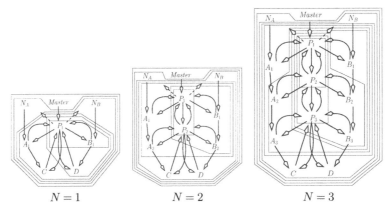

$N = 1$ $N = 2$ $N = 3$

Fig. 6. Opinion poll: Hierarchically partitioned according to rating function $\mathbf{r}^{+}_{pref}$

in the middle and the right table are identical – due to the tree structure, the best rated candidate here is always of size two.

4.2   Opinion Poll Protocol

The second sample problem is meant to demonstrate the behavior of our heuristic in a setting, where there is no obviously preferable choice.

Consider a poll for a public opinion. There is a line of N pollers P_i and two non-connected lines of citizen A_i and B_i, plus two special citizen C and D. Poller P_1 starts raising an issue with a Yes/No question. Let us assume that the way one asks influences the answer. Poller P_1 starts of with an opinion he got from a random source (called *Master*). Poller P_{i+1} is influenced by P_i. The citizen are influenced by one other citizen and the poller who interviews them. For instance, A_{i+1} is influenced by A_i's and P_i's opinions, and A_1 is influenced by a random source N_A and P_1. The communication pattern is indicated by arrows in Fig. 6.

For $N = 1, 2, 3, 4$ we considered three invariants: *(i)* a false property that is easy to falsify, *(ii)* a false property that requires a special scenario (called *bad property* in the following), and *(iii)* a true property.

The experiments compare plain enumerative model-checking and application of the "Next" heuristic, where the preprocessing follows one of the following strategies. *a. 2-merge*: Group any pair with a connection without further preference, i.e., use $sig(\langle\!\langle\rangle\!\rangle)$ as rating function and consider only candidates of size 2, *b. pref*: Partition incrementally according to rating function $\mathbf{r}_{pref}$, and *c. pref+*: Partition incrementally according to rating function $\mathbf{r}^{+}_{pref}$. For the latter, we included the results of the preprocessing in Fig. 6. It is interesting to note that sometimes triples were preferred to pairs.

The quantitative comparison is listed in Table 3. For the false properties *(i)* and *(ii)* the enumerative check is up to five times *slower*, if sophisticated heuristics are applied. Apparently it is more tedious to reach a counter-example scenario here, if more structure is given. For the true property *(iii)*, the enumerative check speeds up by a third, when the "Next" heuristic is applied. For

Table 3. Opinion poll protocol: Run-time comparison of three rating functions

N\Method	plain	2-merge	pref	pref+
colspan for title	(i) false Judgment: System |= (result = DontKnow)			
1	742	2 (partition) 854 (check)	280 (partition) 754 (check)	274 (partition) 861 (check)
2	3˙713	32 (partition) 4˙313 (check)	1˙808 (partition) 6˙862 (check)	1˙886 (partition) 6˙850 (check)
3	32˙181	7 (partition) 26˙330 (check)	1˙790 (partition) 87˙708 (check)	2˙047 (partition) 88˙879 (check)
4	345˙071	22 (partition) 435˙529 (check)	5˙256 (partition) 1˙390˙739 (check)	4˙828 (partition) 1˙351˙527 (check)

N\Method	plain	2-merge	pref	pref+
colspan for title	(ii) bad Judgment: System |= NoNegativeResult			
1	1˙113	2 (partition) 916 (check)	238 (partition) 766 (check)	203 (partition) 886 (check)
2	4˙846	5 (partition) 3˙930 (check)	1˙625 (partition) 7˙130 (check)	1˙667 (partition) 6˙561 (check)
3	32˙580	7 (partition) 29˙324 (check)	1˙788 (partition) 87˙827 (check)	1˙920 (partition) 73˙350 (check)
4	385˙951	20 (partition) 375˙977 (check)	5˙476 (partition) 1˙665˙765 (check)	6˙458 (partition) 1˙306˙961 (check)

N\Method	plain	2-merge	pref	pref+
colspan for title	(iii) true Judgment: System |= ~((result = DontKnow) & (result = Yes))			
1	30˙565	2 (partition) 23˙689 (check)	290 (partition) 24˙369 (check)	292 (partition) 24˙423 (check)
2	610˙131	5 (partition) 454˙089 (check)	1˙787 (partition) 482˙600 (check)	2˙148 (partition) 482˙214 (check)
3	8˙488˙532	17 (partition) 6˙392˙536 (check)	2˙301 (partition) 5˙920˙255 (check)	2˙357 (partition) 5˙865˙170 (check)
4	93˙557˙192	23 (partition) 60˙934˙073 (check)	5˙733 (partition) 57˙762˙294 (check)	5˙068 (partition) 57˙165˙981 (check)

larger N the more sophisticated clustering techniques *pref* and *pref+* perform slightly better than *2-merge*.

5   Conclusion

We developed a notion of hierarchical partitions and introduced a method to compare different structures by means of a cost function. This is applicable, whenever the relationship of entities can be adequately described via hyperedges. For the presented cost function, the problem of determining an optimal hierarchical partition is *NP*-complete [12].

We presented a scalable greedy method to compute approximately good hierarchical partitions based on a heuristic rating function. We argued that – in order to achieve a good result – this function should be based on four criteria: number of covered hyperedges, size, number of occurring hyperedges, and structural depth. This is corroborated by qualitative and quantitative data. We implemented our algorithm in an experimental version of the MOCHA model checking tool and measured its performance on small and medium sized examples.

It should be noted that our proposed method gives no guarantee on how the obtained result compares to an optimal solution. Since we apply a variation of local search, it is to be expected that our algorithm can get caught in local optima. Moreover, optimality in the sense of least cost does not necessarily imply minimal time- or space-consumption when running a model checking algorithm. In general, we cannot expect to express such subtle behavioral properties of a system via a simple function, i.e., a function that is fast to evaluate.

Though our case studies suggest that in the rating function both touch and depth of the candidates should be taken into account, it remains open, *how* this should be reflected. The values for the parameters ε_1 and ε_2 in (6) were chosen according to fit the parity computer example. It would be desirable to investigate the impact of parameter changes in general, but we seem to lack apt mathematical means to do so.

Related Work. Hierarchical structures find a wide range of application in design, description and physical organization of both software and hardware. In particular, the decomposition of large circuits in VLSI layout turns out to be a crucial problem and has received a respectable amount of attention [14]. Here the partitions are typically shallow (i.e., of depth two) and mainly motivated by size constraints that single components have to meet. Optimality is typically described as the least number of components with as few as possibly connections.

Similar structures, called classification trees (e.g. [8]), are used as expressive decision trees over large sets of data. The internal nodes are labeled by distinguishing criteria and all leaf nodes are distinguishable. Finding expressive classification trees is computationally hard.

Though various advanced techniques have been developed for these problems, to the best of our knowledge none of them is applicable in the considered case. In our setting, *every* tree-indexing is a feasible solution, there is no constraint satisfaction component and there might well be two leaves that are alike.

Open Problems. We noted that finding an optimal solution with respect to our cost function is *NP*-hard, but this does not preclude the existence of a polynomial approximation scheme. Also, it remains unknown, how the computational complexity compares with respect to other cost functions, like $depth\_cost(\mathcal{T}) := depth(\mathcal{T})$ or $cost(\mathcal{T}) := \sum_e |leaves(\mathcal{T}_e)|$. It is conjectured that the tree-indexing problem remains *NP*-complete in both cases.

Future Work. Our proposed method is not limited to MOCHA, but can be applied in other settings where connected entities have to be structured or re-structured. The parameters can be adjusted accordingly. An interesting mean to make use of the obtained partitions could be to construct property preserving abstractions based on this particular hierarchy.

We believe that other areas of research can benefit from a more analytic quantification of good structures. As an example we list refactoring [15], where structures of software systems are conservatively modified according to a set of schema transformations, design patterns, or hot spots.

Acknowledgments. We thank Bow-Yaw Wang and Radu Grosu for many useful comments and invaluable help on MOCHA implementation details. This research was supported in part by SRC award 99-TJ-688.

References

1. Rajeev Alur, Luca de Alfaro, Radu Grosu, Thomas A. Henzinger, Minsu Kang, Rupak Majumdar, Freddy Y.C. Mang, Christoph Meyer-Kirsch, and Bow-Yaw Wang. MOCHA: A model checking tool that exploits design structure. *Proceedings of 23rd International Conference on Software Engineering*, 2001. See `www.cis.upenn.edu/~mocha/`.
2. Rajeev Alur and Bow-Yaw Wang. "Next" Heuristic for On-the-fly Model Checking. In *Proceedings of the Tenth International Conference on Concurrency Theory (CONCUR'99)*, LNCS 1664, pages 98–113. Springer-Verlag, 1999.
3. Grady Booch, James Rumbaugh, and Ivar Jacobson. *The Unified Modeling Language User Guide*. Addison-Wesley, 1998.
4. Edmund M. Clarke, Jr. and E. Allen Emerson. Synthesis of synchronization skeletons from branching time temporal logic. *Lecture Notes Comp. Sci.*, 131:52–71, 1982.
5. Rance Cleaveland, Joachim Parrow, and Bernhard Steffen. The Concurrency Workbench: A Semantics Based Tool for the Verification of Concurrent Systems. *ACM Transactions on Programming Languages and Systems*, 15(1):36–72, January 1993.
6. Philippe Flajolet. A problem in Statistical Classification Theory, 1997. `http://pauillac.inria.fr/algo/libraries/autocomb/schroeder-html/`.
7. Phillippe Flajolet. Mathematical Methods in the Analysis of Algorithms and Data Structures. Lecture Notes for *A Graduate Course on Computation Theory*, Udine (Italy), Fall 1984. In Egon Börger, editor, *Trends in Theoretical Computer Science*, pages 225–304. Computer Science Press, 1988.
8. Saul B. Gelfand, C. S. Ravishankar, and Edward J. Delp. An iterative growing and pruning algorithm for classification tree design. *IEEE Transactions on Pattern Analysis and Machine Intelligence*, PAMI-13(2):163–174, February 1991.
9. Patrice Godefroid, Doron Peled, and Mark Staskauskas. Using partial order methods in the formal validation of industrial concurrent programs. *IEEE, Transactions on Software Engineerings*, 22:496–507, 1996.
10. David Harel. Statecharts: A visual formalism for complex systems. *Science of Computer Programming*, 1987.
11. Gerard J. Holzmann. The model checker spin. *IEEE Trans. on Software Engineering*, 23(5):279–295, May 1997. Special issue on Formal Methods in Software Practice.
12. M. Oliver Möller and Rajeev Alur. Heuristics for hierarchical partitioning with application to model checking. Research Series RS-00-21, BRICS, Department of Computer Science, University of Aarhus, August 2000. 30 pp, available online at `http://www.brics.dk/RS/00/21/`.
13. Ernst Schröder. Vier combinatorische probleme. *Zentralblatt. f. Math. Phys.*, 15:361–376, 1870.
14. Naveed Sherwani. *Algorithms for VLSI Physical Design Automation – 2nd Edition*. Kluwer Academic Publishers, Norwell, USA, 1995.
15. Lance Tokuda and Don Batory. Evolving Object-Oriented Designs with Refactorings. In *Proceedings of ASE-99: The 14th IEEE Conference on Automated Software Engineering*. IEEE CS Press, October 1999.

Efficient Reachability Analysis and Refinement Checking of Timed Automata Using BDDs

Dirk Beyer

Software Systems Engineering Research Group
Technical University Cottbus, Germany
db@informatik.tu-cottbus.de

Keywords. Formal verification, Real-time systems, Timed Automata

1 Introduction

For the formal specification and verification of real-time systems we use the modular formalism Cottbus Timed Automata (CTA), which is an extension of timed automata [AD94]. Matrix-based algorithms for the reachability analysis of timed automata are implemented in tools like Kronos, Uppaal, HyTech and Rabbit. A new BDD-based version of Rabbit, which supports also refinement checking, is now available.

For the representation of the models we use an *integer semantics* for *closed timed automata*. Using this discretization, we are able to use a unique representation of the discrete state space (given by the locations) and the continuous state space (given by the clocks). We use an estimate-based strategy for variable ordering which dramatically compresses the BDD representation of the transition relation and the reachable configurations and thus leads to much more efficient verification.

The restricted applicability of reachability analysis due to the high time complexity of the analysis for large models leads to the need of refinement checking for verification. We implemented an algorithm for checking the existence of a simulation relation to investigate the opportunities of refinement checking for Cottbus Timed Automata.

Section 2 introduces our notation for modular modeling of real-time systems: we recall the formal definition of timed automata and our integer semantics for closed timed automata. In Sect. 3 we describe our implementation of reachability analysis and, in more detail, in Sect. 4 we define the corresponding refinement checking. In Sects. 3 and 4, we present performance results for some example models.

2 Cottbus Timed Automata

We start with an informal definition of Cottbus Timed Automata (CTA), which is a modeling concept providing means for modular design [BR98]. A formal definition and the complete semantics of CTA are given in [BR01]. A CTA system description consists of a set of modules. One of them is the top module, which models the whole system. The other modules are used as templates. They can be instantiated several times in different modules. Thus, it is possible to express a hierarchical structure of the system, and to define replicated components of a system just once.

T. Margaria and T. Melham (Eds.): CHARME 2001, LNCS 2144, pp. 86–91, 2001.

Each module is named by an identifier. **(1)** The **interface** contains the declarations of *clock variables* and *synchronization labels*, each of them has a *restriction type* to control the access to the component. We distinguish the restriction types INPUT, OUTPUT, MULTIPLY RESTRICTED, and LOCAL. **(2)** A module contains a **timed automaton** as defined below. **(3)** The **initial condition** is a predicate over the module's variables and locations. **(4)** A module may contain **instances** of previously defined modules.

We now define closed timed automata and their integer semantics. *Clock constraints* are allowed as invariants and guards of a timed automaton. Let X be a set of clocks. Atomic clock constraints over X are comparisons of a clock with a time constant from $\mathbb{N}$, the set of natural numbers (including 0). Clock constraints are conjunctions of atomic clock constraints. Formally, the set $\Phi(X)$ of clock constraints over X for closed timed automata is generated by $\varphi := x \leq c \mid x \geq c \mid \varphi \wedge \varphi$, with $x \in X$ and $c \in \mathbb{N}$. For closed timed automata it is sufficient to use only integer clock values for the computation of reachable locations.

The *clock assignments* $Val_I(X)$ of X are the total functions from X into the set of natural numbers $\mathbb{N}$. For a clock constraint $\varphi \in \Phi(X)$, $[\![\varphi]\!]$ denotes the set of all clock assignments of X that satisfy φ. The clock assignment which assigns the value 0 to all clocks is denoted by v^0. For a timed automaton $\mathcal{A}$ with a clock x, $C_{\mathcal{A}}(x)$ denotes the greatest constant to which x is compared within a clock constraint of $\mathcal{A}$. For $v \in Val_I(X)$ and $\delta \in \mathbb{N}$, $v \oplus \delta$ is the clock assignment of X that assigns the value $min\,(v(x) + \delta, C_{\mathcal{A}}(x) + 1)$ to each clock x. For $v \in Val_I(X)$ and $Y \subseteq X$, $v[Y := 0]$ denotes the clock assignment of X that assigns the value 0 to each clock in Y and leaves the other clocks as in v.

A **closed timed automaton** $\mathcal{A}$ is a tuple $(L, L^0, X, \Sigma, I, E)$, where L is a finite set of *locations*, $L^0 \subseteq L$ is a set of *initial locations*, X is a finite set of *clocks*, Σ with $\Sigma \cap \mathbb{N} = \emptyset$ is a finite set of *synchronization labels*, I is a total function that assigns an *invariant* from $\Phi(X)$ to each location in L, $E \subseteq L \times \Sigma \times \Phi(X) \times 2^X \times L$ is a set of *switches*. A switch (l, a, φ, Y, m) represents a transition labeled with synchronization label a from location l to location m. The guard φ has to be satisfied to enable the switch. The switch resets all clocks in Y to the value 0.

The semantics of a timed automaton is defined by associating a transition system with it. Let $\mathcal{A} = (L, L^0, X, \Sigma, I, E)$ be a closed timed automaton. The **integer semantics** $[\![\mathcal{A}]\!]_I$ of $\mathcal{A}$ is the transition system $(L \times Val_I(X), L^0 \times \{v^0\}, \Sigma \cup \mathbb{N}, \rightarrow_I)$ with the following timed and discrete transitions:

- For $(l, v), (m, w) \in L \times Val_I(X)$ and $\delta \in \mathbb{N}$, $(l, v) \xrightarrow{\delta}_I (m, w)$ holds iff $l = m$, $w = v \oplus \delta$, $v \in [\![I(l)]\!]$ and $w \in [\![I(l)]\!]$.
- For $(l, v), (m, w) \in L \times Val_I(X)$ and $a \in \Sigma$, $(l, v) \xrightarrow{a}_I (m, w)$ holds iff there exists an $(l, a, \varphi, Y, m) \in E$ with $v \in [\![\varphi]\!]$ and $w = v[Y := 0]$.

In the following we define runs and reachable configurations for a transition system $\mathcal{T} = (S, S^0, \Sigma_{\mathcal{T}}, \rightarrow)$. Let $(s_0, s_1, ..., s_k)$ be a finite sequence of configurations, $a_0, a_1, ..., a_{k-1} \in \Sigma_{\mathcal{T}}$, $s_0 \in S^0$, and $s_i \xrightarrow{a_i} s_{i+1}$ for all $i \in \{0, 1, ..., k-1\}$. Then $(s_0, s_1, ..., s_k)$ is a **run** of $\mathcal{T}$. $Run(\mathcal{T})$ denotes the set of runs of $\mathcal{T}$. The configuration s_k is **reachable**. $Reach(\mathcal{T})$ denotes the set of reachable configurations (shorter: reachable set) of $\mathcal{T}$.

In our tool implementation we use the integer semantics for closed timed automata as defined above. This integer semantics is equivalent to the usual, continuous one regarding the set of reachable locations. More details and a formal proof are given in [Bey01]. To verify safety properties of a CTA model we provide two techniques: reachability analysis and refinement checking as described in the following two sections.

3   Reachability Analysis

For verification of a safety property, i.e. whether a configuration of a special set of invalid configurations is reachable or not, we compute at first the set of reachable configurations. Secondly, we intersect the reachable set with the set of invalid configurations. If the intersection is empty, then the safety property is fulfilled. Variants of this algorithm are possible for speed-up, e.g. on-the-fly analysis. We extended our existing matrix-based model checker **Rabbit** by the BDD-based reachability analysis of closed timed automata. Experience with finite automata shows that the efficiency critically depends on the choice of several parameters. In this section we sketch how our implementation determines these parameters.

Variable Ordering. We take the pre-order linearization of the CTA model as initial variable ordering. This implies that we consider the modeler's decision to encapsulate some components together within one module, i.e. local components of a module are assigned to neighboring positions within the variable ordering. Then we apply another heuristic to optimize the ordering respecting a size estimate for the BDD of the set of reachable configurations, which is derived from our upper bound for the transition relation as described and proven in [Bey01].

Partial Transition Relations. Usually the transition relation $\rightarrow$ is represented as implicit union of a timed transition relation $\overset{1}{\rightarrow}$ and discrete transition relations $\overset{a}{\rightarrow}$ for each synchronization label a. Experiments have shown that applying such partial transition relations sequentially is more efficient than using the union of these relations as monolithic transition relation.

Order of Transitions. Using several partial transition relations, we have to determine the order of their application. The intermediate sets of reached configurations in the reachability algorithm depend on this ordering, and therefore the size of the intermediate BDDs. A bad ordering of the partial transition relations can result in intermediate BDDs that are much larger than the final BDD of all reachable configurations. To compute the fixed point using only discrete transitions before applying time transitions is a successful strategy to avoid this problem.

Examples. In the following we report some performance results. All the computation times obtained using our tool are given in seconds of CPU time on a SUN Ultra-Sparc 1 with 200 MHz processor and 64 MB memory. The results of the BDD-based version of Kronos are taken from [BMPY97] and also obtained using a SUN Ultra-Sparc-1.

The BDD-based version of Kronos is able to verify 14 processes of Fischer's mutex protocol. Rabbit needs 13.6 s computation time for the same model. The verification of 32 processes needs 208 s. We are able to verify even the model with 128 of Fischer's processes using 512 MB memory in 9168 s.

Rabbit needs 6 s to compute the whole set of reachable configurations of the AND model with 4 inputs as mentioned in [BMPY97]. Kronos needs 324.7 s for this model. Out tool is able to verify the AND model up to 16 inputs in 1209 s. Another example is the little 'two state' example from [BMPY97], for which they report to handle up to 9 automata. Rabbit computes all rechable configurations of 64 two state automata (having $2.2 \cdot 10^{88}$ configurations) in 94 s. We used on-the-fly analysis for this example, which increases dramatically the performance of models consisting of mostly independent components. For more details about the on-the-fly algorithm see [BN01].

4 Refinement Checking

The intuition behind our refinement concept is an assumption/guarantee principle. We describe it with respect to our formalism: A refinement relation ($\mathcal{P}$ refines $\mathcal{Q}$) for CTA modules $\mathcal{P}$ and $\mathcal{Q}$ has to fulfill the following properties ($\mathcal{M}$ denotes a module, $\mathcal{M}.GI, \mathcal{M}.GO, \mathcal{M}.GMR$ denotes the module's sets of synchronization labels declared as INPUT, OUTPUT, MULTREST, respectively): (1) $\mathcal{Q}.GI \subseteq \mathcal{P}.GI$. The occurrence of a synchronization label g as input in a module $\mathcal{M}$ means that $\mathcal{M}$ guarantees that g is not restricted in $\mathcal{M}$. This clearly is a guarantee. Thus each input label of the specification should be an input label of the implementation. (2) $\mathcal{P}.GO \subseteq \mathcal{Q}.GO$. The occurrence of a synchronization label g as output in a module $\mathcal{M}$ means that $\mathcal{M}$ assumes that g is not restricted in the environment. The implementation should not make more assumptions than the specification, thus each output label of the implementation should also be an output label in the specification. (3) $\mathcal{Q}.G - \mathcal{Q}.GL = \mathcal{P}.G - \mathcal{P}.GL$. The synchronization labels of a module $\mathcal{M}$ can be partitioned into a set of interface labels ($\mathcal{M}.GI \cup \mathcal{M}.GO \cup \mathcal{M}.GMR$), and a set of local labels ($\mathcal{M}.GL$). Interface labels are those via which $\mathcal{M}$ can communicate with the environment. (4) $E_\mathcal{P} \subseteq E_\mathcal{Q}$. The external trace set $E_\mathcal{P}$ (defined below) of the labeled transition system generated by $\mathcal{P}$ is a subset of the external trace set $E_\mathcal{Q}$ of the labeled transition system generated by $\mathcal{Q}$. The intuition is that the occurrence of a trace t in $E_\mathcal{Q}$ means that $E_\mathcal{Q}$ *allows* the system behavior t, and the refinement should not allow more behaviors than the specification.

Let $[\![\mathcal{M}]\!]_I = (S, S^0, \mathcal{M}.G \cup \mathbb{N}, \rightarrow_\mathcal{M})$ be the labeled transition system generated by module $\mathcal{M}$ (integer semantics). $\mathcal{M}.G$ is the set of synchronization labels of module $\mathcal{M}$, $\mathbb{N}$ is used for the time values and $\mathcal{M}.G \cup \mathbb{N}$ is the set of transition labels of the labeled transition system $[\![\mathcal{M}]\!]_I$. For the exact definition we refer to [BR01]. Roughly spoken, it is the integer semantics of the parallel composition $\mathcal{A}_\mathcal{M}$ of all the automata contained by $\mathcal{M}$ regarding various compatibility constraints.

A **trace** of a given labeled transition system $[\![\mathcal{M}]\!]_I$ is an infinite sequence $(a_0, a_1, a_2, \dots)$ of elements of $\mathcal{M}.G \cup \mathbb{N}$. For each element a_k of the trace the following must hold: There exists a run $(s_0, s_1, \dots, s_{k+1}) \in Run([\![\mathcal{M}]\!]_I)$ with $s_i \xrightarrow{a_i}_\mathcal{M} s_{i+1}$ for all $i \in \{0, 1, \dots, k\}$.

We use a simulation relation for the algorithmic analysis of refinement within our tool implementation. To define timed simulation we need the notion of external transitions and external traces. After this we can proceed with the algorithm for the simulation check. We use the concept of safety simulation relation as described in [DHWT92].

Let $\tau \in \mathcal{M}.L$ be a local synchronization label. Then a τ-transition $s \xrightarrow{\tau}_{\mathcal{M}} s'$ is called an internal transition. For some configurations s, s' and $a_i \in \mathcal{M}.G \cup \mathbb{N}$ we define $s \xrightarrow{a_1 a_2 \dots a_n}_{\mathcal{M}} s'$ as: $\exists s''$ with $s \xrightarrow{a_1}_{\mathcal{M}} s''$ and $s'' \xrightarrow{a_2 \dots a_n}_{\mathcal{M}} s'$. For $a \in (\mathcal{M}.G \setminus \mathcal{M}.L) \cup \mathbb{N}$, $s \xrightarrow{a}_{\mathcal{M}} s'$ is an **external transition** and for $\hat{\tau}, \hat{\tau}' \in (\mathcal{M}.L)^*$, $s \xrightarrow{\hat{\tau} a \hat{\tau}'}_{\mathcal{M}} s'$ is the sequence of an arbitrary number of internal transitions followed by one external transition followed by another arbitrary number of internal transitions. In the sequel we write $s \xrightarrow{a}_{\mathcal{M}} s'$ for $s \xrightarrow{\hat{\tau} a \hat{\tau}'}_{\mathcal{M}} s'$. Now we can define an external trace as follows: The sequence $(a_0, a_1, a_2, \dots)$ of elements of $(\mathcal{M}.G \setminus \mathcal{M}.L) \cup \mathbb{N}$ is an **external trace**, if for each a_k there exists a sequence $(s_0, s_1, \dots, s_{k+1}), s_0 \in S^0, s_j \in S, 1 \le j \le k+1$ with $s_i \xrightarrow{a}_{\mathcal{M}} s_{i+1}$ for all $i \in \{0, 1, \dots, k\}$. It hides synchronization labels of internal discrete transitions.

A transition system $\mathcal{Q}$ simulates a transition system $\mathcal{P}$ if $\mathcal{Q}$ can match every step of $\mathcal{P}$ by a step with the same label. We define **timed simulation** for labeled transition systems as follows: The labeled transition system $[\![\mathcal{Q}]\!]_I = (S_{\mathcal{Q}}, S_{\mathcal{Q}}^0, \Sigma_{\mathcal{Q}}, \to_{\mathcal{Q}})$ simulates the labeled transition system $[\![\mathcal{P}]\!]_I = (S_{\mathcal{P}}, S_{\mathcal{P}}^0, \Sigma_{\mathcal{P}}, \to_{\mathcal{P}})$, $\Sigma = (\mathcal{P}.G \setminus \mathcal{P}.GL) \cup \mathbb{N} = (\mathcal{Q}.G \setminus \mathcal{Q}.GL) \cup \mathbb{N}$, $\Sigma_{\mathcal{Q}} = \mathcal{Q}.G \cup \mathbb{N}$, $\Sigma_{\mathcal{P}} = \mathcal{P}.G \cup \mathbb{N}$, if:

- there exists a simulation relation $\mathcal{R} \subseteq S_{\mathcal{P}} \times S_{\mathcal{Q}}$ which fulfills
  $\forall a \in \Sigma, \forall (p, q) \in \mathcal{R}, \forall p' \in S_{\mathcal{P}} :$
  $\left(p \xrightarrow{a}_{\mathcal{P}} p' \right) \implies \left(\exists q' : q \xrightarrow{a}_{\mathcal{Q}} q' \land (p', q') \in \mathcal{R} \right)$, and
- all initial configurations of $\mathcal{P}$ are contained within the simulation relation: $S_{\mathcal{P}}^0 \subseteq \{p \in S_{\mathcal{P}} | \exists q : (p, q) \in \mathcal{R}\}$.

The algorithm of the simulation check is shown in Fig. 1. For the composition of $\mathcal{P}$ and $\mathcal{Q}$ we compute the set of reachable configurations. We consider this set of tuples (p, q) as the initial relation for trying to build a simulation relation between $\mathcal{P}$ and $\mathcal{Q}$. Then, in each cycle of a fixed point iteration we assume that it is a simulation relation and we check whether all configurations of the set fulfill the simulation condition from the definition above. Differing from the definition, we use the transition relation of the product $\mathcal{P} \| \mathcal{Q}$, which is already computed in our approach. (The computation of the reachable set needs more time than checking the simulation relation if using $\mathcal{P} \| \mathcal{Q}$.) If there are 'bad' configurations we have to invalidate our assumption that it is already the simulation relation and we eliminate them from the relation. If we reached the fixed point (i.e. our assumption was true) we got the simulation relation. If the algorithm eliminates some of the initial configurations of $\mathcal{P}$ from the relation there cannot exist a simulation relation and the algorithm aborts. If we reached the fixed point (i.e. our assumption was true) we got the simulation relation.

Note. Modules $\mathcal{P}$ and $\mathcal{Q}$ are not allowed to contain variables within their interfaces (shared variables). The simulation check considers only synchronization labels regarding external traces.

Production Cell Example. To validate the practical relevance of our tool using a complex system, we developed a CTA model of a production cell, which is similar to the Lewerentz/Lindner production cell from FZI. The system consists of 20 machines and belts with 44 sensors and 28 motors. We modeled the system as modular composition of several belts, turntables and machines, including 45 timed automata containing 22

Input: labeled transition system $[\![\mathcal{P}]\!]_I = (S_{\mathcal{P}}, S_{\mathcal{P}}^0, \Sigma_{\mathcal{P}}, \rightarrow_{\mathcal{P}})$
    as integer semantics of module $\mathcal{P}$,
    labeled transition system $[\![\mathcal{P}||\mathcal{Q}]\!]_I = (S_{\mathcal{P}||\mathcal{Q}}, S_{\mathcal{P}||\mathcal{Q}}^0, \Sigma_{\mathcal{P}||\mathcal{Q}}, \rightarrow_{\mathcal{P}||\mathcal{Q}})$
    as integer semantics of the composition $\mathcal{P}||\mathcal{Q}$ with $\Sigma = (\mathcal{P}.G \setminus \mathcal{P}.GL) \cup \mathbb{N}$.
Output: $true$, iff $\mathcal{P}$ simulates $\mathcal{P}||\mathcal{Q}$
$R_{\mathcal{P}||\mathcal{Q}} := Reach([\![\mathcal{P}||\mathcal{Q}]\!]_I)$
do
    $R'_{\mathcal{P}||\mathcal{Q}} := R_{\mathcal{P}||\mathcal{Q}}$
    forall $a \in \Sigma$
    if $S_{\mathcal{P}}^0 \not\subseteq \{p \in S_{\mathcal{P}} | \exists q : (p,q) \in R_{\mathcal{P}||\mathcal{Q}}\}$ **then return** false

$$R_{\mathcal{P}||\mathcal{Q}} := R_{\mathcal{P}||\mathcal{Q}} \cap \left\{ (p,q) \in R_{\mathcal{P}||\mathcal{Q}} \middle| \begin{array}{c} \forall p' : \left(p \xrightarrow{a}_{\mathcal{P}} p' \right) \Rightarrow \\ \left(\begin{array}{c} \exists q' : (p,q) \xrightarrow{a}_{\mathcal{P}||\mathcal{Q}} (p',q') \\ \wedge \ (p',q') \in R_{\mathcal{P}||\mathcal{Q}} \end{array} \right) \end{array} \right\}$$

while $R_{\mathcal{P}||\mathcal{Q}} \neq R'_{\mathcal{P}||\mathcal{Q}}$
return true

Fig. 1. Algorithm for checking a simulation relation

clocks. For the measurement of the throughput, i.e. how long does a piece need to go through the production cycle, we modeled each belt to be able to measure the time of transportation using a clock. For the verification process we can fade out some details of the machines. To verify a safety property, e.g. 'the drilling machine must be off if the transport belt is not off', we verify at first that the timed version of the transport belt implements an untimed version by checking the existence of a simulation relation. Now we can verify the safety property of the model using that smaller untimed version for transport belts. The analysis of the safety property of the system using a timed model for the sensor instances needs 1098 s. Using an untimed version for the transport belts, the same task needs only 556 s. It shows that an abstraction within one small part of the system has a big impact on the computation time. The computation time for the simulation check is 0.5 s because the belt model is a small part of the whole system.

References

[AD94]   Rajeev Alur and David L. Dill. A theory of timed automata. *Theoretical Computer Science*, 126:183–235, 1994.

[Bey01]   Dirk Beyer. Improvements in BDD-based reachability analysis of timed automata. In *Proc. FME 2001*, LNCS 2021, pages 318–343. Springer-Verlag, 2001.

[BMPY97] M. Bozga, O. Maler, A. Pnueli, and S. Yovine. Some progress on the symbolic verification of timed automata. In *Proc. CAV'97*, LNCS 1254, pages 179–190. 1997.

[BN01]   Dirk Beyer and Andreas Noack. Efficient verification of timed automata using BDDs. In *Proc. Formal Methods for Industrial Critical Systems*. to appear, 2001.

[BR98]   Dirk Beyer and Heinrich Rust. Modeling a production cell as a distributed real-time system with Cottbus Timed Automata. *Proc. FBT'98*, pages 148–159. Shaker, 1998.

[BR01]   Dirk Beyer and Heinrich Rust. Cottbus Timed Automata: Formal definition and semantics. In *Proc. FSCBS 2001*, pages 75–87, 2001.

[DHWT92] David L. Dill, Alan J. Hu, and Howard Wong-Toi. Checking for language inclusion using simulation preorders. In *Proc. CAV'91*, LNCS 575, pages 255–265. 1992.

Deriving Real-Time Programs from Duration Calculus Specifications

François Siewe[1] and Dang Van Hung[2]

[1] Department of Maths. and Computer Science, University of Dschang
P. O. Box 96 Dschang, Cameroon. Fax (237) 45 13 81
fsiewe@uycdc.uninet.cm

[2] The United Nations University International Institute for Software Technology
P. O. Box 3058 Macau. Fax (853) 71 29 40
dvh@iist.unu.edu

Abstract. In this paper we present a syntactical approach for deriving real-time programs from a formal specification of the requirements of real-time systems. The main idea of our approach is to model discretization at state level by introducing the discrete states approximating the continuous ones, and then derive a specification of the control program over discrete states. Then the control program is derived from its specification using an extension of Hoare triples to real-time.

Keywords: Continuous specification, discrete design, real-time program, concurrency, shared variables, Hoare triples

1 Introduction

Real-time control systems usually consist of some physical plant, in permanent interaction with its environment, for which a suitable controller has to be constructed such that the controlled plant exhibits the desired time dependent behaviour.

This paper presents a syntactical approach for deriving real-time programs from a formal specification of the requirements of real-time systems. The approach provides a logical framework that can handle both continuous time and discrete time models in a uniform manner. We consider Duration Calculus (DC for short) [ZHR91] as specification and top level design language, for its effectiveness in reasoning about the design of real-time systems, and the fact that DC is successfully used in many case studies. We denote by DC^* an extension of DC with iteration [HuG99]. We link DC and Hoare logic to reason about both the real-time behaviour and the functional behaviour of real-time programs in a uniform manner.

Our design technique is formulated as follows. A real-time control system is a distributed hybrid system as it comprises continuous components (e.g the plant) and discrete components (e.g the controller). At the first step of the design process a state variables model of the system should be defined. The state model comprises continuous state variables (modelling the behaviour of the continuous

T. Margaria and T. Melham (Eds.): CHARME 2001, LNCS 2144, pp. 92–97, 2001.

components) and discrete state variables (modelling the behaviour of the discrete components). Then the requirement of the system is formalised as a *DC* formula *Req* over continuous state variables. A design decision must be taken as how the requirement of the system will be met and refined into a detailed design *Des* over continuous state variables such that *Des* ⇒ *Req*. Then the discretization step follows. We approximate continuous state variables by discrete ones and formalise the relationship between them based on the general behaviour of the sensors and actuators. Then the control requirement is derived from the detailed design and refined into a *simple DC** formula *Cont* over discrete state variables such that $\mathcal{A} \vdash Cont \Rightarrow Des$, for some assumptions $\mathcal{A}$ about the behaviour of the environment and the relationship between continuous state variables and discrete state variables. The discrete formula *Cont* is the formal specification of the controller. The last step of the design process consists to write a real-time program for the controller and verify its correctness w.r.t. the specification *Cont* using our extended Hoare triples.

In the literature, some works have addressed the problem. Fränzle has developed in [Fra96] a technique for synthesizing controllers from Duration Calculus specifications. His approach is semantical, and so more difficult to use by engineers. Our aim is to provide the designers with syntax-based compositional interface for the design and verification of real-time programs, hiding semantic details. Another advantage of a syntactical approach is that the design process can be assisted by proof tools. In [Hoo94,XuM98,PWX98], some extensions of Hoare triples are proposed for real-time programs verification. However the derivation of a discrete design is not considered.

The remainder of the paper is organized as follows. We give a summary of *DC* in Section 2. Section 3 details our design technique. The program construction technique is presented in Section 4. Section 5 concludes the paper.

2   Duration Calculus with Iteration

In this section we give a brief summary of *DC**. The readers are referred to [HuG99] for more details on the calculus.

A language for DC* is built starting from the following sets of *symbols*: a set of *constant symbols* $\{a, b, c, \ldots\}$, a set of *individual variables* $\{x, y, z, \ldots\}$, a set of *state variables* $\{P, Q, \ldots\}$, a set of *temporal variables* $\{u, v, \ldots\}$, a set of *function symbols* $\{f, g, \ldots\}$, a set of *relation symbols* $\{R, U, \ldots\}$, and a set of *temporal propositional letters* $\{A, B, \ldots\}$.

A DC* language definition is essentially that of the sets of *state expressions* S, *terms* t and *formulae* φ of the language. These sets can be defined by the following BNFs:

$$S \ \hat{=} \ \mathbf{0} \mid P \mid \neg S \mid S \vee S$$
$$t \ \hat{=} \ c \mid x \mid u \mid \int S \mid f(t, \ldots, t)$$
$$\varphi \ \hat{=} \ A \mid R(t, \ldots, t) \mid \neg \varphi \mid (\varphi \vee \varphi) \mid (\varphi ^\frown \varphi) \mid (\varphi^*) \mid \exists x \varphi$$

A state variable P is interpreted as a function $I(P) : \mathbb{R}^+ \to \{0, 1\}$ (a state). $I(P)(t) = 1$ means that state P is present at time t, and $I(P)(t) = 0$ means that P is not present at time t. We assume that a state has finite variability in any finite time interval. A state expression is interpreted as a function which is defined by the interpretations for the state variables and Boolean operators.

For an arbitrary state expression S, its duration is denoted by $\int S$. Given an interpretation I of the state variables and an interval, duration $\int S$ is interpreted as the accumulated length of time within the interval at which S is present. So for any interval $[t, t']$, the interpretation $I(\int S)([t, t'])$ is defined as $\int_t^{t'} I(S)(t)dt$.

A formula φ is satisfied by an interpretation in an interval $[t, t']$ when it evaluates to true for that interpretation over that time interval. This is written as $I, [t, t'] \models \varphi$.

Given an interpretation I, a formula $\varphi \frown \phi$ is true for $[t, t'']$ if there exists a t' such that $t \leq t' \leq t''$ and φ and ϕ are true for $[t, t']$ and $[t', t'']$ respectively.

We consider the following abbreviations: $\ell \,\hat{=}\, \int 1$, $\lceil S \rceil \,\hat{=}\, (\int P = \ell) \wedge (\ell > 0)$, $\Diamond \varphi \,\hat{=}\, true \frown \varphi \frown true$, and $\Box \varphi \,\hat{=}\, \neg \Diamond \neg \varphi$.

3   From Continuous Specifications to Discrete Design

A model of real-time control systems is depicted in figure 1. The *plant* denotes the continuous components of the system, in permanent interaction with the physical environment. The *controller* is a discrete component denoting a program executed by a computer. The *sensors* sample the states of the plant for them to be observable by the controller. The *actuators* receive commands from the controller and control the plant accordingly. The sensors and the actuators constitute the continuous-to-discrete and discrete-to-continuous interfaces respectively.

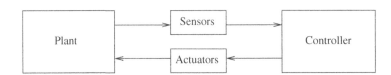

Fig. 1. A model of controlled system

We define three concepts for formalising the relationship between continuous state variables and discrete ones. The first concept is the *stability* of continuous state variables. This property requires that continuous state variables should not change very fast in order to be observable at discrete time.

Definition 1 (Stability). *Given a state variable s and a positive real number δ, we say s is δ-stable iff the following formula is satisfied by any interval*

$$\delta\text{-stable}(s) \,\hat{=}\, \Box(\lceil \neg s \rceil \frown \lceil s \rceil \frown \lceil \neg s \rceil \Rightarrow \lceil \neg s \rceil \frown (\lceil s \rceil \wedge \ell > \delta) \frown \lceil \neg s \rceil)$$

Definition 2 formalises the relationship between a state variable r and a state variable s that always follows r after some time.

Definition 2 (Control state). *Given two state variables r and s, and a non-negative real number δ, we say r δ-controls s iff the following formula is satisfied by any interval*

$$r \triangleright_\delta s \;\widehat{=}\; \Box(\lceil r \rceil \wedge \ell > \delta \Rightarrow (\ell \le \delta)^\frown \lceil s \rceil)$$

The concept of *control state* can be used for formalising the behaviour of actuators. Let r be a state variable modelling a program command, and s a state of the plant. Then the relation $r \triangleright_\delta s$ means that whenever the controller issues the command r, the plant gets into state s within at most δ time units. So the maximum response time is δ time units.

Definition 3 formalises the relationship between a state variable s and a state variable r that observes any change (stable enough) in s.

Definition 3 (Observation state). *Given two state variables r and s, and a non-negative real number δ, we say r δ-observes s iff the following formula is satisfied by any interval*

$$r \overset{\rightrightarrows}{\approx}_\delta s \;\widehat{=}\; (s \triangleright_\delta r) \wedge (\neg s \triangleright_\delta \neg r)$$

The concept of *observation state* can be used for formalising the behaviour of sensors. Let r be a state variable modelling a program variable, and s a state of the environment. Then the relation $r \overset{\rightrightarrows}{\approx}_\delta s$ means that any change (stable enough) in s is observed by the controller within δ time units. So the sampling step is δ time units. Note that the definition says nothing about unstable change of s. We will assume that continuous state variables are stable enough, otherwise there is no way for a digital controller to guarantee safety requirements in general.

The discrete interface is formalised as follows. For any continuous state variable s, we consider a discrete state variable s_c used by the controller to observe s via the sensors. The relationship between s and its sampling s_c is formalised by $s_c \overset{\rightrightarrows}{\approx}_\delta s$, for some non-negative real number δ (representing the sampling step). Similarly, for any state p of the plant we consider a command p_c, a discrete state, for requesting (via the actuators) the plant getting into state p. The relationship between p and p_c is formalised by $p_c \triangleright_\tau p$, for some non-negative real number τ (representing the response time).

We provide rules useful for refining a continuous design towards a discrete design. Following are some examples. More rules and their proofs are given in [SiH00a].

Rule 1 (Transitivity of control)

(a) $\dfrac{(r \triangleright_\delta s)\quad(s \triangleright_\tau t)}{r \triangleright_{(\delta+\tau)} t}$; (b) $\dfrac{(r \overset{\rightrightarrows}{\approx}_\delta s)\quad(s \overset{\rightrightarrows}{\approx}_\tau t)}{r \overset{\rightrightarrows}{\approx}_{(\delta+\tau)} t}$

Rule 2 (Monotonicity)

(a) $\dfrac{r \triangleright_\tau s \quad \tau \le \delta}{r \triangleright_\delta t}$; (b) If $r \Rightarrow s$ then $r \triangleright_0 s$; (c) $\dfrac{r \triangleright_\delta s \quad t \triangleright_\delta u}{(r \wedge t) \triangleright_\delta (s \wedge u)}$

Rule 3 (Sequential (a) and Parallel (b) composition)

$$(a) \ \frac{\psi \Rightarrow \Box\chi \quad \varphi \Rightarrow \Box\chi}{\psi \Rightarrow \neg(\neg\beta^\frown true) \quad \varphi \Rightarrow \neg(true^\frown\neg\alpha) \quad \alpha^\frown\beta \Rightarrow \chi}{\varphi^\frown\psi \Rightarrow \Box\chi} \ ; \quad (b) \ \frac{A \Rightarrow \Box\psi \quad B \Rightarrow \Box\varphi}{A \wedge B \Rightarrow \Box(\psi \wedge \varphi)}$$

4   Real-Time Programs Construction

In this section we present a real-time programming language and our extension of
Hoare triples as an abstract semantics of it. We develop a set of verification and
design rules for a program. The readers are referred to [SiH00b] for the formal
semantics of the language and the proof of the soundness of the verification rules
which are based on Weakly Monotonic Time Duration Calculus ([PaH98]).

In the following BNFs, S ranges over programs, P and Q range over processes, e over arithmetic expressions, g over guards (Boolean expressions), d
over positive natural numbers $(d > 0)$, and x over program variables.

$$P ::= \mathbf{skip} \mid \overline{x} := \overline{e} \mid \mathbf{delay} \ d \mid \mathbf{await} \ g \mid P; Q \mid$$
$$\mathbf{if} \ g \ \mathbf{then} \ P \ \mathbf{else} \ Q \ \mathbf{fi} \mid \mathbf{while} \ g \ \mathbf{do} \ P \ \mathbf{od}$$

$$S ::= P_1 \parallel P_2 \parallel \ldots \parallel P_n, \quad (n \geq 1)$$

For simplicity, we do not include non-determinism in the language. We assume the *true synchrony hypothesis*, viz, computation and communication do not
take time, only the waiting for external synchronisation or explicit delay statements take time. The statement **delay** d delays the process for d time unit(s).
During this period, the process could not do anything. Another timing control
statement is **await** g which makes the process wait until the guard g is fired
by the occurrence of an external event. The statements **delay** and **await** are
the only statements that take time. The other statements conserve their usual
meaning. Processes can share variables. We denote by $\mathcal{WO}(P_i)$ the set of variables written only by process P_i. Of course, local variables of P_i belong to that
set. We also denote by $\mathcal{VAR}(\alpha)$ the set of variables occurring in $\alpha \in \{P, e, g\}$.

We extend the Hoare logic to real-time concurrent programming with shared
variables to reason about terminating as well as non-terminating real-time programs. Unlike computational programs, real-time control programs in general
run forever. Our extension of Hoare triples has two forms. The first form is
$\{p\}[P, \varphi]\{q\}$, meaning that if the precondition p holds in the initial state and
P terminates, then the postcondition q holds in the final state and the execution period of the program satisfies the duration formula φ. The second form is
$\{p\}[P, \varphi]\{\}$, meaning that if the precondition p holds in the initial state and P
does not terminate, then any prefix of the execution period of the program satisfies the duration formula φ. This couple of triples is used to reason about the
properties of real-time programs. Following are some examples of our extended
Hoare triples.

Rule 4 (Assignment)
$$\{q[\overline{x}\backslash\overline{e}]\}[\overline{x} := \overline{e}, \ell = 0]\{q\}$$

Rule 5 (Delay in process P_i)
$$\{q\}[\textbf{delay } d, \ell = d \wedge \lceil q \rceil]\{q\}, \text{ provided } \mathcal{VAR}(q) \subseteq \mathcal{WO}(P_i).$$

Rule 6 (Consequence)
$$(a) \frac{\{p\}[P, \varphi]\{q\} \quad p' \Rightarrow p \quad q \Rightarrow q' \quad \varphi \Rightarrow \phi}{\{p'\}[P, \phi]\{q'\}}; \quad (b) \frac{\{p\}[P, \varphi]\{\} \quad p' \Rightarrow p \quad \varphi \Rightarrow \phi}{\{p'\}[P, \phi]\{\}}$$

5 Conclusion

In this paper we have presented a technique for the design of real-time hybrid systems using Duration Calculus. We provide a set of rules for deriving a discrete design from the specification of a real-time system. Then we extend Hoare triples to real-time and develop rules for designing a real-time program that satisfies the discrete design.

References

[Fra96]   M. Fränzle. Synthesizing Controllers from Duration Calculus Specifications. Proceedings of *Formal Techniques in Real-Time and Fault-Tolerant Systems (FTRFT'96)*, LNCS 1135, Springer-Verlag, 1996.

[Hoo94]   Jozef Hooman. Extending Hoare Logic to Real-Time. *Formal Aspects of Computing*, 6A:801-825, 1994.

[HuG99]   Dang Van Hung and Dimitar P. Guelev. Completeness of a fragment of Duration Calculus with Iteration. Proceedings of Asian Computing Science Conference (ASIAN'99), LNCS 1742, Springer-Verlag, 1999, pp. 139–150.

[PaH98]   Paritosh K. Pandya and Dang Van Hung. Duration Calculus with Weakly Monotonic Time. Proceedings of *Formal Techniques in Real-Time and Fault-Tolerant Systems*, LNCS 1486, pp. 55–64, Springer-Verlag, 1998.

[PWX98]   Paritosh K. Pandya, Wang Hanpin, and Xu Qiwen. Towards a Theory of Sequential Hybrid Programs. Proceedings of the *International Conference on Programming Concepts and Methods* (PROCOMET'98), Chapman & Hall, 1998, pp. 366-384.

[SiH00a]   François Siewe and Dang Van Hung. From continuous specification to discrete design. Proceedings of the *International Conference on Software: Theory and Practice* (ICS2000), Yulin Feng, David Notkin and Marie-Claude Gaudel (eds), Beijing, August 21-24, 2000, pp. 407-414.

[SiH00b]   François Siewe and Dang Van Hung. Deriving Real-time Programs from Duration Calculus Specifications. Technical Report 222, UNU/IIST, P.O. Box 3058, Macau, December 2000.

[XuM98]   Xu Qiwen and Mohalik Swarup. Compositional Reasoning using Assumption-Commitment Paradigm. Technical Report 136, UNU/IIST, P.O. Box 3058, Macau, February 1998.

[ZHR91]   Zhou Chaochen, C.A.R. Hoare, and Anders P. Ravn. A calculus of duration. *Information Processing Letters*, 40(5):269-276, 1991.

Reproducing Synchronization Bugs with Model Checking

Karen Yorav, Sagi Katz, and Ron Kiper

Galileo Technology, Israel
{kareny,sagi,ron}@galileo.co.il

Abstract. In this paper we describe our experience in reproducing synchronization bugs using a model checker. We demonstrate how model checking technology can be utilized for more than just model checking. Synchronization bugs are caused by physical phenomena which cause the actual behavior of a chip to be different than predicted according to the functional model. Traditionally, verification methods such as dynamic simulation and model checking use a synchronous model, whereas the actual behavior is according to an asynchronous model. Because of this, synchronization bugs are very hard to trace. Using a model checker we were able to create a model closer to the actual behavior, and retrace many synchronization bugs. Because model checking allows us to introduce non-determinism when checking a VLSI design, and because of its ability to produce counter examples for specifications that fail, we find that model checking is the ideal tool for reproducing synchronization bugs.

1   Introduction

Finding bugs as early in the design flow as possible is the goal of all types of verification techniques. The main method for functional verification used by the VLSI industry today is *dynamic simulation*. In **Galileo Technology** we use formal verification methods to complement the use of dynamic simulation. The main formal verification method used is (symbolic) *Model Checking* [2]. Model checkers are automatic tools, capable of proving that a design meets its specification. Model checking has two major advantages: it is exhaustive, i.e. it covers all possible combinations of inputs, and when the design does not comply with its specification the model checking tool presents a counter example to prove this. The main limitation of model checking is that it can handle only relatively small parts of a design.

When a functional bug is found in the early stages of design, it is relatively easy to fix. However, sometimes a bug is revealed only after the first prototype is manufactured and tested in the lab. In this case it is far more difficult to identify the cause of the erroneous behavior. An interesting case is when the bug that is seen in the lab was proven to be impossible by the model checking tool. These bugs are caused by physical phenomena which cause the chip to behave differently than expected according to the functional model used for verification.

T. Margaria and T. Melham (Eds.): CHARME 2001, LNCS 2144, pp. 98–103, 2001.

One of the possible causes for such bugs, although not the only one, is the violation of timing constraints on the boundary between asynchronous modules. A typical case is when there are multiple clocks in the design, as in most chips today. Problems arise when a signal (the output of a flip flop) from one clock domain is passed into a different clock domain. Because of certain physical phenomena, two signals that were coordinated in the source clock domain (changed value together) may be passed to the target clock domain with one cycle delay between them. If the design in the target clock domain depends on the coordination between these signals it can cause a **synchronization bug**.

Synchronization bugs are problematic for dynamic simulation, since the model used for dynamic simulation is a deterministic synchronous model, and does not take into account the possibility of the extra delay. To reproduce synchronization bugs we need a non-deterministic model in which each signal crossing the clock domain boundaries is either delayed by one clock cycle or not. It is also necessary to check all possible timings of events, which is impossible with simulation. Although the extra delay is caused by the violation of timing constraints, even if we introduce exact delays on gates and wires into the dynamic simulation, synchronization bugs will not be found, since the delay is caused by a physical phenomena (meta-stable states) which is not simulated.

The method we use is to replace **synchronizer** modules [1] in the design with non-deterministic modules. Synchronizers are modules which are placed on the signals of asynchronous interfaces. Because the space complexity of model checking is very sensitive to the number of variables (flip flops) in the design and the complexity of the behavior, we cannot simply replace all the synchronizers with a sophisticated model that represents the real life behavior. We use several models, each more sophisticated than the previous. We start with the most simple and small, and move to the others only if necessary.

This work is unique in that it shows that it is possible to use model checking in order to verify properties at a level which includes asynchronous physical behaviors, behaviors which do not appear at the register transfer level.

The paper is organized as follows. In Sect. 2 we give a brief overview of model checking of multi-clock systems and explain how synchronizers are used. In Sect. 3 we describe our method for reproducing synchronization bugs, and in Sect. 4 we conclude with some general remarks about our work.

2   Model Checking of Multi-clock Designs

Most real-life designs are multi-clock systems because different parts of the design need to run at different paces. In such systems signals from one clock are driven into flip-flops running on a different clock. Model checking of such systems is either performed using a single clock, which does not represent all the possible behaviors, or using multiple clocks, which is possible only on very small blocks.

A D-flip-flop passes the value from its input (called "d") to its output (called "q") at every transition of the clock from zero to one. The correct behavior of a flip flop is guaranteed only under certain conditions of *setup* and *hold* times on d, which means that d must be stable for a certain amount of time before and after the rising edge of the clock. If these conditions are not maintained it is possible that the flip flop will pass the wrong value to its output, or enter a meta-stable state, which means that the output of the flip-flop is undetermined and may stay this way until the next rising edge of the clock. Any other component of the design that examines the output of a flip flop in a meta-stable state may interpret it as either zero or one.

Assume that the output of flip flop a is the input to the flip flop b. If a is running on a different clock than b then whenever a changes value it may violate setup and hold conditions for b, since the change is not synchronized with b's clock, and b might output the wrong value, or enter a meta-stable state. The design methodology for multi-clock systems is that the interface between clock domains must use a full handshake protocol, so that the design on b's side must acknowledge the change in a's value before a is allowed to change again. This solves the problem of b getting the wrong value because it is guaranteed that by the next cycle a has been stable long enough and b will get the correct value. The only effect here is that b changes value one cycle later than it should have. The solution for the problem of meta-stable states is the introduction of a synchronizer between a and b.

A typical synchronizer [1] decreases the probability of an internal flip-flop entering a meta-stable state by placing two D-flip-flops in a row (Fig. 1). The probability of s_1 entering a meta-stable state because of a change in a's value is very small. If this happens, s_2 may stabilize on zero or one, or enter a meta-stable state itself. The probability of this is even smaller.

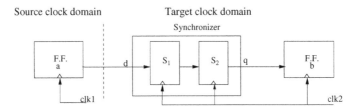

Fig. 1. A typical synchronizer

Each synchronizer creates a delay on its input signal, but it is possible to have different delays on different signals because of slight timing differences. This happens when two different signals that cross clock domains create different effects on their synchronizers - one will pass the change of value in time and the other "misses" a clock cycle (because there was not enough setup time) and passes the change only in the following cycle. Thus, signals that were coordinated in one clock domain are passed to the other clock domain with a delay of one cycle between them. Synchronization bugs occur when the correct behavior of the design inadvertently depends on the coordination of two signals, and this coordination is broken by the additional delay on one of them. This error will not be revealed during verification with dynamic simulation since in simulation the coordination between signals is preserved.

3 Reproducing Synchronization Bugs

Normally, model checking fails to detect synchronization bugs because it ignores the non-deterministic behavior of synchronizers. We show how careful manipulation of the design can create a more accurate model. We present a simple version of the module with which we substitute the synchronizers in the design. We then describe more complicated, and more accurate modules. We need more than one version because using a complicated synchronizer may make the model checking task impractical. We strive to make the behavior of the synchronizer as simple as possible, not to add more variables (flip flops) than necessary, and not to add non-determinism if not necessary, while making sure that the behaviors we add are actually possible, or else we will find false bugs.

3.1 Introducing Early Signals

The situation is that a bug was found in the lab, on real hardware, and needs to be reproduced. A specification ψ claiming that the erroneous behavior is impossible is checked on the design and is proved to be correct. We suspect that this is a synchronization bug. Our goal is to refine the model of the design so that it is closer to the real behavior of synchronizers in a chip.

As explained in Sect. 2, it is possible that the first flip flop inside the synchronizer will not pass a change in value in the cycle that it happens, but only one cycle later. To model this behavior we override the synchronizer code with a non-deterministic environment model. Instead of a 1-cycle delay, the new synchronizer module may pass a change on the input signal 1-cycle early. The effect on the coordination between different signals is the same, since it allows for any two input signals that change value at the same cycle in the source clock domain to be passed to the target clock domain with a difference of 1 cycle between them (even though one of them is early instead of the other being late).

The new synchronizer module is written in the language of the environment, since this language allows for non-determinism. We use IBM's RuleBase model checker [3], and the modules are written in EDL - RuleBase's *Environment Description Language*. The semantics of this language is simple. The **next** clause describes the value of a variable in cycle $n+1$ based on variable values in cycle n. The expression $\{ex_1, \ldots, ex_k\}$ stands for the non-deterministic choice of one of the values $ex_1, \ldots, ex_k$. Following is the code for the *stable_early* synchronizer. Its input is either always passed on time, or always one cycle early.

Module stable_early (/*INPUT*/ d)(/*OUTPUT*/ q)
VAR
    s1, s2 : boolean;
    Status : {Early,no_shift};
ASSIGN

```
    next(s1) := d;
    next(s2) := s1;
    next(Status) := Status;
DEFINE
    q := if (Status=Early) then s1 else s2 endif;
```

The **VAR** section declares the internal variables of this module: $s1$, $s2$, and *Status*, which decides whether the input value d will be early or not. *Status* can have one of two possible values - *Early* or *no_shift*. In the *ASSIGN* section the behavior of the variables is defined so that $s1$ has the value of d delayed by one cycle, and $s2$ has the value of d delayed by two cycles. The value of *Status* never changes, and q is defined according to it.

The initial value for all the variables is left undefined, so *Status* can be either *Early* or *no_shift*. This means that each instance of the synchronizer will either be always early, or always on time. This behavior is reasonable because it is likely that a synchronizer that causes a delay will do so for long periods of time. We do not set an initial value for $s1$ and $s2$ because all of our designs have a *reset* signal, so initial values do not matter.

We replace every synchronizer in the design with the *stable_early* module, and check the same specification ψ. If it fails, the counter example produced by the model checker gives an example of how the different timing of signals passing between clock domains can cause an error. This counter example displays a behavior which, although impossible in the regular model used for model checking, is possible in the actual chip, and is therefore a valid counter example.

3.2   Other Versions of Synchronizer Modules

For a large design we do not want to replace all of the synchronizers with non-deterministic versions, since this can make model checking impossible. However, if we replace only some of the synchronizers then the *stable_early* module is no longer sufficient. If signals a and b pass through synchronizers, but only a's synchronizer is replaced, then there is only the possibility that a is early and not that b is early. For this reason we introduce the *stable_early_and_late* synchronizer, which allows for signals to be either early or late.

We add a variable $s3$ defined by: $next(s3) := s2$. We allow the *Status* variable to take an extra value "Late" so that if $Status = Late$ then $q := s3$. The result is that q can be delayed by either 1, 2, or 3 cycles after d. We can now replace some of the original synchronizers in the design with this version, leaving some of them intact. We restrict our model so that either there are no late synchronizers or there are no early synchronizers (we do mix *Early* and *no_shift*, or *Late* and *no_shift*). This prevents the possibility of two cycles difference between signals that were coordinated, which cannot happen in reality.

In both of the previous versions the delay caused by a single synchronizer was fixed, although in reality it is possible for a synchronizer to add a delay only part of the time. For this reason we have a *dynamic_early_and_late* synchronizer module, in which the *Status* variable can change value in the middle of a run.

We add an input boolean variable called *allow_change*, and define *Status* so that it cannot change when *allow_change* = *False*. When *allow_change* = *True* the *Status* variable can non-deterministically change value (or not), the only limitation being that it is not allowed to change from *Early* to *Late* or visa versa, because such a change is physically impossible.

Most of the time when using this version we allowed *Status* to change only once in each run, because this behavior is closer to the real behavior of the chip (the delay of a synchronizer changes only very rarely). This also reduces the running times of model checking.

4   Our Experience and Conclusions

Galileo Technology develops VLSI chips for the communication markets. Formal verification using RuleBase is an integral part of the verification process of our products. The technique described here was used on several designs, some of which were very large. The design which started the work described here, for example, is a communication system controller, with around 1 Million gates. It is impossible to verify designs of this size using a model checker, so we verify smaller units separately. This unit includes two clock domains, running at different rates. Most of the time the model checking effort was done using the same clock for both domains, and using the *stable_early* or *stable_early_and_late* versions, but other versions were used as well, including complicated ones not described here. All together we were able to find several significant bugs which would have been extremely difficult to find otherwise. It is interesting to note that the main bug in this example was found using the simple *stable_early_and_late* module. In fact, most bugs are found using the simple stable versions, because most of the bugs are conceptual bugs, where the designer did not consider synchronization issues at all. In this case, it is enough to break the coordination between signals once and the design will exhibit wrong behavior.

The experience described in this paper shows that model checking is a very powerful tool. It can be used in many ways, to solve problems which are difficult to solve using other tools. The VLSI design flow is long and complicated, and we believe that there are other such problems where model checking can be applied. For example, we intend to investigate the possibility of using model checking to solve timing problems and to check performance.

References

1. L. Glasser and D. Dopperpuhl. *The design and analysis of VLSI circuits*. Addison-Wesley, 1985.
2. K. L. McMillan. *Symbolic Model Checking: An Approach to the State Explosion Problem*. Kluwer Academic Publishers, 1993.
3. The RuleBase homepage at IBM:
   http://www.haifa.il.ibm.com/projects/verification/RB_homepage/.

Formally-Based Design Evaluation

Kenneth J. Turner and Ji He

Computing Science and Mathematics, University of Stirling, Stirling FK9 4PU, Scotland
kjt@cs.stir.ac.uk, h.ji@reading.ac.uk

Abstract. The paper investigates specification, verification and test generation for synchronous and asynchronous circuits. The approach is called DILL (Digital Logic in LOTOS – the ISO Language Of Temporal Ordering Specification). Relations for (strong) conformance are defined to verify a design specification against a high-level specification. Tools have been developed for automated testing and verification of conformance between an implementation and its specification.

1 Introduction

DILL (Digital Logic in LOTOS [5]) is an approach for specifying digital circuits using LOTOS (Language Of Temporal Ordering Specification [4]). DILL allows formal specification of hardware designs, represented using LOTOS at various levels of abstraction. DILL deals with functional and timing aspects, synchronous and asynchronous design. There is support from a library of common components and circuit designs. Analysis uses standard LOTOS tools. Among Hardware Description Languages, DILL most closely resembles CIRCAL (Circuit Calculus) in that both have a behavioural basis in process algebra. In the authors' experience, DILL can be used successfully at a variety of abstraction levels where CIRCAL appears to be less effective.

LOTOS is a formal language standardised for use with communications systems. DILL, which is realised through translation to LOTOS, is a substantially different application area for this language. LOTOS is neutral with respect to whether a specification is to be realised in hardware or software, allowing hardware-software co-design. LOTOS has well-developed theories for verification and test generation.

The current standard for LOTOS does not support quantified timing, although the authors have developed Timed DILL for hardware timing analysis using ET-LOTOS (Enhanced Timed LOTOS). However, asynchronous circuits are also of interest. Like DILL, other asynchronous verification approaches define relations that judge correctness of a circuit design. However it is not possible to detect deadlocks and livelocks with *conformance* [1] and *decomposition* [2]. Although *strong conformance* [3] can do this, it does not work for non-deterministic specifications. The relations *confor* and *strongconfor* mentioned in this paper resolve these problems.

For validating hardware designs, test cases are in practice manually defined or are randomly generated. More rigorous approaches use traditional software testing techniques or state machine representations. In DILL, tests are derived from higher-level specifications in a novel adaptation of protocol conformance testing theory.

T. Margaria and T. Melham (Eds.): CHARME 2001, LNCS 2144, pp. 104–109, 2001.

2   Synchronous and Asynchronous Circuit Models

DILL supports logic designs at different levels of abstraction, with formal comparison of higher level and more detailed design specifications. A component's ports (e.g. its pins) are represented by LOTOS event gates. To 'wire up' two ports, their LOTOS gates are merely synchronised.

The classical synchronous circuit model has combinational logic to provide the primary outputs and the internal outputs according to the primary inputs and the internal inputs. Internal outputs are then fed into state hold components to produce the internal inputs. DILL incorporates this practice into its synchronous circuit model, assuming that the primary inputs have already been synchronised with the clock signal.

For a synchronous circuit, designers must ensure that the clock cycle is slower than the slowest stage in a circuit. This can be done by analysing the timing characteristics of components used in the circuit. The DILL model is constrained according to the environment in which it operates. If the clock is slow enough to let every signal settle down, it is reasonable to allow the value of each signal to change just once per clock cycle. DILL also requires that there be no cyclic connection within a stage, and storage components have to be specified in the behavioural style.

Some of the better-known asynchronous design methods handle delay-insensitive, quasi delay-insensitive or speed-independent circuits. DILL deals with (quasi) delay-insensitive and speed-independent circuits since they assume unbounded delays that are appropriate for LOTOS. (Quasi) delay-insensitive designs can be easily changed to speed-independent circuits by inserting artificial delay components. Asynchronous DILL concentrates mainly on speed-independent design. Happily this is a good match to the DILL approach since component delays are unbounded, just like the interval between consecutive LOTOS events. If new inputs cannot change any pending outputs, the design is termed semi-modular. Semi-modularity is often used as a correctness criterion for speed independence.

A specification is said to be input-receptive if every input is allowed in every state. In such a case, the DILL model represents the real circuit faithfully. An input quasi-receptive specification can be obtained by adding a choice when there is a potential output. It is not straightforward to transform a LOTOS specification with more than just sequence and choice operators into input quasi-receptive form. Internal events must first be determinised (i.e. internal non-determinism is removed). Outgoing edges are then added to create input quasi-receptive specifications. Since it can be hard to extract an input quasi-receptive environment from a behavioural specification, relations are defined that respect the difference between input and output. These relations do not require a (quasi-)receptive environment or implementation, and are natural criteria for asynchronous circuit correctness.

3   Conformance Testing and Verification

Conformance testing is a term drawn from communications systems to mean evaluating the correctness of an implementation against its specification. To formally define an implementation relation, a test hypothesis is needed that implementations can be expressed by a formal model, DILL describes behaviour using an IOLTS (Input-Output Labelled Transition System) whose actions are partitioned into inputs and outputs.

Several implementation relations have been defined to express conformance of an implementation to its specification. The authors have defined the *confor* (conformance) relation to require that, after a suspension trace of the specification, the outputs that an implementation can produce are included in what the specification can produce. A second implementation relation *strongconfor* (strong conformance) is also defined. This is similar except that output inclusion is replaced by output equality. Normally *confor* is used for a deterministic specification and implementation, while *strongconfor* is used when an implementation is more deterministic than a specification. The verification tool *VeriConf* developed by the authors checks the *(strong)confor* relations.

The *ioconf* relation (input-output conformance) has been defined to reflect the input-output relationship between an implementation and its specification. Suppose that *sp* is a state of the specification and that *im* is the corresponding state in the implementation. If *sp* can produce output *op*, a correct implementation should also produce it. If *sp* cannot produce a certain output, neither should the implementation. However when a specification is allowed to be non-deterministic, it is too strong to require *im* to produce exactly the same outputs as *sp*. A suitable relation should thus require output inclusion instead of output equality. Unfortunately a circuit that accepts everything but outputs nothing would also be qualified as a correct implementation. The overcome this problem, a special 'action' δ is introduced for quiescence, meaning the absence of output. Like any other output, if δ is in the output set of *im* it must be in the output set of *sp* for conformance to hold. That is, *im* can produce nothing only if *sp* can do nothing.

The DILL approach allows a design specification to be formally checked against an abstract specification. The same approach also allows test suites for an implementation to be rigorously derived from its specification. A circuit is specified in LOTOS (whose semantics is given by an LTS – an ordinary Labelled Transition System). The implementation of the same circuit is described by VHDL.

To support the checking of conformance, an intermediate LTS termed a suspension automaton is built from the specification LTS. The suspension automaton of an LTS is obtained by adding self-loops for all quiescent states (δ 'actions' where no output is pending). The resulting automaton is then determinised. Checking conformance is reduced to checking trace inclusion on the suspension automaton. A test case has finite deterministic behaviour that ends with states labelled *Pass* or *Fail* to indicate the verdict of conformance. The test cases generated by DILL have the form of traces to allow easy measurement of test coverage and automatic execution of test cases. The strategy is to cover all transitions in a transition tour that addresses the Chinese Postman problem.

Tests are generated from a suspension automaton by an algorithm that offers three choices in each iteration. The first choice terminates test generation. Since specifications usually have infinite behaviour, test generation has to be stopped at some point. The second choice gives the next input to the implementation. Since inputs are always enabled, this step will never result in deadlock when an input is applied. The third choice checks each possible next output of the implementation. Any implementation producing an unexpected output will result in a *Fail* terminal state, indicating a non-conforming implementation. For all other outputs, test generation may continue.

This test generation algorithm guarantees sound and exhaustive test cases for the *ioconf* relation. The authors have developed the *TestGen* tool to realise this algorithm. Each generated transition tour is a test case and is saved in a test file. A testbench was designed to allow the test cases to be applied and executed against the VHDL description of the circuit.

4   Case Studies

4.1   Bus Arbiter

The Bus Arbiter is a benchmark circuit used to exercise hardware verifiers. Normally the arbiter grants access to the highest priority client: the one with the lowest index number among all the requesting clients. However as requests become more frequent, the arbiter is designed to fall back on a round-robin scheme so that every requester is eventually acknowledged. This is done by circulating a token in a ring of arbiter cells, with one cell per client. Although the Bus Arbiter has been studied by many researchers, as far as the authors know there has not been a formal specification of the arbitration algorithm used in the design. With LOTOS, it is possible to provide such a higher-level specification..

The formulation of properties uses action-based temporal logics ACTL (Action-based Computational Tree Logic [6]) and HML (Hennessy-Milner Logic). The following three properties have to be proved for the circuit: no two acknowledge outputs are asserted in the same clock cycle (safety); every persistent request is eventually acknowledged (liveness); and acknowledge is not asserted without request (safety). To verify the higher-level specification against the temporal logic formulae, the LTS of the specification was produced first. Generation and minimisation of the LTS take a few seconds on a 300 MHz Sun. The temporal logic formulae are then verified against the minimised LTS within a minute.

To check the lower-level specification, the design of the arbiter was divided into pieces – one per cell of the arbiter. An LTS which is safety equivalent to the LOTOS specification of the design was generated in about seven minutes. The two safety properties were verified to be true against this LTS, implying that the design also satisfies these properties. Verification of the formulae took just seconds. However generating an LTS that is branching equivalent to the design took almost one day, after which the liveness property was also verified to be true.

For checking equivalence between the higher-level algorithm and the lower-level design, compositional generation was exploited to generate the LTS for the design. This took about eight minutes to calculate. In fact this LTS is not observationally equivalent to the one representing the higher-level specification. It was found that the circuit does not properly reset the override out signal. This is a fault in the supposedly proven benchmark circuit. The design was modified and then verified to be observationally equivalent to the higher-level algorithmic specification.

4.2   Black-Jack Dealer

The Black-Jack (Pontoon, Vingt-et-Un) Dealer is another verification benchmark circuit [7]. It is a synchronous circuit whose inputs are *Card_Ready* and *Card_Value*. Its outputs are boolean: *Hit* (card needed), *Stand* (stay with current cards) and *Broke* (total exceeds 21). Aces have value 1 or 11 at the choice of the player. Using the authors' *TestGen* program, a test suite for the Black-Jack Dealer was derived. The test suite is able to test 181 different hands of cards that a dealer may hold. The VHDL implementation given in [7] was evaluated against this test suite.

Although the circuit was expected to pass the test suite, a *Fail* verdict was recorded after the dealer was given the following cards: 5, 5, 3, 2, 1, 10. The circuit should initially take an Ace as 11. This should be re-valued as 1 the first time the result would be *Broke*. But the given design continues to re-value the Ace card. Carefully simulating the circuit discovered a problem in the benchmark with one of the flag registers that indicates an Ace should be 11. By slightly modifying the circuit to remove the cause of a short duration pulse, the circuit was enabled to pass the test suite successfully.

4.3 Asynchronous FIFO

As a typical asynchronous circuit, an asynchronous FIFO buffer was specified and analysed. The FIFO has two inputs *InT, InF* and two outputs *OutT, OutF*. Its inputs and outputs use dual-rail encoding in which one bit needs two signal lines. The *Req* input comes from the environment of a stage, indicating that the environment has valid data to transfer. The *Ack* output goes to the environment, indicating that the stage is empty and ready for new data. The implementation uses two C-Elements (transition synchronisers used in asynchronous circuits). To ensure the FIFO works correctly, the environment has to be coordinated. It is convenient to think about the environment in two parts: *EnvF* (front-end) produces data, while *EnvB* (back-end) consumes it.

It was verified that the specification satisfies the following property: if there is an input of *1*, then the output will eventually become *1* (and similarly for input/output of *0*). It was verified that *Spec ≈ Impl || (EnvB ||[· · ·]| EnvF)*, where ≈ denotes observational equivalence.

When speed independence needs to be verified, each building block (including the environment) should be specified in the input quasi-receptive style (*_QR*). It was verified that *Spec ≈ Impl_QR || (EnvB_QR ||[· · ·]| EnvF_QR)*, which gives more confidence in the design of the FIFO. The liveness property is also satisfied by the implementation *Impl_QR || (EnvB_QR ||[· · ·]| EnvF_QR)*.

It was also shown that *Impl_QR || (EnvB_QR ||[· · ·]| EnvF_QR) strongconfor Spec* using the *VeriConf* tool. The *TestGen* tool builds a single test case of length 28:

InF !1	InF !0	OutF !1	InF !1	OutF !0	OutF !1	InF !0
InT !1	OutF !0	InT !0	OutT !1	InT !1	OutT !0	OutF !1
δ	InF !0	OutF !0	InT !1	OutT !1	InT !0	InT !1
OutT !0	OutT !1	δ	InT !0	OutT !0	δ	Pass

4.4 Selector

A selector (an asynchronous design component) allows non-deterministic choice of output. After a change on input *Ip*, output *Op1* or *Op2* may change depending on the implementation. The *TestGen* tool produces a single test case of length 11 for the selector.

This example shows how test branches are marked. After *Ip !1*, the output *Op1 !1* is marked with the current state (*S1) since an implementation may also do *Op2 !1*. A selector that insists on sending its input to *Op1* can follow the first row of steps in the test case below. After the sixth step (*Ip !1*), it cycles back to the second step (*Op !1*) – a loop that the testbench must break.

Ip !1	Op1 !1 (*S1)	Ip !0	Op1 !0 (*S2)	δ	Ip ! 1
Op2 !1 (*S1)	δ	Ip !0	Op2 !0 (*S2)	Pass	

5 Conclusion

An approach to specifying synchronous circuits has been presented. This has allowed standard hardware benchmarks to be verified – the Bus Arbiter and the Black-Jack Dealer in this paper. The authors were pleasantly surprised to find that their approach discovered previously unknown flaws in these circuit designs.

An approach to specifying asynchronous circuits has also been presented. (Quasi) delay-insensitive circuits are transformed into speed-independent designs. Violations of speed-independence (or rather, semi-modularity) are checked using specifications that are input (quasi-)receptive. The *(strong)confor* relations have been defined to assess the implementation of an asynchronous circuit against its specification.

The correctness of a DILL specification can be easily checked by simulation tools. The *TestGen* tool generates test suites using transition tours of automata. This allows automatic generation of test suites for reasonable coverage, and also allows testing of non-deterministic implementations. The *VeriConf* tool was developed to support the *(strong)confor* relations.

References

1. D. L. Dill. *Trace Theory for Automatic Hierarchical Verification of Speed-Independent Circuits.* ACM Distinguished Dissertations. MIT Press, 1989.
2. J. C. Ebergen, J. Segers, and I. Benko. Parallel program and asynchronous circuit design. In G. Birtwistle and A. Davis, editors, *Asynchronous Digital Circuit Design*, Workshops in Computing, pages 51–103. Springer-Verlag, 1995.
3. G. Gopalakrishnan, E. Brunvand, N. Michell, and S. Nowick. A correctness criterion for asynchronous circuit validation and optimization. *IEEE Transactions on Computer-Aided Design*, 13(11):1309–1318, Nov. 1994.
4. ISO/IEC. *Information Processing Systems – Open Systems Interconnection – LOTOS – A Formal Description Technique based on the Temporal Ordering of Observational Behaviour.* ISO/IEC 8807. International Organization for Standardization, Geneva, Switzerland, 1989.
5. Ji He and K. J. Turner. DILL (Digital Logic in LOTOS) project web page. http://www.cs.stir.ac.uk/~kjt/research/dill.html, Nov. 2000.
6. R. D. Nicola and F. Vaandrager. Three logics for branching bisimulation. In *Proc. 5th. Annual Symposium on Logic in Computer Science (LICS 90)*, pages 118–129. IEEE Computer Society Press, 1990.
7. D. Winkel and F. Prosser. *The Art of Digital Design*. Prentice-Hall, Englewood Cliffs, New Jersey, USA, 1980.

Multiclock Esterel

Gérard Berry[1] and Ellen Sentovich[2]

[1] Esterel Technologies, 885 av. J. Lefebvre, 06270 Villeneuve-Loubet, France
`Gerard.Berry@esterel-technologies.com`
[2] Cadence Berkeley Laboratories, 2001 Addison Street, Berkeley, CA
`ellens@cadence.com`

Abstract. We present the Multiclock Esterel language, which extends the synchronous language Esterel to multiple clock zones. While Esterel is good for compact single-clocked hardware or software designs, modern electronic designs are growing rapidly and they can no longer be designed in a monolithic fashion. Problems such as clock distribution, complexity, and power limitations have led designers to construct designs in a modular, multiple clock fashion. Multiclock Esterel is designed precisely to address this design style. It is a natural extension of Esterel, and retains its strong synchronous semantics and internal determinism. Statements driven by different clocks communicate through two special devices called the sampler and the reclocker. Multiclock Esterel should be understood as a preliminary language proposal meant to study multiclocking. It has not yet been validated by large experiments.

1 Introduction

The Esterel synchronous reactive language [5,4,6], which we call Classic Esterel in this paper, is based on a single-clock instantaneous interaction principle. The behavior of a program is defined by a sequence of reactions to input sequences. The execution environment decides when the program is provided an input, and a reaction is viewed as the simultaneous production of an output response to an input event. We call *ticks* the instants in which reactions occur, and *master clock* the sequence of these instants (it is a logical clock and no time regularity is required). This synchronous view has proved useful for a fairly large class of reactive applications such as process or human-machine interface controllers, communication protocols, and hardware circuits. For single-clocked synchronous circuit synthesis [13], the master clock is simply the circuit's clock [2,4]. Then, reaction to an input is not instantaneous, but it is guaranteed to be computed before the next clock tick; this is the best practical approximation of perfect synchrony.

In practice, synchronous single-clocking makes perfect sense for *compact* systems for which it is reasonable to pretend that all the system's component fit within a single circuit or within a single micro-computer for software applications. However, modern electronic designs are characterized by unwieldy size, clock distribution complexity, and power limitation. Their hardware, software,

T. Margaria and T. Melham (Eds.): CHARME 2001, LNCS 2144, pp. 110–125, 2001.
© Springer-Verlag Berlin Heidelberg 2001

or mixed implementation tends to abandon the classical single-clock framework in favor of a multiclock one, each clock controlling a circuit zone or a local processor. A good example of this is described in [12], where a single global clock was abandonned for a design style with multiple local clocks. Communication between clock zones becomes a critical issue and is carefully controlled either by inserting special devices such as latches or by fine timing analysis.

In this paper, we present an extension of Esterel for such multiclock systems, which we call Multiclock Esterel. The new language proposal extends Classic Esterel in two ways. First, it deals with a set of primary clocks instead of a single global clock, each individual reactive module being clocked by one of the primary clocks. Second, it introduces two new communication primitives between clock zones, the *sampler* and the *reclocker*, which make it possible to send an information on a given clock and to receive it on another clock. These extensions are quite minimal, semantically well-defined, and physically reasonable.

Semantically speaking, Multiclock Esterel is still a synchronous language in the sense that clock ticks of different clocks remain comparable in time and information transmission remains instantaneous. We thus retain the strong determinism property which has proved so useful in the synchronous language framework: the behavior of a program is completely determined once the clock timings and the input flows are known. There is a larger amount of external non-determinism (the timings of the clocks), but still no internal non-determinism. Therefore, Multiclock Esterel is not a language for large asynchronous distributed applications for which such notions are not applicable. Since words are heavily overloaded in this field, we shall speak of *single clocked* and *multiclocked* parts of systems, and we shall avoid using the ambiguous word *asynchronous*.

As in [4,3], we concentrate on the kernel Multiclock Esterel calculus, ignoring software engineering issues related to full language development. Our goal is to develop the foundations of multiclock langages, and in particular to study preemption operators in the multiclock framework. In Sect. 2, we study how communication can be organized in a multiclock context, and we define the sampler and the reclocker.

In the rest of the paper, we study the Multiclock Esterel calculi. The syntax defines two kinds of terms: the *clocked reactive statements*, which are basically those of Classic Esterel, and the *multiclocked processes*. A multiclocked process it is either the pair of a reactive statement and of the clock that drives it, or a compound structure involving several of these clocked reactive statements driven by different clocks. The semantics is given by a modeling in Classic Esterel.

In Sect. 3 we begin with the *basic calculus*, where a multiclocked process is a flat network of reactive modules, each controlled by a given clock. In the *full calculus* in Sect. 3.3 we achieve full orthogonality by allowing processes to be recursively launched from within reactive statements and vice versa. We can then define how an arbitrary multiclocked process can be preempted by a reactive preemption statement. We also enrich processes by allowing sequencing and looping of them. We give a simple example of preemption control of a fast process by a slow one. We conclude in Sect. 4.

2   Clocks, Signals, and Communication

2.1   Clocks

We consider a set $C = \{c, c', c_0, c_1, \ldots\}$ of *clocks*. Intuitively, a clock is a primitive object that delivers ticks in sequence. Clocks are logical and need not be periodic in time. For instance, a clock can be generated by a quartz, by the depressing of a button, or by the decision to call a program. Clocks determine *tick sequences*, which can be dealt with in two ways:

- *Discrete time:* A tick sequence is a sequence of *global ticks*, where a global tick is characterized by a set of clock names:

$$c_1, \quad c_2, \quad c_1 \cdot c_2, \quad c_1, \quad c_3, \quad c_2, \quad c_1 \cdot c_2 \cdot c_3, \ldots$$

  Here, $c_1 \cdot c_2$ means that the clocks c_1 and c_2 tick simultaneously. Allowing simultaneity is useful for several reasons: we may want to set two clocks equal in a program, we may want to extract a clock from another one, or we may simply want to deal with coincidental simultaneity of *a priori* independent clocks. When needed, we can require clocks to be non-simultaneous for all of their ticks by asserting an *exclusion relation* of the form $c \# c'$ as we do for signals in Classic Esterel.
- *Continuous time:* here time is continous and a clock is defined by a finite or infinite increasing sequence of real numbers representing the instants at which the clock ticks. Ticks of distinct clocks can coincide in time.

In our framework, both models are essentially equivalent, but one or the other may be more natural in intuitive or formal descriptions. It is easy to go from one model to the other: from discrete time to continuous time, associate any increasing sequence of reals with a discrete tick sequence; conversely, from continuous time to discrete time, extract the clock set sequence from the increasing sequence of real numbers obtained by taking the union of the individual clock sequences.

2.2   Signals and Communication

In Classic Esterel, there is a single implicit clock c for all statements, which makes communication very simple. In Multiclock Esterel, we organize communication between different clock zones by sampling and reclocking the exchanged signals. We explain the semantics of communication and discuss the appropriate software or hardware devices for implementing it.

Communication in Classic Esterel. Two Classic Esterel statements p and q communicate by instantaneously propagating signals: for example, at some tick of c, p emits s and q instantaneously receives s. A *pure signal* is characterized by its broadcast *status* at each tick of c, which is either *present* (*high*) or *absent* (*low*). A pure signal *s* is absent by default, and it is set present at a tick of *c* if and only if at least one "emit *s*" statement is executed at that tick. Simultaneous emission of *s* by concurrent emit statements is allowed and simply sets *s* present.

In addition to the status information, a *valued signal* broadcasts a value belonging to some data type, with the restriction that a value change can occur only when the signal is present. A valued signal is emitted by executing an "emit *s*(*e*)" statement, where *e* is a value expression. The value of *s* is read by the expression '?*s*'. To give a meaning to concurrent emission of a valued signal, one associates with it an associative and commutative *combination function*, which is used to combine all the separately emitted values into one final value. For example, "emit *s*(1) || emit *s*(2)" results in ?*s* = 3 if combination is done by addition.

Pure Communication in Multiclock Esterel. In Multiclock Esterel, we retain the principle of instantaneous reaction to clock ticks and communication by signals, but we may emit and receive signals according to several clocks. Instants become relative to clocks. Consider two processes *A* and *A′* running reactive statements respectively clocked by two different clocks *c* and *c′*. Each process is locally single-clocked: communication within *A* (resp. *A′*) is governed by *c* (resp. *c′*) as in Classic Esterel. To communicate with each other, *A* and *A′* can share an instantaneously broadcast signal *s* without sharing a clock: the emitter *A* will emit *s* at some tick of *c*, and the receiver *A′* will receive *s* at some tick of *c′*. Instantaneous propagation means that *A′* will receive *s* at the very first tick of *c′* that follows the emission tick of *c*, however close these ticks are in time, up to tick equality for which *A′* will also receive *s* as in Classic Esterel.

Which status *s′* of *s* will *A′* receive at *c′*? This question doesn't arise in Classic Esterel, since there is only one clock for the emitters and receivers: communication occurs only on ticks, and statuses outside ticks are irrelevant in the model (they are of course relevant in implementations). In Multiclock Esterel, we view signals as continuous objects and statuses as being held high or low during the emitter's clock cycle according to their emission statuses at the clock tick, as pictured in Fig. 1. We think there are two fundamental choices for *s′*:

1. *Sampling*: *s′* is present at *c′* and set high for the *c′* cycle if and only if *s* is high precisely at *c′*.
2. *Reclocking*: *s′* is present at *c′* and set high for the *c′* cycle if *s* has been high at least once at any time since the previous occurrence of *c′*.

Sampling is the classical operator of Signal Theory. It is like taking a snapshot of the signal at *c′*. Reclocking has a built-in memory, and amounts to taking a long exposure on *s* not to miss any occurrence. This can be necessary if the sender's clock *c* is faster than the receiver's clock *c′*: for instance, if the sender is sending transient alarms, and if the receiver worries about the emission of at least one transient alarm since its last tick, sampling is not enough to catch the required information at the receiver and reclocking is mandatory (another example will be given in Sect. 3.5). In Multiclock Esterel, we make both choices available.

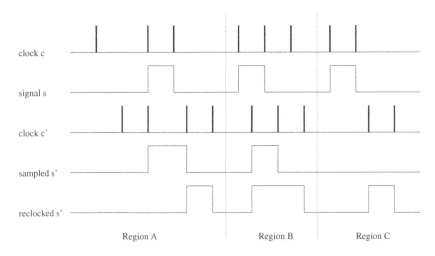

clock c

signal s

clock c'

sampled s'

reclocked s'

Region A   Region B   Region C

Fig. 1. Communication devices in continuous time

As usual with synchronous languages, there are boundary problems to be solved:

- *Sampling:* if the ticks of c and c' coincide, the sampled status is the *new* status of s w.r.t. c, i.e. the one which will be valid for s in M until the next tick of c.
- *Reclocking:* if the ticks of c and c' coincide, the new status of s in M is not considered for s' in M'; reclocking checks that s has been emitted since the *previous tick of c included, current tick excluded.*

The other boundary cases can be obtained simple Boolean combinations of the above ones. If $c' = c$, sampling is simply transparent, while reclocking amounts to taking the previous status of the signal, as for the Classic Esterel pre operator.

In Fig. 1, the following three differences between sampling and reclocking cases are illustrated:

Region A: the reclocked signal can be delayed in its rising transition with respect to the sampled signal when the clocks c and c' coincide on a rising edge of s: reclocked s' cannot see the current value of s until the next tick of c', while sampled s' can.

Region B: the reclocked signal is delayed in its falling transition with respect to the sampled signal: reclocked s' remains high if s was high at any time since the last tick of c', while sampled s' goes low as soon as s is low at a tick of c'. (Region B illustrates this for the case in which c and c' do not coincide for s's falling transition. The same is true when c and c' do coincide, but we have not shown this case.)

Region C: The reclocked signal can detect spurious high values for s that might be missed by the sampled signal.

An example where a reclocker is needed is presented in Sect. 3.5. The extended version of this paper, available from the authors, contains another illustrative example of sampling and reclocking. The example is a video game where a process controls a joystick and another process controls the game proper. The joystick and game processes have different clocks. The joystick process sends position signals to the game, which samples them. The joystick also sends a reset signal to the game. This signal must be reclocked by the game, otherwise it could be lost.

Value Communication in Multiclock Esterel. As far as valued signals are concerned, sampling extends trivially by inputting the current value of a signal in addition to its status at the receiver's clock tick. The value may have changed many times since the previous tick, but we only sample the most recent one, as for the status; this is exactly what sampling means in signal theory. The memory effect of reclocking makes defining the value of a reclocked signal less obvious; a natural extension of the Classic Esterel concept of multiple simultaneous emission would be to combine all the successive values received since the previous tick by an associative-commutative operation. Since we have no really good supporting example, we prefer to postpone the decision and to currently restrict Multiclock Esterel to pure or valued sampled signals and to pure reclocked signals.

Single Clocked and Multiclocked Signals. Given a clock c, we say that a signal is *clocked by* c if its status changes occur only on ticks of c. A signal which is emitted by reactive statements driven by a single clock c is always clocked by c. Sampling and reclocking always generates local views that are clocked by the receiver's clock.

An interesting question is whether or not we want to allow multiclocked signals. A pure multiclocked signal could be handled by merging the results of two different **emit** statements driven by different clocks, saying that s is high at time t if any one of the outputs of the **emit** statements is high at that time. The device to do this is a *wired or*. However, multiclocked merging does not extend trivially to valued signals. We said that multiply emitted values should be combined by an appropriate combination function. We should also combine the values if the ticks of two distinct emitter clocks happen to be simultaneous. This would be semantically meaningful, but we lack magical tricks for the implementation. Therefore, we choose to postpone this problem and to forbid multiclocked signals for the time being. We do not see this as a limitation: in most hardware systems, each signal is computed from a single clock zone.

2.3   Sampling and Reclocking Devices

Having seen that for Multiclock Esterel have two fundamental choices in reading a signal that passes from one clock zone to another, and having defined what sampling and reclocking semantically mean, we now study devices that can do the job for different styles of implementation. We name them generically the *sampler* and the *reclocker*. To give an operational semantics of Multiclock Esterel, we first encode them within Classic Esterel. We then discuss hardware and software realizations.

Classic Esterel Encoding. To model Multiclock Esterel in Classic Esterel, we can write single-clocked programs where the tick is that of a fictitious global clock of the Multiclock Esterel programs, which is faster than all of the actual clocks. A Multiclock Esterel clock can then be modeled as an ordinary Esterel signal. (This is *modeling*, not programming, since this global clock will be inaccessible from within Multiclock Esterel programs.) The obtained Classic Esterel program is deterministic, which means that Multiclock Esterel has no internal non-determinism. However, the fact that Multiclock Esterel clocks have become free input signals leaves room for a large amount of external non-determinism.

We take the freedom of considering C' and S' as valid Esterel identifiers, using primes for the receiver side. The sampler takes as input the signal S, the receiver clock C', and it returns as output the local receiver view S' clocked by C':

```
module Sampler :
input S;
input C'; % the clock
output S';
loop
   present S then
      sustain S'
   end present
each C'
end module
```

Although the module is active on every tick of the fictitious global clock, sampling S only happens on ticks of the clock C'. On any tick of C' in which S is present, S' is emitted, and it keeps being emitted on every global clock tick (i.e. "continuously in discrete time") until the next tick of C' arrives. Notice that loop...each is stongly preemptive, so that when C' occurs, the sustain statement is preempted without being executed, the loop loops, and the sustain statement is instantaneously restarted only if S is present. Therefore, we indeed sample the new value of S as required in the sampler specification.

The reclocker has the same interface. It can be coded as follows:

```
module Reclocker :
input S;
input C'; % the clock
output S';

signal MEM in
   loop
```

```
      await immediate S;
      sustain MEM
   each C'
||
   loop
      present pre(MEM) then sustain S' end
   each C'
end signal
end module
```

The first part of the parallel statement implements a memory to remember the occurrence of S. As soon as S goes high on a global clock tick, the internal signal MEM is set high, where it remains (by being emitted each global clock tick) until a tick of C' occurs. When C' occurs, the previous value of MEM w.r.t. the global clock is tested and S' is sustained for the next cycle if MEM was high. Notice that an S occuring at the same time as a C' is taken into account only in the next tick of C', as required by the reclocker specification.

Software Implementation. A software implementation of Multiclock Esterel can follow the principles of Polis [1]. Clock ticks correspond to module activations, acting as Polis CFSMs. A sampler is a single memory cell in which writers write and readers read, i.e., a Polis 1-place buffer. A reclocker is a 1-place buffer with an additional bit set at each write and reset by the reader. For multiprocessor applications, atomicity conditions on activations, reads and writes are required as usual to provide a consistent view between readers and writers.

Hardware Implementation. In hardware, one can build a sampler and a reclocker using electronic devices such as *latches* and *flip-flops* (unfortunately, the terminology is a little sloppy in this field). For instance, the sampler can be built as a *positive edge-triggered D flip-flop* [10]. However, it is impossible to build a *perfect* sampler or reclocker, because of the standard metastability problem. For any sampling device, there is a time interval δ around the clock edge inside of which the sampled signal cannot itself have an edge. Otherwise, the device can enter a metastable state in which it incorrectly drives its output for some amount of time.

The metastability problem is not specific to Multiclock Esterel. It arises in any kind of harware multiclock design. In today's technology for multiclock design, a sophisticated timing analysis program is run to ensure that there are no violations of clock setup and hold times. A similar analysis should be made for a synthesized implementation of an Multiclock Esterel program. In this exploratory paper, we shall ignore these issues since we want to experiment with the technology-independent aspects of language design.

3   The Multiclock Calculus

In this section, we develop the Multiclock Calculus. We begin with some definitions of basic elements and the flat calculus. This is followed by the modeling of the flat calculus in Classic Esterel. We continue by addressing the issues of process sequencing and preemption. Finally, we give an example containing a slow module controlling a fast process.

Terms in Multiclock Esterel are divided into two categories, multiclocked *processes* and single-clocked *reactive modules*. Recall that Classic Esterel contains only reactive modules.

- Processes P, Q, etc., are elementary or compound. A process has a signal interface and a body. Elementary process bodies consist in executing a reactive module with a given clock and a given sampling / reclocking interface specification. Compound process bodies are built from elementary ones by concurrency and signal scoping declaration, and possibly by sequencing and looping in the extended calculi.
- Reactive modules M, N, etc., are as in Classic Esterel. Each reactive module is governed by a single clock, which is implicit in the module. The interface specifies the input / output signals. The body is a reactive statement. All the statements in kernel Classic Esterel are imported in kernel Multiclock Esterel. Only one statement is added, execution of a process.
- A *program* is defined by a set of clocks and a process.

For compound processes, clocks appear only in the leaf elementary processes, which in turn call clocked reactive modules. Therefore, there is no global clock accessible to the executable reactive statements: a reactive statement knows only the clock that runs it.

We first present the *flat calculus*, limited to putting classical reactive modules into a flat multiclocked parallel structure. Then, we present the full calculus, where we can recursively embed multiclocked processes into single-clocked reactive modules and conversely, up to any depth. In this fully orthogonal language, one can deal in a general way with preemption of multiclocked processes. As in [4], we use indifferently a mathematical style or a programming language style syntax.

3.1   The Flat Calculus

In the flat calculus, a process is limited to being composed of reactive modules driven by given clocks and put in parallel.

Processes. A process $P = (I, O).A$ has a list I of input signals, a list O of output signals, and a body. Process bodies are written A, B, etc. Their syntax is as follows, M denoting a reactive module defined below:

$$
\begin{array}{ll}
c * M\,(I^m) & \texttt{run M clock C input sample I \ldots} \\
A \,|\, B & A \,\|\, B \\
A \setminus s & \texttt{signal S in } A \texttt{ end}
\end{array}
$$

There are clock constaints on signals, which will be presented in Sect. 1.

In the $c * M(I^m)$ reactive module run statement, the decorated vector I^m of *input modes* specifies an exponent $m \in \{s, r\}$ for each input signal of M. The exponent defines how the signal is brought into the clock c: s means that the signal is sampled, r means that it is reclocked.

Reactive Modules. A reactive module M is defined by an interface specification and a body:

$$M = (I, O).p$$

The interface specifies the input signal vector I and the output signal vector O. The body is a reactive statement p^1, with syntax that of Classic Esterel:

0		nothing
1		pause
k	(for $k > 1$)	exit t
$!s$		emit s
$s\,?\,p, q$		present s then p else q end
$s \supset p$		suspend p when s end
$p\,;\,q$		$p\,;\,q$
$p\,*$		loop p end
$p\,|\,q$		$p \,\|\, q$
$\{p\}$		trap t in p end
$\uparrow p$		
$p \backslash s$		signal s in p end

Notice that there are two concurrency operators '$|$', one for reactive statements and one for processes, and similarly two local signal declaration operators '$\backslash$'. These operators perform the same kind of operation in both worlds, although the behaviors are technically different. There is no danger in overloading the symbols since one can always determine unambiguously from the syntax whether one is inside a process body or a reactive module body.

Signal and Clock Constraints. Signals are subject to the usual visibility rules. A signal refered to in a reactive statement must be an interface signal or a local signal in the enclosing reactive module. For any reactive module run $c * M(I^m)$ occuring in the body A of a process P, a signal s in M's interface must be declared as an interface signal of P or as a local signal in an enclosing local signal declaration $A \backslash s$.

Signals in a reactive module are always clocked on the module's implicit clock. Input signals are either sampled or reclocked by the module run interface; local and output signals are set by the **emit** statements which only acts on the module's clock.

Signals in a process must obey the following *clock consistency* rules: *all reactive modules that share a signal as output must be clocked by the same clock.*

---

[1] We use P for a process and p for a reactive statement, which may be confusing. The whole Esterel literature uses p for statements, a notation we keep here.

This requirement ensures that each signal is properly clocked by a single clock, as required in Sect. 2.2.

3.2   Modeling Flat Multiclock Esterel in Classic Esterel

We give the semantics of a flat Multiclock Esterel program by modeling its behavior in Classic Esterel. For this, we introduce a fictitious base clock faster than all Multiclock Esterel clocks, as in Sect. 2.3. We consider this fictitious clock as the Classic Esterel model base clock and we view each Multiclock Esterel clock c as a pure signal also named c in the Classic Esterel translation. We define a translation function T which translates any multiclocked process P into a Classic Esterel term $T(P)$.

The interface of $T(P)$ is simply that of P. The process body is translated in a trivial recursive way until reaching a clocked module run statement:

$$T(A \,|\, B) = T(A) \,|\, T(B)$$
$$T(A \setminus s) = T(A) \setminus s$$

We now translate a leaf explicitly clocked module run statement $c * M\,(I^m)$, $m \in \{s, r\}$, with $M = (I, O).p$. The basic idea is to introduce local views I' and O' clocked by c' of the input and output signals vectors I and O, ant to use three components in parallel: the appropriate translation of p, an input handler $TI(I^m, c', I')$, which builds the local view I' of I according to the sampling or reclocking directives m, and an output handler $TO(O', c', O)$ which builds the output signals O from their local views O'.

The input handler puts in parallel individual signal handlers that sample or reclock input signals using the sampler and reclocker defined in Sect. 2.2:

$$TI(I^m, c', I') = TI(i_0{}^{m_0}, c', i'_0) \,|\, TI(i_1{}^{m_1}, c', i'_1) \,|\, \ldots \,|\, TI(i_n{}^{m_n}, c', i'_n)$$
$$TI(i^m, c', i') = \begin{cases} Sample\,(i, c', i') & \text{if} \quad m = s \\ Reclock\,(i, c', i') & \text{if} \quad m = r \end{cases}$$

The output handler puts in parallel individual signal sustainers $Out(o', c', o)$ defined in Classic Esterel as follows:

```
loop
    present O' then
        abort
            sustain O
        when C'
    else
        await C'
    end present
end loop
```

The effect of these sustainers is to clock the output signal on c', in the sense of Sect. 2.2. The definition of TO is:

$$TO(O', c', O) = TO(o'_0, c', o) \,|\, TO(o'_1, c', o_n) \,|\, \ldots \,|\, TI(o'_n, c', o_n)$$
$$TO(o', c', o) = Out\,(o', c', o)$$

To translate p, we use two auxiliary operators. The "await immediate c" operator or $sync(c)$ waits for the first occurrence of c and terminates. The c-trigger operator, written "suspend p when immediate not c" in Esterel and $\bar{c} \supset p$ in short form, triggers p exactly at the instants where c is present, and freezes p at the other instants ($\bar{c}$ is the negation of c). These operators are defined by

$$sync(c) = \{(c\,?\,1\,,\,2)\,*\}$$
$$\bar{c} \supset p = sync(c)\,;\,\bar{c} \supset p$$

Let $p\,[I'/I, O'/O]$ be p where I and O are renamed into I' and O'. The final translation is:

$$T(c * M\,(I^m)) = ((\bar{s} \supset p)\,[I'/I, O'/O]\,|\,TI(I^m\,,\,c'\,,\,I')\,|\,TO\,(O'\,,\,c'\,,\,O))\setminus I', O'$$

Notice that the translation of p starts acting at the first tick of c', first instant included, because of the initial $sync(c)$ in $\bar{c} \supset p$.

3.3 Process Sequencing

In the flat calculus, termination of a reactive module is ignored within the process that holds the module since there is no process sequencing. It is sensible to exploit the existing termination information and to define concurrent process termination as termination of all the concurrent reactive modules they contain. In the Classic Esterel translation, we get termination detection for free: a process body A terminates when $T(A)$ terminates. In hardware, concurrent termination detection can be done using devices such as Muller C-elements [11].

Then, we can define process sequencing $A; A'$. Semantically, we just write $T(A; A') = T(A); T(A')$. As before, the syntactic context disambiguates the '$;$' symbol between processes or reactive statements.

There are many elementary delays involved in process sequencing. Consider the trivial example $(c * 1\,())\,;\,(c' * 1\,())$. One first waits for the first occurrence of c to start the first 1 pause statement. This statement waits for one more c and terminates. Then, one waits for the first occurrence of c' to start the second 1 statement, this instant included. The second 1 terminates at the next occurrence of c'. The final translation of the above process sequence in Classic Esterel is

$$sync(c);\ await(c);\ sync(c');\ await(c')$$

Therefore, to implement process sequencing in practice, we have to implement $sync(c)$. We need a device with an input for the incoming control in and an output for the outgoing control out. The device should immediately set out high if in is high when c occurs, or else reclock in on c. The solution is to use the disjunction of a sampler and a reclocker with input in and clock c.

Having defined process sequencing, we can also define *process looping* $A *$ by $T(A *) = (T(A)) *$, provided that A cannot terminate instantaneously, which is checked as in Classic Esterel.

3.4 Process Preemption: The Full Calculus

The last step to the full Multiclock Esterel calculus is to allow an abritrary multiclocked process P to be launched and preempted from within a clocked reactive module M clocked by c. This is done by adding a new reactive statement that starts P within M with a list of inputs and outputs and samples or reclocks the output signals of P on M's clock c since they become inputs of the body of M. The calculus and language syntax are as follows:

$$\langle P(I, O^m) \rangle \qquad \texttt{process run } P \texttt{ input I}$$
$$\texttt{output reclock O ...}$$

This new reactive statement can appear anywhere where a reactive statement can, including in a preemption context. Of course, the construction is fully orthogonal and recursive: P can be a general process that can itself run synchronous modules that themselves run other multiclock processes, etc.

Signal names in P's interface must exist in the scope of the $\langle P(I, O^m) \rangle$ statement within M's body, and signal binding is by name (in a real language, renaming facilities such as the ones used in the full Classic Esterel language should be added).

However, with the current definition of a reactive module interface, all signals in P's interface would be clocked by M's clock c: inputs come either from M's interface where they are either sampled or reclocked, local signals are directly clocked by c, while outputs are sampled or reclocked for M. Since P is an arbitrary process, it should also be able to view signals which are in the current process scope but not in M's interface. For this, we slightly change the reactive module syntax into $(I, O, H).p$, where H is a set of *hidden signals*. Hidden signals are neither sampled nor reclocked by M, and they cannot be used by M's reactive statements. They can only be passed by M as inputs or outputs to subprocesses. In the call $\langle P(I, O^m) \rangle$, we set $m = \perp$ for outputs of P which are hidden in M. To sample or reclock the non-hidden outputs of P for M, we rename them O' in P to keep the names O in M (beware, this time the prime is on the emitter side). The translation in Classic Esterel is:

$$T(\langle P(I, O^m) \rangle)c = T(P[O'/O]) \mid TO(O'^m, c', O)$$
$$T(o'^m)c = \begin{cases} Sample(o', c, o) & \text{if} \quad m = s \\ Reclock(o', c, o) & \text{if} \quad m = r \\ 0 & \text{if} \quad m = \perp \end{cases}$$

This very simple translation tells everything about the semantics of the construct, and especially about subprocess preemption. For example, consider the Multiclock Esterel reactive statement

$$s \supset \langle P(I, O^m) \rangle$$

within a reactive module M clocked by c. When control reaches the statement, P is started autonomously and it communicates with M through its input /

output ports, where M sees all non-hidden signals as clocked by c. If M receives s from the environment or emits it internally, s is sustained for the whole clock cycle of c in the Classic Esterel translation. Therefore, $T(P)$ is suspended for the whole clock cycle as one expects. If, for some reason, M exits a trap that kills P, then $T(P)$ instantaneously dies as any other Classic Esterel term.

3.5   The Bureaucrat Example

Here is an example where a slow module called Bureaucrat controls the life and death of a fast process Worker. We assume that Worker reads an input flow InFlow, writes an output flow OutFlow, and sends a signal Done when its computation has finished. The bureaucrat wakes up every hour and returns OK and goes home if Worker has finished in the last hour. Of course, by construction, the bureaucrat does not care about the work being done with InFlow and OutFlow, which are hidden to him. After five hours, if Worker has not finished, the bureaucrat kills it, reports FIRED, and goes home. In concrete syntax, the program may look like:

```
clock Hour;
clock WorkClock;

process Global :
input InFlow;
output OutFlow;
signal OK, FIRED in
    run Bureaucrat clock Hour input InFlow
                               output OutFlow, OK, FIRED
end signal
end process


module Bureaucrat :
output OK, FIRED;
hidden InFlow, OutFlow;
signal Done in
   weak abort
       process run Worker input InFlow output OutFlow,
                          reclock Done
   when
     case Done do
        emit OK
     case 5 tick do
        emit FIRED
   end weak abort
end signal
end module
```

assuming that Worker itself involves one or more reactive modules clocked by the fast clock WorkClock.

Notice that bureaucrat termination depends on a **weak abort** preemption statement [5]. Every hour, i.e. on each of its ticks, the bureaucrat checks whether Done has occurred since the previous hour using the **reclock** directive (the worker goes home when done and does not sustain Done, hence the need to reclock this signal). If so, OK is emitted and the bureaucrat terminates. Otherwise, the counter from 5 in the second case is decremented. If the count reaches 0, Worker is killed, FIRED is emitted and Bureaucrat terminates.

3.6   Notes

In full Multiclock Esterel, one should allow direct reactive module inclusion within a module body as in full Classic Esterel. In this case, the clock of the included module is as usual that of the caller. To include a reactive module with a different clock, one needs to use an intermediate process. Process / module strict alternation makes signal usage very clear and is one of the main characteristics of Multiclock Esterel.

The physical hardware implementation of multiclocked preemption is not fully explored yet. The main idea is of course to "gate the clocks" when passing them through Worker, but one has to be careful about the numerous boundary problems which can occur, because of clock simultaneity for instance.

4   Conclusion

We have presented the new Multiclock Esterel language proposal that extends Classic Esterel to multiclock systems. Although the language makes it possible to write very complex behaviors including well-defined multiclocked process preemption, it is technically a simple extension of Classic Esterel. We believe that this is a good sign of semantic soundness. It is too early to claim that the language is really well-suited to real-life multiclock systems and that it can be correctly and efficiently implemented in hardware or software. Experiments are on the way.

We did not discuss causality issues and the handling of combinational cycles [4]. Within each node, they are as in Classic Esterel. Between nodes, there is no extra issue if the clocks are declared exclusive since there is always a positive delay from one clock zone to another one. If clocks ticks can be simultaneous, there can be nasty cycles beteen clock zones, which we have not studied yet.

Similar techniques can be used for other synchronous languages such as Lustre [9] and Signal [8], making their nodes communicate through sampler and reclockers. These languages are much simpler since they do not support preemption. See for example [7] for the use of samplers in distributed continuous control applications.

Acknowledgements. We thank Paul Caspi (Verimag), Michael Kishinevsky (Intel), and the reviewers for their helpful comments and suggestions.

References

1. F. Balarin, M. Chiodo, P. Giusto, H. Hsieh, A. Jurecska, L. Lavagno, C. Passerone, A. Sangiovanni-Vincentelli, E. Sentovich, K. Suzuki, and B. Tabbara. *Hardware-Software Co-Design of Embedded Systems: The POLIS Approach*. Kluwer Academic Publishers, 1997.
2. G. Berry. Esterel on Hardware. *Phil. Trans. R. Soc. London A*, 339:87–104, 1992.
3. G. Berry. Preemption in concurrent systesms. In *Proc. FSTTCS93*, Lecture Notes in Computer Science 761, pages 72–93. Springer-Verlag, 1993.
4. G. Berry. *The Constructive Semantics of Esterel*. Draft book, preliminary version available from `http://www.esterel.org`, 1995-1999.
5. G. Berry. *The Esterel v5 Language Primer*. Available from `http://www.esterel.org`, 2000.
6. G. Berry and G. Gonthier. The Esterel Synchronous Programming Language: Design, Semantics, Implementation. *Science of Computer Programming*, 19(2):87–152, 1992.
7. P. Caspi, C. Mazuet, and N. Reynaud-Parigot. About the design of distributed control systems: the quasi-synchronous approach. In *Proc. Safecomp'01*, September 2001.
8. P. Le Guernic, M. Le Borgne, T. Gauthier, and C. Le Maire. Programming Real-Time Applications with Signal. *Another Look at Real Time Programming, Proceedings of the IEEE, Special Issue*, September 1991.
9. N. Halbwachs, P. Caspi, and D. Pilaud. The Synchronous Dataflow Programming Language Lustre. *Another Look at Real Time Programming, Proceedings of the IEEE, Special Issue*, September 1991.
10. Randy H. Katz. *Contemporary Logic Design*. Benjamin / Cummings, 1994.
11. David E. Muller and W. S. Bartky. A theory of asynchronous circuits. In *Proceedings of an International Symposium on the Theory of Switching*, pages 204–243. Harvard University Press, April 1959.
12. S. Schuster, W. Reohr, P. Cook, D. Heidel, M. Immediato, and K. Jenkins. Asynchronous Interlocked Pipelined CMOS Circuits Operating at 3.3-4.5MHz. In *Proceedings of the ISSCC*, February 2000.
13. E.M. Sentovich, K.J. Singh, C. Moon, H. Savoj, R.K. Brayton, and A. Sangiovanni-Vincentelli. Sequential Circuit Design Using Synthesis and Optimization. In *Proc of the ICCD*, pages 328–333, October 1992.

Register Transformations with Multiple Clock Domains*

Alvin R. Albrecht and Alan J. Hu

Department of Computer Science
University of British Columbia
{albrecht,ajh}@cs.ubc.ca

Abstract. Modern circuit design increasingly relies on multiple clock domains –
at different frequencies and with different phases – in order to achieve performance
and power requirements. In this paper, we identify a special case of multiple
clocking that encompasses typical design styles, and we present a theory enabling
a wide range of register transformations relating to the multiple clock domains. For
example, we can perform pipelining, phase abstraction, and retiming across clock
domain boundaries. We believe our theory will be useful to extend current work
on formal hardware design, synthesis, and verification to multiple-clock-domain
systems.

1 Introduction

Modern circuit design increasingly relies on multiple clocks running at different frequencies and with different phases. The reasons are to attain higher performance, to reduce
power consumption, and to allow integration of diverse components into systems-on-chip.

Higher performance comes mainly through the use of level-sensitive (a.k.a. transparent or flow-through) latches. A level-sensitive latch allows its input to flow combinationally to its output when its clock input is high, but ignores its input and holds the value of
its output when its clock input is low. To prevent combinational loops from fast signals
"racing through" the latch when it is transparent, latch-based designs group the latches
into several distinct, non-overlapping clock phases. Only one phase is transparent at a
time. (See Fig. 1.) This multi-phase design style is more complex than a single-clock-domain design, but allows more relaxed circuit timing constraints, faster clock speeds,
and higher performance (e.g., [5]).

Multiple clock domains are used to reduce power consumption because synchronous
CMOS dynamic power dissipation is linear in the clock speed. For performance reasons,
not all parts of a chip can run at a slow clock speed, but for power savings, not all parts of
the chip should run at full speed, either. Designing with multiple clock domains allows
different parts of a chip to run at their most efficient speeds. More elaborate techniques
involving multiple clock phases can produce substantial additional power savings [16].

An exponential gap is widening between the number of transistors per chip and the
number of transistors that a human designer can generate per unit time. The trend in

---

* This work was supported in part by a research grant from the National Science and Engineering
Research Council of Canada.

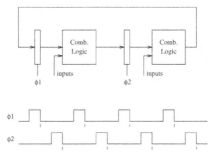

Fig. 1. Basic Two-Phase Clocking: The latches are transparent when their clock input is high. Two non-overlapping clocks ϕ_1 and ϕ_2 prevent both groups of latches from becoming transparent at the same time. The tic marks on the falling edges of the clocks show when the latch closes and holds its current value. This clocking style allows for higher performance designs than single edge-triggered clocking

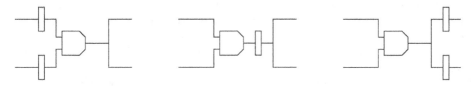

Fig. 2. Retiming allows moving registers (shown as small rectangles) forward or backward through arbitrary combinational logic gates and fanout stems. These three circuits have the same cycle-by-cycle behavior. Moving registers forward in the circuit (left-to-right in the figure) is called *forward retiming*. Moving registers backward is called *backward retiming*

design, therefore, is towards system-on-chip, in which multiple previously-designed IP blocks are integrated into a complete system on a single integrated circuit. Obviously, different blocks will have different performance and clocking needs, necessitating multiple clock domains on a single chip. For example, ARM's AMBA on-chip bus standard explicitly defines a high-performance clock domain and a separate, slower clock domain for peripherals [1]. As system complexity grows, additional clock domains will also be needed for slow, cross-chip communication.

With this increasing reliance on multiple-clock-domain design styles, new formalisms are needed to reason about, transform, and optimize these circuits. In this paper, we present a theory allowing a wide variety of register transformations to be done to multiply clocked circuits.

1.1 Background

Throughout this paper, we make extensive use of retiming, so we give a brief explanation here. Retiming is a behavior-preserving (at the cycle level) circuit transformation that allows moving registers forward and backward through combinational logic gates and fanout stems (Fig. 2). Leiserson and Saxe [12] laid the original foundation for retiming, showing how the problem of optimum retiming could be formulated as a linear program, and finding several important applications including clock period minimization, state minimization and pipelining of unpipelined circuits.

The linear programming formulation of retiming naturally leads to the concept of negative registers. A normal register delays a signal on a wire by one clock cycle. A negative register *accelerates* a signal by one clock cycle. A register/negative-register pair is equivalent to a no-op. Obviously, circuits with negative registers cannot be implemented directly[1], but negative registers are a useful bookkeeping formalism and can be retimed through a circuit just as a normal register.

One subtlety of retiming is that it can change the behavior of the circuit for the first few cycles after reset. For example, in Fig. 2, the rightmost circuit can produce unequal values at its two outputs for the first cycle after reset, whereas the other two circuits cannot. After the first cycle, the circuits behave identically. In general, forward retiming always allows preserving the initial behavior of the circuit, whereas backward retiming might not. In some applications, the initial behavior is not important, and both forms of retiming can be used freely. In other applications, the initial behavior is important, so either we restrict to forward and limited cases of backward retiming, or else we add special reset circuitry to force the retimed circuit to behave properly for the cycles following reset (e.g., [18,8, 17,3]). Similarly, our formalism supports more general register transformations if initial behavior is not important and imposes additional constraints if initial behavior must be preserved.

The initial behavior problem is related to a more general concept: retiming can be viewed as running a circuit forward or backward through time. We introduce a distinguished *host node* that is connected to all inputs and outputs of the circuit. The host node represents the external view of the behavior of the circuit. Retiming a register forward through the host node is equivalent to having the circuit step forward one clock cycle; retiming a register backward through the host node is equivalent to backing up the circuit one clock cycle [18].

1.2 Related Work

The retiming techniques described above can be applied only to circuits with a single type of register clocked from a single global clock, or within a single clock domain in a multi-clocked circuit. With the growing importance of multiple-clock-domain design, a few research results dealing with multiple clock domains have started to appear.

Hasteer et al. [10] concentrate on k-phase, level-sensitive clocking, where there are k non-overlapping clocks, all with the same frequency, and introduce the concept of phase abstraction, in which all but one phase of registers is eliminated. The resulting circuit has $1/k$ as many registers, reducing the state space of the circuit for verification purposes. Baumgartner et al. [2] achieve a similar result for two-phase circuits using a different formalism that preserves the initial behavior of the circuit and allows some different optimization. We will show how both of these works can be understood in terms of register transformations in our theory.

In a pure, well-formed multi-phase circuit, the sequence of register phases along any path is the same. (This restriction also applies to the circuits considered in the above

---

[1] Hassoun and Ebeling propose implementing negative registers via precomputation or prediction [9].

two works.) In that case, the different clock phases never interact in a way that prevents retiming [14,11]. Only very recently has work started to appear regarding more general instances of multiple clock domains [6,7]. In these works, the key idea is that only if all the registers at the inputs (outputs) of a gate are in the same clock domain, can they be retimed forward (backward) across that gate. Our work enables changing the clock domains of registers, thereby creating additional opportunities for these retiming techniques to work.

In general, our work is intended to provide tools necessary to reason about multiple-clock-domain circuits in a unified manner. Our theory supports retiming, phase abstraction, and a number of other transformations not published previously.

The theorem-proving community has long relied on temporal abstraction to relate hardware models with different granularities of time (e.g., [15, Sect. 4.1.3, Chap. 6]. The basic idea is to introduce a monotonic mapping function that takes time values from one model to time values in the other. Typically, the mapping is from time values for an abstract specification model into time values for a concrete implementation model. In contrast, our research deals with multiple granularities of time within a single circuit. Nevertheless, one could imagine embedding our work into the more classical framework by treating our Sampling Formula (Sect. 2.2) as a temporal mapping function, defining the behavior of circuit elements to include temporal abstraction, and then developing the necessary theories analogous to our derived transformations to reason about circuit behaviors involving these temporal abstractions. The main advantage of such an approach is that it may be possible to unite our work with other temporal mapping functions in a single theorem-proving system. Such a formalization is a possible direction for future research.

2   Register Transformations with Multiple Derived Clocks

2.1   The Multiple Derived Clocks Abstraction

In this subsection, we introduce our abstraction for multiple clock domains: the *multiple derived clocks* (MDC) abstraction. Just as the synchronous circuit model is an abstraction of the timing complexities of a single-clocked circuit, our MDC abstraction separates detailed low-level timing verification from higher-level functionality for multiple-clock-domain circuits.

First, we assume that the low-level clock routing and circuit timing analysis has been done properly. In particular, our model ignores clock skew (differences in timing for registers driven by the same clock), and we assume that signals have enough time to stabilize before they are latched. These premises are identical to the standard cycle-based view of a single-clocked, synchronous circuit, and allow separating the problem of timing analysis from issues of higher-level functionality.

If level-sensitive latches are used, we assume that the low-level timing analysis has properly accounted for the clocking scheme – in particular that adjacent latches have non-overlapping transparent periods. This model allows us to treat level-sensitive latches as edge-triggered latches, e.g., the circuit in Fig. 1 would behave exactly the same (assuming low-level timing constraints are met) if the latches were edge triggered at the tic marks. To map from an edge-triggered circuit back to a level-sensitive circuit requires the same low-level timing analysis on the level-sensitive circuit.

The general problem of interactions between multiple clock domains has long been studied (e.g., [5, pp. 473–475]). If clocks can be related arbitrarily, there is no way to eliminate the problem of metastability, and it is impossible to create a more abstract model. The key insight behind our MDC abstraction is that rarely do single-chip systems involve arbitrarily related clock domains. Instead, the clocks almost always maintain fixed, well-defined temporal relationships among themselves.

The reasons for this design style are practical. In order to avoid metastability and other timing problems, designers never want to risk having unknown clock timing. Furthermore, on a single chip, it is impractical to have multiple frequency references (e.g., crystals). As a result, all clocks on the chip are actually derived from a single, fast reference clock (e.g., [13]), using frequency dividers, phase-locked loops, etc.

Our MDC abstraction exactly captures this clocking style. We imagine a master reference clock with a certain clock period. Every clock in the system has a period that is some integer multiple λ of the master clock period and a phase offset θ with $0 \leq \theta < \lambda$ that is an integer number of master clock periods. Every clock tick occurs at the same instant as a master clock tick. Each register triggers instantaneously on the ticks of its clock. Combinational logic is considered to have zero delay.

Accordingly, any clock in the system can be characterized by the pair (λ, θ). We define the convention that for all λ, all $(\lambda, 0)$ clocks tick simultaneously at reset. At all times, a (λ, θ) clock ticks exactly θ master clock periods later than the $(\lambda, 0)$ clock. Registers on a given clock are labeled with the (λ, θ) pair of their clock. If the reset value of a register is relevant, we will label a register with a triple (λ, θ, v), where v is the value of the register immediately after the circuit resets. Wires also belong to clock domains. A wire driven by a register belongs to the same clock domain as the register that is driving it. A wire driven by a gate belongs to the same clock domain as the inputs to the gate, if all the inputs have the same clock labeling. Otherwise, we can conservatively assign the wire to the clock domain $(1, 0)$, indicating that it can change as fast as the master clock. Inputs and outputs can also be assigned a period-phase pair to indicate how often the value changes or is sampled.

For example, in Fig. 1, we can imagine a master clock that ticks at twice the frequency of the ϕ_i, with the master clock ticks lining up with the falling edges (tic marks in the figure) of the ϕ_i. This master clock would be called $(1, 0)$, and ϕ_1 and ϕ_2 would be called $(2, 0)$ and $(2, 1)$ (or vice-versa, depending on which clock ticked at reset). If the inputs have the same period-phase label as the latches to its left, then each combinational logic block will also have the same period-phase.

Our theory does not directly handle gated clocks. In the most common usage, however, the gating function simply enables and disables clocking of some registers. In that case, we can treat a register with a gated clock as the same register with an ungated clock and the enable signal controlling a multiplexer (Fig. 3). While the loop limits retiming, we can still determine that the input is sampled and the output is driven by the (λ, θ) clock domain.

A series of registers will be denoted by the sequence of their period-phase pairs. For example, the expression $(2, 0)(2, 1)$ indicates a $(2, 0)$ register whose output is the

Fig. 3. The most common case of a clock gating can be handled by changing the circuit to one without clock gating. These two circuits are equivalent under a zero-delay timing model

input of a $(2, 1)$ register. We denote a series of n registers of the same type (λ, θ) by the expression $n(\lambda, \theta)$; or if the reset values of the registers are relevant, by the expression $n(\lambda, \theta, \sigma)$, where σ is a string of n reset values, one per register.

Abusing the notation slightly, we will denote a negative $(1, 0)$ register by $(-1, 0)$. A negative register written as a triple $(-1, 0, v)$ indicates that v is the initial value at the input to the register, which it will discard because it is a negative register.

2.2 Sampling Formula

Given a register of type (λ, θ), the following equation relates its input and output as a function of time:

$$\text{Out}[t] = \text{In}\left[\lambda\left\lfloor\frac{t-\theta}{\lambda}\right\rfloor - 1 + \theta\right]$$

where t is time measured in master clock periods, and the floor function $\lfloor x \rfloor$ denotes the greatest integer less than or equal to x.

The circuit is "turned on" at time $t = 0$, when all registers take on their initial values. The equation relating individual register inputs and outputs is defined for all time, but when simulating the initial behavior of a physical circuit after reset, references to values at negative times in the equation should be considered references to the initial state of the register.

Our register transformation theory derives entirely from applying the sampling formula. For example, with two registers in series, the output of the first register drives the input of the second register, so $\text{In}_2[t] = \text{Out}_1[t]$. Let the two registers in series be $(\lambda_1, \theta_1)(\lambda_2, \theta_2)$. Then we can compute the output at time t as a function of the input as follows:

$$\text{Out}_2[t] = \text{In}_2\left[\lambda_2\left\lfloor\frac{t-\theta_2}{\lambda_2}\right\rfloor - 1 + \theta_2\right]$$

$$= \text{Out}_1\left[\lambda_2\left\lfloor\frac{t-\theta_2}{\lambda_2}\right\rfloor - 1 + \theta_2\right]$$

$$= \text{In}_1\left[\lambda_1\left\lfloor\frac{\lambda_2\left\lfloor\frac{t-\theta_2}{\lambda_2}\right\rfloor - 1 + \theta_2 - \theta_1}{\lambda_1}\right\rfloor - 1 + \theta_1\right]$$

In general, I/O equations for series combinations of registers can be found by working from right to left, successively making substitutions for $\text{In}_n[t]$ with $\text{Out}_{n-1}[t]$.

The I/O equations of three common series cases are summarized here:

$$n(1,0) : \text{Out}[t] = \text{In}[t - n]$$
$$n(-1,0) : \text{Out}[t] = \text{In}[t + n]$$
$$n(\lambda, \theta) : \text{Out}[t] = \text{In}\left[\lambda \left\lfloor \frac{t - \theta}{\lambda} \right\rfloor - 1 + \theta - (n - 1)\lambda \right]$$

These relationships are used frequently in order to prove the transformations in the next section.

2.3 Derived Transformations

Using the sampling formula, we have derived several circuit transformations that allow adding registers, removing registers, and changing the clock domain of registers. This is obviously not a complete list of derivable transformations, but simply an illustration of some useful transformations that our theory can generate. Determining the completeness of our theory, finding a small and useful set of derived transformations, and determining how to use the transformations to optimize circuits are all open questions.

Except where noted, the rules are shown below in a more restricted form that preserves the initial behavior of the circuit. If the initial behavior is not important, all of the rules can be considered to omit the specification of initial register values. We use X to denote a don't-care value.

Backward (De)composition allows decomposing/composing faster-clocked registers before a slower-clocked register. Given $\frac{\lambda_1}{\lambda_2} \in I$ and $0 \le \theta_1 < \lambda_1$,

$$\lambda_2(1, 0, \sigma)(\lambda_1, \theta_1, B) \equiv (\lambda_2, \theta_1 \bmod \lambda_2, A)(\lambda_1, \theta_1, B)$$

where σ is a string of λ_2 reset values, with the first θ_1 values from the right are X, then one A, then the rest are X. (If $\lambda_2 < \theta_1$, all of σ are don't-cares.)

Forward (De)composition allows decomposing/composing faster-clocked registers following a slower-clocked register. Given $\frac{\lambda_1}{\lambda_2} \in I$ and $0 \le \theta_1 < \lambda_1$,

$$(\lambda_1, \theta_1, A)\lambda_2(1, 0, \sigma) \equiv (\lambda_1, \theta_1, A)(\lambda_2, \theta_1 \bmod \lambda_2, B)$$

where σ consists of: first $(\theta_1 \bmod \lambda_2 + 1)$ from the right are B, and the rest are A.

Register Transformer changes the types of registers on a clock domain boundary. Given $m\lambda_1 = n\lambda_2$ and $m, n \in I \ge 0$

$$(\lambda_1, \theta_1, A)m(\lambda_1, \theta_1, \sigma_1)(\lambda_2, \theta_2, C) \equiv (\lambda_1, \theta_1, A)n(\lambda_2, \theta_2, \sigma_2)(\lambda_2, \theta_2, C)$$

where σ_1 and σ_2 consisting entirely of the same value B is a sufficient condition to preserve the initial behavior of the circuit.

Common Period Rule decomposes adjacent registers on the same clock period. Given $0 \le \theta_1, \theta_2 < \lambda$, and let

$$n = \begin{cases} \theta_2 - \theta_1 & \text{if } \theta_2 > \theta_1 \\ \theta_2 - \theta_1 + \lambda & \text{otherwise} \end{cases}$$

then
(i)

$$(\lambda, \theta_1, A)(\lambda, \theta_2, B) \equiv n(1, 0, \sigma)(\lambda, \theta_2, B)$$

where σ is: first θ_2 from right are X, then A, rest are X, and
(ii)

$$(\lambda, \theta_1, A)(\lambda, \theta_2, B) \equiv (\lambda, \theta_1, A)n(1, 0, \sigma)$$

where σ is: first $\theta_1 + 1$ from right are B, rest are A.

Phase Abstraction allows elimination or insertion of certain phases in a multiphase clocking scheme. Given $0 < j < k \leq \lambda$ and $0 \leq \theta < \lambda$:
(i) if $[\theta + j] \bmod \lambda < [\theta + k] \bmod \lambda$,

$$(\lambda, \theta, A)(\lambda, [\theta + j] \bmod \lambda, X)(\lambda, [\theta + k] \bmod \lambda, B) \equiv (\lambda, \theta, A)(\lambda, [\theta + k] \bmod \lambda, B)$$

(ii) if $[\theta + j] \bmod \lambda > [\theta + k] \bmod \lambda$,

$$(\lambda, \theta, X)(\lambda, [\theta + j] \bmod \lambda, A)(\lambda, [\theta + k] \bmod \lambda, B) \equiv (\lambda, \theta, A)(\lambda, [\theta + k] \bmod \lambda, B)$$

Minimum Period Register Insertion allows elimination or insertion of $(1, 0)$ registers between two more slowly clocked registers. Given $0 \leq \theta_1 < \lambda_1, 0 \leq \theta_2 < \lambda_2$ and either $\frac{\lambda_1}{\lambda_2} \in I$ or $\frac{\lambda_2}{\lambda_1} \in I$,

$$(\lambda_1, \theta_1, A)(\lambda_2, \theta_2, B) \equiv (\lambda_1, \theta_1, A)n(1, 0, \sigma)(\lambda_2, \theta_2, B)$$

where n is determined from:
(i) if $\frac{\lambda_2}{\lambda_1} \in I$ and $\theta_2 \bmod \lambda_1 \geq \theta_1 + 1$,

$$(\theta_2 \bmod \lambda_1) - \lambda_1 - \theta_1 \leq n \leq (\theta_2 \bmod \lambda_1) - \theta_1 - 1$$

(ii) if $\frac{\lambda_2}{\lambda_1} \in I$ and $\theta_2 \bmod \lambda_1 \leq \theta_1$,

$$(\theta_2 \bmod \lambda_1) - \theta_1 \leq n \leq (\theta_2 \bmod \lambda_1) - \theta_1 - 1 + \lambda_1$$

(iii) if $\frac{\lambda_1}{\lambda_2} \in I$ then for $0 \leq k \leq \frac{\lambda_1}{\lambda_2} - 1$,

$$1 - \lambda_1 + max\,[(k\lambda_2 + \theta_2 - \theta_1 - 1) \bmod \lambda_1] \leq n \leq min\,[(k\lambda_2 + \theta_2 - \theta_1 - 1) \bmod \lambda_1]$$

and σ is defined as follows: if $n < 0$, the first θ_1 values from the left are A, and the rest are X; if $n > 0$, then σ containing all A is sufficient to preserve initial behavior.

Domain Swallow allows neighboring registers in a different clock domain to be absorbed. Domain swallow is contrasted with the register transformer rule by noting that the absorbed register does not have to be sandwiched between two other registers. Although in some cases, it is possible to preserve initial behavior, the case analysis is cumbersome, so we present the transformation only for the steady-state behavior of the circuit. Given $0 \leq \theta_1 < \lambda_1$ and $0 \leq \theta_2 < \lambda_2$,
(i) if $\frac{\lambda_2}{\lambda_1} \in I$ and either $\frac{\theta_2 - \theta_1 - 1}{\lambda_1} \in I$ or $\frac{\theta_2 - \theta_1 - 1}{\lambda_2} \in I$ then

$$(\lambda_1, \theta_1)(\lambda_2, \theta_2) \equiv (1, 0)(\lambda_2, \theta_2)$$

(ii) if $\frac{\lambda_1}{\lambda_2} \in I$ and $\frac{\theta_2-\theta_1}{\lambda_2} \in I$ then

$$(\lambda_1, \theta_1)(\lambda_2, \theta_2) \equiv (\lambda_1, \theta_1)(1, 0)$$

(iii) if $\frac{\lambda_1}{\lambda_2} \in I$ and $\frac{\theta_2-\theta_1-1}{\lambda_2} \in I$ then

$$(\lambda_1, \theta_1)(\lambda_2, \theta_2) \equiv (1, 0)(\lambda_1, \theta_1)$$

Phase Change changes the phase of a register at the expense of more or fewer surrounding minimum period registers.

$$(\lambda, \theta, S) \equiv \begin{cases} n(-1,0)(\lambda, \theta - n, S)n(1,0,S) & \text{if } 0 \le n \le \theta \\ (-1,0,A)(\lambda, \lambda - 1, A)(1,0,S) & \text{if } \theta = 0 \\ n(1,0,X)(\lambda, \theta - n, S)n(-1,0,S) & \text{if } n < 0 \end{cases}$$

Time Passage A forward movement of a $(1,0)$ register past another register is equivalent to a single cycle passage of time on the register passed. A reverse movement of a $(1,0)$ register past another register is equivalent to a single cycle reversal of time on the register passed. The analogous interpretation exists for the movement of $(-1,0)$ registers, but with time reversed.

(a)
$$(1,0,I)(\lambda, \theta, S) \equiv \begin{cases} (\lambda, \lambda - 1, I)(1,0,S) & \text{if } \theta = 0 \\ (\lambda, \theta - 1, S)(1,0,S) & \text{if } \theta \ne 0 \end{cases}$$

(b)
$$(\lambda, \theta, S)(1,0,I) \equiv \begin{cases} (1,0,X)(\lambda, \theta + 1, S) & \text{if } S = I \\ (1,0,S)(\lambda, 0, I) & \text{if } \theta = \lambda - 1 \end{cases}$$

(c)
$$(\lambda, \theta, S)(-1,0,S) \equiv \begin{cases} (-1,0,X)(\lambda, \theta - 1, S) & \text{if } \theta > 0 \\ (-1,0,A)(\lambda, \lambda - 1, A) & \text{if } \theta = 0 \end{cases}$$

(d)
$$(-1,0,A)(\lambda, \theta, S) \equiv \begin{cases} (\lambda, \theta + 1, S)(-1,0,S) & \text{always} \\ (\lambda, 0, S)(-1,0,S) & \text{if } \theta = \lambda - 1 \text{ and } A = S \end{cases}$$

The notion of an environment host node having all outputs of a circuit tied to the input of the node and all inputs to a circuit tied to the output of the node allows forward retiming through the host node to represent forward time passage in the circuit. Likewise, reverse retiming through the host node represents reverse time passage in the circuit. Registers leaving the host node and entering the inputs to the circuit are initialized with the circuit's input values from the environment. Registers entering the host node and leaving the outputs of the circuit carry the output values sampled by the environment. This understanding of time passage in single clock domain circuits is described in [18].

In our general multitple-clock-domain case, $(1,0)$ registers passing through the host node represent time passage in units of the quickest period in the circuit. The time passage rule described here propagates input values on $(1,0)$ registers from the host node, through the circuit and back to the host node where they carry the output value sampled by the environment. A complete traversal of one such register passing through the host node represents one cycle of time passage.

Spontaneous Generation $(-1, 0)$ registers accelerate signals along a wire by one cycle, effectively annihilating preceding $(1, 0)$ registers. A $(-1, 0)$ register followed by a $(1, 0)$ register will not annihilate each other unless they hold the same state. A $(-1, 0)$ register's "state" is the input value on its left that it annihilates. Since it is known what the input value to the left of the negative register is, it is safe to annihilate the same value to the right without changing circuit behavior.

$$wire \equiv \begin{cases} (1, 0, S_1)(1, 0, S_2)...(1, 0, S_n)(-1, 0, S_n)(-1, 0, S_{n-1})...(-1, 0, S_1) \\ (-1, 0, S_1)(-1, 0, S_2)...(-1, 0, S_n)(1, 0, S_n)(1, 0, S_{n-1})...(1, 0, S_1) \end{cases}$$

This rule is best understood in the context of a circuit's environment. Imagine taking a circuit and cutting a particular wire connecting two halves. Then simulate this whole circuit n cycles forward in time in the manner described earlier. In that case, $n(1, 0)$ registers will accumulate on the left side of the cut wire, holding the n input values that would have been sampled on the right side of the cut wire had the cut not been made. On the right side of the wire, introduce the sequence $n(-1, 0)n(1, 0)$ and then retime the positive registers created on the right to the outputs to supply the host node and complete a single cycle time passage. Facing each other across the cut wire are the $n(1, 0)$ registers on the left and the $n(-1, 0)$ registers on the right. Reconnect the wire and using this rule, conditions on the state of the negative registers are found in order for the two sets to annihilate each other. These conditions on the state of the negative registers impose conditions on the state of the positive registers that were retimed toward the output. Effectively, negative registers act as place holders for expected input yet to be seen.

Each of these transformations has been described assuming a certain series of registers existed on some wire of the circuit. However, it is not the existence of registers per se that allows these transformations to occur, but rather the different rate of information supply and sampling that occurs between adjacent registers of different type. If it is known that a wire carries (supplies or is sampled for) a certain type of information, it can take the place of a register in any of the rules listed above, provided that this "phantom" register is not transformed by the application of the rule.

3 Examples

In this section, we give a few short examples showing the application of the above rules. In the first example (Fig. 4), we demonstrate how Hasteer et al.'s [10] phase abstraction can be viewed in our framework. This example circuit is taken from their paper. Based on the information being supplied and sampled, we see that each $(2, 1)$ register is being supplied by a $(2, 0)$ wire, and its output is being sampled by a $(2, 0)$ register. (Alternatively, we can retime the leftmost and rightmost $(2, 0)$ registers inward, and then retime them back out at the end.) Therefore, each wire in the middle of the circuit is $(2, 0)(2, 1)(2, 0)$, and our phase abstraction rule lets us remove the $(2, 1)$ registers.

Our next example (Fig. 5) demonstrates how we can change the clock domain of registers, allowing retiming through a gate that has different register types on its inputs. This is something that was impossible with earlier theories of retiming. The series of registers

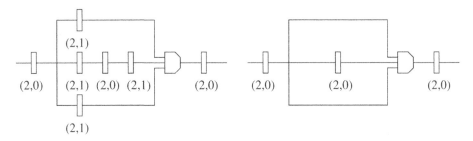

Fig. 4. Our phase abstraction rule converts $(2,0)(2,1)(2,0)$ to $(2,0)(2,0)$, allowing us to eliminate the $(2,1)$ registers

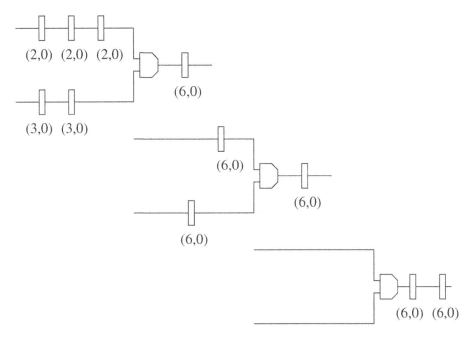

Fig. 5. The backward composition rule converts the $(3,0)(3,0)(6,0)$ register sequence into $(6,0)(6,0)$. Similarly, $(2,0)(2,0)(2,0)(6,0)$ becomes $(6,0)(6,0)$. Once the two inputs have been converted to the same register class, the registers can be retimed forward through the gate

are sampled by a $(6,0)$ register, so we can reason about the registers as $2(3,0)(6,0)$ for the top fanin and $3(2,0)(6,0)$ for the bottom fanin. Applying the backward composition rule converts both fanin branches to being $(6,0)(6,0)$. Now, we can retime the $(6,0)$ registers forward through the gate.

Our last example (Fig. 6) illustrates pipelining through different clock domains, another possibility created by our theory. The circuit shown illustrates incoming information arriving at a faster rate and being summarized into a slower rate output signal (e.g., imagine filtering an input signal). We can pipeline this circuit by inserting slow registers at the output and retiming them backwards through the circuit. The key trans-

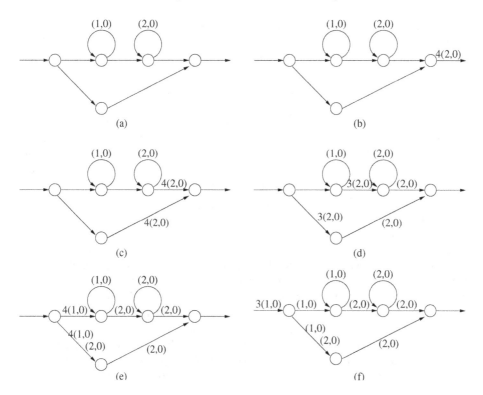

Fig. 6. For brevity, we indicate registers simply by the period-phase notation on edges. The circles denote arbitrary combinational logic. (a) Initially, we have a circuit with two clock domains that we wish to pipeline. (b) The output is being sampled at a $(2, 0)$ rate, so we insert four $(2, 0)$ registers at the output and start retiming them backwards, producing steps (c) and (d). In step (e), the backward decomposition rule allows us to convert $2(2, 0)$ registers into $4(1, 0)$ registers, which we can retime through the $(1, 0)$ clock domain. (f) Finally, we retime $3(1, 0)$ registers back through the first gate, leaving a single $(1,0)$ register behind the combinational node to reduce cycle time

formation is that we can use the backward decomposition rule to convert registers from the slower clock domain into registers in the faster clock domain and then retime them into the faster part of the circuit.

4   Conclusion and Future Work

In this paper, we have introduced the multiple derived clock abstraction for multiple-clock-domain circuits and developed a theory for transforming the registers of these circuits. We have shown how these transformations allow a wide range of reasoning about and manipulation of multiple-clock-domain circuits in a unified framework.

Future work extends in both theoretical and practical directions. The obvious theoretical question is whether our theory is complete, or if we will need to introduce additional circuit transformation primitives.

Our theory suggests behavior-preserving circuit transformations, but it does not say how to use them. Ideally, one could derive optimum algorithms for applications such as minimizing clock period or state space, as has been done for single-clock-domain retiming. More likely, given negative complexity results on generalized retiming problems, we will need to develop good heuristics for using these transformations to optimize circuits.

To be useful in practice, our theory will need tool and language support. At the bare minimum, tools will need to be able to determine the clocking schemes used in a circuit. Possible ways to determine this information range from recognizing idiomatic styles of HDL coding, through user annotation of the HDL code, to new description languages with explicit language support for multiple clocking schemes. Unfortunately, the more elegant solutions will likely require greater methodological changes.

Finally, we hope that our theory will be useful to extend a wide range of formal design, synthesis, and verification techniques to multiple-clock domain circuits. On the design side, it may be possible to design an easy-to-prove-correct circuit, and then apply these transformations in a step-by-step, refinement based design/proof strategy to arrive at a high-performance, multiple-clock implementation. If such a strategy were automatable, our theory could provide additional design-space flexibility to a formal synthesis system. Similarly, our theory can provide new inference rules for applying theorem-proving to multiple-clock circuits. For model checking, modeling multiple-clock systems is a challenging task (e.g., [4]), which our theory could simplify, and our theory could allow more freedom for retiming to minimize the state space. Practical demands are forcing people to design with multiple clock domains; we need to develop the supporting theory.

References

1. ARM Limited. *AMBA Specification (Rev 2.0)*. 13 May 1999.
2. Jason Baumgartner, Tamir Heyman, Vigyan Singhal, and Adnan Aziz. Model checking the IBM gigahertz processor: An abstraction algorithm for high-performance netlists. In *Computer-Aided Verification: 11th International Conference*, pages 72–83. Springer, 1999. LNCS Number 1633.
3. Gianpiero Cabodi, Stefano Quer, and Fabio Somenzi. Optimizing sequential verification by re-timing transformations. In *37th Design Automation Conference*, pages 601–606. ACM/IEEE, 2000.
4. Hoon Choi, Byeong-Whee Yun, and Yun-Tae Lee. Simulation strategy after model checking: Experience in industrial SOC design. In *International High-Level Design, Validation, and Test Workshop*, pages 77–79. IEEE, 2000.
5. William J. Dally and John W. Poulton. *Digital Systems Engineering*. Cambridge University Press, 1998.
6. Klaus Eckl and Christian Legl. Retiming sequential circuits with multiple register classes. In *Design, Automation and Test in Europe*, pages 650–656. IEEE, March 1999.
7. Klaus Eckl, Jean Christophe Madre, Peter Zepter, and Christian Legl. A practical approach to multiple-class retiming. In *36th Design Automation Conference*, pages 237–242. ACM/IEEE, 1999.
8. Guy Even, Ilan Y. Spillinger, and Leon Stok. Retiming revisited and reversed. *IEEE Transactions on CAD*, 15(3):348–357, March 1996.

9. Soha Hassoun and Carl Ebeling. Architectural retiming: Pipelining latency-constrained circuits. In *33rd Design Automation Conference*, pages 708–713. ACM/IEEE, 1996.

10. Gagan Hasteer, Anmol Mathur, and Prithviraj Banerjee. Efficient equivalence checking of multi-phase designs using retiming. In *International Conference on Computer-Aided Design*, pages 557–562. IEEE/ACM, 1998.

11. Alexander T. Ishii, Charles E. Leiserson, and Marios C. Papaefthymiou. Optimizing two-phase, level-clocked circuitry. *Journal of the ACM*, 44(1):148–199, January 1997. An earlier version of this paper appear in *Advanced Research in VLSI and Parallel Systems: Proceedings of the 1992 Brown/MIT Conference*, March 1992.

12. Charles E. Leiserson and James B. Saxe. Retiming synchronous circuitry. *Algorithmica*, 6(1):5–35, 1991.

13. Bill Lin, Steven Vercauteren, and Hugo De Man. Embedded architecture co-synthesis and system integration. In *4th International Workshop on Hardware/Software Codesign*, pages 2–9. IEEE, 1996.

14. Brian Lockyear and Carl Ebeling. Optimal retiming of multi-phase, level-clocked circuits. Technical Report TR-91-10-01, University of Washington, Department of Computer Science and Engineering, 1991.

15. T. Melham. *Higher Order Logic and Hardware Verification*. Cambridge University Press, 1993.

16. Christos A. Papachristou, Mehrdad Nourani, and Mark Spining. A multiple clocking scheme for low-power RTL design. *IEEE Transactions on VLSI Systems*, 7(2):266–276, June 1999.

17. Vigyan Singhal, Sharad Malik, and Robert K. Brayton. The case for retiming with explicit reset circuitry. In *International Conference on Computer-Aided Design*, pages 618–625. IEEE/ACM, 1996.

18. Hervé J. Touati and Robert K. Brayton. Computing the initial states of retimed circuits. *IEEE Transactions on CAD*, 12(1):157–162, January 1993.

Temporal Properties of Self-Timed Rings

Anthony Winstanley and Mark Greenstreet

Department of Computer Science, University of British Columbia, 2366 Main Mall,
Vancouver, BC V6T 1Z4, CANADA
{winstan,mrg}@cs.ubc.ca

Abstract. Various researchers have proposed using self-timed networks
to generate and distribute clocks and other timing signals. We consider
one of the simplest self-timed networks, a ring, and note that for timing
applications, self-timed rings should maintain uniform spacing of events.
In practice, all previous designs of which we are aware cluster events
into bursts. In this paper, we describe a dynamical systems approach to
verify the temporal properties of self-timed rings. With these methods,
we can verify that a new design has the desired uniform spacing of events.
The key to our methods is developing an appropriate model of the timing
behaviour of our circuits. Our model is more accurate than the simplistic
interval bounds of timed-automata techniques, while providing a higher
level of abstraction than non-linear differential equation models such as
SPICE. Evenly spaced and clustered event behaviours are distinguished
by simple geometric features of our model.

1 Introduction

Like many problems in hybrid systems, the problem of analyzing temporal clustering of events in self-timed rings involves a modeling problem. Clustering shows
up in detailed circuit models (e.g. SPICE), but such models are too detailed to
provide insight into the causes of the clustering or possible solutions. At the other
extreme, timed automata tools and logic simulators use bounded delay models
that are too coarse to distinguish the timing modes of self-timed circuits. We
require a model that captures the non-linear timing dependencies of real circuits
while abstracting away most of the details of a low-level circuit model.

We propose a new model of timing behaviour. The time of the output event
of a logic element is expressed as a function of the times of the input events
and the time of the previous output. This model recognizes the fact that gate
delay times depend not only on the time of the last input event, but also on
the time separation of these events. For a two input gate, this model can be
represented as a surface in a three dimension space. In a regular structure such
as a self-timed ring, the times of events are determined by an iterated map on
this surface. We identify the fixpoints of this map and the stability properties
of these equilibria. In particular, we show that self-timed rings exhibit critical
behaviours that lead to phase transitions with hysteresis between evenly spaced
and clustered equilibria. These phase transitions are in the *timing* of events; the

T. Margaria and T. Melham (Eds.): CHARME 2001, LNCS 2144, pp. 140–154, 2001.
© Springer-Verlag Berlin Heidelberg 2001

logical behaviour of the circuit is unaffected. In other words, the timed, discrete event model gives rise to a continuous map with Hopf bifurcations. We present simple, geometric criteria for identifying these phenomena.

Self-timed rings are introduced in Sect. 2. Section 3 describes our timing model, an extension of the "Charlie Diagrams" introduced in [EFS98]. Section 4 presents the main results of this paper; in particular, we derive geometrical criteria for classifying the behaviour of a ring as clustered or evenly spaced. In Sect. 5, we apply our approach to two examples. In the first example, we use a simple, synthetic model that admits easy analysis and provides the motivation for our circuit design. In the second example, we describe a circuit that we have designed and is currently in fabrication in a 0.35μ CMOS process. This latter example shows the value of our model as a design guide while revealing numerical limitations of our current methods for generating Charlie Diagrams from SPICE models.

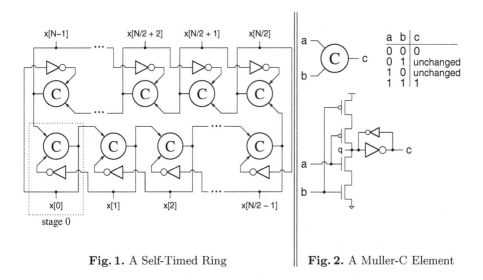

Fig. 1. A Self-Timed Ring   **Fig. 2.** A Muller-C Element

2   Self-Timed Rings

Self-timed circuits use handshake signals to control the sequencing of operations. These handshake signals take the role of clocks in synchronous designs. Figure 1 shows a ring composed of self-timed, handshaking stages (see [Sut89]). Each stage consists of a Muller C-element and an inverter. As illustrated by the transition table in Fig. 2, a C-element drives its output to the value of its inputs when its inputs agree; when the inputs of the C-element have different values, then the output of the C-element retains the value from the last time the inputs agreed.

The schematic in the lower half of Fig. 2 shows the CMOS implementation of the C-element used in this paper.

Figure 3 illustrates the operation of a self-timed ring. Let the output of n stages be represented by the array $x[0..n-1]$. By the operation of inverters and C-elements, stage i is enabled to change the value of its output, $x[i]$, when $x[i]$ differs from $x[i-1]$, and $x[i]$ is the same as $x[i+1]$. We say that a stage holds a "token" if the value of its output is different than that of its successor, and a "bubble" if their output values are the same.   In Fig. 3, stages that

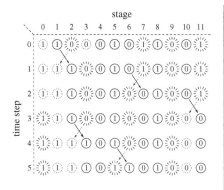

Fig. 3. Operation of Self-Timed Ring ‖ **Fig. 4.** Burst and Evenly Spaced Events

hold tokens are represented with a solid circle, $\bigcirc$; stages that hold bubbles and are not enabled are represented with a dotted circle, $\bigcirc$; and stages that hold bubbles and are enabled are represented with a "star burst", . It is convenient to rephrase the rules for when stages are enabled in terms of tokens and bubbles: stage i is enabled if it holds a bubble and stage $i-1$ holds a token. After stage i performs its action, stage i will hold a token, and stage $i-1$ will hold a bubble. In other words, performing an action exchanges a token and a bubble; tokens move forward around the ring, and bubbles move backward. The number of tokens in the ring is invariant, as is the number of bubbles. For example, if there are 6 tokens and 6 bubbles at time step 1, there will always be 6 tokens and 6 bubbles in the ring. The sum of the number of tokens and the number of bubbles is the number of stages.

In many applications of self-timed rings, tokens and/or bubbles are used to convey data values. The invariants on the number of tokens and the number of bubbles guarantee that data values are neither lost nor duplicated. In our current analysis, we focus on the timing of operations in the ring and do not give further consideration to any data values that may be conveyed with the tokens or bubbles.

Self-timed rings are ubiquitous in self-timed designs (e.g. [Wil91,SS93]), and their performance has been studied in many contexts. For example, [Thi91] analyzed throughput assuming each stage has a fixed time for each operation. Self-timed rings with exponentially distributed processing times were analyzed in [GS90]. Xie and Beerel have developed tools that analyze general networks of self-timed processors for general probabilistic models [XB97,XKB99]. All of these analyses have focused on the long-term throughput of the self-timed network.

Although long-term throughput is an important measure, the details of time separation between consecutive events is also important. In particular, in most self-timed rings, events occur in "bursts" as depicted in Fig. 4. Noting that stages in the ring are idle between bursts, even spacing can produce higher performance. Furthermore, evenly spaced events are desirable if the self-timed network is used to generate or distribute clocks and other timing signals in an otherwise synchronous design. Our work was motivated by these applications where self-timed circuits are used within a synchronous system. Although even spacing is desirable, all designs that we had seen prior to the ones described in Sect. 5 produced bursts of events. In the remainder of this paper, we present methods for analyzing the timing modes of self-timed rings.

3   Models

As with many hybrid systems, the key to understanding the timing properties of self-timed rings lies in finding an appropriate model for the ring. Circuit simulators such as SPICE [Nag75] use non-linear ordinary differential equations (ODEs) to model the circuit, and numerically integrate these equations to predict the circuit's behaviour. These ODE models can be quite accurate, and they correctly predict the burst behaviour that is observed by laboratory measurements. However, the device models are complicated, and even the models for small rings have dozens to hundreds of variables. Thus, whatever their virtues for accuracy, ODE models are too detailed to provide the insight into the causes of burst behaviour and how it can be controlled.

Another approach is to model the system as having discrete values that change at instants in continuous time. This is the approach taken by discrete event simulators (e.g. SHIFT [DGV96]) and timed automata techniques [LPY97, Yov97]. Close to our current problem, Amon and Hulgaard [HBAB95,AH99] have developed algorithms for computing bounds on the separation of events given bounds on operation times. All of these models specify the range of possible event times for each operation with an interval. Such models admit a wider range of behaviours than occur in practice. In particular, they show that bursts and evenly spaced events are both admitted by the models, but they don't predict which behaviour actually occurs.

Typical hardware delay models specify a delay after the *last* input event that enables the change. Using such a model, the time at which the output of a C-element changes, t_c is given by $\max(t_a, t_b) + \delta$, where t_a is the time of the change of the a input; t_b is the time of the change of the b input; and δ is some value with $\delta_{\min} \leq \delta \leq \delta_{\max}$. Let $t_{c,\max}$ be the latest time at which an output may change. For $t_a < t_b$, $\frac{\partial}{\partial t_a} t_{c,\max} = 0$, and for $t_a > t_b$, $\frac{\partial}{\partial t_a} t_{c,\max} = 1$. An equivalent observation holds for $\frac{\partial}{\partial t_a} t_{c,\mathrm{min}}$, and for derivatives with respect to t_b. The ODE models for circuits don't exhibit such discontinuities. To remain consistent with the ODE models, the delay intervals of these traditional hardware models must be fairly large. This is what makes them too imprecise for our purposes.

More accurate delay models account for the effects of closely spaced input events [DM+94,CS96] and intersymbol interference [DP98]. As described in Sect. 3.2, when enabling input events are closely spaced, the delay from the last input event to the resulting output event is greater than when the input events are more widely separated. In [CS96], this phenomenon is modeled with function that applies a correction term to a delay model for a single enabling event. This is similar to the Charlie Diagram model described in Sect. 3.1. However, the model in [CS96] lacks the continuity of the Charlie Diagram. In particular, for large separations, it assumes that the effect of the earlier signal on the output time is negligible. As described in Sect. 5, even very small dependencies can be critical in determining whether a ring has evenly spaced or clustered events.

3.1   The 2D Charlie Diagram

The starting point for our models are the "Charlie Diagrams" first described in [EFS98]. Based on observations like those above, Charles Molnar (after whom the diagrams were named) proposed measuring the output delay from the *average* of the two input arrivals. This delay is parameterized by the half-difference of the arrival times:

$$t_c = m + \mathsf{Charlie}(s)$$
$$\text{where:}$$
$$m = (t_a + t_b)/2$$
$$s = (t_a - t_b)/2$$

(1)

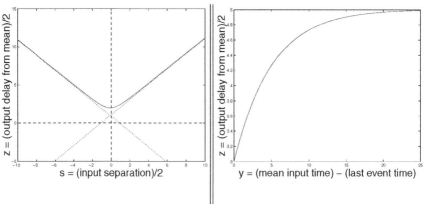

Fig. 5. A Charlie Diagram   **Fig. 6.** "Drafting"

Figure 5 shows a typical Charlie Diagram. The curve of Charlie(s) versus s resembles a hyperbola. For large separations of the input events, the output time approaches the time of the last input plus some constant. Thus, the Charlie Diagram has asymptotes with slopes of ± 1. We assume that all stages in the ring are identical and that their behaviour is causal; these assumptions imply that the asymptotes of the Charlie Diagram intersect the vertical axis at positive values.

Charlie Diagrams can be used to model any two-input gate, and the obvious higher dimensional diagrams can be used to model gates with more inputs. We use Charlie Diagrams to model stages in a ring, where each stage consists of a C-element and an inverter. For stage i, signal x[i-1] is the a input to the stage; signal x[i+1] is the b input of the stage; and signal x[i] is the output of the stage. The forward delay of stage i is the time from receiving an event on signal x[i-1] until producing an event on x[i]. Likewise, the reverse delay is the time to propagate an event on x[i+1] to x[i]. Using the Charlie Diagram notation, we obtain:

$$\begin{aligned}
\delta_F &= t_c - t_a = \mathsf{Charlie}(s) - s, && \text{forward delay} \\
\delta_R &= t_c - t_b = \mathsf{Charlie}(s) + s, && \text{reverse delay}
\end{aligned} \qquad (2)$$

3.2   The Charlie Effect

We now examine how the curve of a Charlie Diagram approaches the asymptotes. Consider a scenario where both a and b make low-to-high transitions, and a changes after b. If a changes a long time after b, then the p-channel transistor controlled by b will be in its cut-off region, and the n-channel transistor controlled by b will be fully conducting as a changes. Furthermore, the node between the two n-channel transistors will be close to ground potential. This allows a relatively fast transition on signal q and therefore on the output c. On the other hand, if a changes only slightly after b, then the transistors controlled by b will both be partially conducting as a changes. This results in a greater delay from the transition of a to the transition of c. Similar effects occur if a changes before b.

Charlie Diagrams provide a simple way to model these effects. If input a changes a long time after b, then s is large and positive. Conversely, if a changes only slightly before b, then s is small and positive. The delay from an event at input a to an event at output c is $t_c - t_a = \mathsf{Charlie}(s) - s$. The dependence of output delay on relative arrival times described above is reflected in the curve of the Charlie Diagram approaching the asymptotes monotonically from above. In general, when enabling events for a multiple input gate arrive at roughly the same time, the delay from the last input event to the output is greater than when the arrival times of the events are widely separated. Because this effect is naturally modeled by Charlie Diagrams, we call this effect the "Charlie Effect".

The curves of the Charlie Diagrams described in [EFS98] monotonically approach their asymptotes from above. This is due to the Charlie Effect described above. As we show in Sect. 4, this monotonicity implies that events are evenly spaced. In practice, most self-timed rings produce bursts of events. The authors of [EFS98] noted this discrepancy – it shows that their Charlie Diagram model neglects phenomena that are crucial to understanding the spacing of events.

3.3   The Drafting Effect

We extend the Charlie Diagram to model the effects of the output capacitance of the gate. Due to this capacitance, output transitions are not instantaneous. Instead, the voltage of the output asymptotically approaches the level of the power supply or ground. If input events are closely spaced, then the output of the gate will still be a significant distance from the power or ground rail when a new transition occurs. This allows subsequent transitions to occur faster than in the case where the output has reached a value closer to the rail. We call this phenomenon "drafting," after the practice of bicyclists to ride in closely spaced lines to reduce wind drag. Just as the lead cyclist reduces the work required of those behind her, the lead token in a burst allows subsequent tokens to propagate with reduced delay. The handshake protocol prevents trailing tokens from overtaking earlier ones (fear serves an equivalent purpose for bicyclists). As an example, Fig. 6 shows the time from an input event to the corresponding output event for our FIFO stage as a function of the input period. In this example, both inputs of the FIFO change simultaneously.

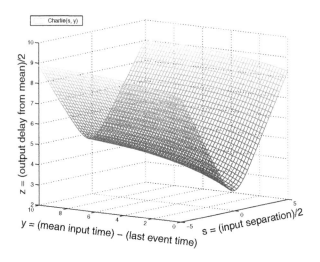

Fig. 7. A Three-Dimensional Charlie Diagram

3.4   The 3D Charlie Diagram

We model drafting by extending the Charlie Diagram to three dimensions. As with the original Charlie Diagram, the time separation of the input events is drawn along one axis of the domain. The other domain axis is the time from the last output event to the mean of the input event times. Figure 7 shows such a Charlie Diagram.

4   Analysis

Consider a burst of events propagating around a self-timed ring. We'll call the leading event in the burst event 1, and number the subsequent events consecutively. In the following analysis, we assume that the spacing of events in the burst remains invariant as the burst travels around the ring. This is intuitively plausible (otherwise, the burst would alternately expand and contract, etc. as it propagated). While we don't have a proof for this conjecture, we observed it in all cases for a wide variety of synthetic models and models from real circuits.

Now consider event 1. By the invariance assumption above, each stage that propagates event 1 has the same separation of input events, call this separation s_1. Likewise, each stage that propagates event 1 has the same time since its last output change, call this y_1. The values s_1 and y_1 specify a point on the 3D Charlie Diagram. These observations about event 1 apply to any event. If there are n_T tokens in the ring, and these tokens form a burst, then the operation of the ring is completely characterized by the values of s for each of the n_T events in the burst, and the n_T values for y. This cycle of n_T points on the 3D Charlie Diagram define an *operating point* for the ring.

In the remainder of this section, we will show how to identify the steady state operating points of a self-timed ring. This identification is based on geometric properties of the 3D Charlie Diagram. In particular, we show that our invariance assumption above implies that the forward latency, δ_F is the same for all event propagations. If the events form a burst, then there are two values of δ_R: one for the leading event in the burst, and the other for all other events. This means that the leading event of the burst corresponds to one point on the Charlie Diagram, and all other events correspond to one other point. If the events are evenly spaced, then all events occur at the same point on the Charlie Diagram.

More abstractly, the operation of the ring defines a mapping between points on the Charlie Diagram. We are interested in limit cycles of this mapping. In the rest of this section, we show how to locate and characterize these limit cycles.

4.1   The Constant δ_F Assumption

The conjecture stated above that the spacing of events in a burst remains invariant as the burst travels around the ring is equivalent to assuming that δ_F is the same for all events. We would like to find sufficient conditions on the Charlie Diagram than guarantee this conjecture. For the time being, we note that we have never observed a counter-example and will assume a fixed value for δ_F in the remainder of our analysis. We write δ_F^* to denote this common forward latency.

4.2   The $z = y$ Constraint

An enabling of stage i requires two input events: one from stage $i - 1$ and the other from stage $i + 1$. The $z = y$ constraint says that for any limit cycle of the Charlie Diagram, the midpoint of the input arrival times coincides exactly with the midpoint of the time of the last output event and the time of the output event triggered by these inputs. In this section, we show that this condition is implied by the definition of the Charlie Diagram and the handshake rules for the ring.

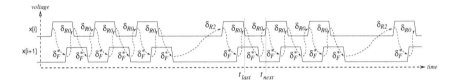

Fig. 8. Forward and Reverse Delays

Let t_{last} and t_{next} be the time of any two consecutive events at some stage i. Let $z = \mathsf{Charlie}(s, y)$ where

$$y = m - t_{last} \tag{3}$$

and m and s are defined as in equation 1. Let $\delta_R \in \{\delta_{R0}, \delta_{R1}\}$ be reverse latency of stage i leading to the event at time t_{next}. We derive:

$$
\begin{array}{rll}
& t_{next} = m + z, & \text{eq. 1} \\
\wedge & m = t_{last} + y, & \text{eq. 3} \\
\wedge & t_{next} = t_{last} + \delta_F^* + \delta_R, & \text{see fig. 8} \\
\wedge\, & \delta_F^* + \delta_R = 2z, & \text{eq. 2} \\
\Rightarrow & y = z &
\end{array}
\tag{4}
$$

Graphically, we can intersect the surface of a three-dimensional Charlie Diagram with the $y = z$ plane to obtain a curve that includes the operating points of any burst or evenly spaced equilibrium. We write $\mathsf{Charlie}_{y=z}(s)$ to denote this curve. In the remainder of this section, we restrict our attention to the $\mathsf{Charlie}_{y=z}(s)$ curve.

4.3   The δ_F^* Constraint

The assumption that δ_F^* is the forward latency for all events means that all points on a limit cycle lie on the intersection of the $\mathsf{Charlie}_{y=z}(s)$ curve with a line corresponding to δ_F^*. We call this the δ_F^* constraint.

More specifically, the points of a limit cycle lie at the intersection of

$$z = \mathsf{Charlie}(s, y),$$
$$z = y, \tag{5}$$
$$\text{and } z = s + \delta_F^*$$

If $\partial\mathsf{Charlie}_{y=z}(s)/\partial s < 1$ at the point of intersection, then that intersection is stable – in other words, if the ring is operating at a point near that intersection, it will converge to that stable point. On the other hand, if $\partial\mathsf{Charlie}_{y=z}(s)/\partial s > 1$, then the point of intersection is an unstable equilibrium.

Noting that $\partial\mathsf{Charlie}_{y=z}(s)/\partial s$ approaches -1 as s approaches $-\infty$ and $+1$ as s approaches $+\infty$ we classify five scenarios as exemplified in Figs. 9 and 10. In the "low drafting" case, $\partial\mathsf{Charlie}_{y=z}(s)/\partial s$ approaches $+1$ monotonically as s goes to infinity. For any value of δ_F^* greater than the asymptotic forward delay, the $z = s + \delta_F^*$ line intersects the $\mathsf{Charlie}_{y=z}(s)$ curve at a single point. As there is only one solution to equation 5, all events have the same operating point, and therefore the same reverse latency. Thus, the first scenario has evenly spaced events.

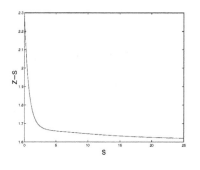 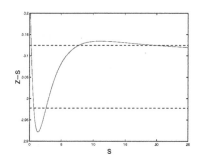

Fig. 9. $z = \mathsf{Charlie}_{y=z}(s)$, low drafting   **Fig. 10.** $z = \mathsf{Charlie}_{y=z}(s)$, high drafting

In the second scenario, $\partial\mathsf{Charlie}_{y=z}(s)/\partial s$ takes on values greater than one for some range of s. In this range, the drafting effect dominates the Charlie Effect, and forward delay increases with increases of the arrival time of the input token. If δ_F^* is large enough, then equation 5 is satisfied at three points: the leftmost and rightmost of these points are stable, and the inner one is unstable. The upper, dashed line in Fig. 10 illustrates this scenario. This solution gives rise to the burst of events: δ_{R0} corresponds to the right intersection, and δ_{R1} corresponds to the left intersection.

The third, fourth, and fifth scenarios arise for smaller values of δ_F^* as shown by the lower, dashed line in Fig. 10. This can either be a stable, evenly spaced mode where all events take place at the left intersection; an unstable, evenly spaced mode where all events take place at the right intersection; or an unstable, burst mode that exits to the stable burst mode or the evenly spaced mode.

4.4   Computing δ_F^*

Not all values of δ_F^* are feasible for any particular number of tokens and bubbles. Every time a stage performs an action, it exchanges a token and a bubble. This imposes a balance in the rates of token and bubble propagation which in turn determines the feasible values for δ_F^*.

Let the number of tokens be n_T, and the number of bubbles be n_B. We derive another constraint by considering a token that makes one complete orbit of the ring. This token is processed by n stages, with a delay of δ_F^* at each stage. The total time for the orbit is $n\delta_F^*$. For any i, stage i processes each of the n_T tokens during the same interval. After processing one token, the stage waits δ_F^* for the bubble to come back from stage $i + 1$, and then δ_R additional time to process the bubble. For the first token in a burst, $\delta_R = \delta_{R0}$. For the remaining $n_T - 1$ tokens, $\delta_R = \delta_{R1}$. Setting the time for one token to go around the ring equal to the time for one stage to process n_T tokens yields:

$$n\delta_F^* = (n_T)\delta_F^* + \delta_{R0} + (n_T - 1)\delta_{R1}$$
$$\Rightarrow \quad 0 = n_B\delta_F^* - \delta_{R0} - (n_T - 1)\delta_{R1} \tag{6}$$

For any value of δ_F^* we compute the intersection of $z = s + \delta_F^*$ and $z = \text{Charlie}_{y=z}(s)$ to find the possible values of s and thereby the values of δ_{R0} and δ_{R1}. Thus, the right side of equation 6 is a function of δ_F^*; the roots of this function are the feasible values of δ_F.

In evenly spaced mode, $\delta_{R0} = \delta_{R1}$. In this case, equation 6 can be rewritten as $n_B\delta_F = n_T\delta_R$. Substitutions for equation 2 yield:

$$z = \frac{n_T + n_B}{n_T - n_B}s \tag{7}$$

4.5   Classification

Given a three-dimensional Charlie Diagram, and values for n_T and n_B, we classify by the following procedure:

1. Find the curve $z = \text{Charlie}_{y=z}(s)$ by intersecting the surface of the Charlie Diagram with the $z = y$ plane.
2. Compute the intersection of $z = \text{Charlie}_{y=z}(s)$ with $z = ((n_T + n_B)/(n_T - n_B))s$. This is the evenly spaced solution. If $\partial\text{Charlie}_{y=z}(s)/\partial s < 1$ at the intersection, then the evenly spaced solution is stable (i.e. an attractor). If $\partial\text{Charlie}_{y=z}(s)/\partial s > 1$, then the evenly spaced solution is unstable.
3. Determine $\max(\partial\text{Charlie}_{y=z}(s)/\partial(s))$. If this value is greater than one, then burst behaviours are possible. If it is less than one, then burst behaviours are excluded.
4. If burst behaviours are possible, find the feasible values of δ_F^* by solving equation 6. If there is a solution with $\delta_{R0} \neq \delta_{R1}$ where $\partial\text{Charlie}_{y=z}(s)/\partial(s) < 1$ at both points, then the ring has stable, burst behaviours.

5   Examples

We have applied the analysis of the previous section to a small handful of examples; both real circuits and synthetic models. We start with our synthetic modeling because it permits succinct description.

5.1   Synthetic Example

Our synthetic model consisted of a ring with 30 stages modeled by the family of 3D Charlie Diagrams described below. We analyzed the ring using the methods described in Sect. 4 and observed the predicted phase transitions between clustered and evenly-spaced events using event driven simulation.

The two-dimensional Charlie Diagram from Fig. 5 is given by:

$$\mathsf{Charlie}(s) = 1 + \sqrt{1 + (s + 0.1)^2} \tag{8}$$

To simulate drafting, we added a delay that converges exponentially to a limit value:

$$\begin{aligned} u(s) &= 1 + \sqrt{1 + (s + 0.1)^2} \\ \mathsf{Charlie}(s, y) &= u(s) + b(1 - e^{-a(y + u(s))}) \end{aligned} \tag{9}$$

Note that $y + u(s)$ is the time from the previous firing until the gate would fire next without drafting. The exponential convergence to the asymptotic delay corresponds to a simple, first-order, RC delay. Figure 7 corresponds to this model with $a = 0.2$ and $b = 3.0$.

Varying b in equation 9 varies the strength of the drafting effect. Figure 9 corresponds $b = 0.5$, and Fig. 10 corresponds to $b = 2.0$ (both drawn with $a = 0.2$). For $a = 0.2$, evenly spaced events are the only stable behaviour for $b < 0.63166476$. For $0.63166476 < b < 1.18011777$, both evenly spaced and burst spacings are stable. As b increases in this interval, the intersection of the $z = \mathsf{Charlie}_{y=z}(s)$ curve with the $z = ((n_T + n_B)/(n_T - n_B))s$ line moves to the right. When it reaches the local minima of slope equals 1 point of the curve, the evenly spaced solution becomes unstable. For $b > 1.18011777$, only the burst behaviour is stable.

5.2   Circuit Example

Following the example of [EFS98], we initially considered an entire FIFO stage as a two-input module consisting of a C-element and an inverter. Based on our analysis, we assumed that drafting was the cause of event clustering. A signal that bounced away from the rail should generate a negative drafting effect and produce evenly spaced events. We tried this by adding small, delayed negative feedback circuitry to the output (c) of each stage in our ring.

This did not work as expected: events in the ring clustered into bursts regardless of the feedback that we applied. To understand this outcome, we attempted to construct 3D Charlie Diagrams for the stage using HSPICE, but this effort was hindered by the limited numerical resolution of HSPICE. At the operating point of the ring, the Charlie Diagram is flat to within the resolution of HSPICE.

Elementary continuity arguments imply that the surface must have some residual curvature, but we cannot observe this curvature directly. After further consideration, we realized that the bounce had a side effect of reducing the Charlie Effect. Consider a scenario where the output of stage i is initially low and the forward input (from stage $i-1$) goes high before the reverse input (from stage $i+1$) goes low. While waiting for the event from stage $i+1$, the output of stage $i-1$ *decreases* towards its asymptotic value. This *increases* the delay from the subsequent transition of stage $i+1$. Thus, while the negative feedback may have counteracted the drafting as desired, it also counteracted the Charlie Effect, resulting in a ring that continued to generate bursts of events.

To obtain the desired outcome, me moved the feedback point to the internal node, q of the C-element. This required some care to avoid spurious transitions. In the final solution, we use an N-channel pull-up to fight the keeper's pull-down and get good device matching. Likewise, we fight the keeper's pull-up with another P-channel device. Our successful solution is shown in Fig. 11. We included

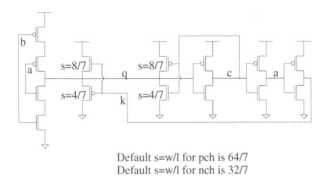

Default s=w/l for pch is 64/7
Default s=w/l for nch is 32/7

Fig. 11. A Ring Stage With Feedback

current mirrors (not shown in the figure) in series with the feedback devices to allow external control of the strength of the feedback for testing purposes. According to HSPICE simulations, this circuit can be started in burst-mode, and forced to evenly-spaced mode by a control input. Circuit considerations limit the range of our control input, and we can't force the ring back into bursting behaviour. Again, we would like to construct detailed Charlie Diagrams to better understand the behaviour of the circuit, but are hindered by numerical issues.

6   Conclusions and Future Work

We have presented a new model for reasoning about the timing behaviour of digital circuits: the three-dimensional Charlie Diagram. We have shown that the temporal properties of self-timed rings can be understood by examining fixpoints of iterated maps of this model. Simple geometric properties of the model distinguish evenly spaced and bursting behaviours. Three-dimensional Charlie Diagrams model behaviours that cannot be distinguished by traditional, timed-automata style models. At the same time, these Charlie Diagrams provide a much higher level of abstraction than circuit-level differential equation models (such as SPICE). This higher level of abstraction makes verification of timing properties practical.

While traditional timing analysis focuses on estimating delay values for gates and paths, we have presented an analysis that focuses on the *dynamics* of the timing. To this end, our models not only estimate the delays of logic elements but also the derivatives of these delays with respect to event times at the gate's inputs and outputs. We have shown that self-timed circuits have distinct timing modes, with phase transitions between these modes that exhibit hysteresis. Understanding these timing modes allows us to design circuits with timing properties not previously achieved.

We have used these techniques to design a new circuit for a self-timed ring stage. We are currently fabricating a chip based on this circuit to obtain experimental measurements and validate our models.

We currently generate three-dimensional Charlie Diagrams for our circuits through iterated SPICE simulations. In the future, we plan to explore ways of combining analytical and numerical techniques for generating these models to reduce the model generation time and improve model accuracy. In particular, by integrating SPICE's Jacobian matrix along paths between events, we should be able to compute an accurate value for the tangent plane of the Charlie Diagram at each point in spite of errors in the exact value of the point. We are also interested in ways to apply these techniques to other design, verification, and simulation problems.

References

[AH99]   Tod Amon and Henrik Hulgaard. Symbolic time separation of events. In *Proceedings of the Fifth International Symposium on Advanced Research in Asynchronous Circuits and Systems*, pages 83–93. IEEE, April 1999.

[CS96]   V. Chandromouli and K.A. Sakallah. Modeling the effects of temporal proximity of input transitions on gate propagation delay and transistion time. In *Proceedings of the 33th ACM/IEEE Design Automation Conference*, pages 617–622, June 1996.

[DGV96]   A. Deshpande, A. Gollu, and P.P. Varaiya. SHIFT, a formalism and programming language for dynamic networks of hybrid automata. In *Proceedings of the Eigth Conference on Computer Aided Verification*, pages 113–133, 1996.

[DM⁺94]   Florentin Dartu, Noel Menzes, et al. A gate delay model for high-speed CMOS circuits. In *Proceedings of the 31th ACM/IEEE Design Automation Conference*, pages 576–580, June 1994.

[DP98]   William J. Dally and John W. Poulton. *Digital Systems Engineering*. Cambridge University Press, 1998.

[EFS98]   Jo C. Ebergen, Scott Fairbanks, and Ivan E. Sutherland. Predicting performance of micropipelines using Charlie Diagrams. In *Proceedings of the Fourth International Symposium on Advanced Research in Asynchronous Circuits and Systems*, pages 238–246, April 1998.

[GS90]   Mark R. Greenstreet and Ken Steiglitz. Bubbles can make self-timed pipelines fast. *Journal of VLSI and Signal Processing*, 2(3):139–148, November 1990.

[HBAB95]   Henrik Hulgaard, Steven M. Burns, Tod Amon, and Gaetano Borriello. An algorithm for exact bounds on the time separation of events in concurrent systems. *IEEE Transactions on Computers*, 44(11):1306–1317, November 1995.

[LPY97]   Kim G. Larsen, Paul Petterson, and Wang Yi. UPPAAL: Status and developments. In *Proceedings of the Ninth Conference on Computer Aided Verification*, pages 456–459. Springer, June 1997. LNCS 1254.

[Nag75]   L.W. Nagel. SPICE2: a computer program to simulate semiconductor circuits. Technical Report ERL-M520, Electronics Research Laboratory, University of California, Berkeley, CA, May 1975.

[SS93]   Jens Sparsø and Jørgen Staunstrup. Delay-insensitive multi-ring structures. *INTEGRATION*, 15(3):313–340, October 1993.

[Sut89]   Ivan E. Sutherland. Micropipelines. *Communications of the ACM*, 32(6):720–738, June 1989. Turing Award lecture.

[Thi91]   Lothar Thiele. On the analysis and optimization of self-timed processor arrays. *INTEGRATION*, 12(2):167–187, December 1991.

[Wil91]   Ted E. Williams. *Self-Timed Rings and their Application to Division*. PhD thesis, Stanford University, May 1991.

[XB97]   Aiguo Xie and Peter A. Beerel. Symbolic techniques for performance analysis of timed systems based on average time separation of events. In *Proc. International Symposium on Advanced Research in Asynchronous Circuits and Systems*, pages 64–75. IEEE Computer Society Press, April 1997.

[XKB99]   Aiguo Xie, Sangyun Kim, and Peter A. Beerel. Bounding average time separations of events in stochastic timed Petri nets with choice. In *Proc. International Symposium on Advanced Research in Asynchronous Circuits and Systems*, pages 94–107, April 1999.

[Yov97]   Sergio Yovine. Kronos: A verification tool for real-time systems. *International Journal of Software Tools for Technology Transfer*, 1(1/2), October 1997.

Coverability Analysis Using Symbolic Model Checking

Gil Ratzaby, Shmuel Ur, and Yaron Wolfsthal

IBM Haifa Research Laboratory, Israel
{rgil,ur,wolfstal}@il.ibm.com

Abstract. In simulation based verification of hardware, as well as in software testing, one is faced with the challenge of maximizing coverage of testing while minimizing testing cost. To this end, sophisticated techniques are used to generate clever test cases, and equally sophisticated techniques are employed by engineers to determine the quality – a.k.a. coverage – attained by the tests. The latter activity is called Test Coverage Analysis.

While it is an essential component of the development process, test coverage can only be analyzed late in the design cycle when the tested entity and the test harness are both stable. To address this serious restriction, we introduce the notion of coverability, which intuitively refers to the degree to which a model can be covered when subjected to testing. We also show an implementation of coverability checking using Model Checking. The notion of coverability highlights a distinction between (1) whether a model has been covered by some test suite and (2) whether the model can ever be covered by any test suite. Coverability Analysis can be performed as soon as the hardware or software are written, before the test harness has been written.

1   Introduction

State machines are a simple and powerful modeling means used in a variety of areas, including hardware [VHDL93] and software design [Biezer90] [Marick95], protocol development and other applications [Hol91]. As a normal part of the modeling process, state machine models need to be analyzed and reasoned with regard to their function, performance, complexity and other properties. Traditionally, functional simulation has been a key vehicle to analyzing state machine models. The model is simulated against its expected real world stimuli and the simulated results are compared with the expected results. Simulation coverage analysis is normally applied to determine the thoroughness and quality of simulation. For example, the most common coverage metric used in the industry is *statement coverage* which checks that each statement has been executed at least once during simulation.

While test coverage analysis is an essential component of the design verification process, it can only be used late in the cycle when the code is stable and simulation environment is running. This is inherent limitation of test coverage analysis.

In this paper, we introduce the notion of coverability. Formally, a *coverability model* is defined by creating a *coverability goal* for every coverage goal in the coverage model of interest. The coverability goal is met if and only if a test that covers the corresponding coverage goal exists. Informally, coverability is a property of a state-machine model and

T. Margaria and T. Melham (Eds.): CHARME 2001, LNCS 2144, pp. 155–160, 2001.

refers to the degree to which the model can be tested using simulation. Reasoning about a model's coverability is a powerful analysis technique as it obviates the aforementioned limitation. Thus, a tool for determining coverability can help assess whether a given fragment of HDL code contains dead code, whether all branches of a particular control-flow statement can be taken, etc.

To implement coverability analysis, we build on symbolic model checking techniques, which provide a framework for reasoning about finite-state systems [McM93]. In recent years, Symbolic Model Checking, and formal verification in general, has been successfully used in the verification of communication protocols, as well as software and hardware systems [BBEL96]). Our work may be viewed as an extension of the recent research trend to bring together simulation-based verification and formal verification. Some recent works, for example, have focused on improving the quality of simulation by using formal verification methods to generate test sequences that ensure transition coverage [CFSM, LLU96].

To contrast coverability analysis and coverage analysis, consider the implementation of statement coverage. Here, a coverage tool typically implements statement coverage by adding a counter after every statement and initializing the counter to zero. Every time a test is simulated, some of the counters are modified. The coverage tool outputs all the counters that remain zero, as they are indicative of either dead-code or holes in the test plan. In coverability analysis, a rule for every statement would be automatically generated to check that it can be reached. These rules are executed by the model checker on the program (or instrumented program) and a warning on the existence of dead-code is created for every statement that cannot be reached.

Our approach is based on two key observations. First, as described above, a coverage model is composed of coverage goals, each of which is mappable to a corresponding coverability goal. The second observation is that a state-machine model can be instrumented with control variables and related transitions; these, on one hand, retain the original model behavior as reflected on the original state variables, and on the other hand can be used for coverability analysis of the model. The analysis is carried out by formulating special rules on the instrumented model and presenting these rules (with the instrumented model) to a symbolic model checker.

The rest of this paper is organized as follows: in Sect. 2 we define the terms used. In Sect. 3, we demonstrate a few coverability models and show their implementation. In Sect. 4, we explain how CAT – our Coverability Analysis Tool – was implemented. In Sect. 5, we discuss our experience using CAT and then we discuss our conclusions.

2   Definitions

A *coverage goal* is a binary function on test patterns. This function specifies whether some event occurred when the test pattern is simulated against the state-machine model. A coverage model is a set of coverage goals. The statement coverage model, for example, is a model containing a goal for every statement and indicates if this statement has been executed in simulation.

Every coverage model has a corresponding *coverability model*. A coverability model is defined by creating, for every coverage goal in the coverage model, a coverability goal which is a binary function on the state-machine model. The coverability goal is true if there exists a test on the state-machine model for which the corresponding coverage goal is true.

3   Coverability Analysis via Model Checking

This section describes and exemplifies the concepts of instrumentation and generation of auxiliary rules in our implementation of coverability analysis. We focus on coverability models relating to values of variables and to dead code analysis, and we show sample results in figure 1.

3.1   Attainability of All the Values of a Variable

This coverability model checks whether all variable values in a code fragment are attainable. To this end, for each variable declaration: `TYPE:var`, where `var` is a variable of interest, a collection of auxiliary rules of the form !EF $(var = V_i)$ – one for each value V_i of `var` – is created. The conjunction of these rules is a property requiring all relevant variables to take their respective values. This rule is presented to the underlying model checker, which in turn decides on the attainability of the value. When the formula passes, an example is also produced, demonstrating how the value is attained. Checking for this kind of coverability does not require instrumentation; it needs only the information about the variable declaration that enables the creation of the auxiliary rule.

3.2   Statement Coverability Analysis

This coverability model checks whether all statements can be reached. To this end, the program is instrumented separately for each statement S_i in the following way:

- Create an auxiliary variable V_i and initialize it to 0.
- Replace statement S_i with the statement "$V_i = 1$".

The model checker is then presented with the following rule: " !EF$(V_i = 1)$, which indicates whether S_i could be reached.

Statement coverability analysis changes the program's behavior. It may now contain dead-locks and behave in an unpredictable way. However, the program's behavior remains the same until the first execution of the replaced statement. As reachability analysis is performed, the behavior of the program after the first execution of the statement is immaterial. In this type of instrumentation, only one statement at a time may be checked. We are currently working on checking more efficient implementation.

4   CAT – Our Coverability Analysis Tool

We have created a coverability tool called CAT (Coverability Analysis Tool) which is very simple to use. It receives two parameters: the name of the program to be tested and the coverability models to be used. CAT outputs a list of all the coverability tasks,

```
module example (A,B,clock,F);
  input A,B;
  input [0:7] clock;
  output F;
  wire W1,W2;
  reg reg1;
  reg [0:7] arr;
  integer int1,int2;
  initial
    begin
        int2 = 1;
        int1 = 0;
        arr[0:7] = 0;
        reg1 = 0;
    end
  assign W2 = ! clock[6];
  assign W1 = reg1;
  always @(clock)
  begin
20int2 = 2;
21if(clock[3]>clock[2] || W2==1)
22 for(int1=0;int1<5;int1=int1+1)
23   begin
24    int2 = int2 + 1;
25    if(int1 > int2)
26        arr[4:5] = 2;
27    if(int1 < int2)
28        arr[4:5] = 3;
29   end
```

```
30 else
31   if(W2 == 1)
32       reg1 = 1;
33end
endmodule
```

Attainability Results:

```
Wire W1:
 Value 0 is attainable.
 Value 1 cannot be attained!
Wire W2:
 Value 0 is attainable.
 Value 1 is attainable.
```

Statement Coverability Results:

```
ASSIGNMENT in line 20: ok
IF in line 21:  ok
ASSIGNMENT in line 24: ok
IF in line 25:  ok
ASSIGNMENT in line 26:
      -- can never execute!
IF in line 27:  ok
ASSIGNMENT in line 28: ok
IF in line 31:  ok
ASSIGNMENT in line 32:
      -- can never execute!
```

Fig. 1. Verilog program with analysis

indicating whether each task is coverable or not. CAT works in the following way: for every coverability goal, CAT instruments the original program with the needed auxiliary statements, and creates a corresponding temporal rule. The rule is then checked by using a model checker on the instrumented program and the result of the run is reported. For example, if we want to find whether a line can be reached, we add an instrumentation that marks this line so it can be referred to by the rule. The model checker then checks the attainability of the marked line and CAT extracts and reports the answer.

CAT uses RuleBase, a symbolic Model Checker developed by the IBM Haifa Research Lab ([BBEL96]), as its underlying engine. RuleBase can analyze models formulated in several hardware description languages, including VHDL and Verilog. The basis for CAT is RuleBases's Verilog parser, Koala. CAT parses the input Verilog design, extracts the information needed in the current coverability goal, and constructs the auxiliary CTL rules on demand. CAT then transforms the design so that it will include the relevant auxiliary statements, and presents the instrumented program to RuleBase.

CAT supports default free-behavior environments, as well as user-defined environments. The user may choose between the two modes of environment modeling – default or user-defined. For example, in the application of CAT for dead-code analysis, default free-behavior environments are used. If a statement cannot be covered with free inputs, it can never be reached under any circumstances.

5 Experience

The coverability report for the sample program of Fig. 1 was generated in less than one minute. Figure 1 shows a Verilog program and Fig. 2 shows the issued report that indicates which variable values are not attainable and which statements cannot be reached. Working on benchmark PCI local BUS ([AZ97]), statement-reachability rules for a 1500-line Verilog program were evaluated at the rate of one rule per minute. To test CAT's industrial applicability, we used CAT to analyze the coverability of some customer code. It took two hours to complete a dead-code analysis report for a 1200-line Verilog module of a relatively simple control structure. The rules looking for dead code (e.g., "line 312 is never reached") execute quickly since each rule induces a relatively small cone-of-influence, which is amenable to the many reductions supported by RuleBase, our underlying model checker. Working on another Verilog file of comparable size (1300 lines) but with a significantly more complex control structure, a report was created at a rate of ten statement coverability rules per hour. This appears to be excessive for the developers, and we therefore are currently evaluating some of the optimization techniques.

6 Discussion

In this paper we introduced the concept of coverability analysis and described how a number of coverability metrics, corresponding to some commonly-used coverage metrics, can be implemented via Symbolic Model Checking. The same ideas can be used to implement many other coverability metrics (e.g., define-use, mutation, and loop [Marick95][Kaner95]).

The presented technique integrates ideas from traditional simulation and formal verification. It is somewhat similar to ideas seen in fault grading [KPKR94]. In fact, the technique derives its strength from its use of the exhaustiveness of formal verification to improve the planning of simulation.

A possible critique of the coverability approach might be that if a state machine is small enough to measure its coverage, it would also be small enough to be formally verified. However, we believe the merit of the coverability approach does in fact complement – and extend – the capabilities of formal verification due to the following reasons:

1. Today's model checkers are complicated tools that require the users to write complex rules in some form of temporal logic. The application suggested in this paper is fully automatic and enables naive users to take advantage of the power of model checkers with a "one button" interface for coverability analysis. We believe that adding this capability to existing model checkers will increase their utilization and applicability in the hardware design community.

2. Our approach can be used in the debugging stage (as soon as code is being written), before formal verification or simulation are used.
3. Formal verification does not guarantee that the design is correct. Even units that were formally verified need to be tested via simulation in which coverability can help.

References

[AZ97]   Adnan Aziz, Example of Hardware Verification Using VIS, The benchmark PCI local BUS, http://www-cad.eecs.berkeley.edu/Respep/Research/vis/texas-97/.

[BBEL96]   I. Beer, S. Ben-David, C. Eisner, A. Landver, "RuleBase: an Industry-Oriented Formal Verification Tool", Proc. DAC'96, pp. 655–660.

[Beizer90]   Boris Beizer, Software Testing Technique, 2/e. New York: Van Nostrand Reinhold, 1990.

[CGP99]   E.M. Clarke, O. Grumberg. D.A. Peled. Model Checking, MIT Press, 1999.

[CFSM]   D. Geist, M. Farkas, A. Landver, Y. Licthenstein, S. Ur and Y Wolfsthal, Coverage Directed Test Generation using Symbolic Techniques, FMCAD 96: Int. Conf. on Formal Methods in Computer-Aided Design, November 1996.

[Hol91]   G. J. Holtzman, Design and Validation of Computer Protocols, Prentice Hall, 1991.

[Kaner95]   C. Kaner, Software Negligence & Testing Coverage, Software QA Quarterly, Vol 2, #2, pp 18, 1995.

[KN96]   M. Kantrowitz, L. M. Noack, "I'm Done Simulating; Now What? Verification Coverage Analysis and Correctness Checking of the DECchip 21164 Alpha Microprocessor", Proc. DAC'96.

[KPKR94]   S. Kajihara, I. Pomerantz, K. Kinoshita and S. M. Reddy, "Cost Effective Generation of Minimal Test Sets for Stack-At Faults in Combinatorial logic Circuits", 30th ACM/IEEE DAC, pp. 102-106, 1993.

[LLU96]   D. Levin, D. Lorentz and S. Ur, "A Methodology for Processor Implementation Verification", FMCAD 96: Int. Conf. on Formal Methods in Computer-Aided Design, November 1996.

[Marick95]   B. Marick, The Craft of Software Testing: Subsystem Testing Including Object-Based and Object-Oriented Testing, Prentice-Hall, 1995.

[McM93]   K. L. McMillan, Symbolic Model Checking, Kluwer Academic Publishers, 1993.

[Mil90]   Raymond E. Miller, Protocol Verification: The first ten years, the next ten years; some personal observations, in Protocol specification, Testing, and Verification X, 1990.

[RB]   RuleBase User Manual V1.0, IBM Haifa Research Laboratory, 1996.

[TCE]   I. Beer, M. Dvir, B. Kozitsa. Y. Lichtenstein, S. Mach, W.J. Nee, E. Rappaport. Q. Schmierer, Y. Zandman, VHDL TEST COVERAGE in BDLS/AUSSIM Environment, IBM HRL Technical Report 88.342, December 1993.

[VHDL93]   D. L. Perry, VHDL Second Edition, McGraw-Hill Series on Computer Engineering, 1993.

[Weyuker94]   E. Weyuker, T. Goradia and A. Singh, Automatically Generating Test Data from a Boolean Specification, IEEE Transaction on Software Engineering, Vol 20, No 5 May 1994.

Specifying Hardware Timing with ET-LOTOS

Ji He and Kenneth J. Turner

Computing Science and Mathematics, University of Stirling, Stirling FK9 4LA, Scotland
h.ji@reading.ac.uk, kjt@cs.stir.ac.uk

Abstract. It is explained how DILL (Digital Logic in LOTOS) can specify and analyse hardware timing characteristics using ET-LOTOS (Enhanced Timed LOTOS –
the ISO Language Of Temporal Ordering Specification). Hardware functionality
and timing characteristics are rigorously specified and then validated.

1 Introduction

DILL (Digital Logic in LOTOS [2]) is an approach for specifying digital circuits using
LOTOS (Language Of Temporal Ordering Specification [1]). DILL allows formal speci-
fication of hardware designs, represented using LOTOS at various levels of abstraction.
DILL deals with functional and timing aspects, synchronous and asynchronous design.
There is support from a library of common components and circuit designs. Analysis
uses standard LOTOS tools.

LOTOS is a formal language standardised for use with communications systems. DILL,
which is realised through translation to LOTOS, is a substantially different application
area for this language. LOTOS is neutral with respect to whether a specification is to
be realised in hardware or software, allowing hardware-software co-design. LOTOS has
well-developed theories for verification and test generation. The paper uses ET-LOTOS
(Enhanced Timed LOTOS [3]). Because ET-LOTOS tools are currently under development, the authors have also used TE-LOTOS (Time Extended LOTOS [5]). Although these
LOTOS variants adopt different semantic models, the equivalence between them has been
established [4].

ET-LOTOS is a timed LOTOS that allows the modelling of time-sensitive behaviour.
The delay operator *delta (time)* means that the subsequent behaviour will be delayed
by *time*. A time value is relative to the instant when the previous action occurs. The
time measurement operator $@t$ is used to measure the time elapsed between the instant
when an event is offered and the instant when it occurs. The time value is stored in t.
The life reducer operator has different semantics when applied to internal events (**i**) and
observable events (**e**). **i** $\{d\}$ means that **i** *must* occur non-deterministically within the
next d time units. In the case of observable behaviour $e \{d\}; B$, the event *may* happen
within d time units. If so the behaviour evolves to B, otherwise the process deadlocks.
The default life reducer for internal events is *0*, while for observable events it is the
maximal value of the time domain. ET-LOTOS adopts maximal progress, i.e. if a hidden
action can occur it must happen at once (unless an alternative action occurs).

The input-output timing relationship is normally called *delay*. A timing relationship
among inputs is called a *timing constraint*, meaning that the digital circuit can work
correctly only if the constraints are met. In an *integrated method* for describing delay,

T. Margaria and T. Melham (Eds.): CHARME 2001, LNCS 2144, pp. 161–166, 2001.

a digital component is specified in one process that deals with both functionality and timing. Although the integrated method may result in compact specifications, it is not a 'structural' method and is hard to apply. Untimed behaviour should also be a special case of timed behaviour. *Combined methods* for specifying delay are thus preferred. These separate the functionality and the timing characteristics into different processes.

The approach selected for Timed DILL is called the *parallel-serial* model. Functionality is assumed to be specified with zero delay. Timing constraints are placed in parallel with the inputs of the functional specification to check if input requirements are met. Delays are placed in series with outputs of the functional specification to describe overall delay. Error events are introduced to discover violation of timing constraints; they have no counterpart in a real physical component.

If the timing constraints are void and the delays are between zero and arbitrarily large, the timed model is equivalent to the untimed model. An untimed specification is thus just a special case. Component functionality is supposed to have zero delay. This can be easily obtained from the untimed functionality. To change an untimed specification to one with zero delay, a life reducer {0} is appended to each output event offer.

2   Delays and Timing Constraints

Suppose the delay of a digital component is D. If a component has *pure delay*, all input changes will have an effect on output. If a component has *inertial delay*, output will respond only to input changes that have persisted for time D. Sometimes, the delay of a component has a more general form. There may exist a threshold $T < D$ such that the component absorbs input pulses whose width is less than T. However output follows input if the pulse width is more than T. In DILL this is termed *general delay*. In fact, it can be considered as inertial delay T cascaded with a pure delay $D - T$.

The DILL library supports non-deterministic delays ranging from *MinDel* (minimum) to *MaxDel* (maximum). For general delay, *MinWidth* corresponds to the threshold T. Timed DILL also handles high-to-low, low-to-high and pin-to-pin delays.

A naive attempt at specifying inertial delay would use the ET-LOTOS generalised life reducer. If the delay is connected to other components in a larger design, an output port might well be hidden. This would mean that the delay time is exactly *MinDel* instead of being a non-deterministic value since ET-LOTOS adopts maximal progress for hidden events. A better specification of inertial delay is given by:

> **process** DelayInertial [Ip, Op] (MinDel, MaxDel: Time) : **noexit** :=
> DelayInertialAux [Ip, Op] (MinDel, MaxDel, 0 **of** Bit, 0 **of** Bit)
> **where**
>   **process** DelayInertialAux [Ip, Op]   (* auxiliary definition *)
>   (MinDel, MaxDel : Time, DataIp, DataOp : Bit) : **noexit** :=
>   Ip ? NewDataIp : Bit;   (* input change *)
>   DelayInertialAux [Ip, Op] (MinDel, MaxDel, NewDataIp, DataOp)
>   []   (* or ... *)
>   [DataIp ne DataOp] $\Rightarrow$ (* output must change? *)
>   **i** {MinDel, MaxDel};   (* allow delay to pass *)
>   Op ! DataIp {0};   (* output changes at once *)

DelayInertialAux [Ip, Op] (MinDel, MaxDel, DataIp, DataIp)
endproc (* DelayInertialAux *)
endproc (* DelayInertial *)

The internal event **i** introduces non-deterministic delay, i.e. output can change at any time between *MinDel* and *MaxDel*. The exact delay is determined by the component itself and not by the environment. Moreover, even if the component is connected to other components, the delay is still non-deterministic since only hidden events are urgent.

Because delay is assumed non-deterministic rather than fixed, the pure delay specified below exhibits sequences like *Op ! 0; Op ! 0; Op ! 1* where the second *Op ! 0* overtakes *Op ! 1* and results in two consecutive *Op ! 0* events. The phenomenon of *catch-up* arises if a later input change takes less time to reach the output than an earlier input change. Catch-up may exhibit various forms in real hardware if delays vary significantly. However, digital components generally operate in a stable environment so the variation in delays is in a narrow range. Thus the catch-up condition is rarely met in practice. The phenomenon exists in any delay model that is based on pure delay.

```
process DelayPure[Ip, Op] (MinDel, MaxDel : Time) : noexit :=
   DelayPureAux [Ip, Op] (MinDel, MaxDel, 0 of Bit, 0 of Bit)
   where
      process DelayPureAux [Ip, Op]                         (* auxiliary definition *)
         (MinDel, MaxDel : Time, DataIp, DataOp : Bit) : noexit :=
      Ip ? NewDataIp : Bit;                                 (* input change *)
      (
         [NewDataIp eq DataOp] ⇒                            (* no output change? *)
            DelayPureAux [Ip, Op] (MinDel, MaxDel, NewDataIp, DataOp)
      []                                                    (* or ... *)
         [NewDataIp ne DataOp] ⇒                            (* output must change? *)
         (
            i {MinDel, MaxDel};                             (* allow delay to pass *)
            Op ! NewDataIp {0};                             (* output changes at once *)
            stop                                            (* delay behaviour now done *)
         |||                                                (* interleaved with ... *)
            DelayPureAux [Ip, Op] (MinDel, MaxDel, NewDataIp, NewDataIp)
         )
      )
   endproc (* DelayPureAux *)
endproc (* DelayPure *)
```

The general delay element in DILL is specified such that it can model not only a general delay but also inertial or pure delay by choosing appropriate timing parameters. The specification of general delay is not given here as it is just the combination of inertial and pure delay. The following gives the rules of using the timing parameters. *Inf* is the maximal value of the time domain (taken as arbitrarily large):

$0 < MinWidth < MinDel \leq MaxDel < Inf$ This describes general delay. It is meaningful only when *MinWidth* is a positive number less than *MinDel*.

$MinWidth = 0, MinDel \leq MaxDel < Inf$ This is pure delay. The difference from general delay is that *MinWidth* is 0 so the component does not absorb a narrow pulse.

$0 \leq MinDel \leq MaxDel < Inf, MinWidth > MinDel$ This is inertial delay. It applies if
  MinDel is less than threshold *MinWidth*, often set to *Inf* for inertial delay.
$MinDel = 0, MaxDel = Inf, MinWidth > 0$ This is equivalent to an untimed delay com-
  ponent. Usually *MinWidth* is set to the value *Inf*.

Timing constraints in DILL are used to check if the inputs of a component satisfy
some conditions. Common timing constraint elements have been added to the DILL
library, including those for setup, hold, pulse width and period.

Setup and hold times are always associated with flip-flops. Setup time is the time
interval between a change in data input and the trigger that stores this data. The hold
time is the interval in which input data must remain unchanged after triggering by the
clock. The setup time constraint is specified as follows for a D flip-flop where the active
clock transition is positive-going. A similar approach specifies a hold time constraint,
checks the minimum input pulse width, or checks the period of clock signals.

process SetupDel [D, Ck, Err] (SetupTime : Time) : **noexit** :=
  D ? NewDataIp: Bit;   (* new data input *)
  AfterD [D, Ck, Err] (SetupTime, SetupTime)   (* check setup time *)
[]   (* or ... *)
  Ck ? NewClock : Bit;   (* new clock input *)
  SetupDel [D, Ck, Err] (SetupTime)   (* no setup time to check *)
endproc (* SetupDel *)

process AfterD [D, Ck, Err] (SetupTime, SetupRem : Time) : **noexit** :=
  delta (SetupRem) **i**;   (* enforce min. setup time *)
  SetupDel [D, Ck, Err] (SetupTime)   (* restart setup check *)
[]   (* or ... *)
  Ck ? NewClock : Bit @ t;   (* new clock input *)
  (
    [NewClock eq 0] $\Rightarrow$ (* negative-going clock? *)
      AfterD [D, Ck] (SetupTime, SetupTime - t)   (* check setup time left *)
  []   (* or ... *)
    [NewClock eq 1] $\Rightarrow$ (* positive-going clock *)
      Err ! SetupError;   (* min. setup time violated *)
      SetupDel [D, Ck, Err] (SetupTime)   (* restart setup check *)
)
[]   (* or ... *)
  D ? NewDataIp: Bit;   (* new data input *)
  AfterD [D, Ck, Err] (SetupTime, SetupTime)   (* restart setup check *)
endproc (* AfterD *)

3   Timed DILL Example: 2-to-1 Multiplexer

As an example, a 2-to-1 multiplexer will be specified and analysed. A selection input S of 0
or 1 chooses input A or B, which appears at C after some delay. A higher level specification
defines the required behaviour and timing performance. A lower level specification gives
a component design that implements the higher level. The behavioural specification of
the 2-to-1 multiplexer in DILL is as follows:

```
define(MinDel, 10)                              # min. delay value
define(MaxDel, 15)                              # max. delay value
include(dill.m4)                                # include DILL library
circuit(                                        # circuit description
  Multiplexer2to1_BB [A, B, S, C],              # circuit name and ports
  hide InC in                                   # internal gate to delay
    Multiplexer2to1_BB_0 [A, B, S, InC]         # multiplexer instance
  |[InC]|                                       # sync with delay
    Delay [InC, C] (Inf, MinDel, MaxDel)        # delay instance
  where                                         
    Multiplexer2to1_BB_0_Decl                   # multiplexer from library
)
```

DILL provides a veneer on top of LOTOS – mainly a library of components that can be combined using LOTOS operators. The **circuit** declaration names the overall specification and its parameters. It then gives a LOTOS behaviour expression for the whole circuit. Library components are declared and automatically included by giving their names (*Component_Decl*). In the above, *Multiplexer2to1_BB_0* is a 2-to-1 multiplexer in behavioural style (*BB* = black box) that exhibits zero delay (*0*).

The behavioural specification was validated using the TE-LOLA simulator. Basically, the behaviour of the multiplexer is simulated for each input combination to see if it is as expected. The results of simulation are regarded as the criteria against which simulation of the lower-level specification should be judged. The design of the 2-to-1 multiplexer uses the selection signal to gate one or other input. Although this design might be found in a standard textbook, it was found that it contains hazards. The corresponding DILL specification is:

```
define(DelayData, Inf, 5, 5)               # MinWidth, MinDel, MaxDel
include(dill.m4)                           # include DILL library
circuit(                                   # circuit description
  timed,                                   # declare timed design
  Multiplexer2to1 [A, B, S, C],            # circuit name and ports
  hide AIn, BIn, SIn in                    # internal gates
    Inverter [S, SIn]                      # inverter instance
  |[S, SIn]|                               # sync with selection signals
    (
      And2 [SIn, A, AIn]                   # two-input and instance
    |||
      And2 [S, B, BIn]                     # two-input and instance
    )
  |[AIn, BIn]|                             # sync with inputs
    Or2 [AIn, BIn, C]                      # two-input or instance
  where                                    
    Inverter_Decl                          # inverter from library
    And2_Decl                              # two-input and from library
    Or2_Decl                               # two-input or from library
```

The delay is fixed at 5 and is inertial because *MinWidth* is *Inf.* The first parameter of the **circuit** declaration is optional. In this example it is **timed**; the default value is **untimed**, which appends *Inf, 0, Inf* to every instantiation of a basic logic gate.

Timed behaviour was investigated using the *TestExpand* function of TE-LOLA that automatically explores a test in parallel with a specification. If the test process can be followed for all executions of the composed specification, the result of testing is *must pass*. If the test process can be followed only for some executions, the result is *may pass*. Otherwise the test is considered to be *rejected*.

Firstly, the functionality of the multiplexer was tested. Secondly, there were tests to see if the design had a timing hazard (static or dynamic). Input transitions were checked with tests that deliberately risked hazards. Unfortunately 6 of the 56 transitions pass the tests, i.e. they exhibit hazards when the delays of each gates are fixed (transitions $000 \rightarrow 101$, $010 \rightarrow 101$, $011 \rightarrow 100$, $011 \rightarrow 110$, $111 \rightarrow 100$, $111 \rightarrow 110$). By simulating a passed test sequence, it becomes obvious that hazards are due to inputs following different lengths of path to reach the output. One solution is to introduce delay elements to equalise input-output path lengths.

4   Conclusion

DILL allows formal specification and analysis of digital hardware. It has extended the experience with LOTOS in the communications field. Timed DILL offers a number of important benefits. It can check whether timing requirements are respected by a design, making use of timing constraint components. Potential timing errors like hazards can be discovered, as in the multiplexer example. Timed DILL can also be used to analyse performance such as minimum/maximum delays and timing on critical paths. Although the paper has deliberately been illustrated with only a small example, the approach is applicable to much larger problems. A future goal is support of Timed DILL with verification based on KRONOS, HYTECH or timed automata.

References

1. ISO/IEC. *Information Processing Systems – Open Systems Interconnection – LOTOS – A Formal Description Technique based on the Temporal Ordering of Observational Behaviour.* ISO/IEC 8807. International Organization for Standardization, Geneva, Switzerland, 1989.
2. Ji He and Kenneth J. Turner. DILL (Digital Logic in LOTOS) project web page. http://www.cs.stir.ac.uk/~kjt/research/dill.html, November 2000.
3. Luc Léonard and Guy Leduc. An enhanced version of timed LOTOS and its application to a case study. In Richard L. Tenney, Paul D. Amer, and M. Ümit Uyar, editors, *Proc. Formal Description Techniques VI*, pages 483–500. North-Holland, Amsterdam, Netherlands, 1994.
4. Luis Llana and Gualberto Rabay Filho. Defining equivalences between time/action graphs and timed action graphs. Technical report, Department of Telematic Systems Engineering, Polytechnic University of Madrid, Spain, December 1995.
5. Gualberto Rabay Filho and Juan Quemada. TE-LOLA: A timed LOLA prototype. In Zmago Brezocnik and Tatjana Kapus, editors, *Proc. COST 247 International Workshop on Applied Formal Methods*, pages 85–95, Slovenia, June 1996. University of Maribor.

Formal Pipeline Design

Tiberiu Seceleanu and Juha Plosila

University of Turku, Dept of Applied Physics, Lab. of Electronics and Information Technology, FIN-20014 Turku, Finland, tel: +358-2-3336954, fax: +358-2-3336950, {Tiberiu.Seceleanu,Juha.Plosila}@utu.fi

Abstract. The action systems formalism has recently been applied to the area of asynchronous and synchronous VLSI design. In this paper, we study formal aspects of synchronous pipelining. We show how the framework of synchronous action systems can be used to derive a pipelined structure from a non-pipelined specification in a correctness-preserving manner.

1 Introduction

Pipelining is a very common technique used in digital systems design in order to increase parallelism. It has an important impact on the rate at which data is produced and consumed, a pipelined version allowing a higher speed than the non-pipelined version of the same circuit [7].

Formal methods of concurrent programming become increasingly important in design of complex VLSI systems. In our earlier work, we have shown how the *action systems formalism* [1] can be applied in design of both self-timed [8] and synchronous circuits [9,10]. In this paper, we continue our work on synchronous modeling by focusing on the formal aspects of clocked pipelining. The actions composing a synchronous action system are high-level representations of circuit functionality. One single action may actually map on a very complex hardware implementation, in terms of the time required to perform. Therefore, the device should operate on a large clock period, situation which, often, does not represent a solution. Thus, a further splitting of the execution of the initial action becomes necessary, reflected further in a pipelined realization of the digital device. Here, we present a stepwise procedure by which a non-pipelined synchronous action system specification is transformed into a pipelined form, in a correctness-preserving manner.

2 Synchronous Action Systems

The action systems formalism is based on an extended version of Dijkstra's language of *guarded commands* [5]. A *synchronous action system* [9] $\mathcal{A}$ has the form **sys** $\mathcal{A}(g)$ :: $|[$ **var** l; **init** $g,l := g_0,l_0$; **do** $A_1 \nabla A_2$ **od** $]|$, where g and l are lists of global and local variables initialised to g_0 and l_0, respectively, and $A_1 \nabla A_2$ is the *synchronous composition* of the actions A_1 and A_2 which access the variables g and l.

T. Margaria and T. Melham (Eds.): CHARME 2001, LNCS 2144, pp. 167–172, 2001.

Any of the actions of the synchronous system $\mathcal{A}$ is defined as a non-deterministic assignment, $x := x'.Q$. Here, Q is a boolean relation which determines the value(s) x' assigned to the (list of) variable(s) x. For example: $Q \;\widehat{=}\; x' = x + y$, where y is another variable. The synchronous composition operator '∇' between the actions A_i in the above system $\mathcal{A}$ indicates that the write variables of the involved actions are updated simultaneously in an atomic manner. At a lower abstraction level, a synchronous composition consists of a *read phase* and a *write phase* which are executed sequentially one after another.

The hardware representation of a synchronous action is represented by a set of registers (D flip-flops), each associated with one of the write variables of the action, and the corresponding combinational logic (specified, under all input situations, by the predicate Q).

Two synchronous action systems $\mathcal{A}_1$ and $\mathcal{A}_2$ can be composed using the operator '∇'. The composition $\mathcal{A}_1 \nabla \mathcal{A}_2$ is defined to be a synchronous action system which is composed of the actions of the constituent systems. This system merges the global variables of the components $\mathcal{A}_i$ keeping the local variables distinct.

Actions and action systems are intended to be developed in a stepwise manner within the *refinement calculus* [2]. A *correct* transformation of a given action A into the action A' is denoted as $A \leq A'$. Refinement of synchronous action systems is mainly based on the theory of *trace refinement* [3]. A *trace*, or a sequence of values of the global variables, represents an observable behaviour of a system. An abstract action system $\mathcal{A}$ is said to be *(trace) refined* by the concrete system $\mathcal{C}$, denoted $\mathcal{A} \sqsubseteq \mathcal{C}$, if the traces of $\mathcal{A}$ and $\mathcal{C}$ are equivalent. Notice that, in a trace, several successive equivalent states are considered as a single one.

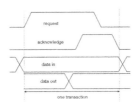

Fig. 1. The four phase handshake

The design of synchronous action systems [10] is based on correctly deriving synchronous representations from asynchronous action systems. An asynchronous system interacts with its environment via an asynchronous *communication (handshake) channel*, which is composed of two boolean variables [6], denoted *req* and *ack* (request, acknowledge), and *data variables*. The handshake variable *req* is updated by the *master* system which is the active party of communication requesting tasks from the *server* system which is the passive party of communication. The variable *ack* is updated by the server whenever it has completed a task requested by the master. In our approach, the synchronous system is the server component of an asynchronous description and a system that models the environment acts as the master. We build our design flow on the four-phase signalling protocol [4] (Fig. 1).

3   Pipelined Synchronous Action Systems

In order to obtain a pipelined version of an initially non-pipelined system, we have to introduce / eliminate local variables of the specific system. We do this according to the methods presented in [11], accommodating for our purposes the rules presented in [2]. We continue with the introduction of the pipelining procedure. For performing the translation from an asynchronous to a synchronous representation, the system under analysis has to comply with a set of *synchronization requirements* [10]. Thus, one of the requirements specifies that it is possible to transform a sequence (for instance $A_1; \ldots; A_n$) into a simultaneous assignment $(A_1 \star \ldots \star A_n)$, and from here, into a synchronous composition, if actions in the sequence do not read what other actions, positioned earlier in the sequence, have just updated. The process preserves the existence of the communication variables *req* and *ack* in the synchronous description.

Pipelining Procedure. Let us consider now a synchronous action system, $\mathcal{A}$, obtained from an initial asynchronous system. Thus, $\mathcal{A}$ contains the action ACK that updates the acknowledge signal *ack* :

$$
\begin{aligned}
&\textbf{sys } \mathcal{A} \, (req, ack : bool; x, y, a, b : data) :: \\
&|[\textbf{ var } l; \textbf{ init } l, x, \ldots, b := l_0, x_0, \ldots, b_0; req, ack := false \\
&\textbf{do } A \nabla B \nabla ACK \textbf{ od} \\
&]|,
\end{aligned}
\tag{1}
$$

$$
\begin{aligned}
A &\;\hat{=}\; a := a'.Q_A, & Q_A &\;\hat{=}\; (req \Rightarrow Q_a) \wedge (\neg req \Rightarrow a' = a), \\
B &\;\hat{=}\; b := b'.Q_B, & Q_B &\;\hat{=}\; (req \Rightarrow Q_b) \wedge (\neg req \Rightarrow b' = b), \\
ACK &\;\hat{=}\; ack := ack'.Q_K, & Q_K &\;\hat{=}\; (req \Rightarrow ack' = true) \wedge (\neg req \Rightarrow ack' = false)
\end{aligned}
$$

Let us now assume that the direct circuit implementation of the actions A and B of the system $\mathcal{A}$ is estimated to be so complex that it cannot be operated with the specified clock frequency. Then the computations of A and B have to be divided into n $(n > 1)$ phases each of which is simple enough to be executed in one clock cycle. A set of $n - 1$ new local variables is introduced for every action we split, in order to store the results of each phase. The combinational logic that updates the variable *ack* is a very simple construct. Consequently, the update is possible to be completed in a single clock step. Therefore, the update on the variable *ack* is only delayed but not split over n stages.

Pre-analysis Step. Intuitively, the first step in dividing the action A into n pipelined actions is to transform it locally into the atomic n-element sequence:

$$
\begin{aligned}
A &\;\leq\; A', A' \;\hat{=}\; A_0; \ldots; A_{n-1}, \\
A_j &\;\hat{=}\; a_j := a'_j.Q_a^j, j = 0, \ldots, n-1, a_{n-1} = a
\end{aligned}
$$

The above refinement requires that

$$
(Q_a^0 \wedge Q_a^1[a'_0/a_0] \wedge \cdots \wedge Q_a^{n-1}[a'_{n-2}/a_{n-2}]) \;\equiv\; Q_A
$$

As each action of the sequence A' reads the values of the variables updated by the previous action in the sequence, it is not possible to obtain a synchronous representation of the action A' (for instance $A_0 \nabla \ldots \nabla A_{n-1}$). However, this study is necessary as it indicates what the relations Q_a^j, $j = 0 \ldots n-1$ are (in other words, it shows *how* to split the execution of action A). They will be used in the following steps, when we describe the transformation of the system $\mathcal{A}$ into an equivalent pipelined system. Observe that a similar procedure is to be applied to action B, too.

Step 1. Buffering. We start the transformation of the action system (1) by introducing $3 \cdot (n-1)$ variables. Consequently, we also have $3 \cdot (n-1)$ new actions. Initially we only modify the original actions so that they have as input the new local variables, instead of the initial global variables:

$$\textbf{sys } \mathcal{A}^1 \ (req, ack : bool; x, y, a, b : data) ::$$
$$|[\ \textbf{var } l, a_0, \ldots, a_{n-2}, b_0, \ldots, b_{n-2}, req_0, \ldots, req_{n-2};$$
$$\textbf{init } l, x, \ldots, b := l_0, x_0, \ldots, b_0; req, ack, req_0, \ldots, req_{n-2} := false;$$
$$a_0, b_0, \ldots, a_{n-2}, b_{n-2} := a_{0_0}, b_{0_0}, \ldots, a_{n-2_0}, b_{n-2_0};$$
$$\textbf{do } A' \nabla A_0 \ldots \nabla A_{n-2} \nabla B' \nabla B_0 \ldots \nabla B_{n-2}$$
$$\nabla ACK' \nabla ACK_0 \ldots \nabla ACK_{n-2} \ \textbf{od}$$
$$]|,$$

$$\begin{aligned}
A' &\mathrel{\hat{=}} A[a_{n-2}, req_{n-2}/x, req], & B' &\mathrel{\hat{=}} B[b_{n-2}, req_{n-2}/y, req], \\
A_0 &\mathrel{\hat{=}} a_0 := x, & B_0 &\mathrel{\hat{=}} b_0 := y, \\
ACK' &\mathrel{\hat{=}} ACK[req_{n-2}/req], & ACK_0 &\mathrel{\hat{=}} req_0 := req,
\end{aligned}$$

$$S_j \mathrel{\hat{=}} s_j := s_{j-1}, s \in \{a, b, req\}, S \in \{A, B, ACK\}, j = 1 \ldots n-2$$

The hardware implementation of the system $\mathcal{A}^1$ is shown in Fig. 2.

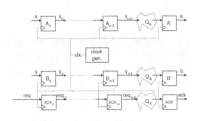

Fig. 2. Intermediate stage in the pipelining process

The result of **Step 1** is the refinement $\mathcal{A} \sqsubseteq \mathcal{A}^1$. Observe that until $req_{n-2} = true$, the system $\mathcal{A}^1$ does not modify its observable state (it stutters). Also, after req becomes $false$, until $req_{n-2} = false$, $\mathcal{A}^1$ stutters again.

Step 2. Auxiliary Actions. Next, we introduce $2 \cdot n$ additional actions $D_0, \ldots, D_{n-1}$ and $C_0, \ldots, C_{n-1}$, that synchronously update the new local variables $d_0, \ldots, d_{n-1}$ and $c_0, \ldots, c_{n-1}$, respectively. We have

$$\begin{aligned}
D_0 &\mathrel{\hat{=}} d_0 := d_0'.Q_D^0, & Q_D^0 &\mathrel{\hat{=}} Q_a^0[d_0'/a_0'], \\
C_0 &\mathrel{\hat{=}} c_0 := c_0'.Q_C^0, & Q_C^0 &\mathrel{\hat{=}} Q_b^0[c_0'/b_0']
\end{aligned}$$

For $j = 0 \ldots n - 1$, the actions D_j and C_j update the corresponding output variables as specified by the predicates selected in the pre-processing step:

$$D_j \;\widehat{=}\; d_j := d'_j.Q^j_D, \; Q^j_D \;\widehat{=}\; (req_{j-1} \Rightarrow Q^j_a[d'_j/a'_j]) \wedge (\neg req_{j-1} \Rightarrow d'_j = d_j),$$
$$C_j \;\widehat{=}\; c_j := c'_j.Q^j_C, \; Q^j_C \;\widehat{=}\; (req_{j-1} \Rightarrow Q^j_b[c'_j/b'_j]) \wedge (\neg req_{j-1} \Rightarrow c'_j = c_j)$$

At the system level we have **sys** $\mathcal{A}^1 \sqsubseteq$ **sys** $\mathcal{A}^2$, where $\mathcal{A}^2$ is described as

> **sys** $\mathcal{A}^2 \, (req, ack : bool; x, y, a, b : data) ::$
> $|[$ **var** $l, a_0, b_0, req_0, d_0, c_0 \ldots, a_{n-2}, b_{n-2}, req_{n-2}, d_{n-1}, c_{n-1};$
> **init** $l, x, \ldots, b := l_0, x_0, \ldots, b_0; req, ack, req_0, \ldots, req_{n-2} := false;$
> $\quad a_0, b_0 \ldots, a_{n-2}, b_{n-2} := a_{0_0}, b_{0_0} \ldots, a_{n-2_0}, b_{n-2_0};$
> $\quad d_0, c_0 \ldots, d_{n-2}, c_{n-2} := d_{0_0}, c_{0_0} \ldots, d_{n-2_0}, c_{n-2_0};$
> **do** $A' \bigtriangledown A_0 \ldots \bigtriangledown A_{n-2} \bigtriangledown B' \bigtriangledown B_0 \ldots \bigtriangledown B_{n-2}$
> $\quad \bigtriangledown ACK' \bigtriangledown ACK_0 \ldots \bigtriangledown ACK_{n-2}$
> $\quad \bigtriangledown C_0 \ldots \bigtriangledown C_{n-1} \bigtriangledown D_0 \ldots \bigtriangledown D_{n-1}$ **od**
> $]|$

Step 3. Removal of Auxiliary Variables. The Final Description. Next, notice that

$$req \wedge req_{n-2} \Rightarrow (Q^{n-1}_D[a/d_{n-1}] \equiv Q_A[a_{n-2}, req_{n-2}/x, req])$$
$$\wedge \; (Q^{n-1}_C[b/b_{n-1}] \equiv Q_B[b_{n-2}, req_{n-2}/y, req])$$

The interpretation of the above relation is as follows. The update on variable a, either according to Q_A, or to Q^{n-1}_D, leads to the same result. Therefore, we can write

$$A' \leq A'', \; A'' \;\widehat{=}\; D_{n-1}[a/d_{n-1}],$$
$$B' \leq B'', \; B'' \;\widehat{=}\; C_{n-1}[b/c_{n-1}]$$

Now we can safely eliminate the variables a_{n-2}, b_{n-2} from the system description, together with the actions that update them. We repeat this procedure for the variables a_j, b_j $(j = n-3 \ldots 0)$ and for the corresponding actions. Eventually we come to the system $(\mathcal{A}^2 \sqsubseteq \mathcal{A}^3)$:

> **sys** $\mathcal{A}^3 \, (req, ack : bool; x, y, a, b : data) ::$
> $|[$ **var** $l, req_0, d_0, c_0 \ldots, req_{n-2}, d_{n-1}, c_{n-1};$
> **init** $l, x, \ldots, b := l_0, x_0, \ldots, b_0; req, ack, req_0, \ldots, req_{n-2} := false;$
> $\quad d_0, c_0 \ldots, d_{n-2}, c_{n-2} := d_{0_0}, c_{0_0} \ldots, d_{n-2_0}, c_{n-2_0};$
> **do** $A'' \bigtriangledown B'' \bigtriangledown D_0 \ldots \bigtriangledown D_{n-2} \bigtriangledown C_0 \ldots \bigtriangledown C_{n-2}$
> $\quad \bigtriangledown ACK' \bigtriangledown ACK_0 \ldots \bigtriangledown ACK_{n-2}$ **od**
> $]|$

The implementation of the system $\mathcal{A}^3$ is shown in Fig. 3.

4   Conclusions

We introduced an action systems-based method to transform a non-pipelined synchronous system description into a pipelined structure within a correctness preserving formal framework. The pipelining procedure was viewed as a four-step refinement which started from atomic sequencing of an action and was completed by splitting the action in question into separate synchronous components each

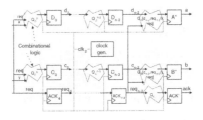

Fig. 3. The system $\mathcal{A}^3$ in hardware

of which constituted a stage of the created pipeline. The work presented in this paper was motivated by importance and generality of pipelining in contemporary VLSI design and by the view that the need for formal design methods is becoming more and more obvious with increasing complexity of digital VLSI systems.

Several issues are still to be addressed in further work studies, such as the elimination of the communication variables and its implication in pipelined-system modelling, as well as the subject of pipeline control.

References

1. R. J. R. Back and K. Sere. Stepwise refinement of action systems. *Structured Programming*, 12:17-30,1991.
2. R. J. R. Back and J. von Wright. *Refinement calculus: A Systematic Introduction.* Springer. April 1998.
3. R. J. R. Back and J. von Wright. Trace refinement of action systems. In B. Jonsson and J. Parrow, editors, *CONCUR'94: Concurrency Theory, Sweden, 1994*, volume 836 of *LNCS*. Springer, 1994.
4. A. Davis and S. M. Nowick. *Asynchronous circuit design: motivation, background and methods.* Asynchronous Digital Circuit Design, G. Birtwistle and A. Davis (eds.), pages 1-49. Springer, 1995.
5. E. W. Dijkstra. *A Discipline of Programming.* Prentice-Hall International, 1976.
6. J. C. Ebergen, J. Segers, I. Benko. *Parallel Program and Asynchronous Circuit Design.* Asynchronous Digital Circuit Design, G. Birtwistle and A. Davis (eds.), Springer, 1995.
7. G. De Micheli. *Synthesis and Optimisation of Digital Circuits.* McGraw-Hill International Editions, 1994.
8. J. Plosila. *Self-Timed Circuit Design – The Action Systems Approach.* Ph.D. Thesis, University of Turku, Dept of Applied Physics, Turku, Finland, 1999.
9. J. Plosila and T. Seceleanu. *Modeling Synchronous Action Systems.* In *Proc. of the 17th NORCHIP Conference*, Oslo, Norway, pages 242-248. November 1999.
10. J. Plosila, T. Seceleanu. *Design of Synchronous Action Systems.* In Proceedings of The 13th International Conference on VLSI Design, January 2000, pages 578-583.
11. T. Seceleanu, J. Plosila. *Synchronous Pipeline Design in Action Systems.* Turku Centre for Computer Science Technical Report No. 403, 2001.
12. J. M. Rabaey. *Digital Integrated Circuits – A Design Perspective.* Prentice-Hall International, 1996.

Verification of Basic Block Schedules Using RTL Transformations*

Rajesh Radhakrishnan, Elena Teica, and Ranga Vemuri

ECECS Department, ML. 30
814 Rhodes Hall, University of Cincinnati
Cincinnati, OH 45221-0030
{rradhakr,eteica,ranga}@ececs.uc.edu, Fax: 513-556-3025

Abstract. We present an approach to aid in debugging/development of scheduling algorithm implementations. Our technique makes use of a sequence of a correctness-preserving RTL transformation called *Register Transfer Split (RTS)*, to collectively perform the same task as that of a scheduler. Violation of the transformation precondition signals an error and the sequence of RTS transformations applied so far forms a trace which can be used for debugging purposes.

1 Introduction

Researchers have addressed the problem of creating bug-free synthesis systems into separately verifying each synthesis stage. In this paper, we specifically deal with the verification of the scheduling stage.

Lock et al [1] captured the scheduling result in the form of a table and checks for correctness inside the HOL theorem prover. Ashar et al [2] used model checking and symbolic simulation to check the signal correspondences between RTL and behavior. Importantly, all arithmetic operations are uninterpreted and loops are handled via identifying loop invariants. In Eveking et al [3], failure to transform the behavioral description towards the scheduled result signals an error. Naren et al [4] proved the correctness of the force-directed list scheduler (FDLS) algorithm in PVS and embedded the correctness conditions developed during the exercise as program assertions in the implementation.

We do not handle constraint violations in the scheduler and hence only consider *legal* schedules and not optimal ones [5]. Also, our approach is applicable only for schedules for sequences of straight-line code or *basic blocks*.

In Sect. 2 we introduce models for a register transfer and the RTS transformation. Section 3 discusses our methodology. The conclusion is presented in Sect. 4.

2 Models

We use the models of a register transfer and the RTS transformation from [6], where a completeness proof for a set of RTL transformations (including RTS)

---

* This work was sponsored by DARPA, monitored by US Army Ft. Huachuca under contract number DABT63-96-C-0051 and NSF (grant number 9634462).

T. Margaria and T. Melham (Eds.): CHARME 2001, LNCS 2144, pp. 173–178, 2001.

is presented. *Completeness* of a set of behavior-preserving RTL transformations means: for any two behaviorally-equivalent RTL designs, by applying a finite sequence of these correctness-preserving transformations we can move from one design to the other.

2.1   Register Transfer

A register transfer maps a set of source registers to a set of destination registers. It denotes the activity performed at a certain part in the data path at a *unique control step*. We use the term *expressions* to collectively refer to operators, registers and their interconnections. Let E refer to the set of expressions, OP to the set of operators in E and REG to the set of registers in E.

The data path consists of a set of operators, a set of registers and the interconnect between them. Using the notations above, we can now define the data path (DP) as the following tuple:

$$DP = (E, OP, REG) \tag{1}$$

The activity inside a data path during a register transfer RT is represented by a subset of expressions from E. The interconnect between these expressions are determined by the computations scheduled to be performed in the data path at the control step defined by RT.

Definition 1 *A register transfer RT associated with a data path is a tuple of the form:*

$$RT = (E_{RT},\ REG_{RT}^{out},\ f_{op}:\ OP \to (E \times E),\ f_{reg}:\ REG_{RT}^{out} \to E) \tag{2}$$

where $E_{RT} \subseteq E$ and $REG_{RT}^{out} \subseteq REG$ is the set of output registers in RT. f_{op} and f_{reg} define the interconnect between expressions of the data path at the control step corresponding to RT as follows: function f_{op} maps an operator to a pair of expressions (for each of the two source expressions of the operator) and function f_{reg} maps an output register to an expression (its input).

2.2   Well-Formed Register Transfer

In our model, we do not allow register transfers to contain combinational cycles, floating inputs for operators and registers or concurrent operations to be performed on the same hardware resource.

To formally define such requirements we first introduce the definition of an *ancestors* set *Anc* for an expression e (operator or register) as the set of all expressions which are connected via a *direct path* to e.

Definition 2 *The ancestors set Anc of an expression e of a data path $= (E, OP, REG)$, with respect to a mapping function $f_{op}: OP \to (E \times E)$, is defined recursively as:*

$$Anc(e) = \begin{cases} \emptyset & e : register \\ Anc(f_{op}(e)`1) \cup\ Anc(f_{op}(e)`2)\ \cup\ \{f_{op}(e)`1, f_{op}(e)`2\} & e : operator \end{cases} \tag{3}$$

where $f_{op}(e)`1$ and $f_{op}(e)`2$ represent the first and second projections of f_{op} respectively.

Definition 3 *A register transfer RT is said to be* **well-formed** *if:*

1. $\forall (e \in E_{RT}) : Anc(e) \subseteq E_{RT}$
   The ancestors set Anc for each operator in E_{RT} must also be present in E_{RT}.
2. $\forall (e \in E_{RT}) : e \notin Anc(e)$
   There are no combinational cycles in an RT.
3. $image(f_{reg}) \subseteq E_{RT}$
   The source of each output register in REG_{RT}^{out} must also be present in E_{RT}.
4. $\forall (e_1, e_2 \in OP) : (e_1 = e_2) \Rightarrow (\ (f_{op}(e_1)\text{'}1 = f_{op}(e_2)\text{'}1) \ \wedge \ (f_{op}(e_1)\text{'}2 = f_{op}(e_2)\text{'}2) \ )$
   Concurrent operations are performed on different hardware resources.

2.3   Register Transfer Split (RTS)

Transformation: **RTS** (RT,*split_set*)
Operation: *Split register transfer RT into two register transfers RT_1 and RT_2 by inserting new temporary registers between them. RT_2 contains the sub-image of RT induced by the operations present in* split_set *and RT_1 contains the rest.*
Precondition:

   i. Dependencies must be preserved as the operators are assigned to different register transfers.

Body:

Step 1. Copy the sub-image in *RT* induced by the operators in *split_set* and place it in a new register transfer RT_2. The remaining sub-image (induced by any remaining operators) in *RT* goes into another new register transfer RT_1.

Step 2. The output of RT_2 is connected to the inputs of the temporary registers. The outputs of the temporary registers are connected to the inputs of components of RT_1. *temp_set* represents the set of temporary registers.

No change is made to RT if split_set has only one operator. The application of the RTS is deemed correct if its precondition, called the *Well-Formedness* precondition, is satisfied.

Well-Formedness Precondition (WFP):

$$WFP(RT, split\_set) \stackrel{\Delta}{=} \forall (e \in split\_set) : Anc(e) \in split\_set \tag{4}$$

$$WFP(RT, split\_set) \Leftrightarrow$$
$$comp\_behavior(CG) = comp\_behavior(RTS(CG, RT, split\_set)) \tag{5}$$

Equation 4 states that the ancestor operator(s) $Anc(e)$ must also be in the *split_set*. This statement corresponds to the first condition for a well-formed register transfer. Equation 5 states that the computational behavior (*comp_behavior*) of the design is preserved if there is no WFP violation.

Thus, the application of RTS is sound if and only if the precondition is satisfied and further, is complete with respect to the scheduling task [6]. Hence the task performed by most scheduling algorithms (excluding those that move code across basic blocks) can be viewed as a sequence of RTS transformations.

3   Methodology

Our methodology is based on the precondition-based correctness of RTS and on the completeness of RTS transformations to perform the scheduling task. Figure 1 illustrates our approach. The primary components of our system are the

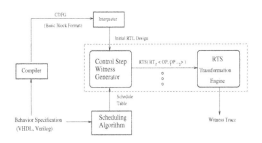

Fig. 1. Verification Methodology

Control Step Witness Generator (CSWG) and the *RTS transformation Engine* (shown in the dotted box in the above figure). The CSWG takes an initial RTL representation of the behavioral design and the *schedule table*. The schedule table is generated by the scheduling algorithm under test and contains a mapping of operations to control steps. For each control step in the schedule table, the CSWG generates an RTS transformation with the operators scheduled at that control step as one of its arguments. All applications of RTS (with its arguments) are written to a log file called the *Witness trace*. If any WFP violation occurs, it too is written to the witness trace along with the offending operators in question. If a violation does occur, the trace file shows the exact steps taken by the scheduler, hence can be used to understand why it scheduled operators that violated the WFP.

An *initial RTL* design of a behavioral specification is obtained by creating a register transfer for every basic block in the behavior. We assign a structural operator/register to each operation/carrier respectively in the behavior. The control flow remains the same. This initial RTL is successively modified by the RTS transformations based on the schedule table. Figure 2 shows an initial RTL being split based on the schedule table using RTS transformations.

3.1   Results

Our system can be used for debugging an existing (or new) implementation of a scheduling algorithm. If the operators are referenced via a special naming scheme in the schedule table, this can be provided as another input to our system (not shown in the Fig. 1).

We used an existing implementation of Force-Directed List Scheduler (FDLS) [5] from DSS [7]. A bug was seeded in at the dependency graph creation routine used by FDLS to get successor and predecessor node information. The following

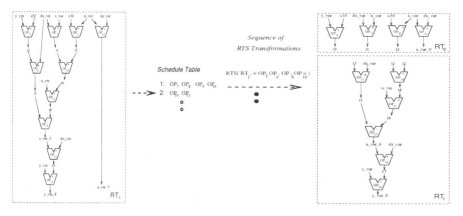

Fig. 2. Sample Flow

examples were run through the scheduler and the resulting schedule table was input to our verification framework. The run times for verifying each design are shown.

The Discrete Cosine Transform (DCT) is the largest example we tried. FFT2D is an 8-point Fast Fourier Transform (FFT) butterfly network composed of two 4-point FFTs (FFT1D). Linear System Solver (LSS) [8] solves a linear system of equations using matrix inversion. The Elliptic Wave Filter (Ellip) and Differential Equation solver (Diffeq) are taken from the High Level Synthesis Workshop Benchmarks'92 [9]. The examples were run on a Sun Enterprise 2. The bug in

Table 1. Information on the synthesis benchmarks used

Design Examples	Total Operations	Schedule Length	CPU Time
1. DCT8x8	1922	72	5.37
2. FFT1D	32	14	0.14
3. FFT2D	104	26	0.39
4. LSS	28	32	0.52
5. Ellip	26	18	0.2
6. Diffeq	10	4	0.08

the scheduler resulted in WFP violations in the Diffeq and Ellip examples. The witness trace file for the Diffeq example is shown below:

Witness Trace File for Diffeq Example:

```
RTS < Blk_3, 6 >
RTS < Blk_4, 7 8 9 11 12 15 16 > RTS WFP Violation: [ 10 13 14 ]
```

From Fig. 2, operation 10 which is the successor of operation 12 was omitted. Also operation 15 was scheduled before operation 14 which in turn required operation 13. There were no WFP violations in the other designs as they were legal schedules, that is, no dependencies were violated.

Though we considered only FDLS, any of the other scheduling algorithms whose output can be captured in the form of a schedule table, could have been used. Our approach cannot be directly applied to algorithms like percolation scheduling [10] which perform scheduling across basic block boundaries. However a schedule table can be extracted after the code motions are performed.

4   Conclusion

We presented an approach to verifying legal, basic block schedules produced by a scheduler in a high-level synthesis system. Based on the schedule table, a sequence of *RTS* transformations is used to perform the same task as the scheduler. The *Well-Formedness* precondition of RTS checks the correctness of the input to RTS. If a precondition is violated then the sequence of transformations applied so far forms a trace and can be used for debugging purposes.

References

[1] T. Lock, M. Mendler, and M. Mutz. Combined Formal Post- and Presynthesis Verification in High Level Synthesis. In *Formal Methods in Computer-Aided Design*, pages 222–236, 1998.

[2] P. Ashar, A. Raghunathan, A. Gupta, and S. Bhattacharya. Verification of scheduling in the presence of loops using uninterpreted symbolic simulation. In *International Conference on Computer Design (ICCD)*, 1999.

[3] H. Eveking, H. Hinrichsen, and G. Ritter. Automatic Verification of Scheduling Results in High Level Synthesis. In *Design, Automation and Test in Europe (DATE)*, 1999.

[4] N. Narasimhan, E. Teica, R. Radhakrishnan, S. Govindarajan, and R. Vemuri. Theorem Proving Guided Development of Formal Assertions in a Resource-Constrained Scheduler for High Level Synthesis. In Andreas Kuehlmann, editor, *International Conference on Computer Design (ICCD)*, Austin, TX, October 1998. IEEE Computer Society.

[5] R. Walker and S. Chaudhuri. Introduction to the scheduling problem. *IEEE Design and Test*, 12(2):60–69, 1995.

[6] E. Teica and R. Vemuri. A Mechanical Proof of Completeness for a Set of Register-level transformations. Technical Report 257/05/01/ECECS, University of Cincinnati, 2001.

[7] J. Roy, N. Kumar, R. Dutta, and R. Vemuri. DSS: A Distributed High Level Synthesis System. In *IEEE Design and Test of Computers*, volume 9, pages 18–32, 1992.

[8] M. Wolfe. *High Performance Compilers for Parallel Computing*. Addison-Wesley Publishers, 1996.

[9] N. Dutt. High Level Synthesis Workshop Benchmarks'92.

[10] G. De Micheli. *Synthesis and Optimization of Digital Circuits*. McGraw-Hill, 1994.

Parameterized Verification of the FLASH Cache Coherence Protocol by Compositional Model Checking

K.L. McMillan

Cadence Berkeley Labs

Abstract. We consider the formal verification of the cache coherence protocol of the Stanford FLASH multiprocessor for N processors. The proof uses the SMV proof assistant, a proof system based on symbolic model checking. The proof process is described step by step. The protocol model is derived from an earlier proof of the FLASH protocol, using the PVS system, allowing a direct comparison between the two methods.

1 Introduction

The verification of cache coherence protocols was perhaps the earliest commercial application of model checking [MS91]. Later, more efficient model checking methods were developed for this application [CD93], and compositional methods were applied to show that a verified protocol was implemented correctly in hardware [Eir98]. However, these techniques were unsound, in the sense that they could be applied only to fixed number N of nodes in the network, whereas in fact N had no useful upper bound. This left open the possibility that a protocol error was missed, which only manifested itself for N greater than the size verifiable by model checking.

Verification for arbitrary N can be accomplished in a number of ways. For example, Park [PD96] applied the general purpose theorem prover PVS [ORS92] to the verify the cache protocol of the Stanford FLASH multiprocessor [KOH$^+$94] for arbitrary N. This is a laborious process, since inductive invariants must be devised, and the theorem prover must be manually guided. It was also shown that in some cases, a protocol could be verified automatically for arbitrary N, using finite state methods with a "symbolic state" abstraction [PD93,ID96]. This abstraction cannot be applied to all protocols, however, and is too coarse to prove liveness. In fact, the first error detected in a cache protocol by model checking was a liveness error.

Thus, there is some interest in finding methods of proof for arbitrary N that are at the same time general, capable of proving liveness, and not unduly time consuming. Here, we consider applying methods of compositional model checking to the problem. This approach has the advantage that parameterized systems can be proved correct without the need to state inductive invariants, since invariant information is obtained by model checking abstract systems. We use the SMV proof assistant [McM99] to verify both safety and liveness of the

T. Margaria and T. Melham (Eds.): CHARME 2001, LNCS 2144, pp. 179–195, 2001.
© Springer-Verlag Berlin Heidelberg 2001

FLASH protocol. This protocol has the advantage that it has been approached previously using a variety of methods, and thus offers a good basis for comparison. In particular, we verify the same model used in Park's PVS proof (with one minor change). Relative to that proof, the proof we obtain is at least an order of magnitude more concise. We will begin in Sect. 2 with some background on the SMV proof assistant and compositional model checking. Then in Sect. 3, the protocol model is described. The next section covers the proof of correctness. Then in Sect. 5, a comparison is made with previous approaches.

2   Background

The SMV proof assistant [McM99] is designed to support the reduction of correctness conditions for unbounded or infinite-state systems to lemmas that can be verified by model checking checking finite state systems. The reduction to finite-state is done by methods of *abstract interpretation*. In effect, we throw away information about the model that is not needed to prove a given property. To make this effective, however, we must first break the desired property into localized properties that rely on only part of the system state. SMV provides two alternatives for this purpose: compositional proof and temporal case splitting.

Compositional Proof. The correctness of an implementation model is specified by a collection of temporal logic properties, usually with respect to a reference model. In our case, the reference model defines the programmer's model of the memory. Temporal specifications relative to a reference model are called *refinement relations*. In the SMV system, these can be specified in one of two ways. We might, for example, write a linear time temporal logic property, such as the following:

p : **assert G** ($cnd \Rightarrow impl\_data = ref\_data$);

This says that, at all times, if condition cnd holds, then $impl\_data$ (some data value in the implementation) must be equal to ref_data (some corresponding value in the reference model). Typically, ref_data will refer to some data computed by the reference model in the past. That is, there is no need for the implementation and the reference model to operate in "lockstep". On the other hand, we might express the same property as a "layer":

layer p: **if**(cnd) $impl\_data := ref\_data$;

This has the same semantics as the property above, but it tells SMV that we wish to use p as an abstraction of variable $impl\_data$ when proving other properties.

The refinement relations are proved by mutual course-of-values induction over time [McM99]. Each refinement relation must be a temporal property of the form $G\phi$, meaning that ϕ is true for all times t. To prove that ϕ is true at time t, we can assume by induction that the other refinement relations hold for all times less than t. This is useful in a methodology based on model checking, because the notion that ψ up to time $t-1$ implies ϕ at time t can be expressed in temporal logic as $\neg(\psi \; \mathcal{U} \; \neg\phi)$. Hence, this proposition can be checked by a model checker. To tell SMV to use property p up to $t-1$ to prove property q at time t, we use the following notation:

using (p) **prove** q;

The parentheses indicate that p is to be used up to time $t-1$ and not t. If p is expressed as a layer, we don't need to use this directive. SMV will determine whether *impl_data* is relevant to property q, using a dependency analysis. If so it will use property p in place of the implementation definition of *impl_data*. In this way, the proof of q can be localized to a small part of the implementation.

This mutually inductive approach is important because it allows us to assume, for example, when proving correctness of data in one cache, that the data in other caches have been correct in the past. Note this technique is quite different from the method of proof by invariant, in which we show that some state property at time $t-1$ implies itself at t. In our case, the properties are temporal, and the inductive hypotheses are assumed for all times less than t, and not just at $t-1$. This is important, since it allows us to avoid writing inductive invariants.

Temporal Case Splitting. An alternative approach to localization is to specialize the properties we wish to prove, so that they depend on only a finite part of the overall state. For example, suppose that we have a state variable v, which is read and written by a collection of processes $1 \ldots n$. We wish to prove a property p of v, of the form $G\phi$. We add an auxiliary state variable w to the model which records a pointer to the most recent process to write the variable v (this is supported by a definitional mechanism in the proof assistant). Now, suppose we can prove for *all* process indices i that $G((w = i) \Rightarrow \phi)$, that is, ϕ holds at all those times when the most recent writer is process i. We can then infer that $G\phi$ holds, since at all times w must have some value. We call this "splitting cases" on the given variable, since it generates a parameterized property with one instance for each value of the given variable. We can tell SMV to do this as follows:

forall(i **in** *PROCS*)
   subcase $p[i]$ **of** p **for** $w=i$;

where *PROCS* is the type of process indices. This generates, for every i, an instance $p[i]$ of p, of the form $G((w = i) \Rightarrow \phi)$. The importance of this approach is that for any given parameter value i, we may be able to verify the property by abstracting away all processes except process i, since the particular case $w = i$ depends directly only on process p_i.

Abstract Interpretation. SMV uses abstract interpretation to reduce the verification of each parameterized property to a tractable model checking problem. The difficulty is that there may be variables in the model with large or unbounded ranges (such as memory addresses) and arrays with a large or unbounded number of elements (such as memory arrays), and the number of instances of the property may be unbounded. We deal with this problem by using abstract interpretation to reduce each data type to a small number of abstract values. For example, suppose that we have a property with a parameter i that ranges over memory addresses. We can reduce the type T of memory addresses to a set containing two values: the parameter value i, and an abstract symbol $T \setminus i$ that represents all values other than i. In the abstract interpretation, accessing a memory array m at location i will produce the value $m[i]$ of that location, whereas accessing the array at $T \setminus i$ produces $\perp$, a symbol representing no information about the value. The net effect of this abstraction is that, for each time we create a parameter by "splitting cases" on a variable of type T, there is one value in the abstract type and one element in each abstracted array indexed by that type. Thus, case splits come at the cost of increased verification complexity.

Normally, a suitable abstract interpretation for each data type can be chosen automatically. However, it can also be specified manually, as in the following:

using $T \Rightarrow \{i,j\}$ **prove** q;

This tells SMV to distinguish the values i and j of type T, but to represent all other values by an abstract symbol. If there are two parameters i and j of type T, then the proof assistant may, for example, split the problem into two cases: one where $i = j$ and one where $i \neq j$. Alternatively, it may consider separately the cases $i < j$, $i = j$ and $i > j$, if information about the order of these values is important to the property. This behavior depends on the declaration of the data type: a *scalarset* type considers only equality, while an *ordset* type considers the relative order of parameter values as well.

The abstractions used by the proof assistant are sound, in the sense that validity of a formula in the abstract interpretation implies validity in the concrete model for *all valuations* of the parameters. Thus, we need not check all parameter valuations separately. Of course, it is possible that the abstraction used may be too coarse to verify the given property (*i.e.*, the truth value of the property in the abstract model may be $\perp$) even though the property is true. Note, however that the user does not need to verify the correctness of the abstractions used, since these are drawn from a fixed set that is built into the proof assistant.

The proof process thus proceeds in the following steps. First, the user specifies refinement relations (and other lemmas, as necessary), which are proved by mutual temporal induction. These properties are parameterized by "splitting cases" on appropriate variables, so that any particular case depends on only a finite part of the system state. Finally, the proof assistant abstracts the model relative to the parameter values, reducing the types with large or unbounded ranges to small finite sets. The resulting proof obligations are discharged by a model checker. In the sequel we will consider how this general methodology can be applied to Park's model of the FLASH cache coherence protocol.

3   The Protocol Model

A FLASH system consists of a collection of N processors, each with a local cache memory. We will refer to a processor/cache pair as a *node*. The nodes act asynchronously, exchanging messages via a network with arbitrary latency. The protocol ensures that multiple cached copies of a given address are consistent, in an appropriate sense. Each memory address has a *home* node. The home node contains the master copy of the address in main memory, and also a directory entry, indicating the set of caches that may hold the given address, and whether the cached data may be modified (*dirty*). In the actual system, this set is maintained in a list structure. In Park's model, there is a variable containing the head of the list, and a set representing the remainder of the list. The model has only one memory address. Generalization to an arbitrary number of memory addresses is trivial, however.

A cache may contain the given address in one of three states: *invalid* (not present), *shared* (readable) and *exclusive* (readable and writable). As the name implies, there can be only one cache in the exclusive state at any time.

The protocol works roughly as follows. A node requiring read access to a given address sends a *Get* message to the home node. If there are no exclusive copies in the system (the directory entry is not marked *dirty*), then the home fetches the value from main memory, and returns it in a *Put* message. The index of the requesting node is added to the sharing list in the directory, and the requesting cache enters the shared state. On the other hand, if the address is dirty, there must be an exclusive copy at some node (the *owner*) and the data in main memory are obsolete. The home thus *forwards* the *Get* request to the owner. The owner is then expected to send the data to the original requester and change to the shared state. Modified data are also returned to the home node via a sharing write-back (*ShWB*) message. This leaves the address clean, with two nodes on the sharing list. While this transaction is in progress, the directory remains in the *pending* state, and will not accept new requests.

When a node requires write access, it sends a *GetX* message to the home, to obtain an exclusive copy. If the address is not dirty, the home sends invalidation messages (*Inv*) to each node on the sharing list. It immediately returns the data in a *PutX* message without waiting for the invalidations to complete (that is, we model the "eager" mode [KOH+94]). The directory then remains in the pending state until acknowledgments (*InvAck*) are obtained for all the invalidations sent. On the other hand, if the address is dirty, the *GetX* is forwarded to the owner, which is expected to send a *PutX* to the requester, enter the invalid state, and send an acknowledgment message *FAck* to the home. Again, the directory stays in the pending state until the *FAck* is received.

The owner may eject the address from its cache at any time, by sending modified data to the home in a write-back message (*WB*). This introduces a race condition, since a forwarded *Get* or *GetX* may be on its way to the owner. If this happens, the former owner will return a negative acknowledgment to the home, aborting the transaction.

Park's model encodes the network state in an unusual way. That is, instead of a uniform collection of message buffers, it has specialized sets of message buffers for each class of messages. For example, *unet* holds the *Put* and *Get* messages (with one entry per requester), while *invnet* holds the *Inv* and *InvAck* messages (one per node) and *wbnet* holds the write-back messages (one global entry). In particular, *unet*[i] holds *Get* messages for requester i, even if the destination of the message is some other node, due to forwarding. This encoding is in some ways convenient for model checking, although we will observe one case in which it is inconvenient.

This model was translated rather directly from PVS into the SMV notation. One notable difference is that the original uses a counter *real* to hold the number of pending invalidations. Since SMV has no reasonable way to deal with the cardinality of sets, this is replaced by the set of processor indices that have pending invalidations. As stated earlier, we would like to prove both coherence and liveness of this model.

4   Proof of Coherence

We begin with the specification. Ideally, a cache coherence protocol would satisfy the condition of *sequential consistency* [Lam79]. That is, there should always exists a global sequence of reads and writes that agrees with the local observations of each processor, and in which every read gets the value of the most recent write to the given address. Our version of the FLASH protocol does not satisfy this condition. However, it does satisfy the weaker condition that reads and writes to a given cache line are sequentially consistent. We will refer to this condition as *coherence*.

In fact, we will prove a considerably stronger condition than coherence. That is, a read at a given node gets the value of the most recent write at the last time the address was *observable* to that node. We require that all values written by a given node be observable to that node. However, other than this, the definition of *observable* is up to the implementer. It is easy to show that any protocol implementing this model is coherent. We do not do that here, however, since this is a generic theorem, not related to the FLASH protocol *per se*. The intuition behind this definition is that if a node is holding a shared copy that is pending invalidation, then the address is not observable. Instead, the node sees the value of the address before the invalidating transaction began.

To express our specification, we add an auxiliary state variable m to the model, which holds the most recent value written to the address:

if(*store*) **next**(m) := *store_data*;

Here, *store* is an implementation variable indicating that a store is occurring, and *store_data* is the data value being stored. This definition says that when *store* is true, m is modified to hold the value of *store_data*. SMV allows such definitions of new variables to be added, so long as the implementation does not depend on them. We now specify correctness of data in the caches relative to this variable. It is most convenient to write the specification as a *layer* in the SMV system, which will cause it to be used by default in the proof of any property involving the cache data. We could, however, write the property as an ordinary temporal assertion. Here is the specification as a layer:

layer *L1*: **forall**(i **in** *Proc*)

if(*readable*[*i*]) *cache*[*i*].*data* := *m* **whenlast** (*obs*[*i*] ∨ *last_writer* = *i*);

That is, the cache sees the value of *m* at the last time it was observable. Note that "*x* whenlast *c*" is a shorthand for the value of *x* at the most recent time that *c* was true (and is undefined if *c* has never been true). The "forall" prefix creates one instance of the property for each node. The property requires that the address must be observable if the most recent writer was *i* (as indicated by the auxiliary variable *last_writer*). Otherwise, observability is determined by the predicate *obs*[*i*]. Note that *obs* is not defined in the specification. We need only show that the specification holds for *some* definition of *obs*. To prove *L1*, we use the following *witness function* for *obs*:

$$obs[i] := \neg(dir.pending \wedge collecting) \vee cache[i].state = exclusive;$$

That is, shared copies are "out of date" when there is a transaction pending in the directory that is collecting invalidation acknowledgments. Here, *collecting* is an auxiliary variable that records whether the current pending transaction sent invalidations. It is needed because the directory does not directly record this fact. Figure 1 shows the structure of the SMV input file at this point.[1]

The first thing we do is to use the model checker to check that the property is true for $N = 3$. To do this, we use the following declaration of the type *Proc*:

typedef *Proc* 0..2;

This check is done to give us confidence that the property is actually true before attempting to prove it for the general case. In fact, several errors in the model were found in this way, resulting from incorrect translation from the PVS description. As a "sanity check", the $N = 3$ case was also checked for absence of deadlock using the model checker. Having some confidence in the correctness of the model, we now attempt to verify the property *L1* for arbitrary N. To do this, we change the declaration of type *Proc* to the following:

scalarset *Proc* **undefined**;

This tells SMV that type *Proc* is used in a symmetric way (which governs the abstraction used for data of type *Proc*) and that the range of the type is unknown.

Our general approach to the proof is to attempt to check the property using an abstraction of the data types. When a counterexample is produced, the likely source is a message produced by some processor that has been "abstracted out" of the model. There are two tactics we can apply at this point. Either we "split cases" on the producer of this message (which will, in effect, add this processor to the abstraction), or we add a lemma that rules out the interfering message. The first approach is simpler for the user, but adds to the complexity of the model checking.

When we first attempt to verify property *L1*, SMV uses an abstraction of type *Proc* that contains one fixed value *i*, the index of the cache we are verifying.

---

[1] The complete file can be obtained at
http://www-cad.eecs.berkeley.edu/~kenmcmil

scalarset *Proc* **undefined**;

/* ... Other type declarations go here ... */

module *main*(){

    /* The reference model */
        abstract *m* : *Data*;
        if(*store*) **next**(*m*) := *store_data*;

    /* ... The implementation model goes here ...*/

    /* coherence specification */
        abstract *obs* : **array** *Proc* **of boolean**;
        abstract *last_writer* : *Proc*;
        if(*store*) **next**(*last_writer*) := *dst*;

        layer *L1*: **forall**(*i* **in** *Proc*)
            if(*readable*[*i*]) *cache*[*i*].*data* := *m* **whenlast** (*obs*[*i*] ∨ *last_writer* = *i*);

    /* witness functions */
        obs[*i*] := ¬(*dir.pending* ∧ *collecting*) ∨ *cache*[*i*].*state* = *exclusive*;
        init(*m*) := *mem*;   /* make sure reference model has correct initial value */

    /* ... The proof goes here ... */
}

Fig. 1. Structure of SMV input file

An abstract symbol is used to represent all of the other processor indices. This abstraction is far too coarse to verify the property, as we can see from the counterexample we obtain. This shows a case in which *i* is the home node, and obtains a readable copy of the data. However, one of the processors that is abstracted away executes a store operation at this point. This is not a possible behavior of the system since only caches in the exclusive state can execute a store. However, the abstraction cannot rule it out, since the state of the abstracted processors is unknown in the abstraction. Here, there are two possible approaches. We can split cases on the most recent processor to execute a store. Or, we might write a lemma saying that only the owner of the cache line (the most recent to receive an exclusive copy) can write. Either will rule out our counterexample.

    Since it is a bit simpler, we try the case splitting approach first. We add the following declaration:

    forall(*j* **in** *Proc*)
        subcase *L1*[*j*] **of** *cache*[*i*].*data*//*L1* **for** *last_writer* = *j*;

where *last_writer* is the auxiliary variable pointing to the most recent node to execute a *Store*. Since the parameter *j* becomes a distinguished value in the abstraction, this node is no longer abstracted out. Thus, we rule out the false counterexample in which an abstracted node corrupts the reference model data by executing a *Store*.

Because *Proc* is a scalarset, we now have two cases to consider: $i = j$ and $i \neq j$. In the latter case, we get a counterexample in which an abstracted processor sends a write-back, thus corrupting the state of main memory. This incorrect value is then loaded into the cache of processor i, violating our property. To fix this problem, we could split cases on the most recent processor to execute a write-back. However, we do not wish to include too many processors in the abstraction. Instead, let us split the problem into two parts by adding a lemma, stating that data in main memory are always correct when there is no exclusive copy in the system. This will rule out the false counterexample, although it increases our proof effort, since we have to prove this property separately. The lemma is as follows:

layer $L2$: **if**$(\neg dir.dirty) mem := m;$

That is, when the directory is not in the dirty state, main memory must have the value of the most recent write. Because it is a layer, SMV will use this specification of *mem* by default when proving *L1* (and *vice versa*), using mutual temporal induction. With this added lemma, we get a counterexample in which the home node has the line in its cache. Node i requests a copy of the cache line, which is then forwarded directly from the home's cache (note the cache in the home node is not the same as the main memory *mem*). Unfortunately, the home node is abstracted in this counterexample. We can tell this because the value of *home* in the counterexample is the abstract symbol representing "other than i or j". At this point, we could add a lemma stating that the home node always sends correct data. This strategy seems unlikely to succeed in the long run, however, since a large part of the control of the protocol depends on whether a given node is the home node or not. Thus, we can expect many other false counterexamples to arise if the home node index is not present in the abstraction. Therefore, let us split cases on the value of *home*:

forall$(k$ **in** $Proc)$
    subcase $L1[j][k]$ **of** $cache[i].data//L1[j]$ **for** $home = k;$

We now have an abstract model with three processors represented: the node i, whose cache we are verifying, the most recent writer j, and the home k. With this model, we get a counterexample in which data is forwarded from an abstracted processor to processor i. Thus, processor i receives incorrect data. Note that in principle, the forwarded value should be correct, since we are allowed to assume that *L1* holds for all nodes up to time $t - 1$. However, by default *L1* is only instantiated for the process indices that are in the abstract model, that is, i, j and k. We need this property to hold for "the processor currently forwarding data", which in the model is indicated by the implementation variable *dst*. This is the index of the processor that is current receiving a request to forward data. Since SMV does not know enough to instantiate property *L1* for $i = dst$, we do it manually, adding the following lemma:

$L1a$: **assert G** $(cache[dst].state = exclusive \Rightarrow cache[dst].data = m);$

Note, we have weakened $L1a$ a bit, omitting the case when the state is *shared*, since we don't need it. This lemma can be trivially proved as a special case of $L1$, so we omit the proof here. We now use this instantiation of $L1$ up to time $t-1$, with the following declaration:

using $(L1a)$ **prove** $cache[i].data//L1[j][k];$

This ensures that good data are always forwarded from other processors. However, it is still possible that a *Store* will occur while the data are in transit. This happens in the next counterexample. Processor j gets an exclusive copy of the data. However, an abstracted processor sends a write-back message to the home, causing it to think that the line is no longer owned by j. As a result, it sends a shared copy to i. Now processor j, which still thinks it has an exclusive copy, writes the line, causing k's data to be incorrect. The problem here is that it is not possible for another processor to have an exclusive copy while j is holding one (hence the write-back is impossible). However, this information is lost by the abstraction. To solve this problem, we might try case splitting on the sender of the write-back, but at this point the abstract model is already nearly too large to handle. Instead, we write a lemma stating when it is possible for a processor to be in the exclusive state. That is, it is possible when the directory thinks that the line is "dirty", and when no other processor holds an exclusive copy. Note that an exclusive copy can be held in a cache, or it can exist in a *PutX* message *en route* to a node. Thus, we write the following lemma:

$L3[i]$: **assert G** $(dst\_dirty \Rightarrow dir.dirty \wedge$
$((cache[i].state = exclusive \vee unet[i].mtype = PutX) \Rightarrow dst = i));$

That is, if the current "destination" node is in the exclusive state (dst_dirty), then the directory must be in the dirty state, and no other processor i may be in the exclusive state, or have a *PutX* message in transit to it. Note, that rather than writing this property for all pairs of nodes, we have written it only relative to the current dst, since we only need to rule out bad write-back messages. If we wrote the lemma for all pairs (i, j) we would then have to instantiate it manually for $dst = j$, as above. To prove this lemma, we clearly must split cases on dst, so that the state of node dst will be present in the abstraction. In addition, we will split cases on *home*, for the reasons mentioned above. As usual with such "non-interference" lemmas, we will assume that the general lemma holds up to time $t-1$ to prove a specific case at time t. Here is the SMV declaration we use:

forall$(j$ **in** $Proc)$ **forall**$(k$ **in** $Proc)\{$
    subcase $L3[i][j][k]$ **of** $L3[i]$**for** $home = j \wedge dst = k;$
    using $(L3)$ **prove** $L3[i][j][k];$
$\}$

From the ensuing counterexamples we discover that in fact there are three other types of messages that represent an "exclusive" copy: a write-back, returning ownership to the directory, a "sharing write-back" message, that is performing the same function but leaving shared copies in the system, and a *Put* message from the owner to home, which has the same effect as a write-back, but is elicited when the home cache needs a copy of the cache line. Thus, we add the following requirements to *L3*, in case *dst* is in the exclusive state:

$$wbnet.mtype{=}Empty \land shwbnet.mtype{\neq}ShWB \land unet[home].mtype \neq Put$$

At this point, in trying to verify *L3*, we run into a problem: the verification does not complete within our limit of 2GB of memory. We now have two choices. We could further decompose the proof into cases or lemmas, or we could help the model checker by adjusting the abstraction manually. Since the verification seems close to finishing, we choose the latter course. Adjustments to the abstraction can be made by "freeing" some variables that we think are irrelevant (*i.e.*, we ignore their definitions and leave them unconstrained). The topic of choosing an appropriate abstraction for model checking is beyond the scope of this article. Suffice it to say for our purposes that making a few educated guesses allows us to reduce the memory consumption of the model checker to about 270 MB, which is well within the capacity of our server, and that *L3* is verified.

Now, having proved our "non-interference lemma", let us return to the proof of *L1* (data correctness in the cache), adding *L3* to the list of properties used to prove *L1*. Now we get a counterexample showing another kind of "interference". That is, node *i* gets a shared copy, after which node *j* requests and obtains an exclusive copy. This causes an invalidation message to be sent to node *i*, and the directory state to be set to "pending", *i.e.*, awaiting invalidation acknowledgments. A new transaction cannot be started until all invalidations are complete. Oddly, then, node *j* (the owner) now receives a forwarded *Get* request for some abstracted node. How does this happen? It is an artifact of the way the network is coded in the model. Even though the forwarded *Get* message for a node *n* can only be sent from the home, it is stored in the message buffer *unet[n]*. This encoding, chosen by Park for his PVS proof [PD96], helps us in some ways, but hurts us here, since *unet[n]* in this case is abstracted. Thus, the abstract model allows this incorrect message to arrive at the owner node while invalidations are pending. The owner sends a *FAck* message back to the home, acknowledging this bogus request. The home, receiving the bogus *FAck*, clears its pending bit. Now node *i* is holding stale data, but the pending bit is not set, violating *L1*.

At this point, we might choose to recode the model of the network in such way that a special buffer is reserved for forwarded *Get* messages. One such buffer would suffice, since only one such message can be in the system at a time. This buffer would always be present in the abstraction, so *Get* messages forwarded from abstracted nodes would not trouble us. However, for comparison purposes, we wish to use the same model that was used in the PVS proof. In any event, we can get the same effect by introducing the proposed message buffer as an auxiliary variable. We then prove that forwarded *Get* messages can only occur when they are found in this imaginary message buffer. This technique of replacing an inconvenient encoding of state information by a more convenient encoding is useful in many contexts. Here is the property we write:

$$L4[i]: \textbf{assert G } (unet[i].mtype \textbf{ in } \{Get, GetX\} \wedge unet[i].proc \neq home$$
$$\Rightarrow (unet[i].mtype = fwd\_get));$$

Here, *fwd_get* is our imaginary message buffer for forward *Get* messages. We say that if there is such a message in any message buffer i, and if the destination is not home (*i.e.*, it is being forwarded to some other node), then it must match our imaginary buffer.

Now we attempt to prove this. Initially, we try splitting cases on just the home node. This produces a counterexample in which a *GetX* is forwarded for node i and entered in *fwd_get*. However, this imaginary message doesn't specify that i was the original sender of the *GetX*. Some abstracted node's message buffer issues the *GetX*, clearing *fwd_get*. Now node i's message buffer violates the property, since it has a forwarded *GetX*, but *fwd_get* is empty. We decide to split cases on the message buffer that issued the bogus *GetX*, so that it will not be abstract in the model. This seems reasonable, since presently there are only two parameters in the property:

subcase $L4[i][j][k]$ **of** $L4[i]$ **for** $fwd\_src = j \wedge home = k$;

Here, *fwd_src* is an auxiliary variable that remembers which message buffer most recently issued a forwarded *Get*. This should solve the problem, since both message buffers i and j cannot hold a forwarded *Get* at the same time. Unfortunately, we now get a counterexample in which two such forwarded messages *do* exist at the same time. In this counterexample, one node requests an exclusive copy, and its *GetX* is forwarded to the owner (some abstracted node). In principle, a new transaction cannot be started until the directory knows that this one has terminated. However, at this point the directory receives in invalidation acknowledgement message (*InvAck*) from some abstracted node. Such messages should never arrive during an ownership transfer (they only occur when changing from shared to exclusive). The directory doesn't know this, however, and responds to the bogus *InvAck* by terminating the transaction. This allows a new transaction to begin, which results in two forwarded *GetX* messages being in the system at the same time. At his point, to avoid another case split, we introduce another non-interference lemma. This one says that *InvAck* messages cannot arrive during an ownership transfer:

$$L5: \textbf{assert G } (src\_invack \Rightarrow dir.pending \wedge collecting);$$

Here *src_invack* says that an *InvAck* is arriving (recall that *collecting* means that invalidations are currently being collected). We assume this lemma up to time $t - 1$, to prove $L4$ at time t. With this assumption, we obtain one more counterexample, in which an abstracted message buffer violates lemma $L4$. This leads to a later violation of $L4$ by message buffer i. This should be ruled out, because we can assume the general lemma up to $t - 1$ when proving case i. However, to do this, we are forced to instantiate $L4$ manually for $src = i$, just as we did above for $L1$. With this addition lemma $L4$ is verified.

This leaves lemma $L5$ to prove, that is, that no unexpected *InvAck* messages arrive. To check this, as usual, we split cases on the sender of the bogus messages, and also on the home location, as follows:

subcase $L5[i][k]$ **of** $L5$ **for** $src = i \wedge home = k$;

This is sufficient to prove the lemma. Note that when proving the non-interference lemmas, we always use all the lemmas up to time $t - 1$. If nothing else, this practice cuts down on the size of the reachable state space (since the lemmas constrain the possible transitions) and may prevent us from looking at bogus counterexamples. Also, we use the same abstraction adjustments for all of the lemmas, except that for $L5$, we cannot abstract out the part of the network model that handles invalidations.

Now we return to property $L1$, data correctness in the caches. Recall that a counterexample to this property caused us to add $L4$, and in turn $L5$. Now we assume these lemmas to prove $L1$. This time the verification succeeds. All that remains is property $L2$, which specifies correctness of the contents of main memory. We use exactly the same strategy as for $L1$, that is, we split cases on the most recent node to execute a *Store*, and on the index of the home node:

subcase $L2[j][k]$ **of** $mem//L2$ **for** $last\_writer{=}j \wedge home = k$;

Using our non-interference lemmas, this property is also verified (layer $L1$ is automatically assumed as the definition of the cache contents).

At this point, we have verified that contents in the caches satisfy our specification. This proof was completed in two days. In fact, an earlier proof of the same model required less time, about 12 hours. However in the present proof, a greater burden was placed on the model checker in order to make the proof simpler for presentation. Thus, most of the two days was spent waiting for the model checker to terminate. The proof of liveness is omitted here for space reasons. It is actually substantially simpler than the coherence proof and required four hours to accomplish.

5   Related Work

As stated earlier, the model of the FLASH protocol that is verified here is adapted from the work of Park [PD96], who verified using the PVS theorem prover that it implements an abstract model. In this model, the cache states are updated globally in a single atomic action. This proof was done using an abstraction function that computes the result on the system state of executing all the "committed" transactions that are currently in the system. This has the effect of emptying the network of messages, producing an abstract state. It is then proved that each action of the implementation model implements some (possibly null) atomic action of the abstract model, modulo the abstraction function. This proof also requires an inductive invariant of the system to be stated and proved, since the abstraction function does not hold for unreachable states.

In fact, such invariants can be quite complex and time-consuming to produce. The lemmas in Park's refinement proof in PVS requires 776 lines (21KB) to state, perhaps half of which is taken up by invariants. This does not include the proof script required to prove the lemmas, which can be quite large. Nor does this include a proof that the atomic model implements some memory model.[2] Park does not report the time required to complete the proof, though we can infer from the 111 lemmas and theorems that it must have been substantial. The safety proof presented here takes 30 lines (1.7KB) in SMV (including the four auxiliary lemmas, case splits, instantiations and abstraction adjustments) plus two auxiliary variable definitions scattered in the model code, and required two days to complete. We should note there is one simplification in the model – a counter is replaced by a set. Since liveness was not proved in Park's work, a comparison of liveness proofs is not possible.

The relative conciseness of the SMV proof appears to stem from two factors. First, when using model checking of abstract models, it is not necessary to have an inductive invariant. Rather, the model checker generates the strongest invariant of the abstract model. Strengthening is required when the abstract model is too coarse (hence, the non-interference lemmas L3, L4 and L5). Note, however, that these lemmas are proved by course-of-values induction over time, rather than by simple induction. The net effect is that the non-interference lemmas are considerably simpler than a global inductive invariant. The second possible factor in the simplicity of the proof is that no intermediate model is used, and hence no abstraction function is required.

Another approach to generating an inductive invariant is to use a technique called *predicate abstraction* [SG97]. In this method, the user provides a collection of predicates on the system state, and a program generates the strongest invariant of the system that can be expressed as a boolean combination of these predicates. This method has been used by Das, Dill and Park [DDP99] to generate some invariants of the FLASH protocol (for example, that there are never two exclusive copies in the system). Invariants generated in this way were subsequently transferred to Park's proof of refinement in PVS.

It appears possible that invariants generated by predicate abstraction could be used to aid the proof presented here. For example, lemma *L3*, stating that an exclusive copy cannot coexist with another exclusive copy, write-back, *etc.*, appears to be the kind of invariant that was generated in [DDP99] (although few details are given). On the other hand, it appears that the manual effort required to arrive at the necessary predicates (a reported five days) was greater than the manual effort of the entire proof presented here, so it is not clear that using predicate abstraction to obtain the non-interference conditions would be worth the effort. On the other hand, the method of [DDP99] can in some cases guess necessary instantiations of universal quantifiers. There were two cases in the above proof in which such instantiations were introduced manually. Possibly by applying the methods of [DDP99], these instantiations could have been found automatically. That is an interesting possibility, since the abstractions generated by SMV can in fact be cast in terms of predicate abstraction. This points to a possible integration of the two methods.

---

[2] For the eager model, a lemma is proved for each low level action, stating that it commutes with some high level action, but two models are not defined *per se*, and coherence is not specified.

Note that the predicate abstraction method by itself cannot prove liveness of the protocol. In fact, no single finite abstraction can do this, because a finite system, if it terminates, must terminate in a bounded number of steps. However, there is no fixed bound on the number of steps required for a transaction to terminate in the FLASH protocol. Rather, this number increases with N.

Verification of cache coherence protocols with an arbitrary number of processors has also been done using so-called *symbolic state* abstractions [PD93]. This is a finite-state approach to systems of many identical processes, in which information about the exact number of processes in each state is abstracted, so that only the cases 0, 1 and "more than one" are distinguished. Pong used this method to verify cache coherence protocols. However, the method required that the protocol be described in a specialized abstract form. This left open the problem of verifying that this model in fact is refined by the more operational description that one actually implements. This problem was solved by Ip [ID96], who showed how such a model could be extracted from a description in the Murϕ language [ID96], under certain restrictions. Not all protocols can be handled in this way. In particular, protocols that pass pointers to nodes can only be handled if pointer chains are bounded. However, in the cases when it does apply, the method is certainly more automated than the one presented here (it is unclear whether it could be applied effectively to FLASH). Note, however, that the abstraction it uses is too coarse to prove liveness, so other methods would have to be used for this purpose.

Finally, an advantage of using a system such as SMV for the verification of cache protocols is that SMV is also capable of verifying that the actual hardware implementation of the protocol conforms to the verified model. This kind of refinement verification was done, for example, in [Eir98], although in that case the protocol was only verified for a fixed number of nodes. A general purpose prover such as PVS is not well suited to this task. On the other hand, the Murϕ system provides no means to verify that a hardware system refines a Murϕ model (this, of course, suggests a possible link between the tools).

6   Conclusion

We have seen an example of formal verification of a cache coherence protocol model, for N processors, using a compositional model checking tool. The protocol was specified with respect to a very simple reference model. The proof that the protocol model implements this reference model was accomplished with four auxiliary lemmas. The first is a "refinement relation", specifying the contents of main memory in terms of the reference model. The other three are "non-interference" lemmas. These were required to rule out false counterexamples that occurred because some nodes in the system were abstracted away.

Our general approach was driven by counterexamples. That is, we attempt naïvely to prove a property, and when a counterexample arises, we diagnose the cause of the error. We then attempt to rule out this cause by one of two strategies. If the cause is reception of an impossible message from an abstracted node, we can "split cases" on the sender of this message, causing the sender to be included in the abstraction. This case splitting can only be done a few times, however, since it leads to state space explosion. The other possibility is to write a "non-interference" lemma, ruling out the bad message. This lemma must be proved separately. It is often useful to pose these lemmas as refinement relations, by which we recode the model state in a more convenient way. The lemmas are then proved by course-of-values induction over time. That is, we can assume all the lemmas up to $t - 1$ when proving a case of any given lemma at time t.

This approach appears to be the most effective one to date for verifying both safety and liveness of cache coherence protocols with an arbitrary number of processes. For example, the proof effort for the FLASH protocol using model checking was clearly much less than the effort for the same model using PVS. The method of predicate abstraction requires considerable user assistance to guess the necessary predicates and cannot prove liveness of the FLASH protocol. Symbolic state methods give a fully automated verification of safety. However, they do not apply to all protocols, and cannot prove liveness. Finally, having proved properties of a protocol model in SMV, it is then possible to refine that model to a formally verified implementation in hardware.

References

[CD93]   C.N. Ip and D.L. Dill. Better verification through symmetry. In D. Agnew, L. Claesen, and R. Camposano, editors, *Computer Hardware Description Languages and their Applications*, pages 87–100. Elsevier, 1993.

[DDP99]   S. Das, D. L. Dill, and S. Park. Experience with predicate abstraction. In *Computer Aided Verification (CAV'99)*, pages 160–171, 1999.

[Eir98]   A. Eiriksson. Formal design of 1M-gate ASICs. In *FMCAD '98*, number 1522 in LNCS, pages 49–63. Springer, 1998.

[ID96]   C.N. Ip and D.L. Dill. Verifying systems with replicated components in murphi. In R. Alur and T. A. Henzinger, editors, *Computer Aided Verification (CAV'96)*, volume 1102, pages 147–158. Springer Verlag, 1996.

[KOH+94]   J. Kuskin, D. Ofelt, M. Heinrich, J. Heinlein, R. Simoni, K. Gharachorloo, J. Chapin, D. Nakahira, J. Baxter, M. Horowitz, A. Gupta, M. Rosenblum, and J. L. Hennessy. The stanford FLASH multiprocessor. In *Proc. of the 21th Annual Int'l Symp. on Comp. Arch. (ISCA'94)*, pages 302–313, 1994.

[Lam79]   Leslie Lamport. How to make a multiprocessor computer that correctly executes multiprocess programs. *IEEE Transactions on Computers*, C-28(9):690–691, 1979.

[McM99]   K. L. McMillan. Verification of an infinite state systems by compositional model checking. In L. Pierre and T. Kropf, editors, *Correct Hardware Design and Verification Methods (CHARME'99)*, volume 1703 of *LNCS*, pages 219–233, 1999.

[MS91]   K.L. McMillan and J. Schwalbe. Formal verification of the gigamax cache consistency protocol. In N. Suzuki, editor, *Proceedings of the International Symposium on Shared Memory Multiprocessors*. MIT Press, 1991.

[ORS92]   S. Owre, J. Rushby, and N. Shankar. PVS: A prototype verification system. In E. Kapur, editor, *Conf. on Automated Deduction (CADE'92)*, number 607 in LNCS. Springer-Verlag, 1992.

[PD93]   F. Pong and M. Dubois. The verification of cache coherence protocols. In *Proc. of the 5th ACM Annual Symp. on Parallel Algorithms and Architectures (SPAA'93)*, pages 11–20, 1993.

[PD96]   S. Park and D.L. Dill. Verification of the FLASH Cache Coherence Protocol by Aggregation of Distributed Transactions. In *8th ACM Symposium on Parallel Algorithms and Architectures*, pages 288–296, Padua, Italy, 1996.

[SG97]   Hassen Saïdi and Susanne Graf. Construction of abstract state graphs with PVS. In Orna Grumberg, editor, *Computer-Aided Verification, CAV '97*, volume 1254, pages 72–83, Haifa, Israel, 1997. Springer-Verlag.

Proof Engineering in the Large: Formal Verification of Pentium®4 Floating-Point Divider

Roope Kaivola and Katherine Kohatsu

Intel Corporation, JF4-451, 2111 NE 25th Avenue, Hillsboro, OR 97124, USA

Abstract. We examine the challenges presented by large-scale formal verification of industrial-size circuits, based on our experiences in verifying the class of all micro-operations executing on the floating-point division and square root unit of the Intel IA-32 Pentium®4 microprocessor. The verification methodology is based on combining human-guided mechanised theorem-proving with low-level steps verified by fully automated model-checking. A key observation in the work is the need to explicitly address the issues of proof design and proof engineering, i.e. the process of creating proofs and the craft of structuring and formulating them, as concerns on their own right.

1 Introduction

Verification of large systems is discussed in an increasing number of published case studies. For many of these, the story-line may be paraphrased by *we used theory X and tool Y to verify system Z*. The verification of a system is considered an accomplishment on its own right, and the fact that it could be achieved at all is a contribution worth reporting. Given the current state of the art, we think this is quite justified.

Rather less has been said about the practice of applying formal verification on a large scale in a system development project [1,5,8]. Producing an isolated proof of correctness differs from such wide-scale application in the same way as writing a program to solve a single problem in a single set of circumstances differs from writing a general software system to solve a class of related problems in a variety of circumstances, evolving over time. Although the solution in both cases is likely to be fundamentally the same, the general case will require attention to issues that can be safely glossed over in the restricted case. In effect, when producing an isolated proof of correctness, the main concern is just that that the proof is provided, whereas in the general case, issues of how the proof is constructed and structured become equally important.

In this paper we examine some of the issues present in large-scale verification work, based on our experiences in verifying the family of all micro-operations executing in the division and square root unit of the Intel IA-32 Pentium®4 microprocessor. Although based on a single extended case study, we believe that many aspects of the work are of a more universal nature. Therefore, we have tried to phrase the discussion on the general level, drawing on the case study to illustrate various points in practice.

Our verification methodology is based on human-constructed, mechanically-checked proofs with completely automatically verified model-checking steps at the lowest level. The aim is to take advantage of automation to mechanise tedious low-level reasoning,

T. Margaria and T. Melham (Eds.): CHARME 2001, LNCS 2144, pp. 196–211, 2001.
© Springer-Verlag Berlin Heidelberg 2001

while retaining the relatively complete freedom of the human verifier to set the overall verification strategy. We set out to perform a fully mechanically checked correctness proof in a single, unified framework, relating the high-level correctness statements all the way down to the actual register-transfer level description of the hardware. Technically the verification work was carried out in the the Forte verification framework, a combined model-checking and theorem-proving system built on top of the Voss system [9]. The interface language to Voss is FL, a strongly-typed functional language in the ML family [16], model checking is done via symbolic trajectory evaluation (STE) [18], and theorem proving is done in the ThmTac proof tool [2].

On a philosophical level, we approach verification much the same way as program construction, by emphasizing the role of the human verifier in decomposing the top-level problem to relatively simple steps amenable to automation, instead of striving at maximizing the amount of automation. Continuing the analogy, we identify two separate, although partly overlapping, aspects of proof construction: *proof design*, concerned with the problem of devising a proof of correctness for a given system in the first place, and *proof engineering*, concerned with the structure and formulation of such a proof.

Probably the most important observation in our work is that in large-scale application of formal verification, conscious attention needs to be paid to the proof design and engineering aspects, in addition to the conceptual argument behind the proof, or the fundamental aspects of the verification framework. In retrospect, this should not be surprising. After all, decades of experience have shown the crucial importance of careful software design and engineering practices for large-scale system development projects. However, in proof development we do not have the same wealth of established models on which to base the work as in software development. In our verification work, we failed to appreciate the need for clear development principles early enough. This resulted in extensive amounts of proof rewriting work later on, when the problems caused by poor choices in proof structuring and formulation became apparent.

We start by looking at the aims and challenges of applying formal verification in the large scale in Sect. 2. Section 3 introduces the Pentium 4 divider circuit. In Sect. 4 we outline our verification methodology, and in Sect. 5 the technical verification framework. Section 6 discusses our approach to proof design, and Sect. 7 gives an overview of the steps involved in the verification of one individual division micro-operation (for more proof details, see [13,14]). Then, in Sect. 8 we examine aspects of proof engineering in some more detail.

2   Large-Scale Verification

Before looking at our case study, let us discuss more generally our experiences regarding the challenges of applying formal verification as a routine part of an active industrial development project, as opposed to a one-off case study illustrating the feasibility of a particular verification approach.

A basic difference between the two is that in a development project, formal verification is not the main concern of the project, but only a fairly small part of it, one tool among others. This is reflected in both the properties and the systems to be verified. On the one hand, the choice of what is to be verified is based more on what is considered to be critical for the project, rather than what happens to suit well to a particular verification technique. Although available technology naturally sets limitations to what can be verified, in principle the verifier should be able to address any correctness issue that may be relevant to the final product. On the other hand, systems are less than perfect regarding the needs of verification. The verifier has little control over them, and cannot massage a system to make the verification problem easier.

In actual fact, in a hardware development project like ours, there is an inherent conflict between the goals of the hardware design and the needs of the formal verification. For design, performance considerations are the most crucial concerns, and simplicity and clarity come distant second. Verification, on the other hand, needs elegance and clarity, for making specifications understandable and any kind of formal reasoning possible. In effect, for formal verification we must create a clear and elegant abstract description of something that is not in itself clear and elegant at all.

The practical problems of formal verification start with the formulation of a precise specification. Written design specifications, if they exist, tend to overlook low-level details necessary for formal verification. Furthermore, in a system under development, current specifications often exist only in the minds of the designers. Therefore, writing a precise specification almost invariably involves some reverse engineering of the system, with the obvious danger that a specification replicates problems of the system.

The largest challenge in industrial formal verification is clearly just carrying out the verification at all. Given the complexity of industrial systems, and the level of support current tools provide, this is often a task requiring great ingenuity. To illustrate the size and complexity of current systems, a print-out of the Pentium 4 divider register-transfer-level source code, the basis of our verification work, is about one inch thick, and the unit is only a small fraction of the whole processor.

However, carrying out the verification as part of an active development project sets additional requirements beyond "just doing it": we have to be able to make plans and promises about the verification before actually carrying it out, and then keep these promises. This means that the verification approach must be sufficiently predictable and well understood to make meaningful advance planning possible.

Probably the largest difference between an individual case study and systematic application of formal verification lies in the verification maintenance aspect. For an isolated case, a proof can be almost write-only, as after the verification has been completed, it will not need to be revisited. For an active development project the situation is quite the contrary: the verification will need to adapt to changes in the underlying system and the specification over the lifespan of a project. As a matter of fact, due to the high initial investment required by formal verification, it is natural to reuse the results in future projects, as well, so the verification is quite likely to outlive the project it was originally part of. In our case, the underlying system model changed sometimes several times a week, and we expect the proofs to be used for five years or more. It is also natural to carry out large verification tasks incrementally, starting at a more restricted set of behaviours and properties, generalising this step by step, which means that the verification needs to be carried out repeatedly, even if the underlying system does not change. All this means that for larger-scale formal verification, the robustness of the verification method and easy modifiability of proofs are extremely important.

While the accuracy of formal verification is naturally important in any setting, in an industrial project it is of special significance, in relation to more traditional testing-based approaches to validation. These methods are likely to be used in parallel with formal verification, and as they typically produce partial results much faster, simple errata appearing frequently are likely to be caught by testing long before formal verification would detect them. Therefore, the value of formal verification lies in its ability to discover the hard-to-find errata that testing would miss. In order to find these subtle problems, it is essential that both the model of the system and the properties to be verified reflect accurately the real system and its intended properties.

An ingredient in the accuracy of verification is the concern of reviewability. The specification of a system should be reasonably clear and crisp to be easily reviewable against informal notions of correctness, without understanding internal details of the system. It should also be easy to find out from a verification what exactly does it prove, how this is proved and, especially, what the underlying, unstated assumptions are.

3 Divider Circuit

To illustrate the Pentium 4 divider unit, consider first the simple iterative division-remainder algorithm sketched in Fig. 1. It takes two normal floating-point numbers N and D as input, and produces the rounded quotient Q of N divided by D. This algorithm is essentially the same as the one taught in school for pen-and-paper division, although in binary instead of decimal. The value of *iteration_count* depends on the required precision of result. The algorithm can be easily modified to compute the remainder R instead of the quotient Q by just switching the entity to be output.

The algorithm of Fig. 1 can also be used to compute square root of N with minor modifications. First, a preprocessing step aligning N is added so that $N_e - bias$, the unbiased exponent, becomes even. Second, both occurrences of D_m inside the loop are replaced with $2 * Q_m[i] + 2^{-i}$ (notice that the value varies between iterations), and third, the final exponent computation is replaced with $Q_e := (N_e - bias)/2 + bias$.

Figure 2 depicts a simplified hardware implementation of this division algorithm. The circuit has inputs for the dividend N, the divisor D and some control signals. Mantissa calculation is done in a feedback loop, one iteration per clock cycle, and exponent calculation is done in a separate subunit. As output, the circuit produces the result W of the required calculation and some control information, such as various flags. Correct behaviour of the circuit can be easily characterised by the formula $r(W) = $ round($r(N)/r(D)$) for division and by $r(W) = $ round($\sqrt{r(N)}$) for square root, where the precise meaning of the function 'round' depends on the intended rounding mode and precision, and the function r maps a floating-point representation to the real number it encodes.

Although similar in principle, current industrial hardware implementations of division algorithms are many magnitudes more complex. For example, they may use redundant or multiple representations of Q and R, produce more than one quotient bit per iteration, or perform speculative calculations [6]. The Pentium 4 divider unit is no exception: it implements a highly optimised double-pumped radix-2 SRT division algorithm, producing two quotient bits per clock cycle, and has over 7000 latches.

input: two normal floating-point numbers $N = (N_s, N_e, N_m)$ and $D = (D_s, D_e, D_m)$
(we view abstractly N_e and D_e as natural numbers and N_m and D_m as fractions below)
variables: floating-point numbers $Q = (Q_s, Q_e, Q_m)$ and $R = (R_s, R_e, R_m)$, integers $imax$ and i

$i := 0;$ $imax := iteration\_count;$
$Q_m[0] := 0;$ $R_m[0] := N_m;$
while $i < imax$ **do**
         /* determine quotient bit $q_i \in \{0, 1\}$ */
         if $R_m[i] < D_m$ **then** $q_i := 0$ **else** $q_i := 1$ **fi**
         /* update quotient and remainder accordingly */
         $Q_m[i+1] := Q_m[i] + 2^{-i} * q_i;$ $R_m[i+1] := 2 * (R_m[i] - q_i * D_m);$ $i := i + 1$
od
$Q_s := N_s \textbf{ xor } D_s;$ $Q_e := N_e - D_e + bias;$ $Q_m := Q_m[imax];$
output (**round**(Q_s, Q_e, Q_m))

Fig. 1. Simple iterative division algorithm

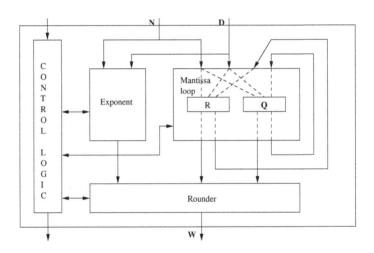

Fig. 2. Simple divider hardware

The Pentium 4 divider unit also supports a number of different variations of the basic division, remainder and square root operations. While the simplest ones differ only with respect to rounding precision, the circuit supports a collection of specialised micro-operations used primarily for microcode flows computing transcendental functions. Additionally, the circuit supports a collection of Single Instruction Multiple Data (SIMD) instructions called SSE (Streaming SIMD Extension) and SSE2, optimised for multimedia applications. For some of these, several passes of the mantissa loop are executed, and for some others, the normal full-width datapath is split into two halves, both effectively executing the same algorithm in parallel. Altogether the Pentium 4 divider unit supports about twenty materially different variants of the basic operations. For more discussion on Pentium 4 micro-architecture, see [10].

4   Verification Methodology

The goal of our verification methodology is to provide a completely machine-verified proof of correctness, with low-level steps justified by fully automated model-checking, relating high-level specifications all the way down to the actual description of the circuit in a unified framework. The four basic principles, mechanised verification, automation in the low level, actual model of the circuit, and uniform framework, are all answers to the challenges of large-scale verification. While we believe these principles to be quite uncontroversial, let us briefly outline the arguments behind them.

Consider mechanised proof-checking first. An alternative view would be to consider proofs as social objects, and trust that the scrutiny of sufficiently observant peers will find any mistakes [17]. Unfortunately, many of the proofs related to formal verification of circuits are rather boring from a mathematical perspective, and do not motivate qualified individuals to delve into them deeply enough. For example, in doing mechanised theorem-proving on our rounding specifications, we discovered a hole that had resisted close scrutiny for a long time. The second principle, automation, is necessary for the sheer size of industrial circuits. While automation does not necessarily need to imply model-checking, we are currently not aware of other sufficiently robust approaches.

Our decision to base the verification on the actual description of the circuit used in the design flow is motivated by the reliability and maintainability of the work. Many actual errata in circuits, e.g. the infamous Pentium FDIV erratum and all errata found in our work, are caused by low-level details. Furthermore, a separate high-level description would need to be constantly updated to reflect changes to the actual design.

Regarding the unified framework, some verification case studies use a variety of tools, e.g. the results of model-checking are transferred from one system to another for theorem-proving purposes. In our opinion this approach leaves room for error in the form of unstated or poorly understood assumptions underlying the translation of statements from one formalism to another. A single, tightly integrated environment also helps in making the verification more manageable and reviewable, as assumptions, qualifications and verified statements can be expressed uniformly.

Underlying our verification methodology is the philosophical belief that verification of systems should be an activity analogous to programming. We view programming as the human activity of organising individual primitive instructions, each of which can be mechanically executed by a computer in an efficient, dependable and predictable fashion, to a larger pattern to perform the intended high-level task. In the same way, we view verification as the human activity of organising individual primitive proof steps, each of which can be mechanically verified by a computer in an efficient, dependable and predictable fashion, to a larger pattern to establish the intended high-level specification.

Based on this view, we tend to emphasise the proof decomposition aspect over automation. We are naturally not in any way against the use of sophisticated algorithms: they can help verification just like a subroutine library can help programming. Nevertheless, our trust in our ability to carry out a given programming task is usually based more on our programming skills, rather than on being lucky enough to find a tool that already happens to perform the task. In the same way, in our opinion, our trust in being able to carry out a given verification task should be based primarily on our decomposition skills and the robustness of the underlying primitive verification steps.

5 Technical Verification Framework

Our verification framework consists of a collection of definition and theorem libraries built in the Forte environment, a combined model-checking and theorem-proving system. The interface and scripting language to Forte is FL, a strongly-typed functional language in the ML family [16]. It includes binary decision diagrams (BDDs) as first-class objects and symbolic trajectory evaluation as a built-in function.

Symbolic trajectory evaluation, based on traditional notions of digital circuit simulation, is an efficient method for determining the validity of a restricted class of temporal properties over circuits. It allows statements of the form $\models_{ckt}[ant\Longrightarrow cons]$, where the *antecedent* (*ant*) gives an initial state and input stimuli to the circuit *ckt*, while the *consequent* (*cons*) specifies the desired response of the circuit. Formally, the meaning of the statement is: all sequences in the language of the circuit satisfying the antecedent will also satisfy the consequent. Antecedents and consequences are formed by conjunction from basic formulae of the form $N^t(node$ is $value$ when $guard)$, where t is an integer, *node* is a signal in the circuit and *value* and *guard* are Boolean expressions. The meaning of a basic formula is "if *guard* is true then at time t, *node* has value *value*".

The efficiency of trajectory evaluation is based on built-in support for data abstraction via a lattice of simulation values. The simulation model used by Forte extends the conventional Boolean domain to a lattice by adding a a bottom element X and a top element ⊤. Intuitively the value X denotes lack of information: the signal could be either T or F. The essential relation between such four-valued and Boolean sequences is that any assertion verified over a sequence containing Xs will hold for sequences with Xs replaced with either T or F [4,3].

Theorem proving in Forte is done in the ThmTac proof tool, an LCF-style implementation of a higher-order classical logic. Its principal aim is to enable seamless transitions between model checking, where we *execute* FL functions, and theorem proving, where we *reason* about the behaviour of FL functions [2]. Roughly speaking, if a term does not include any free variables, contains quantification over Boolean domains only, is evaluatable within the computational resources available, and evaluates to true, we can turn it into a theorem and use it for reasoning.

As the restricted language used for trajectory evaluation is too weak to allow expression of many interesting properties, we use a variant of the traditional pre-postcondition framework (see e.g. [7]) for formulating temporal aspects of our specifications. In our approach specification statements are of the form $\{\phi_{in}\}(tr_{in}, ckt, tr_{out})\{\phi_{out}\}$ where a trajectory assertion $tr_{in}(x)$ binds a vector x of Booleans to some input signals of ckt at the time the input is intuitively read by the circuit, trajectory assertion $tr_{out}(y)$ binds a vector y similarly to some output signals, a formula $\phi_{in}(x)$ expresses the precondition the input is supposed to meet, and $\phi_{out}(x, y)$ the postcondition the circuit is supposed to produce. Formally, this statement is shorthand for the formula:

$$\forall in.\phi_{in}(in) \Rightarrow (\exists out.(\models_{ckt}[tr_{in}(in)\Longrightarrow tr_{out}(out)])) \;\wedge$$
$$(\forall out.(\models_{ckt}[tr_{in}(in)\Longrightarrow tr_{out}(out)]) \Rightarrow \phi_{out}(in, out))$$

Intuitively the formula states that for any vector of values x satisfying the precondition $\phi_{in}(x)$, 1) there is some output vector y such that for every execution e, if $tr_{in}(x)$ is true of e, then so is $tr_{out}(y)$, and 2) for every vector y for which 1 holds, the postcondition property $\phi_{out}(x, y)$ holds, or more loosely, that whenever precondition ϕ_{in} is satisfied, the circuit guarantees that postcondition ϕ_{out} is also satisfied.

The validity of a pre-postcondition statement in the form above can, in principle, be determined by direct evaluation, although computational resource requirements mean that in practice this is feasible only in limited circumstances. To allow reasoning about the flow of computation in a structured way, our proof framework includes general reasoning rules for pre-postcondition statements, such as precondition strengthening, postcondition weakening, conjunction, sequential composition and bounded iteration [13].

When dealing with arithmetic circuits, both specifications and reasoning are often naturally expressed in terms of arithmetics. As model-checking techniques using BDD based representations can only deal with bit-vector operations, our proof framework includes a library of provably correct bit-vector arithmetic operations, which have an exact correspondence with integer operations.

To support verification of floating-point operations, our proof framework includes a general-purpose theorem library for floating-point numbers and rounding [14]. Analogous to the case of integer arithmetics above, the library supports floating-point numbers and rounding at the bit-vector level for model-checking, and at the mathematical level for reasoning. As currently our framework does not support reals, only integers, we have adopted the work-around of multiplying all entities by a sufficiently big number 2^{BN} so that every real number that is relevant for our proofs maps to an integer.

6 Proof Design

Finding a proof for a given property and system is naturally always a heuristic process. Nevertheless, just as in program design, it is worth while to articulate general strategies for finding a solution, and to impose some structure on the process. In addition to offering guidelines for construction of future proofs, spelling out design principles gives a vocabulary for communicating and comparing solution strategies.

For low-level STE model-checking work, the methodology discussed in [1] gives us the basic structure for finding out circuit interfaces, describing them abstractly, and carrying out trajectory evaluation runs. On a higher level, our decomposition strategy is based on looking at the abstract algorithm the circuit is intended to compute.

We start by partitioning the algorithm to regions in such a way that the computation within each region in isolation can, in principle, be efficiently carried out for symbolic initial values using BDD's. For example, a region involving only addition and subtraction, or only shifts, is rather likely to have a concise BDD representation. For the division algorithm, the body of the loop forms a good candidate for a region.

For each region, we then try to locate the computations corresponding to the region in the circuit, find boundaries separating the computations, and signals corresponding to the variables of the abstract algorithm. At this point, we may notice that the algorithmic description of the circuit is too coarse to allow an adequate correspondence. For example, when trying to map the loop body of the division algorithm to the actual circuit, we notice that the circuit uses auxiliary entities, effectively different approximate representations of Q and R. In this case, we will need to refine the abstract algorithm. When mapping entities of the abstract algorithm to the circuit, the relation between the levels is not necessarily one-to-one: a single abstract variable may correspond to a non-trivial function of a collection of signal vectors, with different timing characteristics.

Once we have located the regions and boundaries in the circuit, we verify that each region in the circuit implements correctly the corresponding region of the algorithm. As the original partitioning to regions was chosen so that calculations within each abstract region could be efficiently carried out using BDD's, there is a fairly good likelihood that we can also efficiently simulate the behaviour of the circuit region with STE. While it may naturally happen that an intermediate value in a region is not concisely representable by BDD's even if the boundaries are, we have so far never encountered such a case.

After having model-checked each region separately, the corresponding relations are combined using theorem proving, in the way described by the abstract algorithm to yield a proof of the top-level correctness statement. In other words, our proof design approach is top-down, whereas proof construction takes place bottom-up.

The decomposition of the abstract algorithm to the regions used in verification does not need to coincide with the decomposition used in implementing the algorithm in the circuit. In fact, one perhaps surprising observation in this work has been that this is hardly ever the case: the regions and boundaries used for verification rarely bear much resemblance to the module structure, nor to the latch boundaries of the circuit. On second thought, this may not be so surprising: modules of the circuit are more tied to the physical area, and especially in later stages of a project, circuit logic tends to be moved from one module to another, or from one side of a latch to another, with fairly little regard to its conceptual position in the overall computations. Consequently, circuit modules do not usually have a clear algorithmic characterization useful to us.

We believe that this proof design approach is quite widely applicable to various kinds of datapath-oriented circuits. As neither the proof nor the verifier needs to know the exact way computations within a circuit region are actually carried out, concrete proof plans can be made in advance. The approach also appears to be robust regarding changes to the design: most changes are likely to take place at a local level, so while they may require adjusting the boundaries or the description of the intended computation within a region, the overall proof structure does not need to change.

7   Divider Verification Outline

An informal specification of the circuit's correctness is quite easy to come up with:

> *IF a division operation is started AND the input values N and D are within the range handled by hardware AND the environment behaves according to the expected protocol AND the circuit is internally in normal operating state, THEN at the time the circuit produces output W, the equation $r(W) = round(r(N)/r(D))$ holds.*

When formalising this, the part concerning the relation of input and output data values is straightforward, as it follows from the IEEE specification on floating-point arithmetics [11], although formalisation of the standard itself is non-trivial. However, the problem lies at the left side of the implication: what are "normal internal operating state" and "expected environment protocol"? Characterising these very circuit-dependent aspects required a fair amount of investigative work. To increase confidence in the correct characterisation of the environment assumptions, we also used an existing test suite to check for their validity in a variety of circumstances using traditional test-based validation. In principle the environment assumptions could have been verified in the context of the whole processor, but we did not have the resources for this.

Further, mentioning the internal operating state in the specification violates the principle of external visibility. Therefore, we needed to strengthen the statement by proving separately that *whenever a division operation can be started, the circuit is internally in normal operating state*, which allows us to discharge the last conjunct of the antecedent. This proof was carried out in a fairly traditional temporal-logic-based framework. However, as it is separate from the main datapath proofs, we shall not discuss it here.

The top-level correctness statement can then be formalised by

$$\{IN\} \ \ (tin, ckt, tout) \ \ \{IO\} \tag{1}$$

where the precondition *IN* formalises the four conjuncts of the antecedent of the informal specification, and the postcondition *IO* is defined by

$$\begin{aligned} IO = \exists Q.\, (\ ri(W) \ &= \ \mathrm{round}_{ri}(Q)\,) \ \wedge \\ (\ Q * ri(D) \ &\leq \ ri(N) * 2^{BN} \ \leq \ (Q + \epsilon) * ri(D)\,) \end{aligned} \tag{2}$$

and where trajectory function *tin* binds N and D to input data signals at the start of the operation, and *tout* binds W to output signals at the time the output is ready. The formula *IO* is slightly more complex than the intuitive specification: the extra entities Q, intuitively denoting the unrounded quotient, and ϵ, denoting some fixed small value, are needed because of the lack of real numbers in our current framework. The function round_{ri} is a rounding function working on the integer representation of reals.

As the algorithm and the hardware are iterative in nature, the verification is based on a loop invariant for the mantissa calculation. At the top level, there is a natural mathematical invariant MI_i relating the quotient and remainder mantissas $Q_m[i]$ and $R_m[i]$ to the input numbers D and N, derived from the defining equation of division:

$$MI_i = (N_m = Q_m[i] * D_m + 2^{-i} * R_m[i]) \ \wedge \ (R_m[i] < 2 * D_m) \tag{3}$$

Due to the multiplication operation in this invariant, it is not amenable to verification by direct model-checking. Therefore, the problem is further decomposed into verification of MI_1 after first iteration, and verification of an equation MR_i between current and previous loop values for each subsequent iteration. The equation MR_i is based on the recurrence relation the loop is supposed to compute. Further, to verify the relation MR_i by model-checking, two bit-vector relations are introduced: a bit-vector recurrence relation BR_i that coincides with the mathematical relation MR_i, and a low-level bit-vector invariant BI_i expressing a consistency constraint on loop data.

Using this decomposition, verification of the mantissa computation in the circuit consists of the following steps:

$$\{IN\} \ (tin, ckt, tl_0) \ \{BI_1 \wedge MI_1\} \tag{4}$$

$$\forall i. \ (0 \le i < imax) \ \Rightarrow \ (\{BI_i\} \ (tl_i, ckt, tl_{i+1}) \ \{BI_{i+1} \wedge BR_i\}) \tag{5}$$

$$\forall i. \ (0 \le i < imax) \ \Rightarrow \ (BR_i \Rightarrow MR_i) \tag{6}$$

$$\forall i. \ (0 \le i < imax) \ \Rightarrow \ (MI_i \wedge MR_i \Rightarrow MI_{i+1}) \tag{7}$$

where trajectory function tl_i binds R_m, Q_m and other data items to corresponding signals for iteration i. Statements 4 and 5 are verified directly by model-checking. Considering our proof design strategy, the binding functions tl_i express the boundaries of the regions for model-checking. Statement 6 involves reasoning about the correspondence between bit-vector operations and their arithmetic counterparts, and statement 7 relies on pure arithmetic reasoning. Using pre-postcondition reasoning, steps 4–7 can then be combined to a correctness statement for the complete mantissa computation:

$$\{IN\} \ (tin, ckt, tl_{imax}) \ \{BI_{imax} \wedge MI_{imax}\} \tag{8}$$

The correctness of the final rounding stage can be expressed by the formula:

$$MRND = (\ ri(W) \ = \ \text{round}_{ri}(ri(s, e, m)) \ ) \quad \text{where} \tag{9}$$
$$s = sgn(N) \ XOR \ sgn(D)$$
$$e = exp(N) - exp(D) + bias$$
$$m = Q_m[imax]$$

To verify the rounding stage by model-checking, two further bit-vector relations are needed: a bit-vector version $BRND$ of the mathematical relation $MRND$, and an auxiliary relation AUX, which expresses constraints on the final loop output necessary for the proper behaviour of the rounder. Then, verification of the rounding stage consists of the following steps:

$$\{BI_{imax} \wedge AUX\} \ (tl_{imax}, ckt, tout) \ \{BRND\} \tag{10}$$

$$BRND \Rightarrow MRND \tag{11}$$

$$BI_{imax} \wedge MI_{imax} \Rightarrow AUX \tag{12}$$

$$MI_{imax} \wedge MRND \Rightarrow IO \tag{13}$$

Statement 10 is verified by direct model-checking. This is a good example of the differences between boundaries used for verification and those of the circuit units: the starting boundary tl_{imax} is not at the rounder input in the circuit, but inside the mantissa loop. Statement 11 reasons about the correspondence between bit-vector and mathematical versions of rounding, and statements 12 and 13 involve mostly arithmetic reasoning. Using pre-postcondition reasoning, steps 10–13 then yield:

$$\{BI_{imax} \wedge MI_{imax}\} \ (tl_{imax}, ckt, tout) \ \{IO\} \tag{14}$$

Finally, statements 8 and 14 can be joined by sequential composition to show the top-level correctness statement 1. For more proof details, see [13].

While this discussion has concentrated on the division operation, the proof for square root is analogous. The most crucial issue in the model-checking part of the verification was determining the boundary tl_i and the invariant BI_i exactly: some parts are easy, like the location or the expected ranges of data values in the loop, but some are extremely implementation-dependent.

8   Proof Engineering

Once we had completed a proof for the first micro-operation, effectively the proof outlined in Sect. 7 and reported in [13], we believed that the largest body of work was behind us, and the proofs for the remaining cases would fall out easily. We were wrong. The length of the basic proof for one micro-operation, excluding library code, is about 15 000 lines, about 85% of which is related to the theorem-proving effort, and 15% to the model-checking. The naive solution of replicating the code for each micro-operation and then making necessary changes would have resulted in 20 * 15 000 = 300 000 lines of code, clearly an organisational and maintenance nightmare.

Consequently, we had to reformulate the proofs avoiding code duplication, while accounting for the differences between micro-operations. These differences come in various flavours: The loop body of the division and square root operations is different, although the general structure is the same. Mantissa loop is executed for a different number of times for different precision modes. For certain SIMD operations, the data-path is split to two halves, and both halves implement the same algorithm as the full width datapath, except for certain details. The two halves are nearly, but not exactly symmetric. Flag and fault behaviour for customised micro-operations differs from standard ones. For some SIMD operations, multiple passes of mantissa loop are executed, and for some others, fault behaviour reflects several pieces of data. The rounder depends on different constraints on the loop output for different micro-operations.

It turned out that most of the theorems we proved for the first completed proof were not sufficiently general. Various values, assumptions and definitions were hard-coded into the proofs, although they vary between micro-operations. This lack of generality caused an extensive amount of proof rewriting. Actually, we spent more time rewriting proofs than writing them in the first place. Although some consolation is provided by anecdotal evidence in the literature [15,12] that we were not the only ones encountering the problem, this clearly is not a preferable state of affairs. We basically made the same mistake as starting a software project by rushing to write program code, without precise planning of the overall structure of the system. Moreover, we did not write our original proofs in a fashion that would have supported modifications and maintenance.

So, how do we write robust, understandable, and maintainable proofs? It appears to us that theorem-proving is often used in a fairly static setting, where maintainability is not a crucial concern, so we did not seem to have too many models. While we can naturally learn from principles used to enhance software maintainability, we cannot just simply equate the problems of software and proof maintenance. As a matter of fact, we would argue that the latter is considerably harder than the former. First, for software only the semantics of objects matter: we can freely reformulate a definition as long as its denotation does not change. For proofs, on the other hand, syntactic structure of terms matters as well. Secondly, the rigour required for formal reasoning leaves much less leeway for sticky-tape solutions than in the case of software. Thirdly, the arduousness of theorem proving means that proof reformulation is harder than program reformulation. This conspires against maintainability in two ways: once a proof has been created, no matter how imperfect, there is a great temptation to leave it as it is, but when we will need to modify it later, the penalty is even higher.

Many specific questions emerge in relation to writing robust and maintainable proofs. For example, if we classify objects related to proofs to term language definitions, claims regarding such definitions, and proofs of claims, what principles are relevant for formulation of objects in each group? How to structure a theorem hierarchy? How to formulate the proof of an individual theorem? How to represent proof hierarchies which are similar except for some details? While we cannot claim to have the best possible solutions, we were forced to explicitly address these issues during our verification work.

Let us start from the question of maintainable definitions. We adopted the practice of formulating definitions as hierarchies of definition layers, with each layer concentrating on a separate aspect. The layers are very thin, usually a single level of a definition is less than five lines long. For example, consider the definition of *MRND* (equation 9 in Sect. 7). As written there, it contains the aspects of sign and exponent calculation, mantissa definition, conversion from floating-point representation to a number and rounding. To be able to reason about these aspects separately, we define each of them as a separate layer. Then we can easily change parts of a definition, e.g. to reuse parts of *MRND* for square root proofs by changing sign and exponent calculation.

The layering is repeated in the formulation of claims. For each layer present in a definition, we write a separate claim, with the intention that it can be proved on the basis of the definition of the current layer and claims relating to the layer immediately below. This induces a natural theorem hierarchy. We also try to state claims always in terms of relation names, instead of the actual relations. So, instead of *if $a < 2^{23}$ then ...* we write *if (in_bounds a) then ...* where *in_bounds $x = x < 2^{23}$*. In this way, even if the actual bound used in *in_bounds* will change, subsequent theorems will be unaffected as long as the current theorem remains true. This isolates effects of changes and improves maintainability. On the negative side, the approach may lead to a proliferation of claims.

Given a hierarchy of definitions and a claim relating to some layer, how do we write a maintainable and understandable proof for the claim? A proof composed of many simple manually crafted steps is more likely to need editing, even for small modifications in the claim, than a proof with fewer, more automated steps. However, when modifications are needed, they are likely to be easier to make in the former case than in the latter, as it is easier to recreate the conceptual argument behind the proof from its code representation. Thus, in our experience, for small changes in the claim, highly automated proofs are superior, but for larger changes, manually crafted proofs are more maintainable. Unfortunately, without foreseeing what kinds of changes will be required, it is hard to plan for the best outcome.

The core issue in avoiding code duplication is proving each argument only once. However, it is often easier to prove n instances of a general theorem, than the theorem itself, so there is a tradeoff between maintainability and ease of proving. We usually opted for proving the general case for any $n > 2$, unless the general proof was fundamentally more difficult.

To improve understandability and manageability, we organised all our proofs in modules that, conceptually, are given a set of objects and theorems concerning those objects, and provide another set of objects and theorems. In modules we followed a principle of locality, and packaged all proofs requiring access to the internal details of a definition together with the definition itself. This greatly increases maintainability of code, as changes to a definition only require changes to the local module, as long as externally visible theorems are not affected. Modularity also allows us to represent proof hierarchies differing only in some details without code duplication, by using alternate modules for the differing aspects of the proof. This way, we managed to capture all the different proof variations in only about 45 000 lines of proof code.

We found the layering of definitions, claims and proofs to be indispensable for creating maintainable proofs. However, splitting definitions to minute steps has a negative effect on reviewability. To alleviate this problem, we experimented with proofs using two sets of definitions: monolithic ones, used in the top-level specifications for reviewability, and layered ones, used for the rest of the proof. The transition from layered to monolithic definitions then takes place just below the top level.

Technically our proof development environment was rather austere. In ThmTac a user writes down a textual representation of proof steps, and a proof is an FL term like any other, so no special proof capture or replay mechanism is needed. For interactive proof development we used a ThmTac interface for stepping through a proof, and a mechanism for assuming theorems without having to evaluate their proofs. We used no special module, version or proof consistency management tools. While they might have helped on some occasions, the real problem we faced was not in these aspects, but in deciding how to write down the proof. To guarantee that all pieces fit together after a round of changes, we revalidated the whole proof hierarchy from scratch overnight. What would have made our work easier would have been a mechanism for supporting and enforcing good code writing practices, such as module visibility rules. Now, our work environment did not support modules directly and we had to emulate them by conditional load sequences. Another item in our wish-list would have been tool support for incremental proof changes, such as analysing the precise point in which an old proof and an attempted new proof diverge.

Looking back on the lessons learned during our verification work, we would advocate the strategy of starting the verification by a quick, semi-informal decomposition of the top-level property to model-checkable portions, carrying out the model-checking, and then planning the complete formal proof structure in great detail before starting any theorem-proving. The proof planning stage should result in a documented description of the modules, definitions, claims, required proofs and their relations. Precision in this stage is crucial, as the high cost of proof changes makes it important to get the proof structure right at the first attempt.

9 Conclusion

We have examined verification methodology, proof design and proof engineering aspects relevant to large-scale industrial application of formal verification, based on our experiences in verifying the Pentium 4 divider unit, to our knowledge one of the largest industrial hardware verification case studies. The verification took about two and a half person-years of work in total, and it was carried out in parallel with later stages of the circuit design, before silicon was produced. No errata were found in the original design, but applying the proof suite to a proliferation project caught a few rather tricky errata caused by unintended interactions between micro-operations executing in parallel.

Considering the technical requirements of industrial verification, in our experience it is more important that tools are robust and dependable than that they are technologically the most advanced ones: all the basic techniques that we are using have been well known for years. In a combined theorem-proving and model-checking approach, we found tight integration of the techniques to be necessary: theorem-proving is used in all stages of our verification work, from justifying model-checking optimizations in the low level, to very abstract reasoning in the high level. We also found the open-endedness of general theorem-proving indispensable for the work.

During the course of our work we identified a number of solutions to practical problems arising in proof design and engineering. Nevertheless, we would like the message of the paper to be not so much of a solution but of a problem statement: How do we write large proofs in a manageable way? If we are to apply formal verification, especially formal theorem-proving, as a routine part of industrial design flow, we must have models and principles addressing this issue, as otherwise the work will become infeasible.

References

1. M. D. Aagaard, R. B. Jones, T. F. Melhan, J. W. O'Leary, and C.-J. H. Seger. A methodology for large-scale hardware verification. In *FMCAD*, volume 1954 of *LNCS*, pages 263–282. Springer, 2000.
2. M. D. Aagaard, R. B. Jones, and C.-J. H. Seger. Lifted-fl: A pragmatic implementation of combined model checking and theorem proving. In *TPHOLs*. Springer, 1999.
3. M. D. Aagaard, T. F. Melham, and J. W. O'Leary. Xs are for trajectory evaluation, Booleans are for theorem proving. In *CHARME*. Springer, 1999.
4. C.-T. Chou. The mathematical foundation of symbolic trajectory evaluation. In *CAV*. Springer, 1999.
5. P. Curzon. The importance of proof maintenance and reengineering. In J. Alves-Foss, editor, *HOL-95: B-track Proceedings*, pages 17–31, Sept. 1995.
6. M. D. Ercegovac and T. Lang. *Division and Square Root, Digit-Recurrence Algorithms and Implementations*. Kluwer Academic, 1994.
7. D. Gries. *The Science of Programming*. Springer-Verlag, 1981.
8. I. Hayes. Applying formal verification to software development in industry. In I. Hayes, editor, *Specification case studies*, pages 285–310. Prentice-Hall, 1987.
9. S. Hazelhurst and C.-J. H. Seger. Symbolic trajectory evaluation. In T. Kropf, editor, *Formal Hardware Verification*, chapter 1, pages 3–78. Springer Verlag; New York, 1997.
10. Hinton, G., Sager, D., Upton, M., Boggs, D., Carmean, D, Kyker, A. and Roussel, P. The microarchitecture of the Pentium 4 processor. *Intel Technology Journal*, Q1, Feb. 2001.
11. IEEE. *IEEE Standard for binary floating-point arithmetic*. ANSI/IEEE Std 754-1985, 1985.
12. M. Jones and G. Gopalakrishnan. A PCI formalization and refinement. In *TPHOLs 2000, Supplemental Proceedings*, OGI Technical Report CSE 00-090, pages 115–124, 2000.
13. R. Kaivola and M. D. Aagaard. Divider circuit verification with model checking and theorem proving. In *TPHOLs*, volume 1869 of *LNCS*, pages 338–355. Springer, 2000.
14. R. Kaivola and N. Narasimhan. Multiplier verification with symbolic simulation and theorem proving. Submitted for publication, 2001.
15. A. Mokkedem and T. Leonard. Formal verification of the Alpha 21364 network protocol. In *TPHOLs*, volume 1869 of *LNCS*, pages 443–461. Springer, 2000.
16. L. Paulson. *ML for the Working Programmer,*. Cambridge University Press, 1996.

17. D. M. Russinoff. A mechanically checked proof of IEEE compliance of the floating point multiplication, division and square root algorithms of the AMD-K7 processor. *London Mathematical Society Journal of Computational Mathematics*, 1:148–200, 1998.
18. C.-J. H. Seger and R. E. Bryant. Formal verification by symbolic evaluation of partially-ordered trajectories. *Formal Methods in System Design*, 6(2):147–189, Mar. 1995.

Towards Provably-Correct Hardware Compilation Tools Based on Pass Separation Techniques

Steve McKeever and Wayne Luk

Department of Computing, Imperial College, 180 Queen's Gate, London, UK
{swm2,wl}@doc.ic.ac.uk

Abstract. This paper presents a framework for verifying compilation tools based on parametrised hardware libraries expressed in Pebble, a simple declarative language. An approach based on pass separation techniques is described for specifying and verifying Pebble abstraction mechanisms, such as the loop statement. We show how this approach can be used to verify the correctness of the flattening procedure in the Pebble compiler, which also results in a more efficient implementation than a non-verified version. The approach is useful for guiding compiler implementations for Pebble and related languages such as VHDL; it may also form the basis for automating the generation of provably-correct tools for hardware development.

1 Introduction

Advance in integrated circuit technology leads to an increasing emphasis on building designs from hardware libraries. A single parametrised library can be used to generate many implementations supporting multiple architectures, variable bit widths and trade-offs in speed and size. Such libraries enable effective hardware utilisation by exploiting technology-specific features whenever desirable, allowing designs with optimal performance and resource usage while minimising the need for knowing low-level details.

This paper describes an approach for developing provably-correct compilation tools for the Pebble language [12], which has been used to produce hardware libraries in VHDL, an industry standard language. While it is desirable to have hardware libraries in industry standard languages, there are, however, two major difficulties with developing VHDL libraries. First, VHDL is a versatile but complex language, and it takes much effort to write good parametrised code and to check its behaviour by simulation or other means – even when the subset used for realistic hardware libraries is small [13]. Second, most vendors have their own VHDL dialect; for instance not all VHDL tools support multi-dimensional vectors. It is unattractive to develop and maintain the same set of library elements in various vendor-specific dialects.

Our aim for Pebble is to enable application builders and library developers to work at a higher level of abstraction than that provided by VHDL, while ensuring that the resulting libraries are as flexible and efficient as those produced by hand. Our Pebble compiler targets various description formats, including parametrised and flattened VHDL and EDIF. It enables designers to compose and instantiate library elements, and it has been used to develop many designs for applications such as speech processing, data compression, and special video and graphics effects in augmented reality [14].

T. Margaria and T. Melham (Eds.): CHARME 2001, LNCS 2144, pp. 212–227, 2001.
© Springer-Verlag Berlin Heidelberg 2001

An important component of the Pebble compiler is the flattening procedure, which produces flattened descriptions from hierarchical descriptions. Flattened descriptions are required by many tools, such as place-and-route programs for design implementation, and model checkers for design verification [18]. Interestingly, our proof of the flattening procedure not only offers users greater confidence in its correctness, it also leads to a more efficient implementation. Although this paper focuses on a specific proof, recent work indicates that similar techniques can be applied to verify other abstraction mechanisms, such as polymorphic variables and higher-order functions.

Pass separation [11] provides a framework for verifying abstraction mechanisms for generic descriptions, such as hierarchical blocks and GENERATE-FOR loops in Pebble, with respect to a Structural Operational Semantics [16]. For instance, the key to our proof is an environment invariant – Equation (1) – inspired by pass separation (Sect. 3.2). Such proofs are rare, but we feel that they are well-suited to verifying development tools for domain-specific languages containing multiple evaluation phases.

While much has been published on formal methods and tools for hardware design, it appears that most researchers focus on tools for producing correct designs [2],[18] rather than on the correctness of the tools themselves. Some researchers study embedding of synthesis algorithms in a theorem prover [4], or correctness condition generators for designs generated by synthesis tools [15]. Our work is complementary to their efforts, and is in a similar spirit to research on verifying compiler correctness for imperative descriptions for hardware [6],[7] and software [8] implementations.

2 Overview of Pebble

Pebble can be regarded as a much simplified variant of Structural VHDL. It provides a means of representing block diagrams hierarchically and parametrically [12]. A Pebble program is a block, defined by its name, parameters, interfaces, local definitions, and its body. The block interfaces are given by two lists, usually interpreted as the inputs and outputs. An input or an output can be of type WIRE, or it can be a multi-dimensional vector of wires. A wire can carry integer, boolean or other primitive data values.

A primitive block has an empty body; a composite block has a body containing the instantiation of composite or primitive blocks in any order. Blocks connected to each other share the same wire in the interface instantiation. For hardware designs, the primitive blocks can be bit-level logic gates and registers, or they can, like an adder, process word-level data such as integers or fixed-point numbers; the primitives depend on the availability of corresponding components in the domain targeted by the Pebble compiler. The GENERATE-IF statement enables conditional compilation and recursive definition, while the GENERATE-FOR statement allows the concise description of regular circuits.

Pebble has a simple, block-structured syntax. As an example, Fig. 2 describes the multiplexor array in Fig. 1, provided that the size parameter n is 4. In more complex descriptions, the parameters in a Pebble program can include the number of pipeline stages or the pitch between neighbouring interface connections [12]. Different network structures, such as tree- or butterfly-shaped circuits, can be described parametrically by indexing the components and wires.

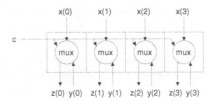

Fig. 1. An array of multiplexors described by the Pebble program in Fig. 2.

```
BLOCK muxarray (n:GENERIC)
        [c:WIRE, x,y:VECTOR (n-1..0) OF WIRE]
        [z:VECTOR (n-1..0) OF WIRE]
   VAR i
BEGIN
   GENERATE FOR i = 0..(n-1)
   BEGIN
     mux [c,x(i),y(i)] [z(i)]
   END
END;
```

Fig. 2. A description of an array of multiplexors (Fig. 1) in Pebble. The external input c is used to provide a common control input for each mutiplexor.

The semantics of Pebble depends on the behaviour of the primitive blocks and their composition in the target technology. Currently a synchronous circuit model is used in our tools (Sect. 3), and special control components for modelling run-time reconfiguration are also supported [12]. However, other models can be used if desired. Indeed Pebble can model any block-structured systems, not just electronic circuits.

Advanced features of Pebble include support for annotations and for modules. Such features improve design efficiency and reusability, and facilitate interface to components in other languages, including behavioural descriptions. Discussions about these features are beyond the scope of this paper.

3   Program Staging and Pass Separation

This section introduces a framework in which abstraction mechanisms for Pebble can be specified and verified. Our approach consists of three steps. The first is to provide a semantics for a flattened version of Pebble. The second is to characterise an abstraction mechanism in two ways: (a) specify how designs exploiting the abstraction mechanism can be transformed into flattened Pebble, and (b) provide the semantics of the abstraction mechanism directly. The third is to show that the semantics of a design produced by (a) is consistent with (b).

In the following, we specify and verify two Pebble abstraction mechanisms: hierarchical blocks, and GENERATE-FOR loops. The insight is to recognise that pass separation provides a framework for the above approach, so that the correctness of the two abstraction mechanisms can be demonstrated with respect to a Structural Operational Semantics [16] for Pebble. Only elementary understanding of such semantics is required to follow our work. We shall first present an overview of pass separation, and then explain how it can be used in verifying Pebble abstraction mechanisms. The key to our proof is an environment invariant – Equation (1) – inspired by pass separation.

Consider a repeated computation, part of whose input context remains invariant across all repetitions. Program staging is a technique which improves performance by separating the computation into two phases. We follow this approach to separate the task of interpreting a Pebble program on a variety of inputs: an early phase flattens the parametrised description into a collection of primitive gate calls, and a late phase completes the task given the varying inputs.

Two popular methods for separating a computation into stages are partial evaluation and pass separation [11]. We have used both methods in our study of Pebble and in tool development; in the following we shall focus on pass separation as a means to study abstraction mechanisms for Pebble.

Pass separation constructs, from a program p, two programs p_1, p_2 such that:

$$\llbracket p \rrbracket (x, y) = \llbracket p_2 \rrbracket (\llbracket p_1 \rrbracket x, y)$$

for all x and y, where $\llbracket p \rrbracket$ is the function mapping the program p to its meaning. The equation indicates that $\llbracket p \rrbracket$ can be split into two stages: computing $v = \llbracket p_1 \rrbracket x$ and $\llbracket p_2 \rrbracket (v, y)$. The intention in performing pass separation is to "move" as many computations from p to p_1 as possible, given only input x. In our framework, let p be the semantics ($\mathcal{PS}$) of Pebble, x be a parametrised circuit description C, and y be some input data. Then p_1 corresponds to the abstraction mechanism, which in this case can be described using a flattening procedure ($\mathcal{FP}$). Similarly p_2 corresponds to the semantics ($\mathcal{FS}$) of the flattened description on the data as shown in Fig. 3.

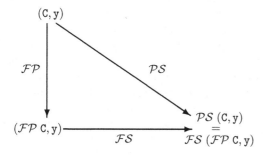

Fig. 3. Commuting diagram describing the pass separation equation.

We shall restrict our attention to a subset of Pebble which does not include vectors or GENERATE-IF statements. We begin by presenting the semantics of Flattened Pebble before adding the necessary structure to create Hierarchical Pebble. We then display a procedure for instantiating the generics. We show that for a set of input values, a flattened description produces the same results as those from the hierarchical description.

3.1 Flattened Pebble

In its most basic form, a circuit consists of a collection of primitive block calls mapping input wires to output wires. Intermediate wires link these primitive block calls together. A circuit description is enclosed within a main block and primitive block identifiers are denoted by $id_{\mathcal{P}}$, as shown in the following syntax:

$$
\begin{aligned}
circuit ::=\ &\texttt{BLOCK main} \\
&[id_{in_1} : type_{in_1}, \dots, id_{in_n} : type_{in_n}] \\
&[id_{out_1} : type_{out_1}, \dots, id_{out_m} : type_{out_m}] \\
&dec_1;\ \cdots\ ; dec_j \\
&\texttt{BEGIN } stmts \texttt{ END} \\
type\ ::=\ &\texttt{WIRE} \\
dec\ ::=\ &\texttt{VAR } id : type \\
stmts\ ::=\ &stmt_1;\ \cdots\ ; stmt_k \\
stmt\ ::=\ &id_{\mathcal{P}}\ [id_1, \dots, id_n]\ [id_1, \dots, id_m]
\end{aligned}
$$

Note that the language is applicative, as each wire is given an attribute only once. Input wires are defined and given the appropriate values; the purpose of the semantics is to find suitable definitions for the output wires. The semantic domain for wires is parametrised by the metavariable a, so that primitive objects can be of any type; it allows us to deal with both bit-level descriptions and word-level descriptions.

$$\textbf{data } \text{wire a} = \text{Undefined} \mid \text{Defined a}$$

To deal with sequential circuits, the datatype a can be lifted to the stream domain [9].

The Structural Operational Semantics rules for Flattened Pebble, defined by $\rightarrow$ transitions [16], are given in Fig. 4. A local environment ρ maps identifiers to their wires. Primitive logic operators that map boolean pairs to booleans, such as xor, are held in an environment labeled δ. Such operators can only be applied to wires that are defined.

Two rules provide the meaning of primitive gate calls. If one or more inputs are Undefined, then the statement is returned unevaluated, as the gate call cannot be completed. The second rule applies the primitive function to the gate's parameters. Statements can be evaluated in any order; those that complete update the local environment ρ. When all statements have been reduced, the final environment is returned.

The output and intermediate wires of the main block are initially declared as Undefined. The block's statements are evaluated to calculate the final environment ρ', from which the output wires are extracted.

3.2 Hierarchical Pebble

Parametrised designs require the addition of a separate parameter list for generic values. Blocks other than main can receive values that define the bounds of loops or that can

$$\frac{\exists j \cdot 1 \le j \le n \land (\rho \; id_j) = \mathsf{Undefined}}{\begin{array}{c} \delta \vdash \langle id_{\mathcal{P}} \; [id_1, \ldots, id_n] \; [id'_1, \ldots, id'_m], \rho \rangle \\ \rightarrow_{stmt} \langle id_{\mathcal{P}} \; [id_1, \ldots, id_n] \; [id'_1, \ldots, id'_m], \rho \rangle \end{array}}$$

$$\frac{\begin{array}{c} (\rho \; id_1) = \mathsf{Defined} \; v_1 \land \cdots \land (\rho \; id_n) = \mathsf{Defined} \; v_n \\ (\delta \; id_{\mathcal{P}}) \; (v_1, \ldots, v_n) = (v'_1, \ldots, v'_m) \end{array}}{\begin{array}{c} \delta \vdash \langle id_{\mathcal{P}} \; [id_1, \ldots, id_n] \; [id'_1, \ldots, id'_m], \rho \rangle \\ \rightarrow_{stmt} \rho \oplus \{id'_1 \mapsto \mathsf{Defined} \; v'_1, \cdots, id'_m \mapsto \mathsf{Defined} \; v'_m\} \end{array}}$$

$$\frac{\delta \vdash \langle stmt_i, \rho \rangle \rightarrow_{stmt} \rho'}{\begin{array}{c} \delta \vdash \langle stmt_1; \cdots stmt_{i-1}; stmt_i; stmt_{i+1}; \cdots stmt_k, \rho \rangle \\ \rightarrow_{stmts} \langle stmt_1; \cdots stmt_{i-1}; stmt_{i+1}; \cdots stmt_k, \rho' \rangle \end{array}}$$

$$\frac{\delta \vdash \langle stmt_i, \rho \rangle \rightarrow_{stmt} \langle stmt_i, \rho \rangle}{\begin{array}{c} \delta \vdash \langle stmt_1; \cdots stmt_{i-1}; stmt_i; stmt_{i+1}; \cdots stmt_k, \rho \rangle \\ \rightarrow_{stmts} \langle stmt_1; \cdots stmt_{i-1}; stmt_i; stmt_{i+1}; \cdots stmt_k, \rho \rangle \end{array}}$$

$$\delta \vdash \langle [\,], \rho \rangle \rightarrow_{stmts} \rho$$

$$\frac{\begin{array}{l} \rho_1 = \{id_{in_1} \mapsto \mathsf{Defined} \; v_1, \cdots, id_{in_n} \mapsto \mathsf{Defined} \; v_n\} \\ \rho_2 = \{id_{out_1} \mapsto \mathsf{Undefined}, \cdots, id_{out_m} \mapsto \mathsf{Undefined}\} \\ \rho_3 = \{id_1 \mapsto \mathsf{Undefined}, \cdots, id_j \mapsto \mathsf{Undefined}\} \\ \rho = \rho_1 \oplus \rho_2 \oplus \rho_3 \\ \delta \vdash \langle stmts, \rho \rangle \rightarrow_{stmts} \rho' \end{array}}{\delta \vdash \left\langle \left(\begin{array}{l} \texttt{BLOCK main} \\ [id_{in_1}:\texttt{WIRE}, \ldots, id_{in_n}:\texttt{WIRE}] \\ [id_{out_1}:\texttt{WIRE}, \ldots, id_{out_m}:\texttt{WIRE}] \\ \quad \texttt{VAR} \; id_1:\texttt{WIRE}; \; \ldots \; \texttt{VAR} \; id_j:\texttt{WIRE} \\ \texttt{BEGIN} \; stmts \; \texttt{END} \end{array} \right), [v_1, \ldots, v_n] \right\rangle}$$
$$\rightarrow_{main} [(\rho' \; id_{out_1}), \ldots, (\rho' \; id_{out_m})]$$

Fig. 4. Semantics of Flattened Pebble, based on $\rightarrow$ rules for *main*, *stmts* and *stmt*.

be passed to subsequent gate calls as defined in the syntax for Hierarchical Pebble:

$$
\begin{array}{lll}
circuit & ::= & main; \; block_1; \; \cdots \; ; \; block_i \\
main & ::= & \texttt{BLOCK main} \\
& & \quad [id_{in_1}:type_{in_1}, \ldots, id_{in_n}:type_{in_n}] \\
& & \quad [id_{out_1}:type_{out_1}, \ldots, id_{out_m}:type_{out_m}] \\
& & \quad dec_1; \; \cdots \; ; \; dec_j \\
& & \texttt{BEGIN} \; stmts \; \texttt{END} \\
block & ::= & \texttt{BLOCK} \; id \; (id_{gen_1}, \ldots, id_{gen_q}) \\
& & \quad [id_{in_1}:type_{in_1}, \ldots, id_{in_n}:type_{in_n}] \\
& & \quad [id_{out_1}:type_{out_1}, \ldots, id_{out_m}:type_{out_m}] \\
& & \quad dec_1; \; \cdots \; ; \; dec_j \\
& & \texttt{BEGIN} \; stmts \; \texttt{END}
\end{array}
$$

$$
\begin{aligned}
type &::= \texttt{WIRE} \\
dec &::= \texttt{VAR}\ id : type \mid \texttt{VAR}\ id : \texttt{NUM} \\
stmts &::= stmt_1 ; \ \cdots\ ; \ stmt_k \\
stmt &::= id_{\mathcal{P}}\ [id_1, \ldots, id_n]\ [id_1, \ldots, id_m] \\
&\quad \mid\ id\ (exp_1, \ldots, exp_q)\ [id_1, \ldots, id_n]\ [id_1, \ldots, id_m] \\
&\quad \mid\ \texttt{GENERATE FOR}\ id = exp_1\ ..\ exp_2 \\
&\quad\quad \texttt{BEGIN}\ stmts\ \texttt{END} \\
exp &::= id \mid n \mid exp_1\ bop\ exp_2 \mid uop\ exp
\end{aligned}
$$

The semantic rules for Hierarchical Pebble statements, defined by $\Rightarrow$ transitions, are given in Fig. 5. Two new environments are introduced: Γ maps block names to their bodies, while σ maps generic variables and loop indices to their values. Arithmetic expressions are evaluated by the valuation function $\mathcal{E}$ in an appropriate value environment. The rules for primitive gate calls and statement lists remain essentially unchanged except for the additional environments.

$$
\frac{\mathcal{E}_\sigma[\![exp_1]\!] > \mathcal{E}_\sigma[\![exp_2]\!]}{\Gamma,\delta,\sigma \vdash \left\langle \left(\begin{array}{l} \texttt{GENERATE FOR}\ id_{index} = exp_1 ..\ exp_2 \\ \texttt{BEGIN}\ stmts\ \texttt{END} \end{array} \right), \rho \right\rangle \Rightarrow_{stmt} \rho}
$$

$$
\frac{\begin{array}{c} \mathcal{E}_\sigma[\![exp_1]\!] \le \mathcal{E}_\sigma[\![exp_2]\!] \\ \Gamma,\delta,\sigma \oplus \{id \mapsto \mathcal{E}_\sigma[\![exp_1]\!]\} \vdash \langle stmts, \rho \rangle \Rightarrow_{stmts} \rho' \\ \Gamma,\delta,\sigma \vdash \left\langle \left(\begin{array}{l} \texttt{GENERATE FOR}\ id_{index} = (exp_1+1) ..\ exp_2 \\ \texttt{BEGIN}\ stmts\ \texttt{END} \end{array} \right), \rho' \right\rangle \Rightarrow_{stmt} \rho'' \end{array}}{\Gamma,\delta,\sigma \vdash \left\langle \left(\begin{array}{l} \texttt{GENERATE FOR}\ id_{index} = exp_1 ..\ exp_2 \\ \texttt{BEGIN}\ stmts\ \texttt{END} \end{array} \right), \rho \right\rangle \Rightarrow_{stmt} \rho''}
$$

$$
\vdots
$$

$$
\frac{\begin{array}{c} (\Gamma\ id) = \left(\begin{array}{l} \texttt{BLOCK}\ id\ (id_{gen_1}, \ldots, id_{gen_q})\ [id_{in_1}:\texttt{WIRE}, \ldots, id_{in_n}:\texttt{WIRE}] \\ \qquad\qquad\qquad\qquad\qquad\qquad [id_{out_1}:\texttt{WIRE}, \ldots, id_{out_m}:\texttt{WIRE}] \\ \quad \texttt{VAR}\ id_{index_1} : \texttt{NUM}; \ \ldots\ \texttt{VAR}\ id_{index_j} : \texttt{NUM}; \\ \quad \texttt{VAR}\ id_{local_1} : \texttt{WIRE}; \ \ldots\ \texttt{VAR}\ id_{local_k} : \texttt{WIRE} \\ \texttt{BEGIN}\ stmts\ \texttt{END} \end{array} \right) \\ \sigma_1 = \{id_{gen_1} \mapsto \mathcal{E}_\sigma[\![e_1]\!], \ldots, id_{gen_q} \mapsto \mathcal{E}_\sigma[\![e_q]\!]\} \\ \rho_1 = \{id_{in_1} \mapsto (\rho\ id_1), \ldots, id_{in_n} \mapsto (\rho\ id_n)\} \\ \rho_2 = \{id_{out_1} \mapsto \mathsf{Undefined}, \ldots, id_{out_m} \mapsto \mathsf{Undefined}\} \\ \rho_3 = \{id_{local_1} \mapsto \mathsf{Undefined}, \ldots, id_{local_k} \mapsto \mathsf{Undefined}\} \\ \rho' = \rho_1 \oplus \rho_2 \oplus \rho_3 \\ \Gamma,\delta,\sigma_1 \vdash \langle stmts, \rho' \rangle \Rightarrow_{stmts} \rho'' \end{array}}{\begin{array}{c} \Gamma,\delta,\sigma \vdash \langle id\ (e_1, \ldots, e_q)\ [id_1, \ldots, id_n]\ [id_1', \ldots, id_m'], \rho \rangle \\ \Rightarrow_{stmt}\ \rho \oplus \{id_1' \mapsto (\rho''\ id_{out_1}), \cdots, id_m' \mapsto (\rho''\ id_{out_m})\} \end{array}}
$$

Fig. 5. Semantics of Hierarchical Pebble. The $\Rightarrow$ rules dealing with primitive statements are similar to the corresponding $\rightarrow$ rules and are not shown.

Two rules are required for loops. The first rule deals with the situation when the loop has terminated. A loop has terminated when the expression representing the lower bound denotes a value that is greater than the upper bound. In such cases the current ρ environment is returned unchanged. Alternatively, the loop's statements are evaluated in a value environment where the loop index is bound to the lower bound value. Once completed, the loop is re-evaluated with the updated ρ' and a lower bound expression that has been incremented by one. The rule for enacting new block calls retrieves the block definition from the environment Γ; it creates a value environment σ_1 by mapping the generic variable names to their values calculated in the outer block's value environment σ; it creates an initial wire environment ρ' by mapping the input variable names to their wire values extracted from ρ and coalesced with Undefined bindings for output and intermediate variable names; and it evaluates the called block's statements to create a final environment ρ'', from which the output wires are extracted. The rule for the main block creates the initial environment ρ in much the same manner as with Flattened Pebble descriptions, and is shown in Fig. 6. The blocks statements are evaluated in an initial empty value environment.

A Hierarchical Pebble description can be flattened by unfolding both the generic variables and the GENERATE-FOR loops. The block environment Γ and the local variable environment σ support the abstraction mechanisms, and do not affect the underlying evaluation mechanism. Hence we can instantiate generic variables prior to the application of input wires, enabling block definitions to be flattened and incorporated into the main block. Flattened Pebble, itself a subset of Pebble, is used as the output language of this flattening process to facilitate its proof. To avoid parameters and local wire names overwriting each other when instantiating block calls, we rename all such variables beforehand using the function $\alpha :: \textit{block} \to \textit{block}$ (Fig. 7).

To model the static behaviour of the wire environment ρ in hierarchical descriptions with local bindings, we introduce a local environment μ which behaves like a symbol table mapping local parameter names to their original definitions, be they inputs to the circuit or local variable definitions. This leads to the invariant equation below, where ρ is the environment for modelling wire values in a hierarchical description, while ρ_d is a dynamic environment for flattened descriptions:

$$\forall id \cdot \rho\ id = \rho_d\ (\mu\ id) \tag{1}$$

This invariant equation will be used extensively in the correctness proof of the flattening procedure for Hierarchical Pebble in Sect. 4. The flattening procedure itself, defined by $\Downarrow$, is given in Fig. 7. Flattening a single statement will result in a pair of lists consisting of primitive gate calls and intermediate wire declarations. The statement list represents those primitive calls required to implement the statement derived from unfolding subsequent parametrised blocks, while the declarations are for the local wire definitions belonging to each unfolded block. The intention is to create the flattened main block from these two lists.

Primitive calls are simply returned with their parameter lists updated with their original variable definitions held in μ. Parametrised gate calls create a new instance of the retrieved block using the function α. Generic variables are bound to their values in σ_1. A static environment μ' is created by mapping the parameter names to their original

$$\rho_1 = \{id_{in_1} \mapsto \text{Defined } v_1, \cdots, id_{in_n} \mapsto \text{Defined } v_n\}$$
$$\rho_2 = \{id_{out_1} \mapsto \text{Undefined}, \cdots, id_{out_m} \mapsto \text{Undefined}\}$$
$$\rho_3 = \{id_1 \mapsto \text{Undefined}, \cdots, id_j \mapsto \text{Undefined}\}$$

$$\frac{\Gamma, \delta, \{\} \vdash \langle stmts, \rho_1 \oplus \rho_2 \oplus \rho_3 \rangle \Rightarrow_{stmts} \rho'}{\begin{array}{l} \Gamma, \delta \vdash \left\langle \begin{array}{l} \texttt{BLOCK main} \\ [id_{in_1}\texttt{:WIRE}, \ldots, id_{in_n}\texttt{:WIRE}] \\ [id_{out_1}\texttt{:WIRE}, \ldots, id_{out_m}\texttt{:WIRE}] \\ \quad \texttt{VAR } id_1\texttt{:WIRE; } \ldots \texttt{ VAR } id_j\texttt{:WIRE} \\ \texttt{BEGIN } stmts \texttt{ END} \end{array} \right\rangle, [v_1, \ldots, v_n] \\ \Rightarrow_{main} \quad [(\rho'\ id_{out_1}), \ldots, (\rho'\ id_{out_m})] \end{array}}$$

Fig. 6. Semantics of the `main` block in Pebble.

$$\frac{\begin{array}{c} \Gamma, \mu, \sigma \vdash \langle id_{\mathcal{P}}\ [id_1, \ldots, id_n]\ [id_1', \ldots, id_m'] \rangle \\ \Downarrow_{stmt}\ ([id_{\mathcal{P}}\ [\mu\ id_1, \ldots, \mu\ id_n]\ [\mu\ id_1', \ldots, \mu\ id_m']], [\,]) \end{array}}{\Gamma, \mu, \sigma \vdash \langle id_{\mathcal{P}}\ [id_1, \ldots, id_n]\ [id_1', \ldots, id_m'] \rangle}$$

$$\begin{array}{l} (\alpha\ (\Gamma\ id)) = \left(\begin{array}{l} \texttt{BLOCK } id\ (id_{gen_1}, \ldots, id_{gen_q})\ [id_{in_1}\texttt{:WIRE}, \ldots, id_{in_n}\texttt{:WIRE}] \\ \qquad\qquad\qquad\qquad\qquad [id_{out_1}\texttt{:WIRE}, \ldots, id_{out_m}\texttt{:WIRE}] \\ \quad \texttt{VAR } id_{index_1}\texttt{ : NUM; } \ldots \texttt{ VAR } id_{index_j}\texttt{ : NUM} \\ \quad \texttt{VAR } id_{local_1}\texttt{ : WIRE; } \ldots \texttt{ VAR } id_{local_k}\texttt{ : WIRE} \\ \texttt{BEGIN } stmts \texttt{ END} \end{array} \right) \\ \mu_1 = \{id_{in_1} \mapsto (\mu\ id_1), \ldots, id_{in_n} \mapsto (\mu\ id_n)\} \\ \mu_2 = \{id_{out_1} \mapsto (\mu\ id_1'), \ldots, id_{out_m} \mapsto (\mu\ id_m')\} \\ \mu_3 = \{id_{local_1} \mapsto id_{local_1}, \ldots, id_{local_k} \mapsto id_{local_k}\} \\ \sigma_1 = \{id_{gen_1} \mapsto \mathcal{E}_\sigma[\![e_1]\!], \ldots, id_{gen_q} \mapsto \mathcal{E}_\sigma[\![e_q]\!]\} \\ \Gamma, \mu_1 \oplus \mu_2 \oplus \mu_3, \sigma_1 \vdash \langle stmts \rangle \Downarrow_{stmts} (stmts', locals') \end{array}$$

$$\frac{}{\Gamma, \mu, \sigma \vdash \langle id\ (e_1, \ldots, e_q)\ [id_1, \ldots, id_n]\ [id_1', \ldots, id_m'] \rangle}$$
$$\Downarrow_{stmt}\ (stmts', [\texttt{VAR } id_{local_1}\texttt{ : WIRE; } \cdots \texttt{; VAR } id_{local_k}\texttt{ : WIRE}] \mathbin{+\mkern-8mu+} locals')$$

$$\frac{\mathcal{E}_\sigma[\![exp_1]\!] > \mathcal{E}_\sigma[\![exp_2]\!]}{\Gamma, \mu, \sigma \vdash \left\langle \left(\texttt{GENERATE FOR } id_{index}=exp_1 \mathbin{..} exp_2 \texttt{ BEGIN } stmts \texttt{ END} \right) \right\rangle\ \Downarrow_{stmt}\ ([\,], [\,])}$$

$$\frac{\begin{array}{c} \mathcal{E}_\sigma[\![exp_1]\!] \leq \mathcal{E}_\sigma[\![exp_2]\!] \\ \Gamma, \mu, \sigma \oplus \{id_{index} \mapsto \mathcal{E}_\sigma[\![exp_1]\!]\} \vdash \langle stmts \rangle \Downarrow_{stmts} (stmts', locals') \\ \Gamma, \mu, \sigma \vdash \left\langle \begin{array}{l} \texttt{GENERATE FOR } id_{index}=(exp_1+1) \mathbin{..} exp_2 \\ \texttt{BEGIN } stmts \texttt{ END} \end{array} \right\rangle \Downarrow_{stmt} (stmts'', locals'') \end{array}}{\begin{array}{c} \Gamma, \mu, \sigma \vdash \left\langle \begin{array}{l} \texttt{GENERATE FOR } id_{index}=exp_1 \mathbin{..} exp_2 \\ \texttt{BEGIN } stmts \texttt{ END} \end{array} \right\rangle \\ \Downarrow_{stmt} (stmts' \mathbin{+\mkern-8mu+} stmts'', locals' \mathbin{+\mkern-8mu+} locals'') \end{array}}$$

$$\frac{\Gamma, \mu, \sigma \vdash \langle stmt_1 \rangle \Downarrow_{stmt} (stmts_1, locals_1) \cdots \Gamma, \mu, \sigma \vdash \langle stmt_n \rangle \Downarrow_{stmt} (stmts_n, locals_n)}{\Gamma, \mu, \sigma \vdash \langle stmt_1; \cdots; stmt_n \rangle \Downarrow_{stmts} (stmts_1 \mathbin{+\mkern-8mu+} \cdots \mathbin{+\mkern-8mu+} stmts_n, locals_1 \mathbin{+\mkern-8mu+} \cdots \mathbin{+\mkern-8mu+} locals_n)}$$

Fig. 7. Transition rules for flattening Pebble statements, based on $\Downarrow$ rules for *stmts* and *stmt*.

names held in μ. Local variables are bound to themselves, but are returned as declarations so that they are properly declared at run time. The blocks statements are flattened and returned with the local wire declarations. The two rules for loops apply the unfolding procedure at compile time to their subterms by incrementing the loop bound, creating primitive gate calls to implement the loop at run time. The rule for statements flattens each statement, and collects the intermediate calls and declarations together.

The flattening rule for the main block is shown in Fig. 8. An initial static environment μ is created binding input, output and local wire names to themselves as these will be their run-time names. The body of the block is flattened with an empty value environment. The returned list of primitives gate calls forms the body of the flattened main block and the derived intermediate wires are declared local to this block.

$$\mu_1 = \{id_{in_1} \mapsto id_{in_1}, \cdots, id_{in_n} \mapsto id_{in_n}\}$$
$$\mu_2 = \{id_{out_1} \mapsto id_{out_1}, \cdots, id_{out_m} \mapsto id_{out_m}\}$$
$$\mu_3 = \{id_1 \mapsto id_1, \cdots, id_j \mapsto id_j\}$$
$$\mu = \mu_1 \oplus \mu_2 \oplus \mu_3$$
$$\Gamma, \mu, \{\} \vdash \langle stmts \rangle \Downarrow_{stmts} (stmts', [\text{VAR } id'_1 : \text{WIRE}; \ldots \text{VAR } id'_k : \text{WIRE}])$$

$$\Gamma \vdash \left\langle \begin{pmatrix} \text{BLOCK main} & [id_{in_1} : \text{WIRE}, \ldots, id_{in_n} : \text{WIRE}] \\ & [id_{out_1} : \text{WIRE}, \ldots, id_{out_m} : \text{WIRE}] \\ \text{VAR } id_1 : \text{WIRE}; & \ldots \text{VAR } id_j : \text{WIRE} \\ \text{BEGIN } stmts \text{ END} \end{pmatrix} \right\rangle$$

$$\Downarrow_{main} \begin{pmatrix} \text{BLOCK main} & [id_{in_1} : \text{WIRE}, \ldots, id_{in_n} : \text{WIRE}] \\ & [id_{out_1} : \text{WIRE}, \ldots, id_{out_m} : \text{WIRE}] \\ \text{VAR } id_1 : \text{WIRE}; & \ldots \text{VAR } id_j : \text{WIRE}; \\ \text{VAR } id'_1 : \text{WIRE}; & \ldots \text{VAR } id'_k : \text{WIRE} \\ \text{BEGIN } stmts' \text{ END} \end{pmatrix}$$

Fig. 8. Transition rules for flattening the `main` block.

To illustrate how blocks are flattened and local variables are renamed, we shall use the following example which creates a row of two not-gates:

```
BLOCK notrow (n) [vin:WIRE] [vout:WIRE]
    VAR inter:VECTOR (n..0) OF WIRE
    VAR i:NUM
BEGIN
    connect [vin] [inter(0)];
    GENERATE FOR i=0..n-1
    BEGIN
        not [inter(i)] [inter(i+1)]
    END;
    connect [inter(n)] [vout]
END;
```

```
BLOCK main [x:WIRE] [y:WIRE]
BEGIN
  notrow (2) [x] [y]
END;
```

The circuit is defined in terms of two primitive components: connect which links two wires together, and the not gate. Figure 9 demonstrates how a block call is flattened given previously established environments. The called block is initially renamed; the environment σ is created for the generic values; the environment μ maps parameter names to their original names; the block's statements are then flattened to create a list of primitive calls, and returned along with the distinct local wire definitions.

$$\vdots$$

$$\Gamma, \mu', \sigma_1 \vdash \left\langle \begin{array}{l} \text{connect [vin1] [i1(0)];} \\ \text{GENERATE FOR} \cdots \\ \quad \vdots \\ \text{connect [i1(n1)] [vot1]} \end{array} \right\rangle \Downarrow_{stmts} ([\text{connect [x] [i1(0)];} \\ \qquad\qquad \text{not [i1(0)] [i1(1)];} \\ \qquad\qquad \text{not [i1(1)] [i1(2)];} \\ \qquad\qquad \text{connect [i1(2)] [y]],} \\ \qquad\qquad []) $$

$\sigma_1 = \{n1 \mapsto 2\}$
$\mu_1 = \{vin1 \mapsto x\}$ $\mu_2 = \{vot1 \mapsto y\}$ $\mu_3 = \{i1 \mapsto i1\}$
$\mu' = \mu_1 \oplus \mu_2 \oplus \mu_3$

```
(α(Γ notrow)) = BLOCK notrow (n1) [vin1:WIRE] [vot1:WIRE]
                VAR i1:VECTOR (n1..0) OF WIRE;
                VAR i2:NUM
              BEGIN
                connect [vin1] [i1(0)];
                GENERATE FOR i2=0..n1-1
                BEGIN   not [i1(i2)] [i1(i2+1)]   END;
                connect [i1(n1)] [vot1]
              END;
```

$\Gamma, \{x \mapsto x, y \mapsto y\}, \{\} \vdash \langle \text{notrow (2) [x] [y]} \rangle$
$\qquad\qquad \Downarrow_{stmt} ([\text{connect [x] [i1(0)];}$
$\qquad\qquad\qquad \text{not [i1(0)] [i1(1)];}$
$\qquad\qquad\qquad \text{not [i1(1)] [i1(2)];}$
$\qquad\qquad\qquad \text{connect [i1(2)] [y]],[])}$
$\qquad\qquad\qquad [\text{VAR i1:VECTOR (2..0) OF WIRE}])$

Fig. 9. Fragment of the proof tree for flattening notrow.

4 Correctness Proof

We can now present the main correctness theorem for flattening hierarchical blocks and GENERATE-FOR loops. This result, given by Equation (2), relies on Lemmas (3), (4) and (5). Each lemma is an instance of the commuting diagram given in Fig. 3, and involves the environment invariant given by Equation (1). At each syntactic level, they show how the sequence of outputs generated by the hierarchical definition can be calculated by first flattening the term and then using the simplified semantics. With the addition of GENERATE-IF statements, recursive block definitions can result in non-terminating programs. In these cases the flattening procedure will also fail to terminate.

The main theorem states that, from a given Hierarchical Pebble description consisting of a main block, a block environment Γ, a primitive gate environment δ, and a sequence of input wires $[v_1, \ldots, v_n]$, one can calculate the sequence of outputs derived from the circuit's proof trees by first unfolding the description to create one large main block, to which the flattening rules can be applied:

$$
\begin{aligned}
\Gamma, \delta \vdash \langle main, [v_1, \ldots, v_n] \rangle \Rightarrow_{main} [v'_1, \ldots, v'_m] \\
\implies \Gamma \vdash main \Downarrow_{main} main' \\
\wedge \quad \delta \vdash \langle main', [v_1, \ldots, v_n] \rangle \rightarrow_{main} [v'_1, \ldots, v'_m]
\end{aligned}
\tag{2}
$$

This result can be proved by structural induction on the rules for the main block (Figs. 4, 6 and 8). We establish the invariant on wire environments initially, and we use Lemma (3) below to show how it holds on completion of the block's statements so that the same final values are derived.

From a given Hierarchical Pebble statement list, a block environment Γ, a primitive gate environment δ, a local value environment σ and a local wire environment ρ, we can calculate the wire bindings ρ' derived from a successful completion of the statements, by staging the computation in two. The first stage flattens the statements into a list of primitive calls, where local wire names are mapped to their original definitions using the static environment μ, and a distinct list of local wire declarations is created. The second stage applies the Flattened Pebble rules to the primitive gate call list using the dynamic wire environment ρ_d. An environment ρ'_d can be derived that will contain the same bindings as those for the hierarchical statements. This implication requires the invariant, given by Equation (1), to hold for wire environments:

$$
\begin{aligned}
\Gamma, \delta, \sigma \vdash \langle stmts, \rho \rangle \Rightarrow_{stmts} \rho' \\
\implies \Gamma, \sigma, \mu \vdash stmts \Downarrow_{stmts} (stmts', locals') \\
\wedge \quad \forall id \cdot \rho\, id = \rho_d(\mu\, id) \\
\wedge \quad \forall id \cdot \rho'\, id = \rho'_d(\mu\, id) \\
\wedge \quad \delta \vdash \langle stmts', \rho_d \rangle \rightarrow_{stmts} \rho'_d
\end{aligned}
\tag{3}
$$

This lemma can be proved by induction on the length of derivation sequences using Lemma (4) below; it completes the presentation of the main theorem.

The next lemma deals mainly with GENERATE-FOR loops (Figs. 4, 5 and 7). From a given Hierarchical Pebble statement, a block environment Γ, a primitive gate environment δ, a local value environment σ, and a local wire environment ρ, one can calculate a set of wire bindings ρ' derived from the successful completion of the statement, by first flattening the statement using the static environment μ, and then executing the derived statements in the dynamic wire environment ρ_d:

$$
\begin{aligned}
\Gamma, \delta, \sigma \vdash \langle stmt, \rho \rangle \Rightarrow_{stmt} \rho' \\
\Longrightarrow \Gamma, \sigma, \mu \vdash stmt \Downarrow_{stmt} (stmts', locals') \\
\wedge \quad \forall id \cdot \rho \ id = \rho_d(\mu \ id) \\
\wedge \quad \forall id \cdot \rho' \ id = \rho'_d(\mu \ id) \\
\wedge \quad \delta \vdash \langle stmts', \rho_d \rangle \rightarrow_{stmt} \rho'_d
\end{aligned}
\tag{4}
$$

This result can be proved by structural induction on statements: primitive gate calls, parametrised gate calls and loops. The first two cases are straightforward, once the invariants of the environments have been established. The third case, however, requires Lemma (5) to show that the appropriate final environment can be derived after staging:

$$
\begin{pmatrix}
\Gamma, \delta, \sigma_1 \vdash \langle stmts_1, \rho \rangle \Rightarrow_{stmts} \rho' \\
\wedge \ \Gamma, \delta, \sigma_2 \vdash \langle stmts_2, \rho' \rangle \Rightarrow_{stmts} \rho''
\end{pmatrix}
\\
\begin{aligned}
\Longrightarrow \Gamma, \sigma_1, \mu \vdash stmts_1 \Downarrow_{stmts} (stmts'_1, locals'_1) \\
\wedge \quad \Gamma, \sigma_2, \mu \vdash stmts_2 \Downarrow_{stmts} (stmts'_2, locals'_2) \\
\wedge \quad \forall id \cdot \rho \ id = \rho_d(\mu \ id) \\
\wedge \quad \forall id \cdot \rho'' \ id = \rho''_d(\mu \ id) \\
\wedge \quad \delta \vdash \langle stmts_1 \mathbin{+\!\!+} stmts_2, \rho_d \rangle \rightarrow_{stmts} \rho''_d
\end{aligned}
\tag{5}
$$

This lemma states that reducing a statement list $stmts_1$ in the value environment σ_1 with wire bindings ρ, followed by reducing a second statement list $stmts_2$ in σ_2 yielding a final wire environment ρ'', can be derived by first flattening the two statement lists and then executing the concatenation of the two primitive gate call lists. The lemma can be proved by induction on the length of derivation sequences.

5 Compiler Development

This section reflects on the implications of our approach for compiler development. Natural semantic rules, as used in specifying Pebble, rely on notions of pattern matching, inference rules and operational semantics. They can be captured in a theorem prover [2], [17] or translated into Horn clauses via a metalanguage such as Typol [3]. Since the transition rules for flattening Hierarchical Pebble descriptions permit a left-right and top-down construction of the proof tree with no backtracking, we can replace a resolution engine by a functional evaluator based on pattern matching [1] to improve efficiency.

In practice, an implementation in a functional language of the core flattening procedure (Fig. 10) follows naturally from the rules in Fig. 7. For a particular language construct, a function definition is created that pattern matches its goal and obtains the result from the intermediate transitions by means of a where clause. This technique offers a way of automatically producing a functional implementation of the compilation tools directly from their specifications.

It is educational to compare the original, hand-developed implementation of the flattening procedure, and the new version in Fig. 10 which results from the specification and proof exercises. The new version is better than the original version in all aspects:

```
data Exp  = Number Int | Var String
            | Binop (Exp,Bop,Exp) | Unop (Uop,Exp)
data Type = WIRE
data Dec  = VARW (String,Type) | VARN String
data Stmt = PrimCall (String,[String],[String])
            | BlkCall (String,[Exp],[String],[String])
            | Loop (String,Exp,Exp,[Stmt])
data Blk  = Block (String,[String],
              [(String,Type)],[(String,Type)],[Dec],[Stmt])

fetch :: [(String,a)] -> String -> a
...
eval_exp :: Exp -> [(String,Int)] -> Int
...

flatten_stmt :: ([(String,Blk)],
  [(String,String)],[(String,Int)]) -> Stmt -> ([Stmt],[Dec])
flatten_stmt (gamma,mu,sigma) (PrimCall (pnm,args1,args2))
  = ([PrimCall (pnm,map (fetch mu) args1,
                     map (fetch mu) args2)],[])
flatten_stmt (gamma,mu,sigma) (BlkCall (bnm,gens,args1,args2))
  = (stmts',[(VARW d) | (VARW d) <- decs] ++ locals')
    where
          (Block (nm,gennms,parms1,parms2,decs,stmts))
                   = rename (fetch gamma bnm)
          sigma1 = zip gennms
                     (map (\ e -> eval_exp e sigma) gens)
          mu1    = zip parms1 (map (fetch mu) args1)
          mu2    = zip parms2 (map (fecth mu) args2)
          mu3    = [(id,id) | (VARW (id,WIRE)) <- decs]
          mu'    = mu1 ++ mu2 ++ mu3
          (stmts',locals')
                   = flatten_stmts (gamma,mu',sigma1) stmts
flatten_stmt (gamma,mu,sigma) (Loop (nm,e1,e2,stmts))
  | n1>n2        = ([],[])
  | otherwise    = (stmts'++stmts'',locals'++locals'')
    where
      n1 = eval_exp e1 sigma
      n2 = eval_exp e2 sigma
      (stmts',locals')
          = flatten_stmts (gamma,mu,(nm,n1):sigma) stmts
      (stmts'',locals'')
          = flatten_stmt (gamma,mu,sigma)
                   (Loop (nm,Binop (e1,Add,Number 1),e2,stmts))
flatten_stmts :: ([(String,Blk)],
  [(String,String)],[(String,Int)]) -> [Stmt]-> ([Stmt],[Dec])
flatten_stmts (gamma,mu,sigma) = unzip . map flatten_stmt
```

Fig. 10. Flattening Pebble in Haskell.

it is clearer, more concise, more robust and more efficient. The main reason is that, based on the formal development, the new version separates variable renaming from the flattening procedure. The original implementation, in contrast, attempted the former without properly considering the latter: the mingling of the two leads to situations where the renaming process is deeply nested within the unfolding procedure, leaving little scope for further optimisations. The new version is amenable to further optimisations, such as the use of de Bruijn indices to avoid the costly rename function [5]. Further refinements would lead to a highly efficient imperative implementation.

Our experience shows that deriving provably-correct compiler implementations can benefit their efficiency, in addition to increasing the confidence in their correctness. Furthermore, it appears possible that the verification of such implementations for domain-specific languages such as Pebble can be mechanised using custom proof engines [2]. Further work, however, is needed to explore this possibility thoroughly.

6   Concluding Remarks

While the version of Pebble described in this paper does not include advanced abstraction mechanisms, current work involves extending Pebble with polymorphic variables, records and higher-order functions. These features enable a combinator style of development [10] that tends to simplify the hardware design process.

Our extended compilation strategy infers the types of the polymorphic variables, unfolds the record definitions and instantiates higher-order functions prior to compile time to create a Hierarchical Pebble description. The correctness proof for Polymorphic Pebble is very similar to that for Hierarchical Pebble. An intermediate environment mapping polymorphic variables to types is used to create distinct blocks, and it leads to an invariant equation similar to Equation (1). Higher-Order Pebble enables nested function calls which require lambda lifting before the calls can be unfolded. In this way the ability to generate correct parametrised VHDL will be maintained.

The combinator style of description facilitates the formulation of correctness-preserving algebraic transformations for design development [10]. The proposed extensions of Pebble take us a step closer to providing, for instance, a generic transformation rule [13] which can be used to derive pipelined designs from a non-pipelined design. Further work will generalise our approach to deal with relational descriptions [10].

Our research contributes to insights about abstraction mechanisms and their validated implementations. It also provides a useful foundation on which further work, such as verifying tools for pipeline optimisation [19], can be based. We believe that provably-correct tools will have a profound impact on understanding the scope and effectiveness of hardware synthesis algorithms and their implementation.

Acknowledgements. Our thanks to the many Pebble developers, users and advisors, particularly Chris Booth, Ties Bos, Arran Derbyshire, Florent Dupont-De-Dinechin, Mike Gordon, He Jifeng, Tony Hoare, Danny Lee, James Rice, Richard Sandiford, Seng Shay Ping, Nabeel Shirazi, Dimitris Siganos, Henry Styles, Tim Todman and Markus Weinhardt, for their patience, help and encouragement. Thanks also to the comments and suggestions of anonymous reviewers of this paper. We acknowledge the support of

UK Engineering and Physical Sciences Research Council, Celoxica, and Xilinx. This work was carried out as part of Technology Group 10 of UK MOD's Corporate Research Programme.

References

1. I. Attali, J. Chazarain and S. Gilette, "Incremental evaluation of natural semantics specifications", *Proc. 4th Int. Symp. on Programming Language Implementation and Logic Programming*, LNCS 631, Springer, 1992.
2. L.A. Dennis et. al., "The PROSPER toolkit", *Proc. 6th Int. Conf. on Tools and Algorithms for the Construction and Analysis of Systems*, LNCS 1785, Springer, 2000.
3. T. Despeyroux, "Executable specification of static semantics", *Semantics of Data Types*, LNCS 173, Springer, 1984.
4. D. Eisenbiegler, C. Blumenroehr and R. Kumar, "Implementation issues about the embedding of existing high level synthesis algorithms in HOL", *Theorem Proving in Higher Order Logics*, LNCS 1125, Springer, 1996.
5. C. Hankin, *Lambda Calculus, A guide for Computer Scientists*, Oxford University Press, 1994.
6. J. He, G. Brown, W. Luk and J.W. O'Leary, "Deriving two-phase modules for a multi-target hardware compiler", *Designing Correct Circuits*, Springer Electronic Workshop in Computing, 1996.
7. J. He, I. Page and J.P. Bowen, "Towards a provably correct hardware implementation of occam", *Correct Hardware Design and Verification Methods*, LNCS 683, Springer, 1993.
8. C.A.R. Hoare, J. He and A. Sampaio, "Normal form approach to compiler design", *Acta Informatica*, Vol. 30, pp. 701–739, 1993.
9. G. Jones and M. Sheeran, "Timeless truths about sequential circuits", *Concurrent Computations: Algorithms, Architectures and Technology*, S.K. Tewksbury et. al. (eds.), Plenum Press, 1988.
10. G. Jones and M. Sheeran, "Circuit design in Ruby", *Formal Methods for VLSI Design*, J. Staunstrup (ed.), North-Holland, 1990.
11. U. Jørring and W. Scherlis, "Compilers and staging transformations", *Proc. ACM Symp. on Principles of Programming Languages*, ACM Press, 1986.
12. W. Luk and S.W. McKeever, "Pebble: a language for parametrised and reconfigurable hardware design", *Field-Programmable Logic and Applications*, LNCS 1482, Springer, 1998.
13. W. Luk et. al., "A framework for developing parametrised FPGA libraries", *Field-Programmable Logic and Applications*, LNCS 1142, Springer, 1996.
14. W. Luk et. al., "Reconfigurable computing for augmented reality", *Proc. IEEE Symp. on Field-Programmable Custom Computing Machines*, IEEE Computer Society Press, 1999.
15. N. Mansouri and R. Vemuri, "A methodology for completely automated verification of synthesized RTL designs and its integration with a high-level synthesis tool", *Formal Methods in Computer-Aided Design*, LNCS 1522, Springer, 1998.
16. H.R. Nielson and F. Nielson, *Semantics with Applications*, John Wiley and Sons, 1992.
17. F. Pfenning and C. Schrmann, "System description: Twelf – a meta-logical framework for deductive systems", *Proc. Int. Conf. on Automated Deduction*, LNAI 1632, Springer, 1999.
18. M. Sheeran, S. Singh and G. Stalmarck, "Checking safety properties using induction and a SAT-solver", *Proc. Int. Conf. on Formal Methods in CAD*, LNCS 1954, Springer, 2000.
19. M. Weinhardt and W. Luk, "Pipeline vectorization", *IEEE Trans. CAD*, Vol. 20, No. 2, 2001, pp. 234–248.

A Higher-Level Language for Hardware Synthesis

Richard Sharp and Alan Mycroft

Computer Laboratory, Cambridge University
New Museums Site, Pembroke Street, Cambridge CB2 3QG, UK
{rws26,am}@cl.cam.ac.uk

Abstract. We describe SAFL+: a call-by-value, parallel language in
the style of ML which combines imperative, concurrent and functional
programming. Synchronous channels allow communication between parallel threads and π-calculus style channel passing is provided. SAFL+
is designed for hardware description and synthesis; a silicon compiler,
translating SAFL+ into RTL-Verilog, has been implemented.
By parameterising functions over both data and channels the SAFL+
fun declaration becomes a powerful abstraction mechanism unifying a
range of structuring techniques treated separately by existing HDLs.
We show how SAFL+ is implemented at the circuit level and define the
language formally by means of an operational semantics.

1 Introduction

In 1975 a single Integrated Circuit contained several hundred transistors; by
1980 the number had increased to several thousand. Today, designs fabricated
with state-of-the-art VLSI technology often contain several million transistors.

The exponential increase in circuit complexity has forced engineers to adopt
higher-level tools. Whereas in the 1970s transistor and gate-level design was the
norm, during the 1980s Register Transfer Level (RTL) Hardware Description
Languages (HDLs) started to achieve wide-spread acceptance. Using such languages, designers were able to express circuits as hierarchies of components (such
as registers and multiplexers) connected with wires and buses. The advent of this
methodology led to a dramatic increase in productivity since, for some classes
of design, time consuming place-and-route details could now be automated

More recently, *high-level synthesis* (sometimes referred to as *behavioural synthesis*) has started to have an impact on the hardware design industry. In the
last few years commercial tools have appeared on the market enabling high-level,
imperative languages (referred to as *behavioural languages* within the hardware
community) to be compiled directly to hardware. Although these techniques un-
doubtedly offer increased levels of abstraction over RTL specification there is
still room for even higher level HDLs, particularly when it comes to specifying
interfaces between separate components (see Section 1.1). Since current trends
predict that the exponential increase in transistor density will continue throughout the next decade, investigating higher-level tools for hardware description
and synthesis will remain an important research area.

T. Margaria and T. Melham (Eds.): CHARME 2001, LNCS 2144, pp. 228–243, 2001.
© Springer-Verlag Berlin Heidelberg 2001

We present a language designed for hardware synthesis which, we argue, is higher level than existing behavioural synthesis packages. This paper builds on previous work in which we use SAFL [14] (Statically Allocated Functional Language) for circuit description. To reflect this we choose to call our new language SAFL+. Our optimising SAFL silicon compiler [16] has been extended to handle SAFL+ and the resulting system has been tested on a number of small designs. The contributions of this paper are:

- We extend SAFL with synchronous channels and assignment and argue that the resulting combination of functional, concurrent and imperative styles is a powerful framework in which to describe a wide range of hardware designs.
- Channel passing in the style of the π-calculus [12] is introduced. By parameterising functions over both data and channels the SAFL+ **fun** declaration becomes a powerful abstraction mechanism unifying a range of structuring techniques treated separately by existing HDLs (Section 2.3).
- We show how SAFL+ is implemented at the circuit-level (Section 3) and define the language formally by means of an operational semantics (Section 4).

1.1   The Motivation for Higher-Level HDLs

Register Transfer Level HDLs (e.g. RTL Verilog) describe hardware as a set of *blocks* parameterised over input and output ports. Once defined, the blocks can be instantiated (possibly multiple times) and explicitly connected with wires and buses. Although behavioural languages provide higher-level primitives for describing block internals, the block remains the primary abstraction mechanism used to structure large designs. For example, at the top level, a Behavioural Verilog program still consists of **module** declarations and instantiations albeit that the modules themselves contain higher-level constructs such as assignment, sequencing and while-loops.

Experience has shown that the notion of block is a useful syntactic abstraction, encouraging structure by supporting a "define-once, use-many" methodology. However, as a *semantic abstraction* it buys one very little; in particular: (*i*) any part of a block's internals can be exported to its external interface; and (*ii*) inter-block synchronisation mechanisms must be coded explicitly on an ad hoc basis.

Point (*i*) has the undesirable effect of making it difficult to reason about the global (inter-module) effects of local (intra-module) transformations. For example applying small changes to the local structure of a block (e.g. delaying a value's computation by one cycle) may have dramatic effects on the global behaviour of the program as a whole. We believe point (*ii*) to be particularly serious. Firstly, it leads to low-level implementation details scattered throughout a program—e.g. the definition of explicit control signals used to sequence operations in separate modules, or (arguably even worse) reliance on unwritten inter-module timing assumptions. Secondly, it inhibits compiler analysis: since inter-block synchronisation mechanisms are coded on an ad hoc basis it is very difficult for the compiler to infer a system-wide ordering on events. Based on these observations, we argue that structural blocks are not a high-level abstraction mechanism.

Through our previous work on SAFL we demonstrated that these problems can be alleviated by structuring code as a series of function definitions. The properties of functions make it easier to reason about the effects of local transformations. As a result we are able to make extensive use of source-to-source program transformation to assist with architectural exploration [15, 14]. The "invoke and wait for result" interface provided by functions removes the burden of explicitly specifying ad hoc inter-module synchronisation mechanisms. Furthermore our compiler is able to automatically infer a system-wide partial-ordering on events thus increasing the scope for global analysis and optimisation. For example consider the SAFL expression `f(g(3),h(5))`. From this (and the call-by-value property of SAFL) we can infer that g and h will be invoked in parallel, after which f will be invoked. Our SAFL compiler exploits a number of analyses and optimisations based on these event orderings (e.g. Soft Scheduling [17] and Data Validity Analysis [16]).

However, although SAFL is well-suited to describing certain types of hardware design, the facility for I/O is lacking. In addition we sometimes find the "call and wait for result" interface to be a little too restrictive. By extending SAFL with channel-communication, channel passing and assignment we intend SAFL+ to be a truly general purpose hardware description language. In Section 2.3 we show that SAFL+ supports a programming style which relaxes many of SAFL's restrictions without sacrificing analysibility.

Related Work Many parallels can be drawn between SAFL+ and the HardwareC [8] language since both provide synchronous channels and allow function definitions to be treated as shared resources. The major differences are: (i) whereas HardwareC offers purely imperative features, SAFL+ also supports a functional style (we find SAFL+'s let-construct for declaring immutable bindings to be particularly useful for describing data-dominated hardware); (ii) the expressivity of SAFL+ is greater due to our less restrictive scheduling policy [17]; and (iii) HardwareC provides a `block` primitive for structural-level declarations. In contrast SAFL+ only allows function declarations.

An interesting observation is that, as a direct result of SAFL+'s channel passing facility, all four of HardwareC's structuring primitives (`block`, `process`, `procedure` and `function`) can be seen as special cases of SAFL+ `fun` declarations (see Section 2.3). Since SAFL+ only requires a single structuring primitive it yields a simpler semantics.

Hoe and Arvind [5] describe TRAC: a hardware synthesis system which generates synchronous hardware from a high-level specification expressed as a term-rewriting system. Broadly speaking, terms correspond to states and rules correspond to combinatorial logic which calculates the next state of the system. Restrictions imposed on the structure of rewrite rules facilitate the static allocation of storage. This closely corresponds to the tail-recursion restriction imposed on SAFL+ programs to achieve static allocation [14].

Previous work on compiling declarative specifications to hardware has centred on how functional languages can be used as tools to aid the *structural* description of circuits—e.g. muFP [18], Lava [3] and the DDD algebra [7]. Functional programming techniques (such as higher order functions) are used to express concisely the regular, repetitive structures that often appear in hardware circuits. In this framework, different interpretations of primitive functions correspond to various operations (e.g. behavioural simulation and netlist generation). Our work differs from this in that we adopt a *behavioural* approach abstracting circuit-level details as far as possible. For example, when expressing a design in Lava, a programmer must explicitly ensure that design-level constraints (e.g. gate fan-out limits) are satisfied. In contrast we consider this to be a low-level detail: ensuring a circuit conforms to low-level design-rules is the job of our SAFL+ compiler. We take SAFL+ constructs (rather than gates) as primitive. Although this restricts the class of circuits we can describe to those which satisfy certain high-level properties, it increases the scope for high-level analysis and optimisation.

A number of languages have been developed which provide structural abstractions similar to those available in the Lava/muFP/DDD framework. For example HML [9] is one such language based on Standard ML [13]; Jazz [1] combines a polymorphic type-system with object oriented features; Hawk [11], like Lava, is embedded in Haskell, but focuses on simulation rather than synthesis.

2   SAFL+ Language Description

In this section we present the syntax of SAFL+ and informally describe its semantics. The language semantics are defined formally in Section 4.

SAFL+ is a concurrent, first-order, call-by-value language which, in the style of ML [13], supports a combination of functional and imperative programming. Function call arguments and **let**-definitions are evaluated in parallel; synchronous channels allow parallel threads to communicate with each other.

Function declarations take the form:

$$\textbf{fun } f(x_1, \ldots, x_k) \, [c_1, \ldots, c_j] = e \qquad \text{(where } k, j \geq 0)$$

We make a syntactic distinction between arguments used to pass data, $x_1, \ldots, x_k$, and arguments used to pass channels $c_1, \ldots, c_j$. Iteration is provided in the form of self-tail-recursive calls. As with SAFL, general recursion is forbidden to permit static allocation of storage [14]. Programs have a distinguished function, **main**, which represents an external world interface—at the hardware level it accepts values on an input port and may later produce a value on an output port. (In the current version of the language, **main** does not have channel parameters).

In the following a ranges over primitive functions (such as +, * etc.), f ranges over user-defined functions and l ranges over record field labels. We use r for array variables, c for channel variables, x for other variables, and i for integer constants. A vector of parameters, $x_1, \ldots, x_k$, is sometimes abbreviated to $\vec{x}$. The abstract syntax of SAFL+ programs, p, is presented in Fig. 1. The **static**

$$
\begin{aligned}
e \;\leftarrow\;\; & x \;\mid\; i \;\mid\; () && \text{(Variable, Integer constant, Unit constant)}\\
\mid\; & \{l_1 = e_1, \ldots l_k = e_k\} \;\mid\; e.l && \text{(Record creation/selection)}\\
\mid\; & r[e] \;\mid\; r[e] := e && \text{(Array read/write)}\\
\mid\; & c? \;\mid\; c\,!\,e && \text{(Channel read/write)}\\
\mid\; & a(e_1, \ldots, e_k) && \text{(Call to primitive function)}\\
\mid\; & f(e_1, \ldots, e_k)[c_1, \ldots, c_j] && \text{(Call to user-defined function)}\\
\mid\; & \texttt{if } e_1 \texttt{ then } e_2 \texttt{ else } e_3 && \text{(Conditional)}\\
\mid\; & \texttt{let } \vec{x} = \vec{e} \texttt{ in } e_0 && \text{(Parallel let)}\\
\mid\; & \texttt{static } p \texttt{ in } e && \text{(Local declarations)}\\
\mid\; & e \parallel e \;\mid\; e; e && \text{(Parallel/sequential composition)}\\[6pt]
d \;\leftarrow\;\; & \texttt{fun } f(x_1, \ldots, x_n)[c_1, \ldots, c_n] = e && \text{(Function declaration)}\\
\mid\; & \texttt{channel } c && \text{(Channel declaration)}\\
\mid\; & \texttt{channel external } c && \text{(I/O Channel declaration)}\\
\mid\; & \texttt{array } [i]\, r && \text{(Array declaration)}\\[6pt]
p \;\leftarrow\;\; & d \;\mid\; d\, p
\end{aligned}
$$

Fig. 1. The abstract syntax of SAFL+ programs, p

construct, used to introduce local definitions, is provided purely for syntactic convenience. It is borrowed from the C language as a way of providing top-level definitions which are only accessible locally. It is not to be confused with the kind of *dynamic* channel-creation present in the π-calculus.

Our existing compiler provides a number of simple syntactic sugarings: the declaration `array [1] r` can be written `reg r`—when accessing such arrays one writes `r` instead of `r[0]`; functions without channel parameters can omit their square brackets completely (both in definition and calls); and `case`-statements are translated into nested conditionals in the usual way.

2.1 Resource Awareness

Our approach is to model hardware as a fixed set of communicating and (possibly) shared resources. As can be seen from Fig. 1, a program consists of a series of resource declarations. There are three different types of resource, each of which addresses a key element of hardware design:

Resource type	Purpose	Hardware Representation
Function	Computation	General Purpose Logic
Channel	Communication	Buses, Wires and Control Logic
Array	Storage	Memories or Registers

We say that SAFL+ is *resource-aware* since each declaration, d, (be it a function, channel or array declaration) corresponds to a *single* hardware block, H_d. Multiple references to d at the source-level (e.g. multiple calls to a function or multiple assignments to an array) correspond to the sharing of H_d at the circuit-level.

A call, $f(\vec{x})[\vec{c}]$, corresponds to: (i) acquiring mutually exclusive access to resource, H_f; (ii) passing data $\vec{x}$ and channel-parameters $\vec{c}$ into H_f; (iii) waiting for H_f to terminate; and (iv) latching[1] the result from H_f's shared output.

For a concrete example of SAFL+, Fig. 3 describes a lock shared between functions f1 and f2. Synthesising this example leads to three resources: H_{f1}, H_{f2} and H_{lock} with H_{lock} shared between resources H_{f1} and H_{f2}.

Sharing issues, such as ensuring mutually exclusive access to resources, are dealt with automatically by our compiler. In [15] we describe how arbiters are generated to protect shared resources from concurrent accesses. A global analysis is presented which allows redundant arbiters to be optimised away.

Resource-awareness means that, although a SAFL+ compiler is free to optimise the internals of fun definitions, it must respect the circuit structure specified by the programmer (i.e. one declaration = one hardware-level resource). We apply source-to-source program transformation as a pre-compilation phase to express resource sharing/duplication and other area-time tradeoffs [14]. Although such transformations are applied manually at the moment, tools to assist with the transformation process and automatically explore the design space are currently being developed.

2.2   Channels and Channel Passing

SAFL+ provides synchronous channels to allow parallel threads to synchronize and transfer information. Channels can be used to transfer data locally within a function, or globally, between concurrently executing functions.

Our channels generalise Occam [6] and Handel-C [4] channels in a number of ways: SAFL+ channels can have any number of readers and writers, are bidirectional and can connect any number of parallel processes. As in the π-calculus, if there are multiple readers and multiple writers all wanting to communicate on the same channel then a single reader and a single writer are chosen non-deterministically.

At the hardware level a channel is implemented as a many-to-many communications bus supporting the atomic transfer of single values between readers and writers (see Section 3). No language-support is provided for *bus-transactions* (e.g. lock the bus for 20 cycles and write the following sequence of data values onto it). In Section 2.3 a SAFL+ code fragment is presented which shows how such transactions can be implemented by using explicit locking.

Channels declared as external are used for I/O: writing to an external channel corresponds to an output action; reading an external channel corresponds to reading an input. There is no synchronisation on external channels although writes are guaranteed to occur under mutual exclusion. For example, for an external channel c, the only two possible output sequences occurring as a result of evaluating expression (c!2 || c!3) are $\langle 2, 3 \rangle$ or $\langle 3, 2 \rangle$. (See Section 4).

---

[1] A data-flow analysis is used to optimise these latches away under certain circumstances [16].

The code in Fig. 2 illustrates channel-passing in SAFL+ by defining two resources parameterised over channel parameters: `Accumulate` reads integers from a channel, returning their total when a 0 is read; and `GenNumbers` writes a decreasing stream of integers to a channel, terminating when 0 is reached. The function, `sum(x)` calculates the sum of the first `x` integers by composing the two resources in parallel and linking them with a common channel, `connect`. (Note that the parallel composition operator, | |, waits for both its components to terminate before returning the value of the rightmost one.)

```
fun Accumulate(state) [c] =
  let val read_value = c?
  in if read_value=0 then state
                     else Accumulate(state+read_value)   end

fun GenNumbers(state) [c] =
  c!state; if c=0 then () else GenNumbers(state-1)

fun sum(x) =
  static channel connect
  in GenNumbers(x) [connect] || Accumulate(0)[connect]    end
```

Fig. 2. Example showing SAFL+ channel-passing

Channel parameters are not passed on recursive calls. Once a function resource, f, has been acquired by means of an *external* (i.e. non-recursive) call, $f(\vec{x})[\vec{c}]$, f's channel parameters remain bound to $\vec{c}$ until f terminates. See the operational semantics presented in Section 4 for a more precise description.

2.3   The Motivation for Channel Passing

By parameterising functions over both data and channel parameters, the SAFL+ `fun` definition becomes a powerful abstraction mechanism, encapsulating a wide range of structuring primitives treated separately in existing HDLs:

- Pure functions can be expressed by omitting channel parameters:
    `fun f(x,y) = ...`
- Structural-level blocks (*cf*. Verilog's `module` construct) can be expressed as non-terminating `fun` declarations parameterised over channels:
    `fun module() [in1,in2,out] = ...; module()`
- HardwareC `process` declarations can be expressed as non-terminating `fun` definitions (possibly without channel or data parameters):
    `fun process() = ...; process()`
- HardwareC `procedures` can be expressed as `fun` declarations that return a unit result:
    `fun procedure(x,y) = ...; ()`

As well as unifying a number of common abstraction primitives, SAFL+ also supports a style of programming not exploited by existing HDLs. Recall the definition of `Accumulate` in Section 2.2. The `Accumulate` function can be seen as a hybrid between a structural-level block (since it is parameterised over a port, c) and a function (since it terminates, returning a result). More generally, by passing in locally defined channels, a caller, f, is able to synchronise and communicate with its callee, g, during g's execution. For example, consider the SAFL+ code in Fig. 3 which declares a lock shared between functions `f1` and

```
fun lock()[acquired, release] = acquired!(); release?

fun f1() = static channel go    channel done
           in (lock()[go,done] ||
               (go?;   (* f1's critical region *) done!()) )    end

fun f2() = static channel go    channel done
           in (lock()[go,done] ||
               (go?;   (* f2's critical region *) done!()) )    end
```

Fig. 3. Mutual exclusion implemented by `lock`

`f2` to implement mutual exclusion from the critical regions. The `lock` function is parameterised over two channels: `acquired` is signalled as soon as `lock` starts executing, indicating to the caller that the lock has been acquired; `release` is used by the caller to signal that it has finished with the lock (at which point `lock` terminates). Recall that resource-awareness means that `lock` represents a single resource shared by functions `f1` and `f2`: the compiler ensures that only one caller can acquire it at a time. By passing in locally defined channels, functions `f1` and `f2` are able to communicate with `lock` during its execution.

3   Translating SAFL+ to Hardware

In [16] we describe in detail how we translate the functional subset of SAFL+ into synchronous hardware. The basic principle involves translating each function definition into a single hardware block consisting of logic to serialise concurrent accesses and registers to latch arguments. Tail-recursive calls[2] are translated into feedback loops at the circuit level.

Here we extend this by showing how the non-functional features (i.e. channels and arrays) can be integrated into our existing framework. As in [16] we adopt the graphical convention that thick lines represent data-wires and thin lines represent control signals.

A channel is translated into a shared bus surrounded with the necessary control logic to arbitrate between waiting readers and writers. Fig. 4 shows

---

[2] The only form of recursion allowed is tail-recursion.

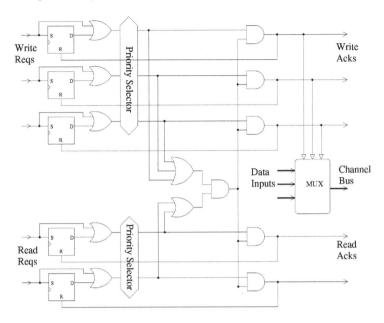

Fig. 4. A Channel Controller. The synchronous RS flip-flops (R-dominant) are used to latch pending requests (represented as 1-cycle pulses). Static fixed priority selectors are used to arbitrate between multiple requests. The 3 data-inputs are used by the three writers to put data onto the bus.

channel control circuitry in a case where there are two readers and three writers. Since we are primarily targeting FPGAs we choose to multiplex data onto the bus rather than using tri-state buffers. To perform a read operation the reader signals its read-request and blocks until the corresponding read-acknowledge is signalled. We extend the convention of our previous compiler [16] that the read-acknowledge line remains high for one cycle during which time the reader samples the data from the channel. To perform a write operation the writer places the data to be written onto a channel's data-input and signals the corresponding write-request line; the writer blocks until the corresponding write-acknowledge is signalled. Our current compiler synthesises static fixed-priority arbiters to resolve multiple simultaneous read requests or multiple simultaneous write requests. However, since the SAFL+ semantics do not specify an arbitration policy, future compilers are free to exploit other selection mechanisms.

Our SAFL+ compiler performs a static flow-analysis to determine which *actual channels* (those bound directly by the **channel** construct) a given formal-channel-parameter may range over. This information enables the compiler to statically connect each channel operation (read or write) to every possible actual channel that it may need to access dynamically. At the circuit level channel values are represented as small integers which are passed as additional parameters on a function call.

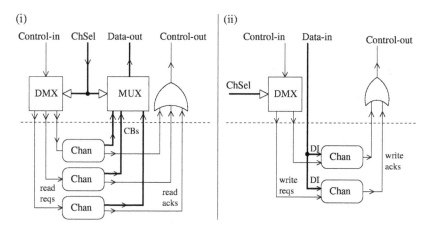

Fig. 5. (*i*) A READ node connected to 3 channels; (*ii*) A WRITE node connected to 2 channels. Each of the boxes labelled 'Chan' is a channel (as in Fig. 4). Although each such channel may well have other readers/writers these are not shown in the figure. The data-wires labelled 'CB' are the channel buses, those labelled 'DI' are channels' data-inputs (multiplexed onto the channel buses—see Fig. 4). 'ChSel' is the channel-select-input. Note that (although not shown in this figure) channel buses may be shared among many readers. The dotted line represents the boundary between the resource performing the channel operation and the channels themselves.

Our intermediate code [16] is augmented with READ and WRITE nodes representing channel operations. In cases where our flow-analysis detects that a channel operation may refer to a number of possible actual channels, multiplexers and demultiplexers are used to dynamically route to the appropriate channel. READ nodes have a control-input (used to signal the start of the operation), a control-output (used to signal the completion of the operation), a channel-select-input (used to select which actual channel to read from) and a data-output (the result of the read operation). Similarly WRITE nodes have a control-input, a control-output, a channel-select-input and a data-output. Fig. 5 shows READ and WRITE nodes connected to multiple channels.

We extend the translation of **fun** declarations described in [16] to include extra registers to latch channel-parameters. At the circuit-level channel-parameters are fed into the select lines of the multiplexers and demultiplexers seen in Fig. 5. In this example 'ChSel' would be read directly from the registers storing the enclosing function's channel-parameters.

Arrays are represented as RAMs wrapped up in the necessary logic to arbitrate between multiple concurrent accesses. Our compiler translates array declarations, **array [i] r**, into SAFL+ function definitions with signature:

```
fun r (addr:int, data:int, wr_select:bit) : int
```

Calling **r** always returns the value stored at memory location **addr**. If **wr_select** is 1 then location **addr** is updated to contain **data**. Hence array assignments,

r[e1] := e2, are translated into function calls of the form r(e1,e2,1) and
array accesses, r[e], are translated into calls of the form r(e,0,0). Treating
arrays as SAFL+ functions in this way allows us to use the compiler's existing
machinery to synthesise the necessary logic to serialise concurrent accesses to the
array and latch address lines. The compiler automatically generates the body of
r, which consists solely of RAM.

4   Operational Semantics for SAFL+

In this section we define the meaning of the SAFL+ language formally through
an operational semantics. Although, at first sight, the semantics may seem theoretical and far-removed from hardware-implementation we argue that this is not
the case. It is worth pointing out that many of the symbols in Fig. 9 have a direct
correspondence to circuit-level components. For example, *channel resources*, $\langle v \rangle_c$
(see below), represent channel controller circuits (as shown in Fig. 4) and the
(*Call*) rule (see Fig. 9) corresponds directly to transferring data into the callee's
argument registers (circuits corresponding to this are presented in [16]).

A SAFL+ program consists of a series of function definitions of the form:

$$\mathtt{fun}\ f\ (x_1, \ldots, x_k)\ [c_1, \ldots, c_j] = b_f$$

We write b_f for the body of function, f, $x_1, \ldots, x_k$ for formal parameters and
$c_1, \ldots, c_j$ for channel parameters. For the sake of brevity, we define $\vec{x}$ to mean
$x_1, \ldots, x_k$ and, similarly, $\vec{c}$ to mean $c_1, \ldots, c_j$.

Due to the static nature of SAFL+, we can simplify matters by assuming
that: (*i*) SAFL+ programs have been α-converted to make all variable names
distinct; and (*ii*) scope-flattening has been performed, bringing local declarations
to the top level and eliminating static statements. (Note that bringing a locally
defined function to the top level may require extra arguments to be added to
the function in order to pass in values for its free variables.)

We give the semantics by describing how one *program state*, P, evolves into
another, say Q, by means of a *transition*: $P \xrightarrow{\alpha} Q$, where α represents an optional
I/O action taking one of the following forms:

$\bar{\mathtt{c}}\langle v \rangle$	Output v on external channel $\mathtt{c}$
$\mathtt{c}(v)$	Read a value v from external channel $\mathtt{c}$
$go(\vec{v})$	Pass parameters $\vec{v}$ into the main function
$done(v)$	Read result v from the main function

Note that we use a bold-face $\mathtt{c}$ to range over external channels (in contrast to c,
which ranges over non-external channels).

A program state consists of a parallel composition of *function resources*,
channel resources and *array resources* (see Fig. 6). Our presentation borrows
notation and ideas from Marlow *et al* [10].

Each non-external channel declaration, channel c, corresponds to a channel
resource. When an empty channel resource (written $\langle \rangle_c$) reacts with a waiting

writer a value, v, is transferred and c becomes full (written $\langle v \rangle_c$). On reacting with a waiting reader, the value is consumed and the c enters an acknowledge state (written $\langle \text{Ack} \rangle_c$). The Ack interacts with the writer, notifying it that communication has taken place and returning c to the empty state, $\langle \rangle_c$. The explicit use of Ack models the synchronous nature of SAFL+ channels ensuring that a writer is blocked until its data has been consumed by a reader.

Array resources, $[\mathcal{S}_i]_r$, correspond to array declarations, `array [i] r`. The contents of the array, $\mathcal{S}_i$, is a function mapping indexes $0 \ldots (i-1)$ onto values. We write $\mathcal{S}_i\{j \mapsto v\}$ to denote the function which is as $\mathcal{S}_i$ but maps index j onto value v. Accessing elements outside the bounds of an array leads to undefined behaviour. To reflect this we define $\mathcal{S}_i(j)$ to be an undefined value if $j \geq i$. Furthermore if $j \geq i$ then $\mathcal{S}_i\{j \mapsto v\}$ represents an undefined state mapping indexes $0 \ldots (i-1)$ onto undefined values.

Each SAFL+ function declaration, `fun` f $(\vec{x})$ $[\vec{c}] = b_f$, is represented by a function resource. At any given time a function resource may be *busy* (performing a computation) or *available* (waiting to perform a computation). An available function resource, f, is written $\mathbb{O}_f$, signifying that f is not in use; a busy function resource takes the form $(\!|e|\!)_f$ signifying that f is currently in *evaluation state* e. The syntax of evaluation states (see Fig. 6) is essentially the same as the syntax of SAFL+ expressions augmented with the $\mathcal{W}_g$ construct which represents waiting for a result from function resource g. To save space, conditional expressions, `if` e_1 `then` e_2 `else` e_3, are shortened to $e_1 \triangleright e_2 : e_3$.

As with the Chemical Abstract Machine [2] program states can be viewed as a "solution" of reacting resources. We formalise this notion in the standard way by defining structural congruence, $\equiv$, to be the least congruence which satisfies the (*Comm*) and (*Assoc*) equations of Fig. 7. Rule (*Par*) allows transitions within parallel compositions and (*Equiv*) makes it possible to use the structural congruence relation to bring different parts of the program state together.

SAFL+ is an implicitly parallel language—an expression may contain a number of sub-expressions which can be evaluated concurrently. To formalise this notion we use a *context*, $\mathbb{E}$, to highlight the parts of an evaluation state which can be evaluated concurrently (see Fig. 8). Intuitively a context is an evaluation state $\mathbb{E}[\cdot]$ with a hole $[\cdot]$ into which we can insert an evaluation state, e, to derive a new evaluation state $\mathbb{E}[e]$.

A useful mental model is to consider a frontier of evaluation which is defined by $\mathbb{E}$ and advanced by applying the transition rules (see Section 4.1 and Fig. 9).

4.1   Transition Rules

For clarity, we present the transition rules for SAFL+ in two parts: Fig. 9(a) gives the rules for SAFL+ without channel passing. Section 4.2 explains how the rules are modified to handle channel passing in Fig. 9(b).

Substitution of values, $v_1 \ldots v_n$, for variables, $x_1 \ldots x_n$ in an evaluation state, e, is written, $\{v_1/x_1, \ldots, v_n/x_n\}e$, and for convenience abbreviated to $\{\vec{v}/\vec{x}\}e$.

$$
\begin{array}{llll}
P & \leftarrow & (\!|e|\!)_f & \text{(busy function)} \\
& | & \mathbb{0}_f & \text{(available function)} \\
& | & \langle\rangle_c & \text{(empty channel)} \\
& | & \langle v\rangle_c & \text{(full channel, holding value } v) \\
& | & \langle\text{Ack}\rangle_c & \text{(channel in acknowledge state)} \\
& | & P \mid P & \text{(parallel composition)}
\end{array}
$$

$$
\begin{array}{llll}
e & \leftarrow & v & \text{(value)} \\
& | & \mathcal{W}_g & \text{(awaiting result from } g) \\
& | & \{l_1 = e, \ldots, l_k = e\} \quad | \quad e.l & \text{(as in Fig. 1)} \\
& | & x \quad | \quad f(e, \ldots, e)[c_1, \ldots, c_n] \quad | \quad a(e, \ldots, e) & \cdots \\
& | & \texttt{let } (x_1, \ldots, x_k) = (e, \ldots, e) \texttt{ in } e & \cdots \\
& | & e \triangleright e : e \quad | \quad c\,!\,e \quad | \quad c? \quad | \quad e \parallel e & \cdots \\
& | & e; e \quad | \quad x[e] := e \quad | \quad x[e] & \text{(as in Fig. 1)}
\end{array}
$$

$$
\begin{array}{llll}
v & \leftarrow & i & \text{(integer, } i \in \mathbb{N}) \\
& | & () & \text{(unit)}
\end{array}
$$

Fig. 6. The Syntax of Program States, P, Evaluation States, e, and values, v

$$
\begin{array}{rcll}
P \mid Q & \equiv & Q \mid P & (Comm) \\
P \mid (Q \mid R) & \equiv & (P \mid Q) \mid R & (Assoc)
\end{array}
$$

$$
\frac{P \xrightarrow{\alpha} Q}{P \mid R \xrightarrow{\alpha} Q \mid R} \ (Par) \qquad\qquad \frac{P \equiv P' \quad P' \xrightarrow{\alpha} Q' \quad Q' \equiv Q}{P \xrightarrow{\alpha} Q} \ (Equiv)
$$

Fig. 7. Structural congruence and structural transitions

$$
\begin{array}{ll}
\mathbb{E} \leftarrow & [\cdot] \\
| & f(\mathbb{E}, e_2, \ldots, e_k) \quad | \quad \cdots \quad | \quad f(e_1, \ldots, e_{k-1}, \mathbb{E}) \\
| & a(\mathbb{E}, e_2, \ldots, e_k) \quad | \quad \cdots \quad | \quad a(e_1, \ldots, e_{k-1}, \mathbb{E}) \\
| & \{n_1 = \mathbb{E}, \ n_2 = e_2, \ \ldots, \ n_k = e_k\} \\
& \qquad\qquad \cdots \\
| & \{n_1 = e_1, \ \ldots, \ n_{k-1} = e_{k-1}, \ n_k = \mathbb{E}\} \\
| & \texttt{let } (x_1, \ldots, x_k) = (\mathbb{E}, e_2, \ldots, e_k) \texttt{ in } e \\
& \qquad\qquad \cdots \\
| & \texttt{let } (x_1, \ldots, x_k) = (e_1, \ldots, e_{k-1}, \mathbb{E}) \texttt{ in } e \\
| & r[\,\mathbb{E}\,] \quad | \quad r[\,\mathbb{E}\,] := e \quad | \quad r[\,e\,] := \mathbb{E} \quad | \quad \mathbb{E}.l \\
| & \mathbb{E} \triangleright e_1 : e_2 \quad | \quad \mathbb{E}; e \quad | \quad \mathbb{E} \parallel e \quad | \quad e \parallel \mathbb{E} \quad | \quad c\,!\,\mathbb{E}
\end{array}
$$

Fig. 8. A context, $\mathbb{E}$, defining which sub-expressions may be evaluated in parallel

The rules in Fig. 9 are divided into six categories:

- (*Call*) and (*Return*) deal with interaction between functional resources.
- (*Ch-Write*), (*Ch-Read*) and (*Ch-Ack*) model communication over channels.
- (*Input*) and (*Output*) deal with I/O through external channels.
- (*Ar-Write*) and (*Ar-Read*) handle access to array resources.
- (*Start*) and (*End*) correspond to external call/return of main().
- The remainder of the rules represent local computation within a function.

Note that the left hand side of the (*Tail-Rec*) rule is not enclosed in a context. This reflects the fact that tail recursive calls cannot occur in parallel with any other expressions; hence a context is unnecessary.

4.2   Semantics for Channel Passing

To deal with channel passing, function resources need to store the channel parameters passed from an external call. We use the notation $(\!|\cdot|\!)_f^{\vec{c}}$ to represent a function resource which has been called with actual channel parameters, $\vec{c}$. For convenience we sometimes omit the channel parameters from a rule, defining $(\!|e_1|\!)_f \to (\!|e_2|\!)_g$ to mean $(\!|e_1|\!)_f^{\vec{c}} \to (\!|e_2|\!)_g^{\vec{c}}$.

The (*Call*), (*Ret*) and (*Tail-Rec*) rules are modified for channel passing as shown in Fig. 9(b).

5   Conclusions and Further Work

This paper has introduced and formally defined the SAFL+ language, motivating its use for hardware description and synthesis. We argue that the major advantages of SAFL+ over most existing high-level synthesis languages are:

- The combination of resource-awareness and channel-passing makes SAFL+ **fun** declarations a very powerful abstraction mechanism. Both structural blocks and functions can be seen as special cases of **fun** declarations.
- By structuring programs as a series of function definitions (as opposed to a collection of structural blocks), SAFL+ supports a wide range of analyses and transformations which are not applicable to conventional HDLs.
- SAFL+ has a formally defined semantics.

The project is in its early stages. Although we have implemented a silicon compiler for SAFL+ and tested it on small examples, we have yet to use the system to build a large system-on-a-chip design. This is very high-priority for our future work. Using SAFL+ to construct a large hardware design will test both the expressivity of the language and the efficiency of our compiler.

The translation of SAFL+ to hardware given in this paper (Section 3) outlines one of many possible implementation techniques. In future we plan to investigate the translation of SAFL+ to globally-asynchronous-locally-synchronous (GALS) hardware. This involves mapping function-resources into separate clock domains and extending our compiler to automatically instantiate the necessary inter-clock-domain interfaces.

(a) Rules for SAFL+ without channel passing:

$$(\mathbb{E}[g(v_1, \ldots, v_k)])_f \mid \mathbb{O}_g \longrightarrow (\mathbb{E}[\mathcal{W}_g])_f \mid (\{ \vec{v}/\vec{x} \} b_g)_g \qquad f \neq g \qquad (Call)$$

$$(v)_f \mid (\mathbb{E}[\mathcal{W}_f])_g \longrightarrow \mathbb{O}_f \mid (\mathbb{E}[v])_g \qquad (Return)$$

$$(\mathbb{E}[c \, ! \, v])_f \mid \langle \rangle_c \longrightarrow (\mathbb{E}[\mathcal{W}_c])_f \mid \langle v \rangle_c \qquad (Ch\text{-}Write)$$

$$(\mathbb{E}[c?])_f \mid \langle v \rangle_c \longrightarrow (\mathbb{E}[v])_f \mid \langle \text{Ack} \rangle_c \qquad (Ch\text{-}Read)$$

$$(\mathbb{E}[\mathcal{W}_c])_f \mid \langle \text{Ack} \rangle_c \longrightarrow (\mathbb{E}[()])_f \mid \langle \rangle_c \qquad (Ch\text{-}Ack)$$

$$(\mathbb{E}[c \, ! \, v])_f \xrightarrow{\bar{c}\langle v \rangle} (\mathbb{E}[()])_f \qquad (Output)$$

$$(\mathbb{E}[c?])_f \xrightarrow{c(v)} (\mathbb{E}[v])_f \qquad (Input)$$

$$(\mathbb{E}[r[v_1] := v_2])_f \mid [\mathcal{S}_i]_r \longrightarrow (\mathbb{E}[()])_f \mid [\mathcal{S}_i\{v_1 \mapsto v_2\}]_r \qquad (Ar\text{-}Write)$$

$$(\mathbb{E}[r[v]])_f \mid [\mathcal{S}_i]_r \longrightarrow (\mathbb{E}[\mathcal{S}_i(v)])_f \mid [\mathcal{S}_i]_r \qquad (Ar\text{-}Read)$$

$$\mathbb{O}_{\text{main}} \xrightarrow{go(\vec{v})} (\{ \vec{v}/\vec{x} \} b_{\text{main}})_{\text{main}} \qquad (Start)$$

$$(v)_{\text{main}} \xrightarrow{done(v)} \mathbb{O}_{\text{main}} \qquad (End)$$

$$(\mathbb{E}[a(v_1, \ldots, v_k)])_f \longrightarrow (\mathbb{E}[v])_f \quad \text{where } v = a(v_1, \ldots, v_k) \qquad (PrimOp)$$

$$(\mathbb{E}[\{\ldots, l = v, \ldots\}.l])_f \longrightarrow (\mathbb{E}[v])_f \qquad (RecSelect)$$

$$(\mathbb{E}[0 \triangleright e_1 : e_2])_f \longrightarrow (\mathbb{E}[e_2])_f \qquad (CFalse)$$

$$(\mathbb{E}[n \triangleright e_1 : e_2])_f \longrightarrow (\mathbb{E}[e_1])_f \qquad n \neq 0 \qquad (CTrue)$$

$$(\mathbb{E}[\texttt{let } \vec{x} = \vec{v} \texttt{ in } e])_f \longrightarrow (\mathbb{E}[\{\vec{v}/\vec{x}\}e])_f \qquad (Let)$$

$$(\mathbb{E}[v; e])_f \longrightarrow (\mathbb{E}[e])_f \qquad (Seq)$$

$$(\mathbb{E}[v_1 \parallel v_2])_f \longrightarrow (\mathbb{E}[v_2])_f \qquad (Par)$$

$$(f(v_1, \ldots, v_k))_f \longrightarrow (\{\vec{v}/\vec{x}\} b_f)_f \qquad (Tail\text{-}Rec)$$

(b) Modifications for Channel Passing:

$$(\mathbb{E}[g(v_1, \ldots, v_k)[d_1, \ldots, d_j]])_f^{\vec{c}} \mid \mathbb{O}_g \longrightarrow (\mathbb{E}[\mathcal{W}_g])_f^{\vec{c}} \mid (\{ \vec{d}/\vec{c}, \vec{v}/\vec{x} \} b_g)_g^{\vec{d}} \qquad (Call)$$

$$(v)_f^{\vec{c}} \mid (\mathbb{E}[\mathcal{W}_f])_g^{\vec{d}} \longrightarrow \mathbb{O}_f \mid (\mathbb{E}[v])_g^{\vec{d}} \qquad (Ret)$$

$$(f(v_1, \ldots, v_k))_f^{\vec{c}} \longrightarrow (\{ \vec{c'}/\vec{c}, \vec{v}/\vec{x} \} b_f)_f^{\vec{c}} \qquad (Tail\text{-}Rec)$$

Fig. 9. Transition Rules for SAFL+

Acknowledgement

This work was supported by (UK) EPSRC grant GR/N64256 "A Resource-Aware Functional Language for Hardware Synthesis"; the first author was also sponsored by AT&T Research Laboratories Cambridge.

References

1. The Jazz Synthesis System. See: http://www.exentis.com/jazz.
2. BERRY, G., AND BOUDOL, G. The chemical abstract machine. *Theoretical Computer Science 96* (1992), 217–248.
3. BJESSE, P., CLAESSEN, K., SHEERAN, M., AND SINGH, S. Lava: Hardware description in Haskell. In *Proceedings of the 3rd International Conference on Functional Programming* (1998), SIGPLAN, ACM.
4. CELOXICA (LTD.). Handel-C language datasheet. Available from Celoxica: http://www.celoxica.com.
5. HOE, J., AND ARVIND. Hardware synthesis from term rewriting systems. In *Proceedings of X IFIP International Conference on VLSI* (1999).
6. INMOS (LTD.). *Occam 2 Reference Manual*. Prentice Hall, 1998.
7. JOHNSON, S., AND BOSE, B. DDD: A system for mechanized digital design derivation. Tech. Rep. 323, Indiana University, 1990.
8. KU, D., AND DE MICHELI, G. HardwareC—a language for hardware design (version 2.0). Tech. Rep. CSL-TR-90-419, Stanford University, 1990.
9. LI, Y., AND LEESER, M. HML, a novel hardware description language and its translation to VHDL. *Transactions on VLSI Systems*, 1 (February 2000).
10. MARLOW, S., PEYTON JONES, S., MORAN, A., AND REPPY, J. Asynchronous exceptions in Haskell. To appear. *Proceedings of ACM SIGPLAN 2001 Conference on Programming Language Design and Implementation (PLDI)* 2001.
11. MATTHEWS, J., COOK, B., AND LAUNCHBURY, J. Microprocessor specification in Hawk. In *Proceedings of the IEEE International Conference on Computer Languages* (1998).
12. MILNER, R. The polyadic π-calculus: A tutorial. Tech. Rep. ECS-LFCS-91-180, University of Edinburgh, October 1991.
13. MILNER, R., TOFTE, M., HARPER, R., AND MACQUEEN, D. *The Definition of Standard ML (Revised)*. MIT Press, 1997.
14. MYCROFT, A., AND SHARP, R. A statically allocated parallel functional language. In *Proceedings of the International Conference on Automata, Languages and Programming* (2000), vol. 1853 of *LNCS*, Springer-Verlag.
15. MYCROFT, A., AND SHARP, R. Hardware/software co-design using functional languages. In *Proceedings of TACAS* (2001), vol. 2031 of *LNCS*, Springer-Verlag.
16. SHARP, R., AND MYCROFT, A. The FLaSH compiler: Efficient circuits from functional specifications. Tech. Rep. tr.2000.3, AT&T Laboratories Cambridge, 2000.
17. SHARP, R., AND MYCROFT, A. Soft scheduling for hardware, 2001. To appear. *Proceedings of the 8th International Static Analysis Symposium*.
18. SHEERAN, M. muFP, a language for VLSI design. In *Proceedings of the ACM Symposium on LISP and Functional Programming* (1984).

Hierarchical Verification Using an MDG-HOL Hybrid Tool

Iskander Kort[1], Sofiene Tahar[1], and Paul Curzon[2]

[1] Concordia University, Canada. {tahar,kort}@ece.concordia.ca
[2] Middlesex University, UK. p.curzon@mdx.ac.uk

Abstract. We describe a hybrid formal hardware verification tool that links the HOL interactive proof system and the MDG automated hardware verification tool. It supports a hierarchical verification approach that mirrors the hierarchical structure of designs. We obtain advantages of both verification paradigms. We illustrate its use by considering a component of a communications chip. Verification with the hybrid tool is significantly faster and more tractable than using either tool alone.

1 Introduction

Automated decision diagram based formal hardware verification is fast and convenient, but does not scale well, especially where data paths and control circuitry are combined. Details of the version of the design verified need to be simplified: e.g., considering 1-bit instead of 32-bit datapaths. Finding a model reduction and appropriate abstractions so that verification is tractable with the tool can be time-consuming. Moreover, significant detail can be lost. An alternative is interactive theorem proving. The verification can be done hierarchically allowing large designs to be verified without simplification. Furthermore it is possible to reason about high level abstractions of datatypes. It can however be very time-consuming, requiring significant user interaction and skill.

The contribution of our work is to implement a hybrid tool combining HOL [9] and MDG [4] which provides explicit support for hierarchical hardware verification. In particular, we have provided an embedding of the MDG input language in HOL, implemented a linkage between HOL and MDG using the PROSPER toolkit [7] and implemented a series of HOL tactics that automate hierarchical verification. This means that a hierarchical proof can be performed as it might be done using a pure HOL system. However, the MDG tools can be seemlessly called to perform verification of components that are within its capabilities. We have verified a component of a communication switch using the tool. Verification is shown to be significantly faster and more tractable using the hybrid tool than with either tool individually.

The remainder of this paper is organized as follows. In Sect. 2 we overview briefly the two tools being linked. We present our hybrid tool and the methodology it embodies in Sect. 3. A case study using the tool to verify a component of an ATM switch is described in Sect. 4. Finally, we discuss related work in Sect. 5 and draw conclusions in Sect. 6.

T. Margaria and T. Melham (Eds.): CHARME 2001, LNCS 2144, pp. 244–258, 2001.
© Springer-Verlag Berlin Heidelberg 2001

2   The Linked Tools

Our hybrid tool links the HOL interactive theorem prover and the MDG hardware verification system. HOL [9] is based on higher-order logic. The user works interactively with the system calling SML functions that implement inference rules to apply proof steps. New theorems are created in HOL by applying inference rules—derived rules call a succession of primitive rules, thus the user can have great confidence in the derived theorems. However, HOL also provides functions to create theorems directly without proof. This feature can be used to import results produced by external tools into HOL. We initially used the PROSPER/Harness Plug-in Interface of HOL [7]. This gives a uniform way of linking HOL with external proof tools. It provides the low level client-server communication interface from HOL to various languages within which other tools are integrated. A range of different external proof tools (such as MDG) can act as servers to a HOL client. The interface removes the burden of writing low-level communication tools, leaving the hybrid tool designer to concentrate on higher-level issues. It also tags theorems produced by plug-ins with a label indicating their source. These labels are propagated to any theorem generated from the imported result allowing the pedigree of any result to be later determined.

The MDG system, which is primarily designed for hardware verification, provides verification procedures for equivalence and property checking. The former provides the verification of two combinational circuits or the verification of two state machines. The latter allows verification through invariant checking or model checking. The strength of the MDG system is its automation and ease of use. It has been used in the verification of significant hardware examples [3,16,18]. The MDG system is a decision diagram based verification tool based on Multiway Decision Graphs (MDGs) [4] rather than on BDDs. MDGs overcome the data width problem of Reduced-Order Binary Decision Diagram (ROBDD) based verification tools. An MDG is a finite, directed acyclic graph (DAG). MDGs essentially represent relations rather than functions. They are much more compact than ROBDDs for designs containing a datapath. Furthermore, sequential circuits can be verified independently of the width of the datapath. The MDG tools combine some of the advantages of representing a circuit at more abstract levels with the automation offered by decision-diagram based tools. The input language for MDG, MDG-HDL, supports structural descriptions, behavioral descriptions as Abstract State Machine (ASM) or a mixture of both. A structural description is usually a netlist of components connected by signals, and a behavioral description is given by a tabular representation of the transition/output relation of the ASM. This is done using the Table construct of MDG-HDL: essentially a case statement that allows the value of a variable to be specified in terms of the values of inputs and other expressions

3   The Hybrid Tool and Verification Methodology

In a pure MDG verification, structural and behavioral descriptions are given for the top level design. An automated verification procedure is then applied. If the problem is sufficiently tractable, the verification is completed automatically. If not, ideally the problem would be attacked in a hierarchical fashion by verifying the sub-blocks independently. However, the management of this process cannot be done within the tool, though could be done informally outside it.

In a pure HOL hardware verification, the proof is structured according to the design hierarchy of sub-blocks within the implementation. For each block, including the top level block of the design, a structural specification and behavioral specification are given. Each block's implementation (apart from those at the bottom of the hierarchy) is verified against its specification in three steps. Firstly an intermediate verification result is obtained about the block based on the behavioral descriptions of its sub-blocks. Essentially the sub-blocks are treated as primitive components in this verification. Secondly the process is repeated recursively on the sub-blocks to obtain correctness theorems for them. Finally, the correctness theorems of the sub-blocks are combined with the intermediate correctness theorem of the block itself to give the actual correctness theorem of the block. This is based on the full structural description of the block down to primitive components. The verification follows the natural design hierarchy. If this process is applied to the top level design block, a correctness theorem for the whole design is obtained. The integration of the verification results of the separate components that would be done informally (if at all) in an MDG verification is thus formalized and machine-checked in the HOL approach.

Our hybrid tool supports hierarchical verification, automating the process discussed above, and fits the use of MDG verification naturally within the HOL framework of compositional hierarchical verification. The HOL system is used to manage the proof, with the MDG system called seemlessly to verify those design blocks that are tractable. This removes the need to provide behavioral specifications for sub-blocks and the need to verify them separately. In particular, if the design of any sub-block is sufficiently simple, then the hierarchical approach can be abandoned for that block and the whole block verified in one go in MDG. Furthermore, verifying a block under the assumption that its sub-blocks are all primitive components may also be done using MDG if tractable. If not, a normal HOL proof can still be performed. No information is lost in using MDG via the hybrid tool. To allow the seamless integration of the tools, we use MDG-style behavioral specifications within HOL. This means the specifications must be in the form of a finite state machine or table description. If a higher level abstraction, unavailable in MDG, is required then a separate HOL proof is performed that an MDG style specification meets this abstraction.

3.1   The Hybrid Tool

Our Hybrid tool was written in SML. It consists of five modules: a parsing module, an extraction module, a hierarchical verification support module, a code generation module and an MDG interaction module (cf. Fig. 1). Subgoal management is done using the HOL subgoal manager. This is an advantage of the hybrid approach—the existing HOL infrastructure augments MDG providing a much more powerful interface to MDG.

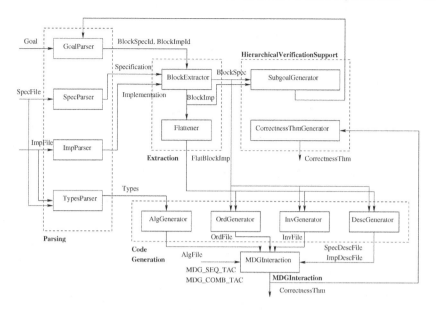

Fig. 1. The hybrid tool's structure

The hybrid tool supports the hierarchical verification process by providing a HOL embedding of the concrete subset of the MDG input language to allow MDG-style specifications to be written in HOL. Three high-level proof tactics that manage the proof process are also provided. A hierarchy tactic, HIER_VERIF_TAC, automates the subgoaling of the correctness theorem of a block by analyzing its structure as outlined in the previous section. It later combines the proven subgoals to give the desired correctness theorem. Where a non-primitive component occurs several times within a block, the tactic avoids duplication, generating a single subgoal that once proved is automatically instantiated for each occurence of that component to prove the correctness of the block. Two other tactics automate the link to the MDG tools: MDG_COMB_TAC attempts to verify a given correctness theorem for a block using MDG combinational equivalence verification; MDG_SEQ_TAC calls MDG sequential equivalence verification to prove the result.

Verification using the hybrid tool proceeds as shown in Fig. 2. An initial goal is set that the top level design's implementation meets its behavioral specification. If the design can be verified using MDG, the appropriate MDG tactic, determined by whether the circuit is sequential, is called. Otherwise, the hierarchy tactic is called to break the design into smaller parts, and the process is repeated. At any point, a HOL proof can be performed directly to prove a goal. MDG verification can fail due to state-space explosion leading to the system running out of memory. In general MDG can fail to terminate, however the current version of the hybrid tool does not do so due to the fact that abstract variables are not yet supported.

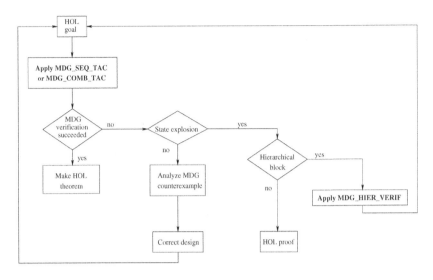

Fig. 2. Using the hybrid tool

3.2   Specifications

The hybrid tool must be supplied with a behavioral specification for each block
in the design that is verified. This is not necessary for sub-blocks within blocks
verified by calls to MDG. The specifications are provided as a normal file of
HOL definitions. However, as these definitions must be analyzed by the tool
and ultimately converted into MDG, they must follow a prescribed form: they
must consist of a conjunction of tables, which input and output arguments must
both be explicitly typed and be in a given order. MDG abstract variables are
not currently supported. The tables are an embedding of MDG tables in HOL
originally defined by Curzon *et. al.* [5] to verify the MDG components in HOL.
The verification of these components increases confidence that the MDG tools
can be trusted when used in the HOL system.

Structural specifications are written in a subset of the HOL logic similar
to that for behavioral specifications. However, the descriptions are not limited
to tables but can include any component of the MDG component library. The
structural specification of a block thus differs from a behavioral specification in
that its body consists of a network of components. A component may be an
MDG built-in component, a functional block, a table or a component previously
defined by the user. The MDG built-in components are an embedding in HOL
of the actual MDG components.

3.3   The Verification Process

The hybrid tool is intended to provide automated support for hierarchical verification and to enable the user to verify some blocks using MDG. We will illustrate
this by refering to the verification of a simple adder circuit. A typical session

HA_i ((x:num → bool), (y:num → bool)) ((z:num → bool), (cout:num → bool)) =
  (MDG_XOR (x,y) z) ∧ (MDG_AND (x,y) cout)

FA_i ((x:num → bool), (y:num → bool),(cin:num → bool))
    ((z:num → bool), (cout:num → bool)) =
  ∃ (z_0:num → bool) ($cout_0$:num → bool) ($cout_1$:num → bool).
        (HA_i (x,y) (z_0,$cout_0$)) ∧ (HA_i (z_0,cin) (z,$cout_1$)) ∧
        (MDG_OR ($cout_0$,$cout_1$) cout)

Fig. 3. A structural specification of an adder

z_TAB ((x:num → bool), (y:num → bool)) (z:num → bool) =
        TABLE [x;y] z [[F; F]; [T; T]] [F;F] T

cout_TAB ((x:num → bool), (y:num → bool)) (cout:num → bool) =
        TABLE [x;y] cout [[F; DONT_CARE]; [T; F]] [F;F] T

HA ((x:num → bool), (y :num → bool)) ((z:num → bool), (cout :num → bool)) =
        (z_TAB (x,y) z) ∧ (cout_TAB (x,y) cout)

Fig. 4. A behavioural specification of a half-adder

with the hybrid tool goes through the following steps. First, the user supplies the tool with a specification file and an implementation file as part of an initialization procedure. These are SML files containing normal SML definitions. The specification file includes the behavioral specifications of the design blocks. The implementation file includes the design structural specification and follows the design hierarchy. Both files may include user defined HOL datatypes. An example of a structural specification for an adder is given in Fig. 3. The behavioral specification of a half-adder in terms of tables is given in Fig. 4. The specification of the full adder is similar. In a table specification, the first list gives the inputs of the table, the next argument is the output. Next is a list of lists giving possible combinations of input values and then a list giving the output values resulting from those combinations. The final argument gives the default value for any combination of inputs not listed. MDG tables are more general than shown in this example in that general expressions can be used as table inputs and variables can appear in the rows. For example, the carry out signal in the half-adder is defined by a table with two inputs x and y and one output cout. If x is False and y is "DON'T CARE" (i.e. anything) then cout is False. Similarly if x is true and y is false then cout is false. The default value for all other combinations of input values is true. The behavioral specifications of the components are similarly defined. The initialization procedure also involves loading the embeddings of the MDG tables and the MDG components in HOL as well as starting a server to the MDG system.

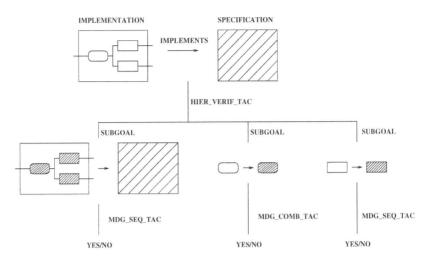

Fig. 5. Hierarchical verification using HIER_VERIF_TAC

Once the tool is initialized, the user sets the correctness goal for the whole design using HOL's subgoal package. This goal states that the design's implementation implies its specification. For example, for our adder, we set the goal:

$\forall$ x y cin z cout. FA_i (x,y,cin) (z,cout) $\Longrightarrow$ FA (x,y,cin) (z,cout)

This correctness goal could then be resolved directly through MDG using MDG_SEQ_TAC or MDG_COMB_TAC. Applying these tactics to complex designs may lead to state explosion. To overcome this, HIER_VERIF_TAC is used. The action of this tactic is summarized in Fig. 5. It automatically generates a correctness subgoal for every immediate sub-block in the design. Where one sub-block is used in several places, only one goal is generated: the hybrid tool generates a general subgoal that justifies its use in each situation. A further subgoal states that the lower level specifications, connected according to the structural specification, imply the current specification.

For example, HIER_VERIF_TAC generates two subgoals for our adder.

$\forall$ x y z cout. HA_i (x,y) (z,cout) $\Longrightarrow$ HA (x,y) (z,cout)

$\forall$ x y cin z cout. FA_i_hl (x,y,cin) (z,cout) $\Longrightarrow$ FA (x,y,cin) (z,cout)

The first is a correctness statement for the half-adder component. Only one general version is generated. This is used to create the two theorems justifying each of the two instances of this component in the design. The second subgoal is a correctness goal for the adder where the half-adder is treated as a primitive component. It contains an automatically generated new structural specification

FA_i_hl, which is in terms of the behavioral specifications of the half-adder submodules rather than their structural specifications:

$\vdash$ FA_i_hl ((x:num $\rightarrow$ bool), (y:num $\rightarrow$ bool),(cin:num $\rightarrow$ bool))
$\quad$ ((z:num $\rightarrow$ bool), (cout:num $\rightarrow$ bool)) =
$\quad\quad$ $\exists$ (z_0:num $\rightarrow$ bool) ($cout_0$:num $\rightarrow$ bool) ($cout_1$:num $\rightarrow$ bool).
$\quad\quad\quad$ (HA (x,y) (z_0,$cout_0$)) $\wedge$ (HA (z_0,cin) (z,$cout_1$)) $\wedge$
$\quad\quad\quad$ (MDG_OR ($cout_0$,$cout_1$) cout)

HIER_VERIF_TAC creates a justification function that given theorems corresponding to the subgoals creates the theorem corresponding to the original goal. The subgoals it produces could be resolved using a conventional HOL proof, by invoking MDG as above or by applying HIER_VERIF_TAC once again. If the subgoals are proved, then the justification rule of HIER_VERIF_TAC will automatically derive the original correctness goal from them. In our example, we apply one of the MDG-based tactics. This circuit is purely combinational so MDG_COMB_TAC is used.

When the MDG-based tactics are applied, the hierarchy in the structural specification is automatically flattened to the non-hierarchical form of primitive components required by MDG (just the next layer down in the case of the second subgoal above). The tool currently generates a static variable ordering for use by MDG though more sophisticated ordering heuristics could be included. Alternatively the tool user can provide the ordering. Each block verified can use a different variable ordering.

The tool analyses the feedback of MDG in order to find out whether the verification succeeded or failed. If the verification fails a counter-example is generated. If it succeeds, the tactic creates the appropriate HOL theorem. For example, for our adder we obtain the theorems:

[MDG] $\vdash$ $\forall$ x y z cout. HA_i (x,y) (z,cout) $\Longrightarrow$ HA (x,y) (z,cout)

[MDG] $\vdash$ $\forall$ x y cin z cout. FA_i_hl (x,y,cin) (z,cout) $\Longrightarrow$ FA (x,y,cin) (z,cout)

The theorem is tagged with an oracle label indicating that it is proved by an external tool. This tag will be passed to any theorem proved using these theorems.

Note also that the theorem proved can be instantiated for any instance. We effectively can prove a single correctness theorem for a block and reuse it for any instance of the block. In our example, there are two instances of the half-adder, but this single theorem is used for both. This process is managed formally and machine-checked within HOL. This contrasts with pure automated tools, where each instance would need a specific theorem to be verified separately or non-machine-checked reasoning to be relied upon. For the half-adder, the subgoals are formally combined using automatic proof by HIER_VERIF_TAC to give the desired theorem about the adder:

[MDG] $\vdash$ $\forall$ x y cin z cout. FA_i (x,y,cin) (z,cout) $\Longrightarrow$ FA (x,y,cin) (z,cout)

The way HOL and MDG are used together is thus that the former manages the compositional aspects of the proof, ensuring duplicated work is avoided. The latter does fast, automated, low-level verification.

4   Case Study: The 4 × 4 ATM Switch Fabric

We have applied the hybrid tool to a realistic example: the verification of a block of the Fairisle ATM (Asynchronous Transfer Mode) switch fabric [13]. The Fairisle switch fabric is a real switch fabric designed and used at the University of Cambridge for multimedia applications. It switches cells of data from 4 input ports to 4 output ports as requested by information in header bytes in each cell.

Curzon [6] formally verified this ATM switching element hierarchically using the pure HOL system. However, this verification was very time-consuming. Verifying the fabric can be done hierarchically following exactly the same structure as the original design using our hybrid tool. However, with the tool, many of the sub-blocks can be verified automatically using the MDG tool, thus saving a great deal of time and effort. Furthermore, HIER_VERIF_TAC automates much of the management of the proof that was previously done manually. Attempting the verification in MDG alone would, on the other hand, be barely tractable taking days of CPU time. This is discussed in more detail below.

The fabric is split into three sub-blocks, namely Acknowledgement, Arbitration and Data Switch. Further dividing the Arbitration sub-module, we have essentially two blocks: the arbiters that make arbitration decisions and a preprocessing block that generates the timing signal and processes the headers of the cells into a form usable by the arbiters (see Fig. 6). We consider the verification of the preprocessor block here (see Fig. 7). The timing block within the preprocessor generates a timing signal for the arbiters from an external frame signal

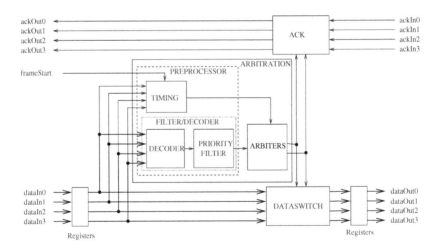

Fig. 6. The vairisle ATM switch fabric

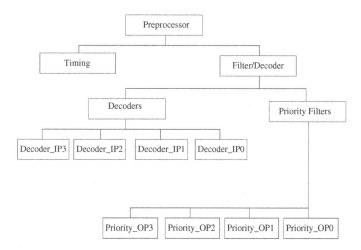

Fig. 7. The preprocessor hierarchy

and from the data stream. The decoder block (made of 4 independent decoders) takes the four cell headers from the data stream and extracts the information about the destinations they are requesting (which is in a binary encoding). For each destination a unary encoding of the cells that are requesting that output is created. The priority filter takes this information together with priority information from the cell headers. If any cell has high priority, then requests from low priority cells are not forwarded to the arbiters.

Setting as goal the correctness statement for the preprocessor, we attack it using HIER_VERIF_TAC. We obtain two subgoals corresponding to the timing block and the filter-decoder block, together with a subgoal that the combined preprocessor is correct on the assumption that its sub-blocks are. As the Timing block is a sequential design, we call MDG_SEQ_TAC to automatically prove the timing unit correctness subgoal. This proves the equivalence of the implementation and its specification, and so proves the implication in our subgoal.

Decoders and *Priority Filters* are purely combinational circuits. Their specifications are the conjunctions of 32 16-input-tables and 16 32-input-tables, respectively. MDG takes 16 hours to verify *Decoders* and it would take days to verify *Priority Filters*. The problem is in finding an efficient variable ordering given that the way the sub-blocks are connected means that the best ordering for one table is bad for another. In order to overcome this problem, we move down one level in the design hierarchy. More specifically, the 32 tables in *Decoders*' specification were partitioned into four 8-table-sub-blocks: *Decoder_IP0* ... *Decoder_IP3*. *Decoder_IPi* is a decoder for input port i, $i = 0..3$. A more efficient variable ordering is then supplied for each of these sub-blocks. Similarly, the 16 tables in *Priority Filters*' specification were partitioned into four 4-table-sub-blocks: *Priority_OP0* ... *Priority_OP3*. *Priority_OPi* is a priority filter for output port i, $i = 0..3$. The preprocessor hierarchy as verified is shown in Fig. 7.

Table 1. Hierarchical verification statistics

Block	CPU Time (sec.)
Preprocessor	495.230
Timing	0.060
Filter/Decoder	488.900
Decoders	45.520
Decoder_IPi	10.050
Priority	437.210
Priority_OPi	107.413

We apply HIER_VERIF_TAC to verify *Decoders* and *Priority Filters* based on this hierarchy. The subgoals associated to *Decoder_IPi* and *Priority_OPi*, $i = 0..3$, are then proved automatically. Note that this still avoids expanding the hierarchy as far as in the original HOL proof—so lower level behavioral specifications do not need to be written.

Table 1 shows the hierarchical verification statistics, including CPU time in seconds. Obviously, using our hybrid tool, the verification of the preprocessor is faster than proving in HOL that the implementation implies the high-level specification. Given the formal specifications, Curzon [6] originally took several days to do the proofs of these blocks using interactive proof whereas the verification is done in minutes using our tool. Verification is also faster than using MDG alone: splitting the decoder block enabled verifying it within less than 1 minute using our hybrid tool instead of 16 hours if only MDG was used. It took a day (approximately 8 hours) to interactively prove the decoder block in HOL. Thus verification is faster using the hybrid tool than with either system on its own as shown in Table 2 which gives approximate times for verifying the decoder block. These times should be treated with caution, as the pure HOL times are not CPU time but that for the human to interactively manage the verification. Times to develop specifications, including those of sub-blocks verified hierarchically rather than directly using MDG, are not included in these times. Though, writing these specifications was straightforward. It therefore is worthwhile additional work, given the overall time improvement. Some extra human interaction time for the verification part is also needed when using the hybrid tool over the bare CPU time. This is needed to call the appropriate tactics. However, this is minimal—a matter of minutes rather than hours, since it follows the existing design hierarchy. The main part that is time consuming is if unsuccessful automated proofs of sub-blocks are attempted. This obviously requires judgement over the limitations of MDG, in knowing when it is worth attempting automated proof, and when it is better to step down a level in the hierarchy.

5 Related Work

Work to combine the advantages of automated and interactive tools falls generally into two areas: hybrid tools in which two existing, stand-alone verification

Table 2. Comparison of verifications of the decoder blocks

HOL (Human Proof Time)	MDG (CPU Time)	Hybrid Tool (CPU Time)
Interactive	Automated	Semi-automated
8 hours	16 hours	1 minute

systems are linked; and systems where external proof packages are embedded as decision procedures for some subset of the logic by an interactive system.

Perhaps the most impressive hybrid verification system to date is the combined Voss-ThmTac System [2]. It combines a simple, specially written LCF style proof system, ThmTac with the Voss Symbolic Trajectory Analysis System. This system evolved out of the HOL-VOSS System [11]. In that system, Voss was interfaced within HOL as a tactic that could be called to perform a symbolic trajectory analysis to verify assertions about sequences of states. The Voss-ThmTac System is thus based on many years of experience combining systems. It has been used to verify a series of real hardware designs including an IA-32 Instruction length decoder claimed to be one of the most complex hardware verifications completed. Much of its power comes from the very tight integration of the two provers allowing the user to interact directly with either tool. This is facilitated by the use of a single language, *fl*, as both the theorem prover's meta-language and its object language.

Schneider and Hoffmann [15] linked SMV (a CTL model checker) to HOL using PROSPER. In this hybrid tool, HOL conversions were used to transform LTL specifications into ω-Automata, a form that can be reasoned about within SMV. These HOL terms are exported to SMV through the PROSPER plug-in interface. On successful model checking the results are returned to HOL and turned into tagged theorems. This allows SMV to be used as a HOL decision procedure. The SMV specification language has also been deeply embedded in HOL, allowing temporal logic specifications to be manipulated in HOL and the model checker used to return a result about its validity.

The use of tightly integrated decision procedures is a major focus of the PVS proof system. Rajan *et al* [14] integrated a BDD-based model checker for the propositional μ-calculus within PVS. An extension of the μ-calculus is defined within higher-order logic and temporal operators then defined as μ-calculus fixpoint definitions. These expressions are converted into the form required by the model checker which can then be used to prove appropriate subgoals generated within PVS. Such results are treated no differently to those created by proof.

An issue with accepting imported results as theorems is whether the external system can be trusted to produce "theorems" that really are host system theorems. This is more of an issue with fully-expansive proof systems such as HOL where the integrity of the system depends on a small core of primitive inference rules. Accepting results from an external package essentially treats that package as one of the trusted primitives. The approach taken by Gordon [8] to minimize this problem in the BuDDy package when integrating BDD based tools is to provide a small set of BDD primitives in terms of which full tools are implemented. In this way only the primitives need to be trusted not the whole package.

Hurd [10] used PROSPER to combine the Gandalf prover with HOL. Unlike other approaches, the system reproves the Gandalf theorems within HOL rather than just accepting the results. The Gandalf proof script is imported into the HOL system and used to develop a fast proof within HOL. The tool is thus used to discover proofs, rather than directly to prove theorems.

The MEPHISTO system [12] was developed to manage the higher levels of a verification, producing first-order subgoals to be proved by the FAUST first order prover. The goals of MEPHISTO are similar to ours: managing the subgoaling of a verification to produce goals that can be proved by another system. The difference is the focus of the way the systems do this and the target system. Our approach is to use the existing design hierarchy, sending to the automated prover (here a hardware verification system itself) subgoals that are correctness theorems about design modules. Thus HIER_VERIF_TAC produces subgoals (and results from failed verification) easily understood by the designer. This approach avoids the problem of the verifier having to inspect goals that bear little relation to the input to the system. MEPHISTO does give some support for hierarchical proof providing a library of preproved modules. However, in our approach such hierarchical verification is explicitly supported by the tactics.

Aagaard *et al* [1] proposed a similar hardware verification management system. They aimed to complete the whole proof within the theorem prover (HOL or Nuprl). As with MEPHISTO, the focus is on producing lemmas to be proved by decision procedures. They developed a series of prototype tactics that could be used to break down subgoals. However, they do not directly support hierarchical verification: the first step proposed is to rewrite with the module specifications.

As in [2] and [15], we integrate a theorem prover (HOL) to an existing hardware verification tool (MDG) rather than embedding a package within the system. We work within the proof system but using the specification style of the automated tool. This is done by embedding the language of the automated verification tool within the proof system. As is done in pure HOL verification, the proof follows the natural design hierarchy embodied in the specifications. This process is explicitly supported by our hierarchy tactic. By working in this way we obtain a seamless integration of the tools. The subgoals automatically generated also have a direct relation to the specifications produced by the designer.

6   Conclusions

We have described a tool linking an interactive theorem prover and an automated decision diagram-based hardware verification system. This builds on previous work [17], where we showed formally how an MDG equivalence proof can be imported to an implication-based correctness theorem in HOL. Our system explicitly supports the hierarchical compositional verification approach naturally used in interactive proof systems, when using an automated tool. The interactive proof system is used to automatically manage the proof as well as complete any proof interactively that is beyond the scope of the automated system. The verification of whole blocks in the hierarchy can however be done automatically. The hybrid tool can be used to verify larger examples than could be done in MDG alone, and these proofs can be done faster than in either system alone.

We used the PROSPER toolkit to perform the linkage of the two tools. This made providing such a linkage relatively easy. However, with the early version of PROSPER used the linkage was slow. An alternative implementation that communicated between the tools directly using files was quicker.

We illustrated the use of the hybrid tool by describing the verification of the preprocessing block of the arbitration unit of an ATM switch. This was done using hierarchical verification with both the combinational and sequential equivalence checking tools of MDG being used. Using the hybrid tool, a verification that originally required many hours of interactive proof work, could be done largely automatically using the hybrid tool.

We intend to extend the capabilities of the tool to increase the automation of the proof management process. For example, we will automate different forms of parameterization. Parameterized circuits must currently be dealt with interactively. A single instance of the parameterized circuit is verified using the hybrid tactics and this theorem used in a pure HOL proof of the parameterized circuit—performing the inductive part of the proof. This process could be automated for a range of common parameterization patterns (see Aagaard *et al* [1]) with a similar tactic to HIER_VERIF_TAC managing the inductive part of the proof. Common abstraction techniques to reduce a model say from 32-bits to 1 bit to make automated verification tractable could also be dealt with in this way. However, MDG provides a better approach: by making fuller use of the abstraction facilities in MDG itself we will remove the need for such abstraction. Currently only bit level designs can be verified using MDG via the hybrid tool. However, a future version of the hybrid tool will allow designs with abstract variables to be exported. This will remove the need to simplify datapath widths to make verification tractable and will enable us to handle data-dependent circuits automatically. We will also extend the hybrid tool to support model checking in MDG. While most of the infrastruture may be reused, ways of translating and composing temporal properties in HOL need to be developed. Finally, we will consider the verification of more complex examples including a full 16 by 16 switch fabric.

Acknowledgments. This work was funded by the NSERC Strategic Grant STP0201836 and EPSRC Research Agreement GR/M45221.

References

1. M.D. Aagaard, M. Leeser, and P. Windley. Toward a super duper hardware tactic. In J.J. Joyce and C.H. Seger, editors, *Higher Order Logic Theorem Proving and Its Applications*, LNCS 780, pages 400–413. Springer-Verlag, 1993.

2. M.D. Aagaard, R.B. Jones, and C-J.H. Seger. Lifted-FL:A Pragmatic Implementation of Combined Model Checking and Theorem Proving. In Y. Bertot, G. Dowek, A. Hirschowitz, C. Paulin, and L. Thery, editors, *Theorem Proving in Higher Order Logics*, LNCS 1690, pages 323–340. Springer-Verlag, 1999.

3. S. Balakrishnan and S. Tahar. A Hierarchical Approach to the Formal Verification of Embedded Systems Using MDGs. In *Proceedings IEEE 9th Great Lakes Symposium on VLSI*, Ann Arbor, Michigan, USA, March 1999, pages 284–287.

4. F. Corella, Z. Zhou, X. Song, M. Langevin, and E. Cerny. Multiway Decision Graphs for Automated Hardware Verification. *Formal Methods in System Design*, 10(1):7–46, 1997.

5. P. Curzon, S. Tahar, and O. Ait-Mohamed. Verification of the MDG Components Library in HOL. In J. Grundy and M. Newey, editors, *Theorem Proving in Higher Order Logics:Emerging Trends*, pages 31–45, Australian National University, 1998.

6. P. Curzon. The Formal Verification of the Fairisle ATM Switching Element. Technical Report 329, Computer Laboratory, University of Cambridge, U.K., 1994.

7. L. A. Dennis, G. Collins, M. Norrish, R. Boulton, K. Slind, G. Robinson, M. Gordon, and T. Melham. The PROSPER Toolkit. In *Proceedings of the Sixth International Conference on Tools and Algorithms for the Construction and Analysis of Systems*. LNCS 1785, Springer Verlag, 2000.

8. M.J.C. Gordon. Combining Deductive Theorem Proving with Symbolic State Enumeration. 21 Years of Hardware Verification, December 1998. Royal Society Workshop to mark 21 years of BCS FACS.

9. M.J.C. Gordon and T.F. Melham. *Introduction to HOL:A Theorem Proving Environment for Higher-Order Logic*. Cambridge University Press, U.K., 1993.

10. J. Hurd. Integrating Gandalf and HOL. In Y. Bertot, G. Dowek, A. Hirschowitz, C. Paulin, and L. Thery, editors, *Theorem Proving in Higher Order Logics*, LNCS 1690, pages 311–321. Springer Verlag, 1999.

11. J.J. Joyce and C.J.H. Seger. Linking BDD-based Symbolic Evaluation to Interactive Theorem Proving. In *Proceedings of the 30th Design Automation Conference*, pages 469–474, Dallas, TX, June 1993.

12. R. Kumar, K. Schneider and T. Kropf. Structuring and Automating Hardware Proofs in a Higher-Order Theorem-Proving Environment. *Formal Methods in System Design*, 2:165–223, 1993.

13. I.M. Leslie and D.R. McAuley. Fairisle:An ATM Network for the Local Area. *ACM Communication Review*, 19(4):327–336, 1991.

14. S. Rajan, N. Shankar, and M.K. Srivas. An Integration of Model-checking with Automated Proof Checking. In Pierre Wolper, editor, *Computer Aided Verification*, Lecture Notes in Computer Science 939, pages 84–97. Springer Verlag, 1995.

15. K. Schneider and D.W. Hoffmann. A HOL Conversion for Translating Linear Time Temporal Logic to ω-Automata. In Y. Bertot, G. Dowek, A. Hirschowitz, C. Paulin, and L. Thery, editors, *Theorem Proving in Higher Order Logics*, LNCS 1690. Springer Verlag, 1999.

16. S. Tahar, X. Song, E. Cerny, Z. Zhou, M. Langevin and O. Ait-Mohamed. Modeling and Verification of the Fairisle ATM Switch Fabric using MDGs. *IEEE Transactions on CAD of Integrated Circuits and Systems*, 18(7):956–972, 1999.

17. H. Xiong, P. Curzon, and S. Tahar. Importing MDG Results into HOL. In Y. Bertot, G. Dowek, A. Hirschowitz, C. Paulin, and L. Thery, editors, *Theorem Proving in Higher Order Logics*, LNCS 1690, 293–310. Springer Verlag, 1999.

18. M.H. Zobair and S. Tahar. On the Modeling and Verification of a Telecom System Block Using MDGs. Technical Report, Concordia University, Department of Electrical and Computer Engineering, December 2000.

Exploiting Transition Locality in Automatic Verification[*]

Enrico Tronci[1][**], Giuseppe Della Penna[1], Benedetto Intrigila[1], and
Marisa Venturini Zilli[2]

[1] Area Informatica, Università di L'Aquila, Coppito 67100, L'Aquila, Italy
{tronci,gdellape,intrigil}@univaq.it
[2] Dip. di Scienze dell'Informazione, Università di Roma "La Sapienza",
Via Salaria 113, 00198 Roma, Italy
zilli@dsi.uniroma1.it

Abstract. In this paper we present an algorithm to contrast *state explosion* when using *Explicit State Space Exploration* to verify protocols. We show experimentally that protocols exhibit *transition locality*.
We present a verification algorithm that exploits transition locality as well as an implementation of it within the Murϕ verifier.
Our algorithm is compatible with all Breadth First (BF) optimization techniques present in the Murϕ verifier and it is by no means a substitute for any of them. In fact, since our algorithm trades space with time, it is typically most useful when one runs out of memory and has already used all other state reduction techniques present in the Murϕ verifier.
Our experimental results show that using our approach we can typically save more than 40% of RAM with an average time penalty of about 50% when using (Murϕ) bit compression and 100% when using bit compression and hash compaction.

1 Introduction

State Space Exploration (*Reachability Analysis*) is at the very heart of all algorithms for automatic verification of concurrent systems. As well known, the main obstruction to automatic verification of *Finite State Systems* (FSS) is the huge amount of memory required to complete state space exploration (*state explosion*).

For protocol verification, *Explicit* State Space Exploration often outperforms *Symbolic* (i.e. OBDD based, [1,2]) State Space Exploration [6]. Since here we are mainly interested in protocol verification we focus on explicit state space exploration. Tools based on explicit state space exploration are, e.g., SPIN [4, 15] and Murϕ [3,10].

In our context, roughly speaking, two kinds of approaches have been studied to counteract (i.e. delay) state explosion: *memory saving* and *auxiliary storage*.

In a memory saving approach essentially one tries to reduce the amount of memory needed to represent the set of visited states. Examples of the memory saving approach are, e.g., in [22,7,8,18,19,5].

[*] This research has been partially supported by MURST project TOSCA
[**] Corresponding Author: Enrico Tronci. Tel: +39 0862 433129. Fax: +39 0862 433180.

T. Margaria and T. Melham (Eds.): CHARME 2001, LNCS 2144, pp. 259–274, 2001.
© Springer-Verlag Berlin Heidelberg 2001

In an auxiliary storage approach one tries to exploit disk storage as well as distributed processors (network storage) to enlarge the available memory (and CPU). Examples of this approach are, e.g., in [16,17,13,21,14].

In this paper we study the possibility of exploiting statistical properties of protocol transition graphs to improve state exploration algorithms. This is quite similar to what is usually done when optimizing a CPU on the basis of program profiling [12].

Our algorithm allows us to reduce the RAM needed to complete state space exploration and exploits auxiliary storage as well. We pay for this memory saving with a longer time to complete state space exploration. Our results can be summarized as follows.

- We show experimentally (Sect. 2) that protocols exhibit *transition locality*. That is transitions tend to be local w.r.t. levels of a Breadth First (BF) visit. We support our claim by measuring transition locality for the set of protocols included in the Murφ verifier distribution. To the best of our knowledge this is the first time that such *profiling* of transition graphs is presented.

- We present a verification algorithm that exploits transition locality as well as an implementation of it within the Murφ verifier (Sect. 3). Essentially our algorithm replaces the hash table used in a BF state space exploration with a cache memory (i.e. no collision detection is done) and uses auxiliary (disk) storage for the BF queue.

  Using a fixed size cache memory we do not incur state explosion. However we may incur nontermination when our cache memory is *too small*. In fact in this case we may visit over and over the same set of states. Note however that upon termination, we are guaranteed that all reachable states have been visited.

  To the best of our knowledge this is the first time that a cache based state space exploration algorithm is presented.

  Note in particular that the approach in [22] is a state compression technique and "no collision detection" there refers to state signatures. That is, a (signature) collision in [22] may lead to declare as visited a nonvisited state. On the other hand, we simply forget a visited state upon a collision, thus declaring as nonvisited a visited state.

  Note that the SPIN verifier can use disk storage for the Depth First (DF) stack. However states are still stored in a hash table which is where state explosion typically occurs.

- Our algorithm is compatible with all well known state compression techniques (e.g. as those in [19,5]). In particular it is compatible with all state reduction techniques present in the Murφ verifier. We show experiments using our algorithm together with the bit compression [10] and hash compaction [18,19] features of the Murφ verifier (Sect. 4).

- Our experimental results (Sect. 4) show that we can verify systems more than 40% larger than those that can be handled using a hash table based approach. Our time penalty is about 50% when using (Murφ) bit compression and 100% when using bit compression and hash compaction.

2   Transition Locality for Finite State Systems

In this section we define our notion of locality for transitions and show experimentally that for protocols most transitions are local. We do this by showing
that for all our benchmark protocols most transitions are indeed local.

We used as benchmark protocols all those available in the Murφ verifier
distribution [10] plus the Kerberos protocol from [20]. This gives us a fairly
representative benchmark set.

For our purposes, a protocol is represented as a *Finite State System*.

Definition 1.   *1. A* Finite State System *(FSS) S is a 4-tuple (S, I, A, R)*
   where: S is a finite set (of states), $I \subseteq S$ is the set of initial states, A is a
   finite set (of transition labels) and R is a relation on $S \times A \times S$. R is usually
   called the transition relation *of S.*
2. *Given states $s, s' \in S$ and $a \in A$ we say that there is a transition from s to*
   s' labeled with a iff $R(s, a, s')$ holds. We say that there is a transition from
   s to s' (notation $R(s, s')$) iff there exists $a \in A$ s.t. $R(s, a, s')$ holds. The set
   of successors of state s (notation $\text{next}(s)$) is the set of states s' s.t. $R(s, s')$.
3. *The set of* reachable states *of S (notation $\text{Reach}(S)$) is the set of states of*
   S reachable in 0 or more steps from I.
   Formally, $\text{Reach}(S)$ is the smallest set s.t.
   1. $I \subseteq \text{Reach}(S)$,
   2. for all $s \in \text{Reach}(S)$, $\text{next}(s) \subseteq \text{Reach}(S)$.

In the following we will always refer to a given system $S = (S, I, A, R)$.
Thus, e.g., we will write Reach for $\text{Reach}(S)$. Also we may speak about the set
of initial states I as well as about the transition relation R without explicitly
mentioning S.

The core of all automatic verification tools is the *reachability analysis*, i.e.
the computation of Reach given a definition of S in some language.

Since the transition relation R of a system defines a graph (*transition graph*)
computing Reach means visiting (exploring) the transition graph starting from
the initial states in I. This can be done, e.g., using a *Depth First* (DF) visit or
a *Breadth First* (BF) visit.

For example the automatic verifier SPIN [15] uses a DF visit. Murφ [10] may
use DF as well as BF, although certain compression options can only be used
with a BF visit.

In the following we will focus on BF visit. As well known a BF visit defines
levels on the transition graph. Initial states (i.e. states in I) are at level 0. The
states in $(\text{next}(I) - I)$ (states reachable in one step from I and not in I) are at
level 1, etc.

Definition 2. *Formally we define the set of states at level k (notation $L(k)$) as*
follows. $L(0) = I$, $L(k+1) = \{s' \mid s \in L(k) \text{ and } R(s, s') \text{ and } s' \notin \cup_{i=0}^{i=k} L(i)\}$.

   Given a state $s \in \text{Reach}$ we define $\text{level}(s) = k$ iff $s \in L(k)$. That is $\text{level}(s)$
is the level of state s in a BF visit of S.

   The set $\text{Visited}(k)$ of states visited (by a BF visit) by level k is defined as
follows. $\text{Visited}(k) = \cup_{i=0}^{i=k} L(i)$.

Informally, *transition locality* means that for most transitions source and target states will be in levels not too far apart.

Definition 3. *Let $\mathcal{S} = (S, I, A, R)$ be an FSS. A transition in $\mathcal{S}$ from state s to state s' is said to be k-local iff $|\mathrm{level}(s') - \mathrm{level}(s)| \leq k$.*

Transition $R(s, a, s')$ is said to be a k-transition iff $\mathrm{level}(s') - \mathrm{level}(s) = k$. Note that for k-transitions, $k \leq 1$ and can be negative.

A 1-transition from state s is a *forward transition*, i.e. a transition leading to a *new state* (a state not in $\mathrm{Visited}(\mathrm{level}(s))$). A k-transition with $k < 0$ is a *backward transition*, i.e. a transition leading to a visited state. A 0-transition from state s just leads to a state s' in the same level of s.

We are interested in the distribution of k-transitions in the transition graph. This motivates the following definition.

Definition 4. *1. We denote with $N(s, k)$ the number of k-transitions from s and with $N(s)$ the number of transitions from s.*

2. We define: $\delta(s, k) = N(s, k)/N(s)$. That is $\delta(s, k)$ is the fraction of transitions from s that are k-transitions, i.e. the probability of getting a k-transition when picking at random a transition from s. Of course if s is at level λ we have: $\sum_{k=-\lambda}^{k=1} \delta(s, k) = 1$.

3. If we consider the experiment consisting of picking at random a state s in Reach and returning $\delta(s, k)$ then we get a random variable that we denote with $\boldsymbol{\Delta}(k)$. The expected value $E\{\boldsymbol{\Delta}(k)\}$ of $\boldsymbol{\Delta}(k)$ is the average value of $\delta(s, k)$ on all reachable states. That is:

$$E\{\boldsymbol{\Delta}(k)\} = \frac{1}{|\mathrm{Reach}|} \sum_{s \in \mathrm{Reach}} \delta(s, k).$$

4. As usual, we denote with $\sigma^2(k)$ the variance of $\boldsymbol{\Delta}(k)$ and with $\sigma(k)$ its standard deviation [11].

We show experimentally that transitions in protocols tend to be *local* w.r.t. levels. We do this by showing that for the set of protocols shown in Fig. 1 most transitions are 1-local. That is, for most transitions $R(s, a, s')$ s' will be either in the same level as s (0-transition) or in the next level (1-transition) or in the previous level (-1-transition).

We want to setup experiments to measure what is the percentage of 1-local transitions in the transition graph of a given protocol.

A 1-local transition can only be a k-transition with $k = -1, 0, 1$. The expected fraction of k-transitions is $E\{\boldsymbol{\Delta}(k)\}$. Thus the expected fraction of 1-local transitions is $SumAvg = E\{\boldsymbol{\Delta}(-1)\} + E\{\boldsymbol{\Delta}(0)\} + E\{\boldsymbol{\Delta}(1)\}$.

When $SumAvg$ is close to 1 almost all transitions in the protocol are 1-local. When $SumAvg$ is close to 0 there are almost no 1-local transitions in the protocol.

Although not strictly needed for our present purposes we are also interested in knowing how 1-local transitions are distributed in the graph. Namely, we want to know whether 1-local transitions are concentrated only in some part of the transition graph or rather they are more or less uniformly distributed in the transition graph.

Protocol	$E\{\delta(-1)\}$	$\sigma(-1)$	$E\{\delta(0)\}$	$\sigma(0)$	$E\{\delta(1)\}$	$\sigma(1)$	Sum Avg
n_peterson.m	0	0	0	0	0.958174	0.0877715	0.958174
adash.m	0.0381775	0.122	0.00558149	0.0436066	0.723393	0.292406	0.76715199
adashbug.m*	0.0376604	0.124403	0.00228899	0.0295028	0.793586	0.270018	0.83353539
eadash.m	0.050598	0.0960145	0.015647	0.0562551	0.765236	0.178076	0.831481
ldash.m	0.0107259	0.0719168	0.0763593	0.138464	0.624139	0.191624	0.7112242
arbiter.m*	0.00848939	0.057543	0.0107366	0.0537135	0.92784	0.168859	0.94706599
cache3.m	0.0401213	0.178884	0.00476603	0.0558438	0.565482	0.381654	0.61036933
cache3multi.m	0.0389835	0.102299	0.00299148	0.0263492	0.831139	0.176004	0.87311398
newcache3.m	0.0360339	0.0991869	0.00912116	0.0533416	0.736229	0.227796	0.78138406
sym.cache3.m	0.041675	0.10478	0.00757049	0.0445772	0.876884	0.17786	0.92612949
down.m*	0	0	0.0776385	0.142608	0.922361	0.142608	0.9999995
kerb.m	0	0	0.18417	0.385163	0.441651	0.494666	0.625821
list6.m	0.0183668	0.0710808	0.0246248	0.0815233	0.844018	0.189053	0.8870096
list6too.m	0	0	0	0	0.988378	0.0621402	0.988378
newlist6.m	1.53327e-05	0.00175109	0	0	0.999586	0.00908985	0.9996
mcslock1.m	0.0128856	0.0675736	0	0	0.881379	0.162002	0.8942646
mcslock2.m	0.0054652	0.0458147	0.00056426	0.0151108	0.921489	0.156949	0.92751846
ns-old.m	0.546833	0.497802	0	0	0.101695	0.302247	0.648528
ns.m	0.585714	0.492598	0	0	0.0969388	0.295874	0.6826528
sci.m	0.215646	0.238654	0.0108192	0.0617188	0.642466	0.265885	0.8689312

Fig. 1. Transition Distribution Table

To some extent this can be done by computing the standard deviation $\sigma(k)$ of $\Delta(k)$. If $\sigma(k)$ is small compared to $E\{\Delta(k)\}$ then for most states $\delta(s,k)$ is *close* to $E\{\Delta(k)\}$.

The above considerations led us to perform the experiments whose results are shown in Fig. 1. For each protocol in Fig. 1 for $k = -1,0,1$, in Fig. 1 we show $E\{\Delta(k)\}$, $\sigma(k)$ and $SumAvg = E\{\Delta(-1)\} + E\{\Delta(0)\} + E\{\Delta(1)\}$. Our findings can be summarized as follows.

Experimental Fact 1. *For all protocols listed in Fig. 1, we have that for most states, more than 60% of the transitions are 1-local. Indeed, for most of the protocols in Fig. 1, we have that for most states more that 75% of the transitions are 1-local.*

In fact, from Fig. 1 we have that for all protocols $SumAvg > 0.6$ and in most cases $SumAvg > 0.75$. This shows that most transitions are 1-local.

Moreover, for many protocols, standard deviations $\sigma(-1)$, $\sigma(0)$, $\sigma(1)$ are *relatively small* compared to $SumAvg$. In such cases, *for most states* the fraction of 1-local transitions is close to $SumAvg$.

Since for most states most transitions are 1-local we have that locality holds uniformly. Hence if we pick at random a state $s \in$ Reach in most cases most transitions from s (say about 75%) will be 1-local.

Remark 1. One may wonder why protocols exhibit locality. We conjecture that this is structural for *human made* systems and, indeed, is a consequence of the techniques used (consciously or unconsciously) to master complexity in the design task.

Remark 2. A very interesting result would be to prove (or disprove) that almost all FSSs exhibit locality. Note however that, a priori, truthness or falsehood of such result would not imply anything on protocols in particular.

E.g. for OBDDs we have already a similar situation (in *reverse mode*). In fact in [9] it is proved that for almost any boolean function f of n variables the smaller OBDD for f has size exponential in n. Nevertheless, it is experimentally known that most of the boolean functions implemented by digital circuits have OBDDs of size polynomial (often about linear) in the number of variables.

We measured locality using the Murφ verifier [10]. This gives us a wide benchmark set for free since many protocols have already been defined using the Murφ input language.

The results in Fig. 1 have been obtained by modifying the Murφ verifier so as to compute $E\{\Delta(k)\}$ and $\sigma(k)$ while carrying out state space exploration.

All the experiments in Fig. 1 were performed using bit compression (Murφ option -b) [10] and disabling deadlock detection (option -ndl).

Deadlock detection has been disabled since some of the protocols (e.g. Kerberos kerb.m) are deadlocked. Murφ stops the state exploration as soon as a deadlock or a bug is found. Since our main interest here is in gathering information about the structure of the transition graph we disabled deadlock detection. This allows us to explore the whole reachable part of the transition graph also for protocols containing deadlocks.

The three protocols with superscript * in Fig. 1 contain bugs. In such cases, state space exploration stops as soon as a bug is found. We included such *buggy* protocols just for completeness.

The data in Fig. 1 depend neither on the used machine nor on the memory available as long as the visit is completed. Of course we would have obtained the same results without bit compression.

3   Cache Based Breadth First State Space Exploration

In this section we give an algorithm that takes advantage of the transition locality (Sect. 2) of protocols.

We modify the standard (hash table based) BF visit as follows. First, we use a *cache memory* rather than a hash table. This means that we do not perform any collision check. Thus, when a state s' is hashed to an entry already holding a state s, we replace s with s' so *forgetting* about s.

Using a fixed size cache rather than a hash table to store visited states is appealing since the cache size does not grow with the state space size. However this approach faces two obvious obstructions.

Queue Explosion. The queue size may get very large. In fact, since we forget visited states, at each level in our cache based BF visit the states *apparently* new tend to be much more than those of a true (hash based) BF visit. This may quickly fill up our queue.

Nontermination. Our state space exploration may not terminate. In fact, because of collisions, we forget visited states. Thus, we may visit again and again the same set of states.

When using a fixed size cache memory to hold visited states no state explosion occurs in the cache. However state explosion may occur in the queue now. For this reason, unlike previous work on usage of auxiliary storage in state space exploration (e.g. in [17]), we use disk storage to hold the queue rather than the hash table. This is good news since keeping a queue on disk is much easier than keeping a hash table on disk.

Nontermination is thus the real obstruction. Here is where *locality* helps us. Since most transitions are local one can reasonably expect that, to avoid looping, it is enough to remember the *recently* visited states rather than *all* visited states. This is exactly what a *cache memory* is for.

Note that a cache memory implements the above discipline only in a *statistical sense*. That is, replaced (i.e. *forgot*) states in the cache are *often* old states, but not always. Fortunately, this is enough for us.

Of course, if the cache memory is *too small* w.r.t. the size of the state space locality cannot help us.

Note that whenever our cache based BF visit terminates it gives the correct answer. That is *all* reachable states have been visited. When we stop our cache based visit to prevent looping we may or may not have visited all reachable states.

In the following we sketch our algorithm and show how we integrated it into the Murφ verifier.

To take advantage of transition locality we modify the standard BF visit as shown in Fig. 2.

We use a cache memory rather than a hash table. This may lead to nontermination since we may visit over and over the same set of states. For this reason we *guarded* the while loop with a check on the collision rate (i.e. <number of collisions in cache>/<number of insertions in cache>).

Since we may forget visited states, our queue tends to contain many more states than the one used in a standard BF visit. For this reason we decided to implement our queue using auxiliary memory. In the present implementation we use disk storage as auxiliary memory. Note however that (the RAM of) another workstation could have been used as well.

The implementation schema for the cache memory is quite standard. In the following we describe our implementation schema for the queue on disk (Fig. 3).

We split the queue Q into two segments: *head queue* and *tail queue*. Both head queue and tail queue are in RAM and each of them can hold ram_queue_size states. States are *enqueued* in the tail queue and *dequeued* from the head queue.

When the tail queue is full, its content is flushed on disk by appending it to swapout_file. Thus, since disks are quite large, our disk queue approximates a potentially infinite queue. It overflows only when our disk is full.

When the head queue is empty, it gets reloaded with states from swapin_file. If swapin_file is empty we swap swapin_file and swapout_file and try to load states from swapin_file again. If swapin_file is still empty then we swap the head queue with the tail queue.

```
Queue Q;        Cache T;        collision_rate = 0.0;
bfs(init_states, next) {
for s in init_states enqueue(Q, s);  /* load Q with initial states */
for s in init_states insert(T, s);  /* mark init states as visited */
while ((Q is not empty) and (collision_rate <= 0.9)) {s = dequeue(Q);
     for all s' in next(s) if (s' is not in T) {insert(T, s'); enqueue(Q, s'); }}}
```

Fig. 2. Cache based Breadth First Visit

```
void enqueue(state s) {if (tail queue is full) swap_out();
/* tail queue is not full */ insert s into tail queue; tail_queue_elements++;}

/* hyp: dequeue() always called on non empty queue */
state dequeue() {if (head queue is empty) swap_in();
/* head queue is not empty now */ head_queue_elements--; return top of head queue;}

void swap_out() { /* flush tail queue on disk and reset counter */
append tail queue to swapout_file; tail_queue_elements = 0;  }

void swap_in() { /* load head queue from disk  */
if (swapin_file is not empty)
   load head queue with at most ram_queue_size states from swapin_file;
else /* swapin_file is empty, we use swapout_file */
   { swap the swapin_file and swapout_file;
     if (swapin_file is not empty)
        load head queue with at most ram_queue_size states from swapin_file;
     else /* also swapout_file is empty, we use tail_queue */
        { swap the head queue and tail queue;
          if (head queue is empty)
             /* underflow error */ { error(''queue is empty''); }}}
head_queue_elements = number of states in head queue; }
```

Fig. 3. Disk Queue Functions

Using pointers, swapping (for files and for queue segments) is immediate (just a few assignments).

Rather than building a tool from scratch we decided to integrate our algorithm into an already existing tool. This gives us at least two advantages. First, we have immediately available many benchmark systems for testing. Second, we can exploit other memory reduction techniques that have already been implemented. We decided to integrate our algorithm within the Murφ verifier [10].

To be consistent with the *standard* (i.e. hash table based) Murφ verifier the value of ram_queue_size (Fig. 3) is such that the head queue and the tail queue together take the same amount of memory as the queue in standard Murφ. Namely, if M is the available memory for verification then gPercentActive*M is the amount of RAM memory used for the queue in standard Murφ as well as in our cache based Murφ.

Integrating our algorithm into the Murφ verifier requires some care and consideration. In fact, to save RAM, Murφ (as well as SPIN for that matter) stores pointers to states rather than states in the queue.

As far as we are concerned, Murφ BF visit can have two behaviours. One when hash compaction is not used and another one when it is used. All other Murφ options have no impact on the hash table or the queue and thus are *transparent* to us.

When no hash compaction is used Murφ stores states in the hash table and pointers to hash table slots (states) in the queue. When hash compaction is used, Murφ stores state signatures in the hash table and pointers to states in the queue.

Let us consider the case without hash compaction first. Using a cache rather than a hash table leads to the following problem. When we overwrite a cache slot (collision) as a side effect we may also change the content of the queue. In fact that slot may be pointed to by the queue. For this reason we modified Murφ queue so that it holds states and not just pointers to states. Of course all Murφ functions depending on this fact have to be changed accordingly. This also fixes the situation for the case in which hash compaction is used.

Of course storing states rather than pointers (to states) in the queue takes more space. However, for the reasons explained above, we are going to use disk storage to implement our queue. Thus having a large queue does not pose any serious problem in our context.

With our approach all optimization strategies implemented within the Murφ verifier (bit compression, hash compaction, symmetry reduction, etc) are also available to us.

Note that the bound on the omission probability for the hash compaction computed "on-line" by Murφ accordingly to [19] is also (a fortiori) valid in our case. In fact at each (BF) level we visit a superset of the states visited by a standard BF visit.

4 Experimental Results

We report the experimental results we obtained using Murφ (modified as described in Sect. 3).

We want to measure how much (RAM) memory we can save by using our cache based approach. To make the results from different protocols comparable we proceed as follows.

First, for each protocol we determine the minimum amount of memory needed to complete verification using the Murφ verifier (namely Murφ version 3.1 from [10]).

Let M be the amount of memory and g (in $[0, 1]$) be the fraction of M used for the queue (i.e. g is gPercentActive using a Murφ parlance). We say that the pair (M, g) is *suitable* for protocol p iff the verification of p can be completed with memory M and queue gM. For each protocol p we determine the least M s.t. for some g, (M, g) is suitable for p. In the following we denote with $M(p)$ such M.

Of course $M(p)$ depends on the compression options one uses. Murφ offers *bit compression* (-b) and *hash compaction* (-c). We determined $M(p)$ when only bit compression is used (-b) and when bit compression and hash compaction are used (-b -c). The results are in Fig. 4.

The meaning of the columns in Fig. 4 is as follows. Column **Bytes** gives the number of bytes needed to represent the state when bit compression is used. With hash compaction 40 bits are used to represent the hash-compressed state. Column **Reach** gives the number of reachable states for the protocol. For pro-

Protocol	Bytes	Reach	Rules	Max Q	Diam	mu -b			mu -b -c		
						M	g	T	M	g	T
n_peterson.m	16	163298	1143086	3775	145	3204	0.0233	269.13	813	0.0233	273.32
adash.m	144	10466	137708	734	37	1516	0.075	60.57	55	0.075	62.98
adashbug.m	144	3742	39619	646	14	544	0.2	18.19	21	0.18125	17.60
eadash.m	376	133491	1786047	9050	47	49571	0.06875	4102.33	688	0.06875	4114.71
ldash.m	144	254986	2647358	14988	64	36930	0.075	4002.97	1307	0.062	3950.50
arbiter.m	8	1103	2365	301	12	15	0.4	0.1	7	0.3	0.1
cache3.m	12	577	2440	102	16	10	0.4	0.15	4	0.4	0.15
cache3multi.m	28	13738	65357	1229	29	435	0.1	34.16	73	0.1	35.11
newcache3.m	52	4357	20201	462	27	240	0.1125	8.01	24	0.15	7.93
sym.cache3.m	28	31433	264758	2877	32	994	0.1039	36.22	167	0.10625	36.03
down.m	4	10957	52315	1313	19	92	0.15	1.35	60	0.15	1.22
kerb.m	80	109282	172111	20523	19	9046	0.191	291.49	615	0.190625	301.86
list6.m	24	23410	99874	1095	53	645	0.05	13.73	119	0.05	14.90
list6too.m	20	1077	11622	64	36	26	0.2	15.55	6	0.1	16.83
newlist6.m	24	13044	53595	631	53	360	0.05	17.23	67	0.05	18.34
mcslock1.m	12	23644	94576	928	69	373	0.04023	16.17	120	0.04375	16.76
mcslock2.m	12	540219	1620657	15655	111	8503	0.0293	234.68	2693	0.02969	237.48
ns-old.m	24	1121	2578	424	11	33	0.4	0.69	8	0.4	0.69
ns.m	24	980	2314	382	11	29	0.4	0.59	7	0.4	0.62
sci.m	56	18193	60455	1175	62	1071	0.066	27.29	94	0.06875	28.17

Fig. 4. Results on a SUN Sparc machine with 512M RAM and SunOS 5.8. Murφ options: -b (bit compression), -c (40 bit hash compaction). All experiments have been carried out with option -ndl (no deadlock detection). Column M gives memory used (in kilobytes) and column T gives CPU time (in seconds).

tocol p, in the following we denote such number with $|\text{Reach}(p)|$. Column **Rules** gives the number of rules fired during state space exploration. This number only depends on the transition graph. For protocol p, in the following we denote such number with $\text{RulesFired}(p)$. Column **Max Q** gives the maximum queue size in a BF visit. This number only depends on the transition graph. Column **Diam** gives the diameter of the transition graph. This number only depends on the transition graph. Column **M** gives the minimum amount of memory (in kilobytes) needed to complete state space exploration. Two cases: -b and -b -c. Column **g** gives the fraction of memory M used for the queue. Two cases: -b and -b -c. Column **T** gives the time (in seconds) to complete state space exploration when using memory M and queue gM. Two cases: -b and -b -c. For protocol p, in the following we denote such numbers with $T_b(p)$ and $T_{bc}(p)$ respectively.

Our next step is to run each protocol p with less and less memory using our cache based Murφ. That is we run p with memory limits $M(p)$, $0.9M(p)$, ... $0.1M(p)$.

This approach allows us to easily compare the experimental results obtained from different protocols. Note that, as in [6], our RAM is not filled up. Thus OS buffers may reduce the time to access the queue. As a result our measured time may be slightly smaller than those obtained with a filled up memory.

The results using option -b (bit compression) are in Fig. 5. The results obtained using options -b -c (bit compression and hash compaction) are in Fig. 6.

We only considered protocols requiring at least 10 kilobytes of RAM (column M of Fig. 4) to complete state space exploration. In the experiments in Figs. 5, 6 the value of g (gPercentActive) is chosen as in Fig. 4.

	Mem	1	0.9	0.8	0.7	0.6	0.5	0.4	0.3	0.2	0.1
n_peterson	states	1.004	1.011	1.028	1.242	4.366*	5.028*	4.036*	3.037$^\infty$	2.039$^\infty$	1.035$^\infty$
	rules	1.004	1.011	1.028	1.242	4.364*	4.994*	3.997*	2.992$^\infty$	1.992$^\infty$	0.993$^\infty$
	time	0.972	0.981	1.004	1.201	4.222*	4.888*	3.914*	2.959$^\infty$	1.975$^\infty$	1.017$^\infty$
	coll	0.062	0.128	0.225	0.437	0.863*	0.901*	0.901*	0.901$^\infty$	0.902$^\infty$	0.904$^\infty$
adash	states	1.006	1.016	1.043	1.124	1.738	5.160*	4.300$^\infty$	3.631$^\infty$	2.389$^\infty$	1.147$^\infty$
	rules	1.007	1.016	1.043	1.125	1.736	4.749*	3.714$^\infty$	2.937$^\infty$	1.819$^\infty$	0.782$^\infty$
	time	1.001	1.010	1.039	1.124	1.747	4.824*	3.792$^\infty$	3.013$^\infty$	1.889$^\infty$	0.818$^\infty$
	coll	0.063	0.132	0.236	0.377	0.655	0.904*	0.907$^\infty$	0.919$^\infty$	0.916$^\infty$	0.918$^\infty$
adashbug	states	1.002	1.006	1.010	1.032	1.055	1.124	1.261	1.477	1.871*	1.336$^\infty$
	rules	1.001	1.001	1.003	1.011	1.021	1.061	1.145	1.275	1.504*	0.965$^\infty$
	time	0.983	0.980	0.982	0.988	0.999	1.045	1.135	1.280	1.540*	1.019$^\infty$
	coll	0.052	0.113	0.199	0.312	0.422	0.545	0.672	0.787	0.904*	0.931$^\infty$
eadash	states	1.007	1.021	1.066	1.382	6.540*	5.251$^\infty$	4.862$^\infty$	3.805$^\infty$	2.292$^\infty$	1.034$^\infty$
	rules	1.007	1.021	1.065	1.384	6.008*	4.440$^\infty$	3.910$^\infty$	2.918$^\infty$	1.649$^\infty$	0.667$^\infty$
	time	1.029	1.043	1.094	1.439	6.354*	4.653$^\infty$	4.133$^\infty$	3.155$^\infty$	1.831$^\infty$	0.775$^\infty$
	coll	0.063	0.135	0.251	0.493	0.908*	0.905$^\infty$	0.918$^\infty$	0.921$^\infty$	0.913$^\infty$	0.904$^\infty$
ldash	states	1.019	1.058	1.295	7.495*	6.114$^\infty$	6.106$^\infty$	4.067$^\infty$	3.753$^\infty$	2.361$^\infty$	1.020$^\infty$
	rules	1.018	1.056	1.287	6.698*	5.084$^\infty$	4.847$^\infty$	3.086$^\infty$	2.726$^\infty$	1.633$^\infty$	0.664$^\infty$
	time	1.043	1.078	1.324	6.758*	5.112$^\infty$	4.984$^\infty$	3.177$^\infty$	2.828$^\infty$	1.703$^\infty$	0.701$^\infty$
	coll	0.069	0.161	0.382	0.907*	0.902$^\infty$	0.918$^\infty$	0.902$^\infty$	0.920$^\infty$	0.915$^\infty$	0.902$^\infty$
arbiter	states	1.000	1.004	1.007	1.021	1.032	1.060	1.067	1.137	1.230	0.907$^\infty$
	rules	1.000	1.000	1.001	1.000	1.010	1.017	1.024	1.078	1.139	0.800$^\infty$
	time	1.000	1.100	1.100	1.200	1.200	1.300	1.300	1.400	1.500	1.200$^\infty$
	coll	0.054	0.133	0.189	0.327	0.405	0.537	0.611	0.755	0.833	0.931$^\infty$
cache3	states	1.002	1.021	1.047	1.192	1.459	3.000	3.466*	1.733$^\infty$	1.733$^\infty$	1.733$^\infty$
	rules	1.001	1.020	1.045	1.180	1.416	2.953	3.064*	1.383$^\infty$	1.291$^\infty$	0.776$^\infty$
	time	1.200	1.200	1.200	1.467	1.467	4.133	3.467*	1.667$^\infty$	1.667$^\infty$	0.800$^\infty$
	coll	0.050	0.127	0.230	0.406	0.586	0.831	0.901*	0.905$^\infty$	0.918$^\infty$	0.906$^\infty$
cache3multi	states	1.009	1.030	1.098	1.291	2.422	5.387*	4.367$^\infty$	3.203$^\infty$	2.184$^\infty$	0.946$^\infty$
	rules	1.011	1.037	1.117	1.328	2.535	5.173*	3.902$^\infty$	2.674$^\infty$	1.725$^\infty$	0.712$^\infty$
	time	1.010	1.037	1.118	1.337	2.564	5.255*	3.974$^\infty$	2.730$^\infty$	1.759$^\infty$	0.728$^\infty$
	coll	0.062	0.141	0.272	0.458	0.752	0.908*	0.909$^\infty$	0.908$^\infty$	0.911$^\infty$	0.900$^\infty$
newcache3	states	1.012	1.029	1.053	1.213	1.895	5.279*	4.131$^\infty$	2.984$^\infty$	2.066$^\infty$	1.148$^\infty$
	rules	1.011	1.027	1.050	1.222	1.927	4.851*	3.446$^\infty$	2.378$^\infty$	1.551$^\infty$	0.754$^\infty$
	time	1.004	1.022	1.049	1.230	1.953	4.905*	3.487$^\infty$	2.401$^\infty$	1.577$^\infty$	0.792$^\infty$
	coll	0.069	0.141	0.242	0.423	0.683	0.907*	0.907$^\infty$	0.902$^\infty$	0.911$^\infty$	0.919$^\infty$
sym.cache3	states	1.008	1.022	1.064	1.213	1.919	5.345*	4.327$^\infty$	3.436$^\infty$	2.672$^\infty$	1.304$^\infty$
	rules	1.010	1.026	1.072	1.233	1.970	5.138*	3.742$^\infty$	2.772$^\infty$	1.987$^\infty$	0.863$^\infty$
	time	1.024	1.043	1.093	1.266	2.035	5.388*	4.011$^\infty$	3.041$^\infty$	2.248$^\infty$	1.045$^\infty$
	coll	0.064	0.137	0.251	0.424	0.688	0.906*	0.908$^\infty$	0.913$^\infty$	0.926$^\infty$	0.924$^\infty$
down	states	1.001	1.006	1.013	1.023	1.038	1.058	1.104	1.202	1.416	1.004$^\infty$
	rules	1.001	1.005	1.011	1.019	1.032	1.050	1.092	1.179	1.365	0.747$^\infty$
	time	0.785	0.822	0.859	0.881	0.911	0.948	1.022	1.126	1.333	0.763$^\infty$
	coll	0.057	0.127	0.219	0.321	0.424	0.527	0.645	0.756	0.862	0.902$^\infty$
kerb	states	1.001	1.001	1.003	1.007	1.015	1.025	1.048	1.097	1.228	1.107$^\infty$
	rules	1.003	1.005	1.012	1.022	1.038	1.057	1.092	1.151	1.285	1.012$^\infty$
	time	0.982	0.986	0.987	0.994	1.004	1.015	1.041	1.090	1.221	0.828$^\infty$
	coll	0.060	0.122	0.207	0.305	0.409	0.512	0.618	0.727	0.837	0.910$^\infty$
list6	states	1.006	1.014	1.044	1.151	1.735	5.041*	4.528$^\infty$	3.460$^\infty$	2.392$^\infty$	1.025$^\infty$
	rules	1.007	1.015	1.045	1.149	1.721	4.531*	3.715$^\infty$	2.682$^\infty$	1.765$^\infty$	0.732$^\infty$
	time	0.999	0.999	1.038	1.157	1.752	4.701*	3.866$^\infty$	2.818$^\infty$	1.886$^\infty$	0.792$^\infty$
	coll	0.063	0.131	0.236	0.392	0.654	0.901*	0.912$^\infty$	0.914$^\infty$	0.917$^\infty$	0.904$^\infty$
list6too	states	1.006	1.030	1.082	1.232	1.516	4.643*	3.714*	2.786$^\infty$	1.857$^\infty$	22.284$^\infty$
	rules	1.006	1.035	1.087	1.240	1.571	4.689*	3.505*	2.459$^\infty$	1.674$^\infty$	6.290$^\infty$
	time	1.024	1.053	1.106	1.266	1.595	4.768*	3.553*	2.489$^\infty$	1.665$^\infty$	0.700$^\infty$
	coll	0.062	0.151	0.291	0.440	0.621	0.903*	0.912*	0.908$^\infty$	0.910$^\infty$	0.915$^\infty$
newlist6	states	1.009	1.022	1.049	1.198	1.830	5.136*	3.987$^\infty$	3.527$^\infty$	2.300$^\infty$	1.227$^\infty$
	rules	1.009	1.023	1.052	1.198	1.816	4.641*	3.328$^\infty$	2.783$^\infty$	1.734$^\infty$	0.888$^\infty$
	time	1.005	1.020	1.052	1.198	1.810	4.656*	3.320$^\infty$	2.777$^\infty$	1.725$^\infty$	0.868$^\infty$
	coll	0.066	0.134	0.239	0.415	0.671	0.903*	0.900$^\infty$	0.915$^\infty$	0.914$^\infty$	0.922$^\infty$
mcslock1	states	1.009	1.020	1.058	1.190	1.900	4.948*	4.187$^\infty$	3.299$^\infty$	2.368$^\infty$	1.100$^\infty$
	rules	1.009	1.020	1.058	1.190	1.900	4.613*	3.709$^\infty$	2.818$^\infty$	1.958$^\infty$	0.895$^\infty$
	time	1.015	1.030	1.074	1.198	1.890	4.547*	3.605$^\infty$	2.678$^\infty$	1.799$^\infty$	0.798$^\infty$
	coll	0.069	0.140	0.248	0.412	0.685	0.904*	0.904$^\infty$	0.911$^\infty$	0.917$^\infty$	0.911$^\infty$
mcslock2	states	1.009	1.017	1.032	1.063	1.178	2.751*	4.008$^\infty$	3.004$^\infty$	2.269$^\infty$	1.061$^\infty$
	rules	1.009	1.017	1.032	1.063	1.178	2.749*	3.742$^\infty$	2.710$^\infty$	1.986$^\infty$	0.903$^\infty$
	time	0.978	0.991	1.009	1.046	1.163	2.734*	3.721$^\infty$	2.692$^\infty$	1.951$^\infty$	0.879$^\infty$
	coll	0.092	0.152	0.236	0.343	0.491	0.818*	0.900$^\infty$	0.900$^\infty$	0.912$^\infty$	0.906$^\infty$
ns-old	states	1.000	1.011	1.000	1.019	1.029	1.103	1.259	2.140	1.784*	0.892$^\infty$
	rules	1.000	1.059	1.000	1.095	1.100	1.315	1.460	2.558	1.770*	0.813$^\infty$
	time	1.014	1.058	1.043	1.072	1.087	1.246	1.406	2.377	1.536*	0.754$^\infty$
	coll	0.047	0.125	0.192	0.302	0.428	0.549	0.682	0.870	0.917*	0.925$^\infty$
ns	states	1.001	1.000	1.010	1.046	1.024	1.057	1.147	2.476	2.041*	1.020$^\infty$
	rules	1.032	1.000	1.066	1.140	1.047	1.128	1.284	2.998	1.974*	0.891$^\infty$
	time	1.085	1.068	1.102	1.169	1.119	1.169	1.305	2.864	1.746*	0.746$^\infty$
	coll	0.042	0.098	0.202	0.326	0.409	0.530	0.659	0.886	0.919*	0.932$^\infty$
sci	states	1.003	1.008	1.020	1.041	1.095	1.369	4.067*	3.243$^\infty$	2.089$^\infty$	1.099$^\infty$
	rules	1.003	1.009	1.022	1.045	1.102	1.382	3.975*	2.919$^\infty$	1.749$^\infty$	0.850$^\infty$
	time	1.017	1.016	1.025	1.048	1.104	1.392	3.970*	2.912$^\infty$	1.752$^\infty$	0.858$^\infty$
	coll	0.062	0.127	0.219	0.328	0.453	0.635	0.902*	0.908$^\infty$	0.906$^\infty$	0.912$^\infty$

Fig. 5. Cache Based BF Visit (bit compression -b)

	Mem	1	0.9	0.8	0.7	0.6	0.5	0.4	0.3	0.2	0.1
n_peterson	states	1.000	1.005	1.024	1.225	4.262*	5.021*	4.036*	2.988$^\infty$	1.990$^\infty$	0.998$^\infty$
	rules	1.000	1.005	1.024	1.225	4.262*	4.988*	3.996*	2.942$^\infty$	1.944$^\infty$	0.957$^\infty$
	time	1.008	1.033	1.079	1.320	5.014*	5.993*	4.797*	3.560$^\infty$	2.374$^\infty$	1.207$^\infty$
	coll	0.006	0.104	0.219	0.428	0.859*	0.901*	0.901*	0.900$^\infty$	0.900$^\infty$	0.900$^\infty$
adash	states	1.000	1.009	1.025	1.133	1.565	5.064*	4.491$^\infty$	3.631$^\infty$	2.293$^\infty$	0.860$^\infty$
	rules	1.000	1.009	1.025	1.133	1.568	4.684*	3.895$^\infty$	2.931$^\infty$	1.763$^\infty$	0.572$^\infty$
	time	0.973	0.994	1.016	1.134	1.593	4.855*	4.046$^\infty$	3.081$^\infty$	1.863$^\infty$	0.609$^\infty$
	coll	0.002	0.103	0.207	0.380	0.611	0.902*	0.911$^\infty$	0.918$^\infty$	0.912$^\infty$	0.900$^\infty$
adashbug	states	1.000	1.006	1.012	1.032	1.061	1.135	1.268	1.532	2.138*	1.336$^\infty$
	rules	1.000	1.000	1.001	1.008	1.021	1.065	1.142	1.299	1.708*	0.961$^\infty$
	time	1.006	1.014	1.020	1.034	1.054	1.110	1.213	1.397	1.872*	1.080$^\infty$
	coll	0.001	0.133	0.235	0.336	0.446	0.567	0.686	0.802	0.912*	0.935$^\infty$
eadash	states	1.000	1.009	1.056	1.365	6.083*	5.124$^\infty$	4.847$^\infty$	3.813$^\infty$	2.285$^\infty$	1.041$^\infty$
	rules	1.000	1.000	1.055	1.366	5.568*	4.344$^\infty$	3.893$^\infty$	2.916$^\infty$	1.645$^\infty$	0.670$^\infty$
	time	1.005	1.014	1.065	1.401	5.870*	4.523$^\infty$	4.078$^\infty$	3.133$^\infty$	1.802$^\infty$	0.761$^\infty$
	coll	0.007	0.108	0.242	0.488	0.902*	0.902$^\infty$	0.918$^\infty$	0.922$^\infty$	0.913$^\infty$	0.906$^\infty$
ldash	states	1.000	1.028	1.284	7.189*	7.047$^\infty$	5.953$^\infty$	4.016	3.722	2.349	1.016
	rules	1.000	1.027	1.278	6.394*	5.861$^\infty$	4.719$^\infty$	3.048$^\infty$	2.705$^\infty$	1.622$^\infty$	0.062$^\infty$
	time	1.013	1.075	1.346	6.851*	6.177$^\infty$	4.987$^\infty$	3.278$^\infty$	2.946$^\infty$	1.762$^\infty$	0.727$^\infty$
	coll	0.007	0.125	0.377	0.903*	0.915$^\infty$	0.916$^\infty$	0.901$^\infty$	0.920$^\infty$	0.915$^\infty$	0.902$^\infty$
cache3multi	states	1.000	1.013	1.089	1.297	2.939	5.168*	4.440$^\infty$	3.421$^\infty$	2.329$^\infty$	0.946$^\infty$
	rules	1.000	1.017	1.104	1.330	3.078	4.929*	3.985$^\infty$	2.862$^\infty$	1.835$^\infty$	0.713$^\infty$
	time	0.978	1.005	1.107	1.363	3.344	5.418*	4.399$^\infty$	3.157$^\infty$	2.026$^\infty$	0.780$^\infty$
	coll	0.003	0.114	0.264	0.457	0.798	0.905*	0.911$^\infty$	0.915$^\infty$	0.918$^\infty$	0.903$^\infty$
newcache3	states	1.000	1.022	1.053	1.313	2.155	5.049*	4.131$^\infty$	3.213$^\infty$	1.836$^\infty$	1.148$^\infty$
	rules	1.000	1.020	1.052	1.330	2.192	4.646*	3.412$^\infty$	2.564$^\infty$	1.347$^\infty$	0.742$^\infty$
	time	1.032	1.061	1.111	1.450	2.483	5.462*	4.026$^\infty$	3.021$^\infty$	1.604$^\infty$	0.919$^\infty$
	coll	0.003	0.136	0.239	0.488	0.725	0.903*	0.908$^\infty$	0.911$^\infty$	0.909$^\infty$	0.936$^\infty$
sym.cache3	states	1.000	1.012	1.058	1.217	1.871	5.027*	4.358$^\infty$	3.500$^\infty$	2.641$^\infty$	1.304$^\infty$
	rules	1.000	1.013	1.065	1.235	1.921	4.756*	3.764$^\infty$	2.814$^\infty$	1.950$^\infty$	0.867$^\infty$
	time	1.020	1.054	1.142	1.384	2.298	6.182*	5.011$^\infty$	3.850$^\infty$	2.730$^\infty$	1.278$^\infty$
	coll	0.005	0.109	0.245	0.427	0.679	0.901*	0.910$^\infty$	0.914$^\infty$	0.925$^\infty$	0.927$^\infty$
down	states	1.000	1.002	1.009	1.016	1.037	1.067	1.102	1.179	1.393	1.369$^\infty$
	rules	1.000	1.001	1.007	1.016	1.031	1.058	1.089	1.160	1.347	1.040$^\infty$
	time	1.098	1.426	1.885	2.221	2.623	3.041	3.475	4.049	5.082	4.270$^\infty$
	coll	0.007	0.100	0.206	0.312	0.419	0.530	0.636	0.745	0.856	0.929$^\infty$
kerb	states	1.000	1.001	1.002	1.006	1.013	1.026	1.047	1.093	1.223	1.116$^\infty$
	rules	1.000	1.003	1.008	1.011	1.034	1.060	1.090	1.146	1.280	1.020$^\infty$
	time	0.983	0.990	1.002	1.011	1.026	1.054	1.078	1.135	1.274	0.891$^\infty$
	coll	0.007	0.101	0.202	0.305	0.407	0.513	0.618	0.726	0.836	0.913$^\infty$
list6	states	1.000	1.009	1.036	1.163	2.015	5.254*	4.742$^\infty$	3.759$^\infty$	2.478$^\infty$	1.111$^\infty$
	rules	1.000	1.009	1.037	1.160	1.997	4.670*	3.868$^\infty$	2.880$^\infty$	1.825$^\infty$	0.779$^\infty$
	time	0.996	1.039	1.121	1.332	2.631	6.946*	5.840$^\infty$	4.406$^\infty$	2.823$^\infty$	1.217$^\infty$
	coll	0.006	0.107	0.229	0.400	0.703	0.906*	0.917$^\infty$	0.922$^\infty$	0.923$^\infty$	0.919$^\infty$
newlist6	states	1.000	1.007	1.050	1.226	1.897	5.443*	4.140$^\infty$	3.373$^\infty$	2.300$^\infty$	1.073$^\infty$
	rules	1.000	1.008	1.053	1.224	1.882	4.863*	3.434$^\infty$	2.674$^\infty$	1.738$^\infty$	0.776$^\infty$
	time	1.046	1.066	1.140	1.360	2.192	6.009*	4.263$^\infty$	3.335$^\infty$	2.155$^\infty$	0.944$^\infty$
	coll	0.002	0.100	0.237	0.434	0.682	0.909*	0.905$^\infty$	0.912$^\infty$	0.917$^\infty$	0.917$^\infty$
mcslock1	states	1.000	1.010	1.046	1.163	1.848	5.160*	4.145$^\infty$	3.341$^\infty$	2.326$^\infty$	1.100$^\infty$
	rules	1.000	1.010	1.046	1.163	1.848	4.817*	3.680$^\infty$	2.849$^\infty$	1.926$^\infty$	0.893$^\infty$
	time	0.982	1.025	1.106	1.284	2.217	6.304*	4.776$^\infty$	3.652$^\infty$	2.410$^\infty$	1.085$^\infty$
	coll	0.006	0.104	0.232	0.395	0.674	0.903*	0.903$^\infty$	0.911$^\infty$	0.915$^\infty$	0.910$^\infty$
mcslock2	states	1.000	1.006	1.022	1.055	1.177	4.444*	4.206$^\infty$	3.038$^\infty$	2.279$^\infty$	1.063$^\infty$
	rules	1.000	1.006	1.022	1.055	1.177	4.427*	3.915$^\infty$	2.734$^\infty$	1.993$^\infty$	0.905$^\infty$
	time	1.023	1.092	1.181	1.301	1.538	6.956*	6.202$^\infty$	4.325$^\infty$	3.043$^\infty$	1.335$^\infty$
	coll	0.009	0.105	0.217	0.337	0.490	0.888*	0.905$^\infty$	0.901$^\infty$	0.912$^\infty$	0.906$^\infty$
sci	states	1.000	1.005	1.013	1.037	1.087	1.313	4.013*	3.133$^\infty$	2.309$^\infty$	1.154$^\infty$
	rules	1.000	1.006	1.014	1.041	1.093	1.325	3.950*	2.793$^\infty$	1.916$^\infty$	0.894$^\infty$
	time	1.027	1.043	1.061	1.105	1.180	1.458	4.544*	3.225$^\infty$	2.226$^\infty$	1.047$^\infty$
	coll	0.005	0.108	0.210	0.331	0.450	0.618	0.902*	0.906$^\infty$	0.917$^\infty$	0.920$^\infty$

Fig. 6. Cache Based BF Visit (bit compression and hash compaction -b -c)

m	SizeC	MemC	SizeQ	MemQ	Visited	Time	Coll	MaxQ	MaxMemQ
200	38,836,153	180	215,756	20	44,780,625	19,349.80	0.17024	16,215,144	1113.40
150	29,127,121	135	161,817	15	46,369,727	19,975.79	0.373777	16,239,606	1115
130	25,243,507	117	140,241	13	48,933,778	21,769.69	0.484372	16,264,901	1116.8

Fig. 7. Cache Based BF Visit for protocol ns with parameters: NumInitiators = 2, NumResponders = 1, NumIntruders = 2, NetworkSize = 2, MaxKnowledge = 10. Results obtained using options -b (bit compression), -c (40 bit hash compaction) and -ndl (no deadlock detection) on a Pentium III 866Mhz machine with 256Mb of RAM.

We give the meaning of rows and columns in Fig. 5. Those in Fig. 6 of course have exactly the same meaning, only they refer to the case in which both bit compression and hash compaction (-b -c) are used.

Column α (with $\alpha = 1, 0.9, \ldots, 0.1$) gives information about the run of protocol p with memory $\alpha M(p)$. Row **States** gives the ratio between the visited states when using memory $\alpha M(p)$ and $|\text{Reach}(p)|$. This is the *state* overhead due to revisiting already visited states. Row **Rules** gives the ratio between the rules fired when using memory $\alpha M(p)$ and $\text{RulesFired}(p)$. This is the *rule* overhead due to revisiting already visited states. Row **Time** gives the ratio between the time (in seconds) to complete state space exploration when using memory $\alpha M(p)$ and option w and $T_w(p)$. Option w can be bit compression ($w = b$) or bit compression and hash compaction ($w = bc$). This is the time overhead. Row **Coll** gives the collision rate. That is the ratio between the number of collisions and the number of states inserted in the cache.

Rows *States, Rules, Time* should have close values. As shown in Figs. 5, 6 this is indeed the case.

Our state space exploration may not terminate. We stop our visit when the collision rate is greater than 0.9. or the max depth exceeds a given threshold.

Note that when our visit is stopped prematurely (for the above reasons) we do not know a priori if all reachable states have been visited. However, for the experiments in Figs. 5, 6 we have such information since they fit in our memory. In Figs. 5, 6 we report such information just to give an idea of the behaviour of our approach when a visit is stopped prematurely.

We mark with a * superscript the data obtained when the following conditions are satisfied: 1. The visit has been stopped because the collision rate exceeded 0.9 or because the max depth exceeded a certain threshold; 2. All reachable states have been visited.

We mark with a ∞ superscript the data obtained when the following conditions are satisfied: 1. The visit has been stopped because the collision rate exceeded 0.9 or because the max depth exceeded a certain threshold; 2. There are reachable states that have not been visited.

The experimental results in Figs. 5, 6 show that our cache based approach typically saves about 40% of RAM w.r.t. to an approach based on hash table. This holds for case b (bit compression) as well as for case bc (bit compression and hash compaction).

For case b time penalty for column 0.6 (40% RAM saving) ranges from 0 to 150% with an average value of about 50%.

For case bc time penalty for column 0.6 (40% RAM saving) ranges from 0 to 234% with an average value of about 100%.

Note that with our cache based approach the amount of memory needed does not increase with the max depth. Instead in a hash table based approach using a max depth limit (e.g. as in SPIN) the memory required increases with the max depth.

We also wanted to test our cache based approach with a large protocol that heavily loads our machine. The results are in Fig. 7. The meaning of the columns of Fig. 7 is as follows.

Column **m** gives the total memory (in Megabytes) given to cached Murφ to complete the verification. Column **SizeC** gives the max number of states that the cache can contain (each state takes 40 bits with hash compaction). Column **MemC** gives the memory (in Megabytes) reserved for the cache (0.9*m). Column **SizeQ** gives the max number of states that the RAM queue can contain (each state takes 68 bytes in the queue). Column **MemC** gives the memory (in Megabytes) reserved for the RAM queue (0.1 * m). Column **Visited** gives the number of states visited by the verifier. Column **Time** gives the time (in seconds) taken to complete the verification task. Column **Coll** gives the collision rate. Column **MaxQ** is the max number of states contained in the BF queue (RAM + disk) during the verification. Column **MaxMemQ** is the max amount of memory (in Megabytes) taken by the BF queue (RAM + DISK).

When **m** = 200 the collision rate is low. Thus the number of visited states is about the number of reachable states. This means that to complete the verification of the protocol in Fig. 7 would require more than 1313MB of RAM using standard Murφ (i.e. 200MB for the hash table, 1113MB for the queue). This is more than our machine can handle (see Fig. 7). However, using our cache based Murφ, we were able to complete verification using only 130MB of RAM.

As shown in Figs. 5, 6, 7 there is a time-space tradeoff when using our cache based approach. However for verification tasks such tradeoff is often acceptable and in any case better than being left with an out of memory message after hours of computation.

5   Conclusions

We showed experimentally (Sect. 2) that protocols exhibit *transition locality*. We supported our claim by measuring transition locality for the set of protocols included the Murφ verifier distribution.

We devised a verification algorithm to exploit transition locality and implemented it within the Murφ verifier (Sect. 3). Essentially our approach replaces the hash table used in a BF state exploration with a fixed size cache (i.e. no collision detection) and uses disk storage for the BF queue.

Our experimental results (Sect. 4) show that, w.r.t. a hash table based approach, our cache based approach typically allows verification of systems more than 40% larger with a time overhead of about 100%.

Our approach is compatible with all (BF) optimization techniques present in the Murφ verifier and, because of its time penalty, it is intended to be used only when such techniques are not enough to avoid running out of memory.

The auxiliary memory used in our algorithm can also be implemented using a NOW (*Network Of Workstations*) rather than with (or together with) disk storage. Moreover, our preliminary analysis shows that our cache memory (holding visited states) can also be implemented using secondary storage (namely disk storage). We are currently working on both such issues.

We think that there are other statistical properties of system transition graphs that can be used to improve performances of state space exploration algorithms. This is a natural next step for our research.

Acknowledgements. We are grateful to our anonymous referees for their helpful comments and suggestions on a preliminary version of this paper.

References

[1] R. Bryant. Graph-based algorithms for boolean function manipulation. *IEEE Trans. on Computers*, C-35(8), Aug 1986.

[2] J. R. Burch, E. M. Clarke, K. L. McMillan, D. L. Dill, and L. J. Hwang. Symbolic model checking: 10^{20} states and beyond. *Information and Computation*, (98), 1992.

[3] D. L. Dill, A. J. Drexler, A. J. Hu, and C. H. Yang. Protocol verification as a hardware design aid. In *IEEE International Conference on Computer Design: VLSI in Computers and Processors*, pages 522–5, 1992.

[4] G. J. Holzmann. The spin model checker. *IEEE Trans. on Software Engineering*, 23(5):279–295, May 1997.

[5] G. J. Holzmann. An analysis of bitstate hashing. *Formal Methods in Systems Design*, 1998.

[6] A. J. Hu, G. York, and D. L. Dill. New techniques for efficient verification with implicitily conjoined bdds. In *31st IEEE Design Automation Conference*, pages 276–282, 1994.

[7] C. N. Ip and D. L. Dill. Better verification through symmetry. In *11th International Conference on Computer Hardware Description Languages and their Applications*, pages 97–111, 1993.

[8] C. N. Ip and D. L. Dill. Efficient verification of symmetric concurrent systems. In *IEEE International Conference on Computer Design: VLSI in Computers and Processors*, pages 230–234, 1993.

[9] Heh-Tyan Liaw and Chen-Shang Lin. On the obdd-representation of general boolean functions. *IEEE Trans. on Computers*, C-41(6), June 1992.

[10] url: http://sprout.stanford.edu/dill/murphi.html.

[11] A. Papoulis. *Probability, Random Variables and Stochastic Processes*. McGraw-Hill Series in System Sciences, 1965.

[12] D. A. Patterson and J. L. Hennessy. *Computer Architecture A Quantitative Approach*. Morgan Kaufmann, 1996.

[13] R. K. Ranjan, J. V. Sanghavi, R. K. Brayton, and A. Sangiovanni-Vincentelli. Binary decision diagrams on network of workstations. In *IEEE International Conference on Computer Design*, pages 358–364, 1996.

[14] J. V. Sanghavi, R. K. Ranjan, R. K. Brayton, and A. Sangiovanni-Vincentelli. High performance bdd package by exploiting memory hierarchy. In *33rd IEEE Design Automation Conference*, 1996.

[15] url: http://netlib.bell-labs.com/netlib/spin/whatispin.html.

[16] U. Stern and D. Dill. Parallelizing the murφ verifier. In *Proc. 9th Int. Conference on Computer Aided Verification*, volume 1254, pages 256–267, Haifa, Israel, 1997. LNCS, Springer.

[17] U. Stern and D. Dill. Using magnetic disk instead of main memory in the murφ verifier. In *Proc. 10th Int. Conference on Computer Aided Verification*, volume 1427, pages 172–183, Vancouver, BC, Canada, 1998. LNCS, Springer.

[18] U. Stern and D. L. Dill. Improved probabilistic verification by hash compaction. In *IFIP WG 10.5 Advanced Research Working Conference on Correct Hardware Design and Verification Methods*, pages 206–224, 1995.

[19] U. Stern and D. L. Dill. A new scheme for memory-efficient probabilistic verification. In *IFIP TC6/WG6.1 Joint International Conference on Formal Description Techniques for Distributed Systems and Communication Protocols, and Protocol Specification, Testing, and Verification*, 1996.

[20] url: `http://verify.stanford.edu/uli/research.html`.

[21] T. Stornetta and F. Brewer. Implementation of an efficient parallel bdd package. In *33rd IEEE Design Automation Conference*, pages 641–644, 1996.

[22] Pierre Wolper and Dennis Leroy. Reliable hashing without collision detection. In *Proc. 5th Int. Conference on Computer Aided Verification*, pages 59–70, Elounda, Greece, 1993.

Efficient Debugging in a Formal Verification Environment

Fady Copty, Amitai Irron, Osnat Weissberg, Nathan Kropp*, and Gila Kamhi

Logic and Validation Technology, Intel Corporation, Haifa, Israel
Microprocessor Group*, Intel Corporation, USA
gila.kamhi@intel.com

Abstract. In this paper, we emphasize the importance of efficient debugging in formal verification and present capabilities that we have developed in order to augment debugging in Intel's Formal Verification Environment. We have given the name the *"counter-example wizard"* to the bundle of capabilities that we have developed to address the needs of the verification engineer in context of counter-example diagnosis and rectification. The novel features of the counter-example wizard are the *"multi-value counter-example annotation,"* *"multiple root cause detection,"* and *"constraint-based debugging"* mechanisms. Our experience with the verification of real-life Intel designs shows that these capabilities complement one another and can considerably help the verification engineer diagnose and fix a reported failure. We use real-life verification cases to illustrate how our system solution can significantly reduce the time spent in the loop of model checking, specification and design modification.

1   Introduction

Verification is increasingly becoming the bottleneck in the design flow of electronic systems. Simulation of designs is very expensive in terms of time, and exhaustive simulation is virtually impossible. As a result, designers have turned to formal methods for verification.

Formal verification guarantees full coverage of the entire state space of the design under test, thus providing high confidence in its correctness. The more automated and therefore the more popular formal verification technique is *symbolic model checking* [2]. While gaining success as a valuable method for verifying commercial sequential designs, it is still limited with respect to the size of the verifiable designs.

The capacity problem manifests itself in an additional obstacle—low productivity. A lot of effort is spent decomposing the proofs into simpler proof obligations on the modules of the design. A global property to be verified is usually deduced from local properties verified on the manually created abstraction [10] of the environment for each module. The abstractions and assumptions needed to get a verification case through increase the chance of getting false failure reports. Further

T. Margaria and T. Melham (Eds.): CHARME 2001, LNCS 2144, pp. 275-292, 2001.
© Springer-Verlag Berlin Heidelberg 2001

more, in case of valid failure reports, back-tracing to the root cause is especially difficult.

Given the inherent capacity and productivity limitation of symbolic model checking, we emphasize in this paper the importance of efficient debugging capabilities in formal verification, and we present capabilities that we have developed in order to augment debugging in a commercial formal verification setting. Our solution provides debugging capabilities needed at three major stages of verification.

- **Specification Debugging Stage.** In the early stages of verification, most of the bugs found are not real design errors, but rather holes in the specification. In this stage, which we call *specification debugging* [4,5], the verifier is in the loop of model checking and specification modification based on the feedback from the symbolic model checker. The turn-around time of the model checker becomes then very critical to the productivity of the verification engineer.
- **Model Debugging Stage.** In order to find the root cause of a bug reported by a formal verification system, one needs intimate knowledge of the design behavior. In order to check whether a bug is spurious or not, one must understand in detail the effect of pruning and environmental assumptions made to verify the design under test. Additionally, once a fix has been made as a result of a *counter-example* report, one must ensure that the previously reported counter-example does not hold any more in the fixed design.
- **Quality Assurance Stage.** A significant problem in industrial-size projects is ensuring that the process of fixing one design error does not introduce another one. In the context of conventional testing this is checked through regression testing[7]. If consecutive test suites check several properties, a failure in one property may require re-testing all the previous suites, once the failure has been rectified. Efficient regression testing clearly requires techniques useful for debugging.

In the case of a failing verification, the output of a formal verification system is a counter-example—a trace illustrating that the design under test does not satisfy a given property. It is especially difficult to diagnose a verification failure reported as a counter-example. On one hand, the verification engineer suffers from too much data and the difficulty to distinguish the relevant data (i.e., the signal values that cause the failure). On the other hand, he has too little information. Rectification of the error displayed by a single counter-example trace does not guarantee the overall correction of the failure. A trace is just one of the many witnesses that demonstrate that the model does not satisfy the given property. In that sense, a counter-example has too little information. Trace analysis is even more difficult in formal verification, where obscure corner cases can produce complex failures and consequently complex counter-examples to debug.

We have given the name *"counter-example wizard"* to the bundle of capabilities that we have developed to address the needs of the verification engineer in context of counter-example diagnosis and rectification during the above stages of verification. The novel features of the counter-example wizard are the *"multi-value counter-example annotation," "multiple root cause detection,"* and *"constraint-based debugging"* mechanisms. Our experience with the verification of real-life Intel designs shows that these capabilities complement one another and can considerably help the verification engineer diagnose and fix a reported failure.

The "multi-value" nature of the counter-example annotation mechanism enables the concise reporting of all the failures (i.e., counter-examples) as a result of one model checking run. Hence, the understanding of more than one root cause of a failing verification facilitates the rectification of the failure. The ability to fix more than one root cause can reduce the number of model checking runs needed to get to a passing verification run. Most importantly, multi-value counter-example reports enable the user to pinpoint the pertinent signal values causing the failure and aid in detecting how to change the values to correct the failure.

"Constraint-based debugging" allows the verification engineer to restrict the set of failures (i.e., counter-example traces) to only those that satisfy a specific sequential constraint. If this subset is empty for a given sequential constraint, this means that the constraint is sufficient to eliminate all counter examples found so far. However, the model checker must still be run again to find out if the constraint is sufficient to resolve all counter-examples of all lengths.

The system solution that we provide reduces the time spent in the loop of model checking, specification and design modification. The usage flow consists of running the model checker, dumping all the model checking data needed to compute all the counter-examples of a given length, and then debugging in an interactive environment by loading the pre-dumped model checking data. The fact that we have taken the model checker out of the "check-analyze-fix" loop reduces the debugging loop to "analyze-fix" and consequently improves the time spent in debugging considerably. The effective usage of secondary memory allows the verification engineer to post-process model checking data and debug without the need to add the model checking run to the verification loop.

This paper is organized as follows. In Sect. 2, we present an overview of the formal verification system with enhanced debugging capabilities. Section 3 depicts in detail the capabilities of the "counter-example wizard" Section 4 explains the algorithms underlying our system solution. In Sect. 5, we illustrate the efficiency of these techniques through verification case studies on Intel's real-life designs. We summarize our conclusions in Sect. 6.

2   System Overview

The formal verification system with the counter-example wizard consists of three major components:

1. A state-of-the-art symbolic model checker which accepts an LTL-based formal specification language
2. An interactive formal verification environment which enables access to all the model checking facilities
3. A graphical user interface which allows the user to display annotated counter-example traces and access interactive model checking capabilities

The usage flow consists of two major stages:

1. **Model Checking.** The model checker is run with the option to dump the relevant model checking data and counter-example information to secondary memory.
2. **Interactive Debugging.** The user loads and interacts with the model checking data to access different counter-examples and perform "what-if analysis" on the existence of counter-examples under specific conditions.

The easy storage and loading of relevant model checking data is due to the "*data persistency mechanism*" of the model checker. At any point of the model checking run, the data persistency mechanism can represent all the relevant, computed model checking information in a well-defined ASCII format which later can be loaded in an interactive model checking environment and analyzed through the usage of a functional language.

The fact that the model checker can dump the relevant information for debugging at any point, enables easy integration of this mechanism into regression testing. When regression test suites are run with the "counter-example data dump" facility enabled, the analysis of the failing verification test cases can be done without rerunning the model checker and regenerating the failing traces, which can be computationally expensive. The computational benefit of the system is also witnessed in the specification , model verification, and modification loop.

3   Counter-Example Wizard

Traditional symbolic model checkers provide a single counter-example as the output of a failing verification. To diagnose a verification failure reported as a counter-example is difficult. On one hand, we suffer from too much data and the difficulty identifying the relevant data. On the other hand, we often do not have sufficient information to find the root cause of the failure and rectify it. In this section, we present the "counter-example wizard" that addresses the counter-example analysis and rectification problem and has three novel capabilities: multi-value counter-example annotation, multiple root cause detection, and constraint-based debugging.

3.1   Multi-value Counter-Example Annotation

We address the counter-example diagnosis problem by introducing a concise and intuitive counter-example data representation. We call this advanced representation "multi-value counter-example annotation." This novel annotation relies on the exhaustive nature of the symbolic model checking and hence the ability to represent all the counter-examples of a given length[1]. The enhanced counter-example reporting classifies signal values at a specific phase of a counter-example trace into three types:

---

[1] The model checker generates counter-examples of the shortest path length. The underlying symbolic model checking algorithms that enable "multi-value counter-example annotation" will be explained in detail in Section 4.

- **Strong 0/1** indicates that in all possible counter-examples that demonstrate the failure, the value of the signal at the given phase of the trace is 0 or 1, respectively.
- **Weak 0/1** indicates that although the value of the signal at the given phase is 0 or 1 respectively for this counter-example, the value of the signal at this phase can be different for another counter-example illustrating the failure. Even though the model checker has some leeway in the choice of a value for this signal, this signal must preserve some relation with other signals at this phase.
- **Weaker 0/1** is similar to "weak" designation, except that weaker values are basically arbitrary, and have little or no influence on the generation of a failure.

The strong values provide the most insight on the pertinent signals causing a failure. For example, if the value of a signal at a certain phase of a counter-example is a strong zero, this means correcting the design so that the value of the signal will be one at that phase will often correct the failure. Hence, the error rectification problem is often reduced to determining how to cause a strong-valued signal to take on a different value.

The counter-example wizard can make use of a waveform display to represent the multi-value counter-example annotation. Figure 2 illustrates a screen shot of the multi-value annotated counter-example graphical display. (Later we will show the use of a text-based display.) The counter-example in the figure demonstrates the violation of the specification *"if X is high, W will be high a cycle later."* The value of Y is clearly not relevant (i.e., weaker); therefore its waveform is shadowed out. Furthermore, the values of X and Z in the second cycle do not affect the failure, so their waveforms in that cycle are shadowed out as well. Examination of the waveform reveals that the cause for the failure is the value of the signal Z, which causes the output of the AND gate to be low.

Our experience shows that strong values alone sometimes provide sufficient information to figure out the root cause of a failure and speed up the debugging. Nevertheless, we have also witnessed many verification cases where the answer to the root cause of failure lay in the weak values (as seen in Fig. 3). The debug of traces with weak values is facilitated by the sequential constraint-based debugging capability which is the second major feature of the counter-example wizard.

3.2 Constraint-Based Debugging

"Sequential constraint-based counter-examples" are traces displaying the failure while obeying some temporal constraint. A sequential constraint in the debugging wizard is described as an associated list of pairs of a Boolean condition and a corresponding phase in which the Boolean condition must hold. In other words, the user specifies a function over the signals in the design and a point in time when the function should hold. The counter-example wizard then looks for traces that satisfy the constraint, and recalculates all weak and strong signal values relative to this subset of traces.

Constraint-based debugging facilitates the rectification and the diagnosis of the root cause of a failure. If no counter-example that holds the constraint is found, then the constraint describes a condition sufficient to make the erroneous design or specification correct. Consequently, "sequential constraint-based debugging" can help significantly in the correction of design or specification errors.

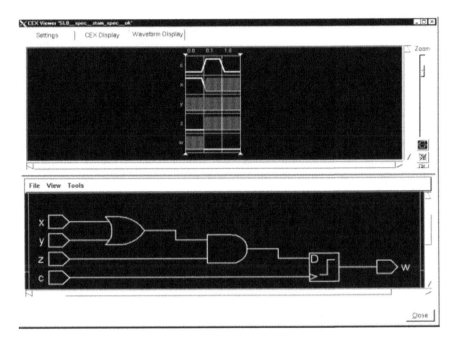

Fig. 2. Graphical counter-example display using multi-value annotation. In the waveform (top), the strong (i.e., significant) signal values are represented by bold lines, while the weaker values are represented by gray boxes

For example, let us assume that some input vector *foo* should be one-hot encoded (exactly one of the bits in the vector is high and the rest are low), but in the counter-example presented it is not encoded as one-hot. In the absence of constraint-based debugging, the user would have to add an environmental assumption that *foo* is one-hot and rerun the model checker to see if the erroneous encoding is indeed the root cause of the failure—a task that can take hours. With constraint-based traces the user can write the environmental assumption as a sequential constraint, and if there are no counter-examples that satisfy the constraint, then the user knows that assumption is sufficient to eliminate all current counter-examples. Thus the user is able to check whether an assumption will cure the current failure without rerunning the model checker.

Additionally, constraint-based debugging allows "what-if analysis" and the ability to investigate the relationships between signals. For example, setting the value of a signal to a constant value at a specific phase and observing other signal values that have consequently become strong, helps the user to understand the relationships between signals over time. Thus, constraint-based debugging refines the information that weak values provide.

Figure 3 illustrates how the usage of two different constraints help debug a failing verification task. The task is to check that the model illustrated in the lower half of the figure satisfies the specification "*if Z is high, W will be high a cycle later.*" On the upper half of the figure, three multi-value annotated counter-example traces are illustrated. The leftmost trace shows all the counter-examples of length three violating this specification. Viewing this trace, we observe that only W, Z and the clock C get strong values in the annotated trace. The signals X and Y have weak values for the first phase and weaker values for all the rest of the phases (indicating that the values of these signals in second and third phases are irrelevant to the failure). In this case, the strong values do not provide enough information; therefore we analyze the weak values (i.e., the relationship between X and Y in the first phase). The middle trace and the rightmost trace demonstrate all the counter-examples of length three, under two constrained values of the signal X in the first phase (one and zero, respectively). When the value of X is high, we can ignore the value of Y. Therefore, the value of Y becomes weaker under this constraint as illustrated in the middle annotated trace. The second constraint, as illustrated by the rightmost trace, assigns X a low value in the first phase. Under this constraint, the signal Y gets a strong one value. Therefore, our conclusion from this constraint-based debugging session is that Y must be high in the first phase to get a violation. Furthermore, to rectify the violation both X and Y need to get a low value in the first phase.

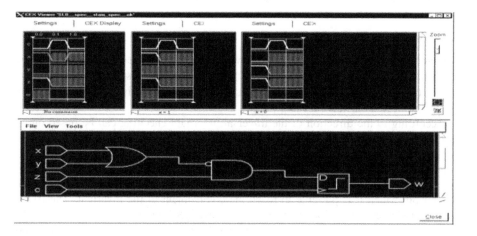

Fig. 3. An illustration of the usage of constraint based debugging. The significant (i.e., strong) signal values are highlighted. The weak values are shadowed out, and weaker values are displayed as gray boxes

3.3   Multiple Root Cause Detection

The user can request a predetermined number of interesting traces that may give clues to several root causes of a failure. The underlying algorithm chooses traces that are unique, and do not share the same prefix and usually share the same postfix. The selection uses a heuristic that selects states to be as different from one another as possible. The purpose is to maximize the parity between states returned by the selection procedure. Usually the design under test has some initialization sequence; therefore, when generating a set of interesting traces, traces with the same prefix but different postfix are chosen. The intuition is that traces with different postfix will give hints on more root causes for the violation.

This mechanism works naturally under the constraint-based debugging mechanism. By request of the user all the interesting traces that comply with a given constraint are produced.

4   Underlying Algorithms

A finite state machine is an abstract model describing the behavior of a sequential circuit. A completely specified FSM M is a 5-tuple $M = (I, S, \delta, \lambda, S_o)$, where I is the input alphabet, S is the state space, δ is the transition relation contained in S x I x S, and $S_o \subseteq S$ is the initial state set. BDDs [2] are used to represent and manipulate functions and state sets, by means of their *characteristic functions*. In the rest of paper, we make no distinction between BDDs and a set of states.

4.1   Background

A common verification problem for hardware designs is to determine if every state reachable from a designated set of initial states lies within a specified set of "good states" (referred to as the *invariant*). This problem is variously known as *invariant verification*,[2] or *assertion checking* [4,5,6].

According to the direction of the traversal, invariant checking can be based on either *forward* or *backward analysis*. Given an invariant G and an initial set of states A, *forward analysis* starts with the BDD for A, and uses BDD functions to iterate up to a fixed point, which is the set of states reachable from A, using the *Img* operator. Similarly, in *backward analysis*, the *PreImg* operator is iteratively applied to compute all states from which it is possible to reach the complement of the invariant. The BDDs encountered at each iteration are commonly referred as *frontiers*.

The *counter-example generation algorithm* [8] is a combination of backward and forward analysis. In Figure 4, we see that counter-example generation starts with the BDD for all the states that do not satisfy the invariant (i.e., F_o) and the *PreImg* operator is applied until a fixed point or a non-empty intersection (i.e, $F_{i+1} \cap S_o \neq \varnothing$) of the last backward frontier with the initial states is reached. A counter-example is then constructed by

---

[2] Although the "debugging wizard" is a valid tool both for invariant and *liveness* property verification, for the sake of simplicity, in this section we will explain the underlying algorithms of the debugging wizard in the context of invariant verification.

applying forward image computation to a state selected from the states in the intersection (i.e., all the states from which there is a path to the states that complement the invariant). Again by iteratively intersecting the forward frontiers with the corresponding backward frontiers and choosing a state from each intersection as a representative for the corresponding phase in the counter-example, the counter-example trace is built.

Target frontiers [4] are the sets of states obtained during backward analysis, starting from the *error states* (states that violate the invariant) and iteratively applying the *PreImg* operator till reaching an intersection with the initial states. More precisely, we define the nth *target frontier*, F_n as the set of states from which one can reach an error state in n (and no less than n) steps.

$$F_0 = \neg \, Invariant \tag{1}$$

$$F_{n+1} = \mathrm{PreImg}(F_n) - \bigcup_{i=1}^{n} F_i$$

In what follows, we denote by N the index of the last target frontier before the fixed-point, such that *Target frontier$_{N+1}$* = *Target frontier$_N$* . The target frontiers are disjoint, and their union, which we denote as *Target* represents all the states from which a state violating the invariant can be reached:

$$Target = \bigcup_{i=1}^{N} F_i \tag{2}$$

As can be seen in the counter-example generation algorithm depicted on the left-hand side of Figure 4, the frontiers F_0, F_1, ..., F_N represent the target frontiers. In order to obtain reachable target frontiers, we need to filter out the states unreachable from $F_N \cap S_0$ in each frontier. The filtering is done by applying forward analysis starting from the states in $F_N \cap S_0$ as seen in the right-hand side of Figure 4. From here on, we will refer to this version of target frontiers as "reachable target frontiers".

The underlying data structure for all the algorithms of the counter-example wizard is the *reachable target frontiers*. Two major characteristics of reachable target frontiers make them very useful for the computations needed for debugging.

- Any trace through the frontiers is guaranteed to be a counter-example.
All possible counter-examples in this verification of length N (when N is the number of frontiers) are included in the frontiers.

Therefore, these frontiers store all the information needed for the querying and extraction of counter-examples in a very concise and flexible way. In our system, the model checker dumps the BDDs representing the reachable target frontiers to secondary memory. Interactive debugging is done by restoring the target frontiers in the interactive environment and querying them through the functional language.

COUNTER-EXAMPLE (δ, S_o, p)	TARGET-FRONTIERS(δ, S_o, p)

i = 0;
Target = F_0 = {pi | pi does not satisfy p};
// F_0 - all the states that do not satisfy p
C = {} // Counter-example initialized, an empty list
While $(F_i \neq \varnothing)$ {
  F_{i+1} = PRE-IMG(δ, F_i);
  if $(F_{i+1} \cap S_0 \neq \varnothing)$ {
    // the property is violated
    $F_{i+1} = F_{i+1} \cap S_0$;
    j = i;
    while (j >= 0) {
      s = SELECT_STATE(F $_{j+1}$);
      $\mathbf{F_j = Img(\delta, s) \cap F_j; j = j - 1;}$
      C = PUSH(s,C);
    }
    return REVERSE(C);
    // return a counter-example
  } //end the property is violated
  i = i + 1;
  F_i = F_i - Target;
  Target= Target U F_i;
} // the property is satisfied
return C; // return an empty list
	i = 0;
	Target = F_0 = {pi | pi does not satisfy p};
	// F_0 - all the states that do not satisfy p
	C = {} // Target frontiers initialized, an empty list
	while $(F_i \neq \varnothing)$ {
	  F_{i+1} = PRE-IMG(δ, F_i);
	  if $(F_{i+1} \cap S_0 \neq \varnothing)$ {
	    // the property is violated
	    $F_{i+1} = F_{i+1} \cap S_0$;
	    j = i ;
	    while (j >= 0) {
	
	      $\mathbf{F_j = Img(\delta, F_{j+1}) \cap F_j; j = j - 1;}$
	      C = PUSH(F$_j$,C);
	    }
	    return REVERSE(C);
	    // return Target Frontiers
	  } //end the property is violated
	  i = i +1;
	  F_i = F_i – Target;
	  Target = Target U F_i ;
	} // the property is satisfied
	return C; // return an empty list

Fig. 5. Classic counter-example generation and target frontiers calculation algorithms

4.2   Annotating the Counter-Example Values

Annotating the counter-example values becomes rather simple, once we have the "reachable target frontiers" at hand. The value of signal x at phase i is

- **Strong**, if $F_{i\,|\,x=0}$ = 0 or $F_{i\,|\,x=1}$ = 1
- **Weak,**   if $F_{i\,|\,x=0} \neq F_{i\,|\,x=1}$
- **Weaker,** if $F_{i\,|\,x=0} = F_{i\,|\,x=1}$ (i.e., x is not in the support of F_i)

when Fi is the reachable target frontier corresponding to phase i. The value of the signal is chosen according to the specific trace at hand.

4.3   Constraint-Based Debugging

As described above, a sequential constraint in the counter-example wizard is described as an associated list of pairs of a Boolean condition and a corresponding phase in which the Boolean condition must hold. In other words, the user specifies a function over the signals in the design and a point in time when the function should hold. The counter-example wizard looks for a trace that leads to a failure and satisfies the constraint, and recalculates all the weak and strong signal values relative to this subset of traces.

Constraints are internally represented as BDDs, and each phase constraint afterwards is intersected with the corresponding reachable target frontier. When a condition is applied to the target frontiers, not every possible trace through the target frontiers is a counter-example any more. States in a frontier that do not comply with the constraint are thrown out leaving some of the traces through the frontiers dangling (i.e., they are not of length N, when N is the number of target frontiers). We remedy the target frontiers by performing an N-step forward propagation followed by an N-step backward propagation through all the frontiers.

The task of calculating a new trace under the constraint now becomes simply finding any trace through the *newly* calculated target frontiers. The multi-value annotation is applied to the new set of target frontiers.

4.4   Computation Penalty

In this section, we present data that compares the CPU time that took our system to compute all the data needed for multi-value annotation and single-value annotation. As seen in the numbers reported in Table 1. based on typical Intel verification cases, multiple-value annotation computation is more costly than single-value annotation computation. Table 1. supports the fact that for average size test cases the multi-value annotation can take 2-3 X more time than single-value annotation. On average the number of single-value counter-example sessions needed to root-cause a failure is more than three. Thus, the computation penalty paid in multi-value annotation is less than the single-value annotation. Additionally each session with a single-value counter-example requires analysis time of the verification engineer. Therefore, our conclusion based on computation data and the utility's deployment in Intel is that although multi-value annotation computation takes more time than single-value annotation, it reduces the overall verification time significantly.

Table 1. Experimental results comparing the model checking time required to compute counter-examples with multi-value annotation and single value annotation making use of eight typical Intel verification test cases

Test case	Multi-value annotation	Single-value annotation
Real 1	31.0 s	21.0 s
Real 2	4.3 s	1.8 s
Real 3	10.7 s	5.2 s
Real 4	292.7 s	89.9 s
Real 5	129.7 s	44.6 s
Real 6	173.0 s	64.2 s
Real 7	8091.8 s	7250.2 s
Real 8	2118.4 s	1421.1 s

5   Experimental Results from the Deployment of the "Counter-Example Wizard"

The counter-example wizard has been used to help debug numerous real-life Intel verification cases. The tool has been found to be beneficial during proof development in reducing the time spent analyzing counter-examples and reducing the number of "check-analyze-fix" iterations, consequently speeding up convergence to a proof. The main advantages of the wizard have been observed to be in determining the root cause of a set of counter-examples and identifying a resolution to the failure.

In this section, we illustrate the benefits of the counter-example wizard through real-life examples. Our productivity claim of the counter-example wizard relies on our experience in its deployment at Intel.

5.1   Determining a Root Cause

5.1.1   Strong versus Weak Values

From our experience, the most important benefit of counter-example wizard has been the ability to home in quickly on the root cause of a counter-example by pinpointing pertinent signal values. The distinction between weak and strong values often indicates which signal values are responsible for a failure and which signal values do not affect the failure (i.e., not related to the root cause). For example, if a signal has weak values in all but one phase, this is a good indication that the one strong phase is the only important phase for this signal. Similarly, if only one of the inputs to a logic gate has a strong value, this is often an indication that the other input signal values can be ignored as being unrelated to the root cause of the failure.

Multi-value counter-example annotation, by reducing the number of relevant signals and phases, reduces the amount of data to be comprehended in the counter-example report. Instead of being distracted by insignificant information, the verification engineer can concentrate on pertinent signals, phases, and values and ignore all the rest. Consequently, it is easier to comprehend the essential behavior of the model, and identify values related to the root cause of a counter-example. In practical use, this characteristic of the counter-example wizard has been its most direct and immediate advantage.

To illustrate how multi-value annotation helps pinpoint the root cause of a failure, we present an example from our experience of the verification of a property involving ten pipe stages of control signals. In one failure report, we observed that almost all the signals in the model had weak values. One signal stood out, however, with a single strong value assignment in a particular phase. This gave us immediate indication that the failure was related to the value of this signal and phase. By tracing the staging of the signal through its several pipe stages, we shortly discovered that at one point the signal was not getting flopped into the next stage (see Fig. 5). Note that the value of the pipelined signal Reg_Rd changes from a strong 1 to a strong 0 at cycle 4. A simple examination of the flip-flop revealed that its clock was not toggling during the phase in which Reg_Rd_s05 took on a strong zero value, illustrating the root cause of the failure. This is a simple case that we could have debugged even with traditional single-value counter-example reporting. Neverthe-less, multi-value signal annotation allowed us to debug much more quickly and without unnecessary trial and error. This is exemplary of the more complex counter-examples with which the wizard has proved helpful.

5.1.2   Sequential Constraints

The "sequential constraint" capability has furthermore helped in root cause determi-nation in our verification problems. There are several cases of when constraints can be useful. First, signals relevant to a counter-example could be related but not con-stant. For example, if two signals are equivalent in the design under test, but their equality is not guaranteed in the verification, and the verification failure is due to this lack of an equivalence guarantee, then the counter-example report would show both signals having weak values. To see if there is a relationship between the signals, the value of one of the signals would need to be constrained. In this example, if the value of one of the signals is constrained to 0, then the other signal will take on a strong 1 value. When this happens, we know there must be some relationship between these signals with respect to this set of counter-examples (namely, that they should be equivalent but are not in the verified model). The advantage that the wizard has pro-vided us in cases like this is the ability to identify relationships between signals in the counter-examples and thus help direct us to the root cause.

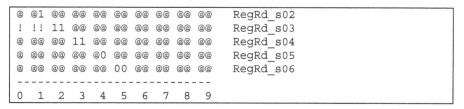

```
@ @1 @@ @@ @@ @@ @@ @@ @@ @@    RegRd_s02
! !! 11 @@ @@ @@ @@ @@ @@ @@    RegRd_s03
@ @@ @@ 11 @@ @@ @@ @@ @@ @@    RegRd_s04
@ @@ @@ @@ @0 @@ @@ @@ @@ @@    RegRd_s05
@ @@ @@ @@ @@ 00 @@ @@ @@ @@    RegRd_s06
-----------------------------
0  1  2  3  4  5  6  7  8  9
```

Fig. 6. In this text-based multi-value annotated counter-example trace, the columns represent phases, and the rows represent the values of each of the signals at each phase. The '@' and '!' symbols represent weak 0 and weak 1 values, respectively, whereas '0' and '1' represent the strong values

Another case for which constraints are useful is when the set of counter-examples is due to multiple root causes. Constraints can be used to partition the set of counter-examples to identify the different root causes. This partitioning is not explicit but hap-pens naturally during analysis while experimenting with constraints and searching for a root cause. Once one root cause is identified, analysis is then switched to the cases not covered by that root cause. This is repeated until the set of counter-examples has been fully covered. The set is thus partitioned according to the various root causes.

Multiple root causes usually cannot be identified when working with a single counter-example at a time. A given counter-example may be due to only one of the root causes; therefore only one root cause could be found from that counter-example. Even when a single counter-example is due to multiple root causes, it is unlikely that all would be identified: When one apparent root cause is found, the verification engineer typically assumes that this is the sole cause. Therefore when multiple root causes exist, the counter-example wizard provides an additional advantage over traditional single-counter-example debugging.

5.2   Identifying a Solution

The counter-example wizard has also been helpful in identifying solutions to a set of counter-examples. In general, once the root cause of a failure is found, the root cause is usually eliminated either by expanding the model to include necessary guaranteeing logic, or by making an environmental assumption about the behavior of signals. The model is then re-verified after the necessary changes have been made. The expectation is that the changes will eliminate any counter-examples, yielding a successfully completed proof.

Analyzing counter-examples one at a time is often a process of trial and error. In a typical verification workflow, a verification run generates a single counter-example, the verification engineer tries to determine the root cause of the failure, a solution is implemented, the model is rebuilt, and the verification is run again. Unfortunately, the root cause may or may not have been correctly identified. Often it results in another counter-example report. When working with only a single counter-example at a time, there is no way to avoid this trial and error process.

The counter-example wizard can eliminate some of this trial and error. Potential solutions can be tested to see if they really do resolve the current set of counter-examples. As noted above, a solution usually takes the form of either an expansion of the model to include necessary guaranteeing logic, or an environmental assumption. Therefore, a solution is essentially a restriction on signal behavior that disallows the behavior observed in the counter-example. To check whether the proposed solution will work, this restriction is given as a sequential constraint to the counter-example wizard. If the wizard determines that there are no counter-examples that satisfy this constraint, then the proposed solution successfully resolves all counter-examples of the given trace length for this verification.

If the wizard does find counter-examples that satisfy the constraint, then the proposed solution does not completely resolve the current set of counter-examples. In this case a different solution can be tried, or the remaining counter-examples can be analyzed to determine why the proposed solution did not resolve them. Possible solutions to verification failures can thus be tested without implementing the solution and rerunning the verification. This can significantly reduce the time spent in the trial and error loop. The wizard gives feedback within seconds, instead of the minutes or sometimes hours it takes to rerun the verification.

In the example in Section 5.1 concerning the flip-flop whose clock failed to toggle, we could quickly test our guess that this was the root cause by specifying to the wizard the constraint that the clock should be high in the phase in which it failed to toggle. We then searched for counter-examples under this constraint, and when none were found, we knew that getting the clock to toggle during the phase in question would be sufficient to resolve all counter-examples of this trace length.

5.3   Evaluating Counter-Example Wizard Capabilities through an Example

In this section, we present the benefits of counter-example wizard by comparing multi-value annotation versus single-value through an example from a real-life formal property verification case.

Let us first briefly explain the inputs of the verification case: the model, the property to be verified, and the design assumptions.

- Property: If a request is killed, then the register that contains the request gets cleared. More specifically, if this "Active Register" holds a request that receives a kill, then it will be clear for the next two cycles.
- Model Behavior:
- Eventually a request is received and is retained in the Holding Register.
- When the Active Register becomes free, the request moves from the Holding Register into the Active Register.
- When the request is finished being serviced, it is cleared out of the Active Register.
- Micro-architectural Assumption :

The same request cannot be made twice.

```
0 00 00 00 00 00 00 00 00 00 01 00 01 00 00  Valid request
0 00 00 00 00 00 00 00 00 00 00 00 01 00 00  Kill request

0 00 00 00 00 00 00 00 00 00 00 11 00 11 00  Holding reg. Valid
0 00 00 00 00 00 00 00 00 00 00 00 11 00 11  Active reg. Valid

0 00 00 00 00 00 00 00 00 00 00 11 00 11 00 00  Holding reg. Write enable
0 11 11 11 11 11 11 11 11 11 11 11 11 11 11  Holding reg.
                                             written/reset
0 00 00 00 00 00 00 00 00 00 00 11 00 11 00  Holding -> Active

0 00 00 00 00 00 00 00 00 00 00 00 00 01 00  Holding - special bit
0 00 00 00 00 00 00 00 00 00 00 00 00 00 00  Active - special bit

---------------------------------------------
0  1  2  3  4  5  6  7  8  9  10 11 12 13 14
```

Fig. 6. Traditional single counter-example trace. The columns represent phases, and the rows represent the values of each of the signals at each phase

```
@ @@ @@ @@ @@ @@ @@ @@ @@ @@ @! @@ @1 @@ @@   Valid request
@ @@ @@ @@ @@ @@ @@ @@ @@ @@ @@ @@ @1 @@ @@   Kill request

@ 00 00 00 00 00 00 00 00 @@ @@ !! @@ 11 @@   Holding reg. Valid
@ @@ @@ @@ @@ @@ @@ @@ @@ @@ @@ @@ 11 00 11   Active reg. Valid

@ @@ @@ @@ @@ @@ @@ @@ @@ @@ !! @@ !1 @@ @@   Holding reg. Write enable
@ 11 11 11 11 11 11 11 11 11 !! !! !! 11 11   Holding reg.
                                             written/reset
@ 00 00 00 00 00 00 00 00 @@ @@ !! 00 !1 00   Holding -> Active

@ @@ @@ @@ @@ @@ @@ @@ @@ @@ @@ @@ @@ @1 @@   Holding - special bit
@ @@ @@ @@ @@ @@ @@ @@ @@ @@ @@ @@ @@ @@ @0   Active - special bit

---------------------------------------------
0  1  2  3  4  5  6  7  8  9  10 11 12 13 14
```

Fig. 7. Multi-value counter-example trace. The columns represent phases, and the rows represent the values of each of the signals at each phase. The '@' and '!' symbols represent weak 0 and weak 1 values, respectively, whereas '0' and '1' represent the strong values

In the above traces two requests (Valid request) arrive one cycle apart. For the first request, the Write Enable for the Holding Register goes high in cycle 10, causing the Holding Register to be valid in the next cycle (cycle 11). Also in cycle 11, the Holding→Active signal goes high, causing the request to move from the Holding Register into the Active Register. Consequently, the Active Register is valid in cycle 12. Furthermore, the trace shows a Kill request arriving in cycle 12. (This should kill the request by clearing the Active Register, which it does: the Active Register is not valid in cycle 13.)

However, the property states that the Active Register should remain clear for two cycles, yet we see that it becomes valid again in the next cycle (cycle 14). This occurs due to the second request, which moves into the Active Register by the same process as the first request. Thus the Active Register becomes valid again after only one cycle, rather than the two cycles specified by the property. Hence the violation of the property and the failure of the proof.

Let us now illustrate how multi-value annotation helps the verification engineer to find the root cause of the failure.

- Multi-value annotation helps narrow the scope of the search for the root cause.
- More strong values are associated with the second request than with the first. Therefore, the focus should be on the second request.
- Multi-value annotation helps identify which logic has to be guaranteed to resolve all counter-examples of the given length.
- Despite the assumption that the same request cannot be made more than once, the first and second requests arrive two cycles apart from one another. A closer examination (not seen in the above trace) of the second request shows that it is indeed identical to the first request but with one exception: its "Special" bit. Multi-value annotation is helpful in identifying this conclusion. With single counter-example debugging, there is no indication that the Special bit is the only exception; it sim-

ply receives a zero or one, just like every other bit that comprises the request. However, with multi-value annotation the Special bit takes on strong values, so it is certain that here the Special bit is the only important component of the request.

- Now that the second request, in particular its Special bit, has been identified as important, the Special bit can be followed from the request through the registers, as shown in the above trace. The Special bit is not getting passed from the Holding Register to the Active Register in cycles 13-14, even though the request is getting passed. Since the Special bit takes on strong values, we know that if the passing of the Special bit from the Holding to the Active Register can be guaranteed, then we will have resolved all counter-examples of this length.

This debugging session helped the verification engineer pinpoint the missing logic that guarantees the transfer of the Special bit. Once that logic was included in the model, the proof succeeded.

The verification case just described is derived from an actual debugging session from our verification work. We have indeed encountered much more complex counter-examples than the one just demonstrated. They include some with multiple root causes that could be identified all at once using multiple counter-example capabilities, but which single counter-example debugging could identify only one at a time. Such examples are quite complex and are beyond the scope of this paper.

6   Conclusions

In this paper we have introduced a novel formal verification debugging aid, the "counter-example wizard." The novelty of the wizard is in its multi-value counter-example annotation, sequential constraint-based debugging, and multiple root cause detection mechanisms. The benefits of counter-example wizard have been observed in an industrial formal verification setting in verifying real-life Intel designs.

The demonstrated advantages of the formal verification system augmented with the counter-example wizard are a shorter debugging loop and unique help in diagnosing and resolving failures. The time saved was due to faster determination of the root cause of a set of counter-examples, and the ability to identify and resolve multiple root causes in a single proof iteration. Furthermore, the wizard allows the verification engineer to test solutions to verification failures and observe if they really do resolve the apparent root cause.

References

[1]   R. Bryant, "Graph-based Algorithms for Boolean Function Manipulations",, IEEE Transactions on Computers,C-35:677-691, August 1986.

[2]   K.L. McMillan. "Symbolic Model Checking", Kluwer Academics, 1993.

[3]   K. Ravi, F. Somenzi, "Efficient Fixpoint Computation for Invariant Checking", In Proceedings of ICCD'99, pp. 467-474.

[4]   R. Fraer, G. Kamhi, L. Fix, M. Vardi. "Evaluating Semi-Exhaustive Verification Techniques for Bug-Hunting" in Proceedings of SMC'99.

[5]   R. Fraer, G. Kamhi, B.Ziv, M. Vardi, L. Fix. "Prioritized Traversal: Efficient Reachability Computation for Verification and Falsification", in Proceedings of CAV'00,Chicago,IL.

[6]   I. Beer, S. Ben-Davis, A. Landver. "On-the-Fly Model Checking" of RCTL Formulas", in Proceedings of CAV'98.

[7]   R.H. Hardin, R. P. Kurshan, K.L. McMillan, J.A. Reeds and N.J.A. Sloane, "Efficient Regression Verification", Int'l Workshop on Discrete Event Systems (WODES '96)

[8]   E. Clarke, O. Grumberg, K. McMillan, X. Zhao, ``Efficient generation of counterexamples and witnesses in symbolic model checking'', in the proceeding of DAC'95.

[9]   B. Kurshan, "Formal Verification in a Commercial Setting", In Proceedings of DAC'97.

[10] J. Jang, S.Quader, M. Kaufmann, C. Pixley, "Formal Verification of FIRE: A Case Study", in Proceedings of Design Automation Conference, 1997, Anaheim, CA

[11] R.K. Brayton, G.D. Hachtel, A. Sangiovanni-Vincentelli, F.Somenzi, A.Aziz, S.T.Cheng, S. Edwards, S. Khatri, Y. Kukimoto, A. Pardo, S.Qadeer, R.K. Ranjan, S. Sarwary, T.R. Shiple, G. Swamy, T. Villa, "VIS: A system for Verification and Synthesis", in Proc. of DAC'94.

[12] I. Beer, S. Ben-David, C. Eisner, A. Landver. "RuleBase: An industry-oriented formal verification tool". In Proc. of Design Automation Conference 1996 (DAC'96)

Using Combinatorial Optimization Methods for Quantification Scheduling[*]

P. Chauhan[1], E. Clarke[1], S. Jha[2], J. Kukula[3], H. Veith[4], and D. Wang[1]

[1] Carnegie Mellon University
[2] University of Wisconsin - Madison
[3] Synopsys Inc.
[4] TU Vienna, Austria

Abstract. Model checking is the process of verifying whether a model of a concurrent system satisfies a specified temporal property. Symbolic algorithms based on Binary Decision Diagrams (BDDs) have significantly increased the size of the models that can be verified. The main problem in symbolic model checking is the *image computation problem*, i.e., efficiently computing the successors or predecessors of a set of states. This paper is an in-depth study of the image computation problem. We analyze and evaluate several new heuristics, metrics, and algorithms for this problem. The algorithms use combinatorial optimization techniques such as *hill climbing, simulated annealing*, and *ordering by recursive partitioning* to obtain better results than was previously the case. Theoretical analysis and systematic experimentation are used to evaluate the algorithms.

1 Introduction

Model Checking and State Explosion. In model checking [CGP00], the system to be verified is represented as a finite *Kripke structure* or *labelled transition system*. A Kripke structure over a set of atomic propositions AP is a tuple $K = (S, R, L, I)$ where S is the set of states, $R \subseteq S \times S$ is the set of transitions, $I \subseteq S$ is the non-empty set of initial states, and $L : S \to 2^{AP}$ labels each state by a set of atomic propositions.

Given a Kripke structure $K = (S, R, I, L)$ and a specification ϕ in a temporal logic such as CTL, the *model checking problem* is the problem of finding all states s such that $K, s \models \phi$ and checking if the initial states are among these. Model checking algorithms usually exploit the fact that temporal operators can

---

[*] This research is sponsored by the Semiconductor Research Corporation (SRC), the Gigascale Research Center (GSRC), the National Science Foundation (NSF) under Grant No. CCR-9505472, and the Max Kade Foundation. One of the authors is also supported by Austrian Science Fund Project N Z29-INF. Any opinions, findings and conclusions or recommendations expressed in this material are those of the authors and do not necessarily reflect the views of GSRC, NSF, or the United States Government.

T. Margaria and T. Melham (Eds.): CHARME 2001, LNCS 2144, pp. 293–309, 2001.
© Springer-Verlag Berlin Heidelberg 2001

be characterized as μ-calculus terms. For example, the set of states where the CTL formula $\mathbf{EF}\phi$ holds, is given by

$$\mathbf{EF}\phi \equiv \mu S.\phi \vee \mathbf{EX}S.$$

Recall that $\mathbf{EF}\phi$ expresses *reachability*, i.e., the existence of a path where ϕ eventually holds. Such fixpoint translations directly correspond to iterative algorithms.

Symbolic Verification. In practice, the systems to be verified are described by programs in finite state languages such as SMV or VERILOG. These programs are then compiled into equivalent Kripke structures. The main practical problem in model checking is the *state explosion problem*: the size of the *state space* of the system is *exponential* in the size of its description. Therefore, even for systems of relatively modest size, it is often impossible to compute their state space explicitly.

In *symbolic verification*, the transition relation of the Kripke structure is not explicitly constructed, but instead a Boolean function representing the transition relation is computed. Sets of states are also represented as Boolean functions. Then, the fixpoint algorithms are applied to the formulas rather than to the Kripke structure. Since the Boolean formula is usually exponentially smaller than an explicit representation, symbolic verification is often able to alleviate the state explosion problem in these situations. Binary Decision Diagrams (BDDs) have been a particularly useful data structure for representing Boolean functions; in addition to their relative succinctness they provide a *canonical* representation for Boolean functions. As a result, equality of two Boolean functions can be easily checked in this representation.

At the core of all symbolic algorithms is *image computation,*[1] i.e., the task of computing the set of successors $\mathbf{Img}(S)$ of a set of states S, where

$$\mathbf{Img}(S) := \{s' : \exists s.R(s, s') \wedge s \in S\}.$$

Image computation is one of the major bottlenecks in verification. Often it is impossible to construct a single BDD for the transition relation R. Instead, R is represented as a *partitioned transition relation*, i.e., as the conjunction of several BDDs, each representing a piece of R. The problem is how to compute $\mathbf{Img}(S)$ without actually computing R.

As the above definition of $\mathbf{Img}$ indicates, the process of image computation involves quantifying state variables. In the BDD representation, this amounts to quantifying over several Boolean state variables. *Early quantification* [BCL91b, TSL$^+$90] is a technique which attempts to reorder the conjuncts so that the scope of each quantifier is minimized. The effect of early quantification is that the evaluation of single quantifiers can be done over *relatively small intermediate BDDs*. An exact definition of early quantification will be given in Sect. 2. The

---

[1] The techniques in this paper also apply to preimage computation. For ease of exposition, we restrict ourselves to image computation.

success of early quantification hinges heavily upon the derivation and ordering of sub-relations. Significant effort has been directed over the last decade to this problem. Since the problem is known to be NP-hard, various heuristics have been proposed for the problem.

In this paper, we propose and analyze several new techniques for efficient image computation using the partitioned representation. The main contributions of the paper are the following:

• We extend and analyze image computation techniques previously developed by Moon et al. [MKRS00]. These techniques are based on the *dependence matrix* of the partitioned transition relation. We explore various *lifetime metrics* related to this representation and argue their importance in predicting costs of image computation. Moreover, we provide effective heuristic techniques to optimize these metrics.

• We show that the problem of minimizing the lifetime metric of [MKRS00] is NP complete. More importantly, the reduction used to prove this result explains the close connection between efficient image computation and the well studied problem of computing the *optimal linear arrangement* for an undirected graph.

• We model the interaction between various sub-relations in the partitioned transition relation as a weighted graph, and introduce a new class of heuristics called *ordering by recursive partitioning*.

• We have performed extensive experiments which indicate the effectiveness of our techniques. By implementing these techniques, we have also contributed to the code base of the symbolic model checker NuSMV [CCGR99].

The main conclusion to be drawn from our analysis is the following: For complicated industrial designs, the effort initially spent on ordering algorithms is clearly amortized during image computation. In other words, the benefits of good orderings outweigh the cost of slow combinatorial optimization algorithms.

The remainder of this paper is organized as follows: in Sect. 2, we introduce notations and definitions used throughout the paper. Section 3 reviews the state of the art for this problem. Section 4 discusses various algorithms that facilitate early quantification. Section 5 describes experimental results. Finally, we conclude in Sect. 6 with some directions for future research.

2 Preliminaries

Notation: Every state is represented as a vector $b_1 \ldots b_n \in \{0, 1\}^n$ of Boolean values. The transition relation R is represented by a Boolean function $T(x_1, \ldots, x_n, x_1', \ldots, x_n')$. Variables $X = x_1, x_2, \ldots, x_n$ and $X' = x_1', x_2', \ldots, x_n'$ are called *current state* and *next state* variables respectively. $T(X, X')$ is an abbreviation for $T(x_1, \ldots, x_n, x_1', \ldots, x_n')$. Similarly, functions of the form $S(X) = S(x_1, \ldots, x_n)$ describe sets of states. We will occasionally refer to S as the *set*, and to T as the *transition relation*. For simplicity we will use X to denote both the set $\{x_1, \ldots, x_n\}$ and the vector $\langle x_1, \ldots, x_n \rangle$. Then the set of variables on which f depends is denoted by $Supp(f)$.

Example 1. [**3 bit counter. (Running Example)**] Consider a 3-bit counter with bits x_1, x_2 and x_3. x_1 is the least significant and x_3 the most significant bit. The state variables are $X = x_1, x_2, x_3$, $X' = x'_1, x'_2, x'_3$. The transition relation of the counter can be expressed as

$$T(X, X') = (x'_1 \leftrightarrow \neg x_1) \wedge (x'_2 \leftrightarrow x_1 \oplus x_2) \wedge (x'_3 \leftrightarrow (x_1 \wedge x_2) \oplus x_3).$$

In later examples, we will compute the image **Img**(S) of the set $S(X) = \neg x_1$. Note that $S(X)$ contains those states where the counter is even.

Partitioned BDDs: For most realistic designs it is impossible to build a single BDD for the entire transition relation. Therefore, it is common to represent the transition relation as a conjunction of smaller BDDs $T_1(X, X')$, $T_2(X, X')$, ..., $T_l(X, X')$, i.e.,

$$T(X, X') = \bigwedge_{1 \leq i \leq l} T_i(X, X'),$$

where each T_i is represented as a BDD. The sequence $T_1, \ldots, T_l$ is called a *partitioned transition relation*. Note that T is *not actually computed*, but only the T_i's are kept in memory.

Example 2. [**3 bit counter, ctd.**] For the 3 bit counter, a very simple partitioned transition relation is given by the functions $T_1 = (x'_1 \leftrightarrow \neg x_1), T_2 = (x'_2 \leftrightarrow x_1 \oplus x_2)$ and $T_3 = (x'_3 \leftrightarrow (x_1 \wedge x_2) \oplus x_3)$.

Partitioned transition relations appear naturally in hardware circuits where each latch (i.e., state variable) has a separate transition function. However, a partitioned transition relation of this form typically leads to a very large number of conjuncts. A large partitioned transition relation is similar to a CNF representation. So as the number of conjuncts increases, the advantages of BDDs are gradually lost. Therefore, starting with a very fine partition $T_1, \ldots, T_l$ obtained from the bit relations, the conjuncts T_i are grouped together into *clusters* $C_1, \ldots, C_r$, $r < l$ such that each C_i is a BDD representing the conjunction of several T_i's. The image **Img**(S) of S is given by the following expression.

$$\mathbf{Img}(S(X)) = \exists X \cdot (T(X, X') \wedge S(X)) \tag{1}$$

$$= \exists X \cdot (\bigwedge_{1 \leq i \leq l} T_i(X, X') \wedge S(X)) \tag{2}$$

$$= \exists X \cdot (\bigwedge_{1 \leq i \leq r} C_i(X, X') \wedge S(X)) \tag{3}$$

Note that in general $\exists x (\alpha \wedge \beta)$ is not equivalent to $(\exists x \alpha) \wedge (\exists x \beta)$. Consequently, to compute **Img**$(S(X))$, formula 3 instructs us to compute first a BDD for $\bigwedge_{1 \leq i \leq r} C_i(X, X') \wedge S(X)$. As argued above, partitioned transition relations have been introduced to *avoid* computing this potentially large BDD.

Early Quantification: Under certain circumstances, existential quantification can be distributed over conjunction using *early quantification* [BCL91b,TSL+90]. Early quantification is based on the following observation: if we know that α *does not contain* x, then $\exists x(\alpha \wedge \beta)$ is equivalent to $\alpha \wedge (\exists x \beta)$. In general, we have l conjuncts and n variables to be quantified. Since loosely speaking, clusters correspond to semantic entities of the design to be verified, it is expected that not all variables appear in all clusters. Therefore, some of the quantifications may be shifted over several C_i's. For a given sequence $C_1, \ldots, C_r$ of clusters, we obtain

$$\mathbf{Img}(S(X)) = \exists X_1 \cdot (C_1(X, X') \wedge \exists X_2 \cdot (C_2(X, X') \ldots$$
$$\exists X_r \cdot (C_r(X, X') \wedge S(X)))) \qquad (4)$$

where X_i is the set of variables which do not appear in $Supp(C_1) \cup \ldots \cup Supp(C_{i-1})$ and each X_i is disjoint from each other. Existentially quantifying out a variable from a formula f reduces $|Supp(f)|$ which usually corresponds to a reduced BDD size. The success of early quantification strongly depends on the order of the conjuncts $C_1, \ldots, C_r$.

Quantification Scheduling. The size of the intermediate BDDs in image computation can be reduced by addressing the following two questions:

Clustering: How to derive the clusters $C_1, \ldots, C_r$ from the bit-relations $T_1, \ldots, T_l$?

Ordering: How to order the clusters so as to minimize the size of the intermediate BDDs?

These two questions are not independent. In particular, a bad clustering results in a bad ordering. Moon and Somenzi [MS00] refer to this combined problem as the *quantification scheduling* problem. The ordering of clusters is known as the *conjunction schedule.*

Our algorithms are based on the concepts of *dependence matrices* (introduced in [MKRS00,MS00]) and *sharing graphs.*

Definition 1 (Moon et al). *The **dependence matrix** of an ordered set of functions $\{f_1, f_2, \ldots, f_m\}$ depending on variables $x_1, \ldots, x_n$ is a matrix D with m rows and n columns such that $d_{ij} = 1$ if function f_i depends on variable x_j, and $d_{ij} = 0$ otherwise.*

Thus, each row corresponds to a formula, and each column to a variable. For image computation, we will associate the rows with the conjuncts of the partitioned transition relation, and the columns with the state variables. For example, $f_m = S(X), f_{m-1} = C_r, \ldots$. Thus, different choices for $f_i, 1 \leq i \leq m$ correspond to different orderings.

We will assume that the conjunction is taken in the order $f_m, f_{m-1}, \ldots, f_2, f_1$, i.e., we consider an expression of the form $\exists X (f_1 \wedge (f_2 \wedge \ldots \wedge (f_{m-1} \wedge f_m)))$. If a variable occurs *only* in f_m, we can quantify it early by pushing it to the right just before f_m.

Example 3. [**3 bit counter, ctd.**] For $f_4 = S(X)$, $f_3 = T_3$, $f_2 = T_2$, $f_1 = T_1$ the dependency matrix for our running example looks as follows:

$$
\begin{array}{c|cccccc}
 & v_1 & v_2 & v_3 & v_1' & v_2' & v_3' \\
\hline
f_1 = T_1 & 1 & 0 & 0 & 1 & 0 & 0 \\
f_2 = T_2 & 1 & 1 & 0 & 0 & 1 & 0 \\
f_3 = T_3 & 1 & 1 & 1 & 0 & 0 & 1 \\
f_4 = S(X) & 1 & 0 & 0 & 0 & 0 & 0
\end{array}
$$

In general, for a variable x_j, let l_j denote the smallest index i in column j such that $d_{ij} = 1$. Analogously, h_j denotes the largest index. We can quantify away the variable x_j as soon as the conjunct corresponding to the row l_j has been considered. The variable does not appear in any conjuncts after h_j. Hence, $h_j - l_j$ can be viewed as the *lifetime* of a variable. Moon, Kukula, Ravi and Somenzi [MKRS00] define the following metric and use it extensively in their algorithms.

Definition 2 (Moon, Kukula, Ravi, Somenzi). *The **normalized average lifetime** of the variables in a dependence matrix $D_{m \times n}$ is given by*

$$
\lambda = \frac{\sum_{1 \leq j \leq n}(h_j - l_j + 1)}{m \cdot n}
$$

Note that the definition of λ assumes that $S(X)$ is given. Therefore, since λ *depends* on $S(X)$, the ordering has to be recomputed in each step of the fixpoint computation. We are considering static ordering techniques here, which are computed independently of any particular $S(X)$, so it is necessary to make assumptions about the structure of $S(X)$. We obtain two lifetime metrics λ_U and λ_L depending on whether we assume $Supp(S) = X$ or $Supp(S) = \emptyset$. It is easy to see that $\lambda_L \leq \lambda \leq \lambda_U$. The terms *average active lifetime* and *total active lifetime* are also used to denote λ_L and λ_U respectively. Moon and Somenzi argue in favour of using λ_L. We will evaluate the effectiveness of each of these metrics to predict image computation costs.

3   Related Work

The importance of the clustering and ordering problem was first recognized by Burch *et al.* [BCL91a] and Touati *et al.* [TSL+90]. Geist and Beer [GB94] proposed a simple heuristic algorithm, in which they ordered conjuncts in the increasing order of the number of support variables. All these techniques are static techniques. Subsequently, the same clusters and ordering are used for all the image computations during symbolic analysis. Since the clustering and ordering problems are not independent, these techniques typically begin by first ordering the conjuncts and then clustering them and finally ordering the clusters again using the same heuristics. The first successful heuristic (commonly known as IWLS95) for this problem is due to Ranjan *et al.* [RAP+95]. They have an elaborate heuristic procedure for ordering the initial conjuncts and the clusters.

The ordering procedure maintains a set Q of conjuncts that are already ordered and a set R of conjuncts that are yet to be ordered. Note that we have used the word conjunct here to mean both the conjuncts before clustering and the clusters in the final ordering phase. The next conjunct in the order is chosen from R using a heuristic score. The score is computed by using four factors: the maximum BDD index of a variable that can be quantified, the number of next state variables that would be introduced, the number of support variables, and the number of variables that will be quantified away. After the ordering phase, the clusters are derived by repeatedly conjoining the conjuncts until the BDD of the cluster grows larger than some partition size limit, at which point a new cluster is started. Bwolen Yang proposed a similar technique in his thesis [Yan99]. However, he introduces a pre-merging phase where conjuncts are initially merged pairwise based on the sharing of support variables and the maximum BDD size constraint. His ordering heuristic is based on six factors which are similar to those used by Ranjan *et al.* [RAP+95]. However, he also takes into account the relative growth in BDD sizes. The clustering algorithm Yang uses is the same as the one used in IWLS95. A recent paper by Moon and Somenzi [MS00] presents an ordering algorithm (henceforth referred to as FMCAD00) based on computing the *Bordered Block Triangular* form of the dependence matrix. Their clustering algorithm is based on the sharing of support variables (affinity). They report large performance gains with respect to the IWLS95 technique.

4   Algorithms for Ordering Clusters

The algorithms we propose also follow the order-cluster-order strategy. The ordering algorithms that we present in this section are used before and after clustering. Our clustering strategy is as in IWLS95. For the sake of clarity of notation, let us assume that the clusters $C_1, C_2, \ldots, C_r$ have been constructed and we are ordering them. But the discussion applies equally well to ordering the initial conjuncts $T_1, \ldots, T_n$.

We present two classes of algorithms. The first one is based on dependence matrix and the other one on *sharing graphs*.

In Sect. 2 we defined a dependence matrix D corresponding to the set of clusters $C_1, \cdots, C_r$. As already pointed out, the number of support variables provides a good estimate of the size of a BDD. Therefore, we seek a schedule in which the lifetime of variables is low. Moon and Somenzi [MS00] provide a method to convert a dependence matrix into bordered block triangular form with the goal of reducing λ_L.

4.1   Minimizing λ is NP-Complete

The main result of this subsection (Theorem 1) motivates the use of various combinatorial optimization methods.

Let λ-OPT be the following decision problem: given a dependence matrix D and a number r, does there exist a permutation σ of the rows of D such that $\lambda < r$? The following theorem shows that $\lambda - OPT$ is NP-complete. The reduction is from the *optimal linear arrangement problem (OLA)* [GJ79, page 200]. Due to space limitations the proof is given in the appendix.

Theorem 1. λ-*OPT is NP-complete.*

The complexity of this problem was not explored by Moon and Somenzi [MS00]. There exists a variety of heuristics for solving the optimal linear arrangement problem and related problems in combinatorial optimization. Some of these heuristics are based on hill climbing and simulated annealing. There are two important characteristics of this class of algorithms. First of all, they all try to minimize an underlying cost function. Second, these heuristics use a finite set of *primitive transformations*, which allows them to move from one solution to another. In our case, the set of swaps of the rows of the dependence matrix constitutes the set of moves and the cost function can be chosen to be either λ_L or λ_U. Our experimental results (Sect. 5) confirm that λ_L correlates with image computation costs much better than λ_U does, in accordance with the claim of [MS00]. Simulated annealing is a more general and flexible strategy than hill climbing.

4.2   Hill Climbing

Hill climbing is the simplest greedy strategy in which at each point, the solution is improved by choosing two rows to be swapped in such a manner as to achieve best improvement in the cost function. This process is repeated until no further move improves the solution. Since the best move is chosen at each point, this strategy is also called *steepest descent hill climbing*. However, this algorithm can easily get stuck in local optima. Randomization is used to alleviate this problem as follows: The best move that improves the solution is accepted only with some probability p, and with probability $1 - p$, a random move is accepted. This allows the algorithms to get out of local optima. Note that with $p = 1.0$, we get the steepest descent hill climbing. The algorithm can be run multiple number of times, each time beginning with a random permutation, and the best solution that is achieved is accepted.

Figure 1 describes the algorithm in exact terms. The hill climbing procedure is repeated *NumStarts* times. In the algorithm, σ denotes a permutation of the rows of the dependency matrix. Hill climbing is performed until no further improvement in λ is possible.

4.3   Simulated Annealing

The physical process of annealing involves heating a piece of metal and letting it cool down slowly to relieve stresses in the metal. The simulated annealing algorithm (introduced by Metropolis *et al.* [MRR+53]) mimicks this process to solve large combinatorial optimization problems [KJV83]. Drawing analogy from the physical process of annealing, the algorithm begins at a high "temperature",

HILLCLIMBORDER(D)

1    $\lambda_{best} = 2$ // any number greater than 1 will do, since λ is always less than 1
2   **for** $i = 1$ to $NumStarts$
3   let σ' be a random permutation of conjuncts.
4   **while** there exists a swap in σ' to reduce λ
5   make the best swap with probability p,
6   or make a random swap with probability $1 - p$ to update σ'.
7   **if** $\lambda' < \lambda_{best}$
8                $\lambda_{best} = \lambda'$
9                $\sigma_{best} = \sigma$
10   **endif**
11   **endfor**

Fig. 1. Hill climbing algorithm for minimizing λ

where the set of moves is essentially random. This allows larger jumps from local to global optima. Gradually, the temperature is decreased and the moves become less random favoring greedy moves over random moves for achieving a global optimum. Finally, the algorithm terminates at "freezing" temperatures where no further moves are possible. At each stage, the temperature is kept constant until "thermal quasi-equilibrium" is reached. While random moves help in the beginning, when the algorithm has a greater tendency to get stuck in local optima, the greedy moves help to achieve a global optimum once the solution is in the proximity of one. In practice, simulated annealing has been successfully used to solve optimization problems from several domains.

The probability of making a move that *increases* the cost function is related to the temperature t_i at the i-th iteration, and is given by $e^{-\Delta\lambda/t_i}$. Thus at higher temperatures, the probability of accepting random moves is high. The gradual decrease of temperature is called the *cooling schedule*. If the temperature is decreased by a fraction r in each stage, we get an exponential cooling schedule. Thus beginning with an initial temperature of t_0, the temperature in the i-th iteration is $t_0 r^i$. It has been shown that a logarithmic cooling schedule is guaranteed to achieve an optimal solution with high probability [B'e92,Haj85]. However, this is an extremely slow cooling schedule and simple cooling schedules like exponential schedules perform well for many problems. Figure 2 describes our algorithm. The parameter $NumStarts$ controls the number of times the temperature is decreased. The parameter $NumStarts2$ controls the number of iterations at a fixed temperature t_i.

4.4   Sharing Graphs and Separators

We build *sharing graphs* as defined below to model interaction between clusters.

SimAnnealOrder(D)
    for $i = 1$ **to** $NumStarts$
1        $t_i \leftarrow t_0 r^i$
2   **for** $j = 1$ **to** $NumStarts2$
3   permute two random rows of D to get D_i
4   **if** $(\lambda_i < \lambda)$ // greedy move
5                $\lambda \leftarrow \lambda_i; D \leftarrow D_i$
6   **else**   // random move
7   with probability $e^{\frac{-(\lambda_i - \lambda)}{t_i}}$, set $\lambda \leftarrow \lambda_i; D \leftarrow D_i$
8   **endif**
9   **endfor**
10 **endfor**

Fig. 2. Simulated annealing algorithm to minimize λ

Definition 3. *A **sharing graph** corresponding to a set of Boolean functions $\{f_1, f_2, \ldots, f_m\}$ is a weighted graph $G(V, E, w_e)$, where $V = \{f_1, f_2, \ldots, f_m\}$, $E = V \times V$ and $w_e : E \to \Re$ is a real-valued weight function.*

We shall use heuristic weight functions to express interaction between clusters. Intuitively, the stronger the interaction between two clusters, the closer they should be in the ordering. IWLS95 and Bwolen Yang's heuristics order the conjuncts based on this type of interaction between conjuncts. We propose to use graph algorithms on sharing graphs to order the conjuncts. We define the weight $w(T_i, T_j)$ of an edge (T_i, T_j) in the sharing graph as

$$w(T_i, T_j) = W_1 \cdot \frac{Supp(T_i) \cap Supp(T_j)}{|Supp(T_i)| + |Supp(T_j)|} + W_2 \cdot \frac{BddSize(T_i \wedge T_j)}{BddSize(T_i) + BddSize(T_j)}$$

The first factor ($W_1 \geq 0$) denotes the relative weight of sharing of support between two conjuncts, while the second factor ($W_2 \leq 0$) denotes the weight of the relative growth in the sizes of BDDs if these two conjuncts are conjoined. Therefore, a higher edge weight between two conjuncts indicates a higher degree of interaction and consequently these conjuncts should appear "close" in the ordering.

A separator partitions the vertices of a weighted undirected graph into two sets such that the total weight of the edges between two partitions is "small". Formally, an *edge separator* is defined as follows:

Definition 4. *Given a weighted undirected graph $G(V, E)$ with two weight functions $w_e : E \to \Re$ and $w_v : V \to \Re$, and a positive constant $\gamma < 0.5$, an edge separator is a collection of edges E_s such that removing E_s from G partitions G into two disconnected subgraphs V_1 and V_2, and $\frac{|\sum_{v \in V_1} w_v(v) - \sum_{v \in V_2} w_v(v)|}{\sum_{v \in V} w_v(v)} < \gamma$.*

Usually, γ is chosen very close to zero so that the size of the two sets is approximately the same. The weight of the edge separator E_s is simply the sum of the weight of the edges in E_s. It has been shown that finding an edge separator of minimum weight is NP-complete [GJ79, pp. 209], in fact finding an approximation is NP-hard, too [BJ92]. The problem of finding a good separator occurs in many different contexts and a wide range of application areas. A large number of heuristics have been proposed for the problem. One of the most important heuristics is due to Kernighan and Lin [KL70]. Variations of this heuristic [FM82] have been found to work very well in practice.

By finding a good edge separator of the sharing graph, we obtain two sets of vertices with a low level of interaction between them. Thus the vertices of these two sets can be put apart in the ordering. A complete ordering is achieved by recursively invoking the algorithm on the two halves. Since this ordering respects the interaction strengths between conjuncts, we expect to achieve smaller BDD sizes.

We use the Kernighan-Lin algorithm for finding a good edge separator E_s. This produces two sets of vertices L and R. A vertex $v \in L$ that has an edge of non-zero weight to a vertex in R is called an *interface vertex*. L_I denotes the set of interface vertices in L. Similarly, R_I denotes the set of interface vertices in R. We invoke the algorithm to recursively order $L \setminus L_I$, L_I, R_I, and $R \setminus R_I$. Finally, the order on the vertices is given by the order on $L \setminus L_I$ followed by the order on L_I, followed by the order on R_I, and followed by the order on $R \setminus R_I$. Figure 3 describes the complete algorithm.

KLinOrder$(G(V, E), W)$

1 Find a separator E_s using
   Kernighan-Lin heuristic

2 Let L and R be two partitions of
   vertices induced by E_s.

3 $L_i \leftarrow Interface(L)$.

4 $R_i \leftarrow Interface(R)$.

5 Recursively call the procedure on
   the subgraphs induced by $L \setminus L_i$, L_i,
   R_i and $R \setminus R_i$.

6 Order the vertices as
   $KLinOrder(L \setminus L_i) \prec KLinOrder(L_i) \prec$
   $KLinOrder(R_i) \prec KLinOrder(R \setminus R_i)$.

Fig. 3. An ordering algorithm based on graph separators

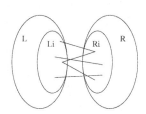

Fig. 4. Kernighan-Lin partition

5   Experimental Results

In order to evaluate the effectiveness of our algorithms, we ran reachability and model checking experiments on circuits obtained from the public domain and industry. The "S" series of circuits are ISCAS'93 benchmarks, and the "IU" series of circuits are various abstractions of an interface control circuit from Synopsys. For a fair comparison, we implemented all the techniques in the NuSMV model checker. All experiments were done on a 200MHz quad Pentium Pro processor machine running the Linux operating system with 1GB of main memory. We restricted the memory usage to 900MB, but did not set a time limit. The two performance metrics we measured are *running time* and *peak number of live BDD nodes*. We provided a prior ordering to the model checker and turned off the dynamic variable reordering option. This was done so that the effects of BDD variable reordering do not "pollute" the result. We also recorded the fraction of time spent in the clustering and ordering phases. The cost of these phases is amortized over several image computations performed during model checking and reachability analysis.

In the techniques that we have described, several parameters have to be chosen. For example, the cooling schedule in the case of simulated annealing needs to be determined. We ran extensive "tuning experiments" to find the best value for these parameters. Due to space constraints, we do not describe all those experiments. However, the choice of lifetime metric to optimize is a crucial one and hence in our first set of experiments, we evaluate the effectiveness of these metrics for predicting image computation costs.

Our algorithms for combinatorial optimization of lifetime metrics can choose to work with either upper or lower approximations of lifetimes. We ran the following experiment to estimate the correlation between the performance, and λ_L and λ_L respectively. We generate various conjunction schedules for a number of benchmarks by different ordering methods and by varying various parameters of the optimization methods. Each schedule gives us different values for lifetime metrics. We measure the running time and the peak number of live BDD nodes used for the model checking or reachability phase. For each circuit, this gives us four scatter plots for running time vs lifetime metric and space vs lifetime metric. A statistical correlation coefficient between runtime/space and lifetime metric indicates the effectiveness of a metric for predicting the runtime/space requirement. The following Table 1 concisely summarizes the correlation results.

It is clear from this data that the active lifetime (λ_L) is a much more accurate predictor of image computation costs than total lifetime (λ_U). Hence, simulated annealing and hill climbing techniques optimize λ_L.

In the following set of experiments (Table 2), we compare our techniques against the FMCAD00 strategy [MS00]. The first column indicates the total running time of the benchmark (including ordering/clustering and model checking/reachability phases), the second column indicates the peak number of live BDD nodes in thousands during the whole computation, the third column indicates time used by ordering phase, the next two columns indicate λ_L and λ_U achieved. From hill climbing and simulated annealing, we only report the results

Table 1. Correlation between various lifetime metrics and runtime/space for a representative sample of benchmarks

Circuit	Runtime		Space	
	λ_L	λ_U	λ_L	λ_U
IU40	0.560	0.303	0.610	0.227
IU70	0.603	0.336	0.644	0.263
TCAS	0.587	0.366	0.628	0.240
S1269	0.536	0.402	0.559	0.345
S3271	0.572	0.350	0.602	0.297

of simulated annealing, as both of them belong to the same class of algorithms. Moreover, we found out that in general, simulated annealing achieves better performance than hill climbing.

The algorithm KLin based on edge separators achieves lower peak live node count for several circuits than FMCAD00. For the 15 large benchmarks for which FMCAD'00 takes more than 100 secs to finish, KLin wins 10 cases in terms of Peak live BDD nodes, and 7 cases in terms of running time. In some cases, the savings in space is 40%.

The result for the simulated annealing algorithm that minimizes λ is shown in Table 2. Again, in comparison to FMCAD00, for the 15 non-trivial benchmarks, simulated annealing wins 14 cases and ties for the other in space, and wins 11 cases in time. In some cases, the savings in space is 55%. The simulated annealing algorithm can also complete 16 reachability steps for the S1423 circuit, which to our knowledge has not be achieved by other techniques. Comparing KLin and simulated annealing, simulated annealing achieves the better results for all the nontrivial benchmarks.

The improvements in execution times are less than the improvements in space, especially for smaller circuits. This is because separator based algorithms spend more time in the ordering phase itself. However, for larger circuits, this cost gets amortized by the smaller BDDs achieved during analysis. An important observation that can be made is that in general, our algorithms spend more time in the initial ordering phase as compared to FMCAD00. This is to be expected since both KLin and simulated annealing are optimization methods.

The last two columns in Table 2 indeed demonstrate that our algorithms improve various λs with respect to FMCAD'00. The main objective of our algorithms was to improve λ_L, though we can see that they also result in better λ_Us in general.

6   Conclusions and Future Work

We have given convincing evidence that variable lifetimes have a crucial impact on the performance of image computation algorithms. We have also presented new algorithms for conjunction scheduling based on hill climbing, simulated annealing, and graph separators and shown the effectiveness of them. The per-

Table 2. Comparing FMCAD00(I), Kernighan-Lin separator (II) and Simulated annealing (III) algorithms. The times are reported in seconds. The peak space is reported by the peak number of live BDD nodes in thousands. (**MOut**)–Out of memory, (†)–SFEISTEL, (*)–8 reachability steps, (**)–14 reachability steps, (#)–13 reachability steps. The lifetimes reported are after the final ordering phase.

Circuit	#FF	#inp.	$\log_2$ of #reach	Total Time I	II	III	Peak space I	II	III	Ordering time I	II	III	λ_L I	II	III	λ_U I	II	III
IDLE	73	0	14.63	159	161	182	289	276	223	2	20	29	0.329	0.293	0.200	0.421	0.515	0.487
GUID	91	0	47.58	14	20	24	137	106	138	4	15	19	0.346	0.220	0.165	0.394	0.452	0.294
S953	29	16	8.98	1	2	3	15	13	15	1	1	3	0.290	0.290	0.271	0.507	0.485	0.410
IU30	30	138	18.07	28	104	63	290	563	290	3	24	34	0.360	0.368	0.324	0.459	0.522	0.634
IU35	35	183	22.49	13	29	11	257	366	202	4	24	6	0.364	0.373	0.304	0.573	0.360	0.308
IU40	40	159	25.85	13	37	14	353	384	232	5	21	5	0.326	0.336	0.302	0.508	0.326	0.334
IU45	45	183	29.82	MOut	11256	165	MOut	3952	483	10	32	39	0.360	0.353	0.300	0.465	0.663	0.569
IU50	50	615	31.57	476	522	540	1627	1599	1602	16	52	77	0.319	0.418	0.133	0.459	0.654	0.403
IU55	55	625	33.94	982	891	870	4683	3358	3298	14	90	84	0.384	0.386	0.324	0.583	0.432	0.515
IU65	65	632	39.32	MOut	1260	1083	MOut	7048	6793	18	81	100	0.389	0.353	0.353	0.659	0.448	0.423
IU70	70	635	42.07	5398	3033	2855	17355	9099	9964	38	95	129	0.303	0.296	0.286	0.424	0.393	0.486
IU75	75	322	46.59	5367	4218	3822	16538	12193	9404	45	115	140	0.398	0.371	0.349	0.731	0.692	0.526
IU80	80	350	49.80	MOut	6586	4824	MOut	22234	17993	49	127	136	0.372	0.335	0.322	0.570	0.628	0.345
IU85	85	362	52.14	MOut	MOut	6933	MOut	MOut	25661	59	141	154	0.332	0.303	0.287	0.623	0.597	0.591
TCAS	139	0	106.87	5058	5285	4598	11931	12376	9140	27	173	165	0.173	0.182	0.227	0.299	0.306	0.261
S1269	37	18	30.07	2109	2466	1875	1440	1736	893	10	19	24	0.584	0.622	0.449	0.659	0.929	0.589
S1512	57	29	40.59	799	1794	651	159	190	135	15	24	30	0.412	0.394	0.386	0.521	0.619	0.714
S5378	179	35	57.71*	18036	MOut	10168	1632	MOut	1279	42	49	67	0.124	0.114	0.099	0.219	0.164	0.152
S4863	104	49	72.35	3565	3109	3013	1124	947	910	38	45	56	0.102	0.103	0.086	0.251	0.109	0.179
S3271	116	26	79.83	4234	3286	3399	8635	6240	6203	33	30	30	0.224	0.185	0.184	0.366	0.306	0.226
S3330	132	40	86.64	23659	19533	24563	12837	9866	11381	69	123	150	0.214	0.217	0.227	0.299	0.335	0.378
SFE†	293	69	218.77	863	916	762	147	146	130	14	84	76	0.383	0.354	0.344	0.554	0.624	0.531
S1423	74	17	37.41**	23325	19265#	35876	65215	27653#	48366	10	17	35	0.486	0.501	0.301	0.622	0.622	0.460

formance of these algorithms was comparable to that of the current best known methods. Our experiments clearly demonstrate that for large circuits, we can achieve savings in memory in the range of 50-60%. Since fine-tuned image computation algorithms are obviously most important for large circuits, we believe that our results are a significant contribution to model checking and reachability analysis. On the implementation side, we have contributed several new image computation algorithms to the NuSMV model checker, and believe that it will be a valuable research tool.

There are some interesting research directions to pursue. First of all, we need to understand the behavior of our algorithms on broader class of systems. The examples were mostly chosen from the circuit domain, and we would like to see the effectiveness of these algorithms on other class of circuits. Secondly, techniques which switch between different conjunction schedules depending on intermediate state sets in the fixpoint computation seem to be promising. We also plan to study in greater depth the effect of various parameters of our methods and automatic ways to tune them.

References

[BCL91a] J. R. Burch, E. M. Clarke, and D. E. Long. Representing circuits more efficiently in Symbolic Model Checking. In *28th ACM/IEEE Design Automation Conference*, 1991.

[BCL91b] J. R. Burch, E. M. Clarke, and D. E. Long. Symbolic Model Checking with partitioned transition relations. In A. Halaas and P. B. Denyer, editors, *Proceedings of the International Conference on Very Large Scale Int egration*, Edinburgh, Scotland, August 1991.

[B'e92] C. J. P. B'elisle. Convergence theorems for a class of simulated annealing algorithms. *Journal of Applied Probability*, 29:885–892, 1992.

[BJ92] T. N. Bui and C. Jones. Finding good approximate vertex and edge partitions is NP-hard. *Information Processing Letters*, 42:153–159, 1992.

[CCGR99] A. Cimatti, E. M. Clarke, F. Giunchiglia, and M. Roveri. NuSMV: A new Symbolic Model Verifier. In N. Halbwachs and D. Peled, editors, *Proceedings of International Conference on Computer-Aided Verification (CAV'99)*, number 1633 in Lecture Notes in Computer Science, pages 495–499. Springer, July 1999.

[CGP00] E. M. Clarke, O. Grumberg, and D. Peled. *Model Checking*. MIT Press, 2000.

[FM82] C.M. Fiduccia and R.M. Mattheyses. A linear time heuristic for improving network partitions. In *19th ACM/IEEE Design Automation Conference*, pages 175–181, 1982.

[GB94] D. Geist and I. Beer. Efficient Model Checking by automated ordering of transition relation partitions. In D. L. Dill, editor, *Sixth Conference on Computer Aided Verification (CAV'94)*, volume 818 of *LNCS*, pages 299–310, Stanford, CA, USA, 1994. Springer-Verlag.

[GJ79] Michael R. Garey and David S. Johnson. *Computers and Intractability: A Guide to the Theory of NP-Completeness*. W. H. Freeman & Co., 1979.

[Haj85] B. Hajek. A tutorial survey of theory and applications of simulated annealing. In *Proc. 24th IEEE Conf. Decision and Control*, pages 755–760, 1985.

[KJV83]   S. Kirkpatrick, C. D. Gelatt Jr., and M. P. Vecchi. Optimization by simulated annealing. *Science*, 220:671–679, 1983.
[KL70]   Brian Kernighan and S. Lin. An efficient heuristic procedure for partitioning graphs. *The Bell System Technical Journal*, pages 291–307, February 1970.
[MKRS00]   In-Ho Moon, James H. Kukula, Kavita Ravi, and Fabio Somenzi. To split or to conjoin: The question in image computation. In *Proceedings of the 37th Design Automation Conference (DAC'00)*, pages 26–28, Los Angeles, June 2000.
[MRR+53]   N. Metropolis, A. W. Rosenbluth, M. N. Rosenbluth, A. H. Teller, and E. Teller. Equation of state calculations by fast computing machines. *Journal of Chemical Phyics*, 21(6):1087–1092, 1953.
[MS00]   In-Ho Moon and Fabio Somenzi. Border-block triangular form and conjunction schedule in image computation. In Warren A. Hunt Jr. and Steven D. Johnson, editors, *Proceedings of the Formal Methods in Computer Aided Design (FMCAD'00)*, volume 1954 of *LNCS*, pages 73–90, November 2000.
[RAP+95]   R.K. Ranjan, A. Aziz, B. Plessier, C. Pixley, and R.K. Brayton. Efficient BDD algorithms for FSM synthesis and verification. In *IEEE/ACM International Workshop on Logic Synthesis*, Lake Tahoe, 1995. IEEE/ACM.
[TSL+90]   H. Touati, H. Savoj, B. Lin, R. K. Brayton, and A. Sangiovanni-Vincentelli. Implicit enumeration of finite state machines using BDDs. In *Proceedings of the IEEE international Conference on Computer Aided Design (ICCAD)*, pages 130–133, November 1990.
[Yan99]   Bwolen Yang. *Optimizing Model Checking Based on BDD Characterization*. PhD thesis, Carnegie Mellon University, Computer Science Department, May 1999.

Appendix A

It is easy to see that for a given permutation σ of rows, we can compute λ in polynomial time $(O(n \cdot m))$ and check if $\lambda \leq r$.

To show that $\lambda - OPT$ is NP-hard, we reduce a known NP-complete problem called *optimal linear arrangement (OLA)* [GJ79, page 200] to $\lambda - OPT$. An instance of OLA consists of a graph $G(V, E)$ and a positive integer K. The question is whether there exists a permutation f of V such that $\sum_{(u,v) \in E} |f(u) - f(v)| \leq K$. The reduction consists of constructing a dependence matrix D and a number r such that $(V, E), K$ is a solution of OLA iff D, r is a solution to $\lambda - OPT$. An example of a reduction is given in figure 5.

Formally, D has $|V|$ rows corresponding to the vertices of $G(V, E)$, and $|E|$ columns corresponding to the edges of $G(V, E)$. For any edge $e_k = (v_i, v_j)$, set $d_{ik} = d_{jk} = 1$ and set all other d_{ij}'s to 0. Thus, in each column there are two occurences of the symbol 1. We set $r = \frac{K+n}{n \cdot m}$. Trivially we obtain the following equivalence:

$$\frac{\sum_{1 \leq j \leq n} (h_j - l_j + 1)}{n \cdot m} \leq r$$

$$\Leftrightarrow \sum_{1 \leq j \leq n} (h_j - l_j + 1) \leq r \cdot (n \cdot m)$$

$$\Leftrightarrow \sum_{1 \leq j \leq n} (h_j - l_j + 1) \leq K + n$$

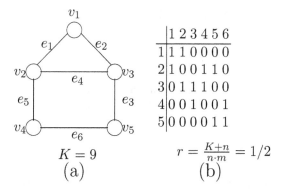

$K = 9$

$r = \frac{K+n}{n \cdot m} = 1/2$

(a)

(b)

Fig. 5. (a) An instance of Optimal Linear Arrangement, (b) its reduction to $\lambda - OPT$. The permutation v_1, v_2, v_3, v_5, v_4 is a solution to both.

Let σ be a permutation of the vertices of V. Note that σ simultaneously is a permutation of the rows of D. We have to show that σ is a solution of $G(V, E), K$ iff σ is a solution of D, r.

The important observation is that because of the construction of D, the only non-zero entries in each column j correspond to the two vertices of the edge $e_j = (u, v)$. Therefore, we conlude that $h_j - l_j = |\sigma(u) - \sigma(v)|$. Continuing the above equivalence we obtain

$$\sum_{1 \leq j \leq n} |\sigma(u) - \sigma(v)| + n \leq K + n$$

$$\Leftrightarrow \sum_{(u,v) \in E} |f(u) - f(v)| \leq K$$

Net Reductions for LTL Model-Checking

Javier Esparza and Claus Schröter

Institut für Informatik, Technische Universität München
{esparza,schroete}@in.tum.de

Abstract. We present a set of reduction rules for LTL model-checking of 1-safe Petri nets. Our reduction techniques are of two kinds: (1) Linear programming techniques which are based on well-known Petri net techniques like invariants and implicit places, and (2) local net reductions. We show that the conditions for the application of some local net reductions can be weakened if one is interested in LTL model-checking using the approach of [EH00,EH01]. Finally, we present a number of experimental results and show that the model-checking time of a net system can be significantly decreased if it has been preprocessed with our reduction techniques.

1 Introduction

In two recent papers [EH00,EH01], Esparza and Heljanko have developed an unfolding technique to the problem of model-checking LTL for concurrent systems. Using the automata-theoretic approach, the model-checking problem is reduced to two simpler problems concerning the infinite executions of a combined system obtained from the original system and from a Büchi automaton for the negation of the property to be checked. Loosely speaking, the unfolding technique exploits the concurrency present in the system to obtain a compact representation of the state space.

Esparza and Heljanko's approach is part of the Model Checking Kit, a collection of programs which allow to model a finite-state system using a variety of modelling languages, and verify it using a variety of checkers, including deadlock-checkers, reachability-checkers, and model-checkers for the temporal logics CTL and LTL. The most interesting feature of the Kit is that, independently of the description language chosen by the user, different checkers can be applied to the same model. The Kit is held together by a program called the Glue. Given a system and a property modelled in one of the description languages supported by the Kit, the Glue (1) translates them into a 1-safe Petri net and a property of the net, described both in a common internal representation language;[1] (2) translates the net and the property into input for the target model checker, and (3) collects the output of the checker and translates it back into the original description language.

---

[1] 1-safe Petri nets are chosen because they are unstructured and have a simple notion of independent actions, which makes them a suitable "assembler language" for concurrent systems.

T. Margaria and T. Melham (Eds.): CHARME 2001, LNCS 2144, pp. 310–324, 2001.
© Springer-Verlag Berlin Heidelberg 2001

Since step (1) is automatic, it may introduce many redundancies, leading to large 1-safe Petri nets. In this paper we introduce some techniques that allow to remove a good number of these redundancies. For that, we exploit well-known Petri net techniques: (sub)invariants [DE95,Rei85], implicit places [CS90], and local net reductions [Ber85,PPP00]. Invariants and implicit places are used to remove places and transitions which do not affect the behaviour of the Petri net. Local net reductions are used to reduce the Petri net while guaranteeing that the result of the analysis on the reduced net will be the same as the result in the original net. Notice that, since we work at the level of the common internal representation language, redundancies can be eliminated independently of the language in which the system was described.

The paper contains two main contributions. First, we show that if we are interested in LTL model-checking using the approach of [EH00,EH01], the conditions for the applications of some net reductions can be *weakened*; therefore, the rules can be applied more often, leading to larger reductions. The second contribution is a number of experimental results on a set of examples modelled using several of the Kit's system description languages.

The paper is organised as follows. Section 2 contains basic definitions about 1-safe Petri nets. Section 3 gives some information on the approach of [EH00, EH01]. Section 4 introduces the reduction rules. (Sub)invariants and implicit places are just taken from the literature, while some local net reductions are improved. Section 5 describes the routine for the application of the rules, and discusses implementation issues. Section 6 presents experimental results on systems modelled using $B(PN)^2$, a high-level programming language integrated in the Kit, and communicating automata.

2   1-Safe Petri Nets

A triple (P, T, F) is a *net* if P and T are disjoint sets and F is a subset of $(P \times T) \cup (T \times P)$. The elements of P are called *places* and the elements of T *transitions*. Places and transitions are generally called *nodes*. We identify F with its characteristic function on the set $(P \times T) \cup (T \times P)$. The *preset* $^\bullet x$ of a node x is the set $\{y \in P \cup T \mid F(y, x) = 1\}$. The *postset* $x^\bullet$ of a node x is the set $\{y \in P \cup T \mid F(x, y) = 1\}$.

A *marking* M of a net (P, T, F) is a mapping $M : P \mapsto \{0, 1\}$. We identify a marking M with the set $P' \subseteq P$ such that $\forall p \in P : p \in P' \Leftrightarrow M(p) = 1$ holds.

A four-tuple $\Sigma = (P, T, F, M_0)$ is a *net system* if (P, T, F) is a net and M_0 is a marking of (P, T, F). M_0 is called the *initial marking* of the net system Σ. A marking M *enables* a transition t if $\forall p \in {}^\bullet t : M(p) = 1$ holds. If t is enabled at M, then t can *occur*, and its occurrence leads to a new marking M' (denoted $M \xrightarrow{t} M'$), defined by $M'(p) = M(p) - F(p, t) + F(t, p)$ for every place p. A sequence of transitions $\sigma = t_1 t_2 \ldots t_n$ is an *occurrence sequence* if there exist markings $M_1, M_2, \ldots, M_n$ such that $M_0 \xrightarrow{t_1} M_1 \xrightarrow{t_2} \ldots M_{n-1} \xrightarrow{t_n} M_n$. M_n is the marking reached by the occurrence of σ, also denoted by $M_0 \xrightarrow{\sigma} M_n$. M is a *reachable marking* if there exists an occurrence sequence σ such that $M_0 \xrightarrow{\sigma} M$.

3   An Unfolding Approach to LTL Model-Checking

The unfolding technique, originally introduced by McMillan [McM92], has been very successfully applied to several verification tasks, e.g. deadlock detection, reachability analysis and LTL model-checking. The 1-safe Petri net is unfolded into an acyclic net until a finite complete prefix is generated. This is a finite acyclic net having exactly the same reachable markings as the original one. Esparza and Heljanko have introduced an unfolding approach to LTL model-checking in [EH00]. This approach makes use of the automata-theoretic approach to model-checking [Var96]. A *synchronized net system* is constructed as the product of the original net system and a Büchi automaton accepting the negation of the property to be checked. Then the model-checking problem is reduced to the problem of detecting illegal ω-traces and illegal livelocks in the synchronized net system. Both problems are solved by constructing finite prefixes of the unfolding of the synchronized net system. The main advantage of this approach with respect to Wallner's approach [Wal98] is its simplicity. Wallner first calculates a complete prefix and then he constructs a graph, but the definition of the graph is non-trivial, and the graph can be exponential in the size of the prefix. The approach of [EH00] avoids the construction of the graph, but unfortunately constructs a larger prefix.

Now, we briefly review the main definitions and results of [EH00]. Given an LTL property φ and a net system $\Sigma = (N, M_0)$ the transitions of the net $N = (P, T, F)$ are divided into two sets V and $T \setminus V$ of *visible* and *invisible* transitions. Let $AP(\varphi) \subseteq P$ be a set of places corresponding to the atomic propositions of φ. Then it holds that $\forall t \in T: t \in V \Leftrightarrow t \in (\bullet p \cup p \bullet)$ for a $p \in AP(\varphi)$. Then a Büchi automaton $A_{\neg\varphi}$ is constructed accepting the negation of the property φ. The Büchi automaton is a tuple $A = (\Gamma, Q, q_0, F, \rho)$ where Γ is an *alphabet*, Q is a finite non-empty set of *states*, $q_0 \in Q$ is an *initial state*, $F \subseteq Q$ is the set of *accepting states*, and $\rho \subseteq Q \times \Gamma \times Q$ is the *transition relation*. A *accepts* an infinite word $w \in \Gamma^\omega$ if some run of A on w visits some state in F infinitely often.

In a next step a *synchronized net system* $\Sigma_{\neg\varphi}$ is constructed as the product of the original net system Σ and the Büchi automaton $A_{\neg\varphi}$. The system Σ and the Büchi automaton are synchronized only by the visible transitions of Σ. For more details we refer the reader to [EH00]. In the following we call the places and transitions of $\Sigma_{\neg\varphi}$ corresponding to the Büchi automaton *Büchi places* and *Büchi transitions*. Büchi transitions and visible transitions have a characteristic feature as described in the following Lemma 1.

Lemma 1. *Property of Büchi transitions and visible transitions*

> *Let $\Sigma_{\neg\varphi}$ be a synchronized net system. Then for all Büchi transitions and visible transitions t it holds that $\bullet t \geq 2$ and $t \bullet \geq 2$.*

Proof: Clear from the construction of the synchronized system $\Sigma_{\neg\varphi}$ (see [EH00]).

Once $\Sigma_{\neg\varphi}$ is constructed, two subsets of Büchi transitions, called *infinite trace monitors* and *livelock monitors*, are defined as follows.

Definition 1. *Infinite Trace Monitors and Livelock Monitors*

- The set I of *infinite trace monitors* contains the transitions t such that there exists in $t^\bullet$ a final state q of $A_{\neg\varphi}$. Loosely speaking, these are the transitions which put a token into a final state of the Büchi automaton.
- The set L of *livelock monitors* contains the transitions t such that there exists in $t^\bullet$ a state q of $A_{\neg\varphi}$ satisfying the following condition: with q as initial state, the automaton $A_{\neg\varphi}$ accepts an infinite sequence of transitions.

∎ 1

With this knowledge we can define a notion of illegal ω-traces and illegal livelocks.

Definition 2. *Illegal ω-Traces and Illegal Livelocks*

Let Σ be a net system where T is divided into two sets of V and $T \setminus V$ of visible and invisible transitions, and T contains the two subsets I and L of transitions as mentioned above.
- An *illegal ω-trace* of Σ is an infinite sequence $M_0 \xrightarrow{\sigma}$ such that σ contains infinitely many I-transitions.
- An *illegal livelock* of Σ is an infinite sequence $M_0 \xrightarrow{\sigma t} M \xrightarrow{\sigma_1}$ such that $t \in L$ and σ_1 contains only invisible transitions.

∎ 2

The main result of [EH00] is Theorem 1.

Theorem 1. *LTL Model-Checking*

Let Σ be a labelled net system and φ an LTL formula. Σ satisfies φ if and only if $\Sigma_{\neg\varphi}$ has no illegal ω-traces and no illegal livelocks.

4   Reduction Rules for LTL Model-Checking

The LTL model-checking approach described in the previous section uses unfolding techniques for detecting illegal ω-traces and illegal livelocks. Since the unfolding grows exponentially in the size of the net system, our aim is to reduce the synchronized net system before unfolding it without changing the conditions of Theorem 2 for the LTL model-checking task.

Theorem 2. *Conditions for LTL Model-Checking*

Let Σ be a labelled net system and φ an LTL formula. Σ violates φ if and only if at least one of the following conditions hold for $\Sigma_{\neg\varphi}$:
(i) there exists an infinite sequence $M_0 \xrightarrow{\sigma}$ containing infinitely many I-transitions, or
(ii) there exists a sequence $M_0 \xrightarrow{\sigma} M$ such that
   (a) there exists an infinite sequence $M|_{A_{\neg\varphi}} \xrightarrow{\sigma_1}$ which goes infinitely often through a final state, and

(b) there exists an infinite sequence $M|_\Sigma \xrightarrow{\sigma_2}$ containing only invisible transitions

Now, with respect to the LTL model-checking approach of [EH00] we only have to guarantee that the two conditions of Theorem 2 hold for a net system if and only if they hold for the reduced net system obtained by applying our reduction rules. Additionally, we restrict ourselves to the fact that we will never remove atomic propositions, Büchi places and Büchi transitions from the net. In the following sections we present some reduction rules which are practicable for the LTL model-checking approach of [EH00].

4.1   Linear Programming Rules

In this section we present two reduction rules which are based on linear programming techniques. First, we briefly introduce some basic notions. When applying linear programming techniques for verification tasks of Petri nets, one of the basic concepts is the so-called marking equation that can be used as an algebraic representation of the set of reachable markings of an acyclic net. Given a marking M reachable from the initial marking M_0 and a place p, the number of tokens of p in M can be calculated as the number of tokens p carries in M_0 plus the difference of tokens added by the input places and removed by the output places. This leads to the following equation: $M(p) = M_0(p) + \Sigma_{t \in {}^\bullet p} \#t - \Sigma_{t \in p^\bullet} \#t$ where $\#t$ denotes the number of occurrences of t in an occurrence sequence $\sigma = t_1 \dots t_m$. Usually this equation is written in the form $M = M_0 + \mathbf{N} \cdot \boldsymbol{\sigma}$, where $\boldsymbol{\sigma} = {}^t(\#t_1, \dots, \#t_n)$ is called the *Parikh vector* of σ and $\mathbf{N}$ denotes the *incidence matrix* of a net N, a $P \times T$ matrix given by $\mathbf{N}(p, t) = F(t, p) - F(p, t)$. Let l_p be the *incidence vector* of a place p. Then we have $M(p) = M_0(p) + l_p \cdot \boldsymbol{\sigma}$.

Dead Transition Rule. In many cases systems that should be checked for LTL properties are specified in a high level programming language, for instance B(PN)^2 which is well-known from the PEP-tool and also an input language of the Kit. The systems have been translated automatically into 1-safe Place/Transition nets. During these translations many redundancies might have been introduced, for instance places which never carry a token. It is clear that also their output transitions never fire. The aim of the dead transition rule is to detect such places and transitions and to remove them from the net system. Applying this rule does not affect the size of the unfolded net prefix which will be calculated during the model checking process of [EH00,EH01] but it surely decreases the prefix construction time. A solution vector $Y \geq 0$ and $Y \neq 0$ for both equations $Y^T \cdot M_0 = 0$ and $Y^T \cdot \mathbf{N} \leq 0$ yields a set of places which never carry tokens. Since the solution is not unique this solution might not give us all such places. Therefore we repeat the application of the dead transition rule. However, a solution of $Y^T \cdot M_0 = 0$ determines a set of places which are not initially marked, and a solution of $Y^T \cdot \mathbf{N} \leq 0$ yields a subinvariant of places for which firing a transition does not increase the number of tokens in these places.

Definition 3. *Dead Transition Rule*

Let $N = (P, T, F)$ be a net. A set $P_d \subseteq P$ of places and a set $T_d \subseteq T$ of transitions satisfy the dead transition rule if and only if
- the following equation system has a solution for Y:
    Variables: Y
    $Y^T \cdot \mathbf{N} \leq 0$
    $Y^T \cdot M_0 = 0$
    $Y \geq 0$
- $\forall p \in P : p \in P_d \Leftrightarrow Y_p > 0$
- $\forall t \in T : t \in T_d \Leftrightarrow \exists p \in P_d : t \in p^\bullet$

The reduced net $N_{red} = (P_{red}, T_{red}, F_{red})$ obtained by applying the dead transition rule on N is defined by
- $P_{red} = P \setminus P_d$
- $T_{red} = T \setminus T_d$
- $F_{red} = F \cap ((P_{red} \times T_{red}) \cup (T_{red} \times P_{red}))$
- $M_0^{red} = M_0|_{P_{red}}$

■ 3

Implicit Place Rule. A place is called *implicit* if it never restricts the firing of its output transitions. Loosely speaking, an implicit place is never the only reason that a transition of its postset cannot fire. Colom and Silva have published a method for detecting implicit places of bounded Place/Transition nets with linear programming techniques [CS90]. Our rule is based on this approach. Figure 1 shows an example for the implicit place rule. Place p can be seen as a linear combination of the places p_1 and p_2 since $M(p) = M(p_1) + M(p_2)$ holds. It is clear that p never restricts the firing of transition t and therefore place p can be removed without changing the firing sequences of the net.

Definition 4. *Implicit Place Rule*

Let $N = (P, T, F)$ be a net. A place $p \in P$ satisfies the implicit place rule if and only if the following linear program has a solution for Y and μ:
    Variables: Y, μ
    Minimize $Y^T \cdot M_0 + \mu$ such that
    $Y^T \cdot \mathbf{N} \leq l_p$
    $Y^T \cdot pre(t_i) + \mu \geq 1, \forall t_i \in p^\bullet$
    $Y^T \cdot M_0 + \mu \leq M_0(p)$
    $Y \geq 0$
    where the vector $pre(t)$ is defined as follows:
    $\forall 1 \leq j \leq |P| : pre_j(t) = 1$, if $p_j \in {}^\bullet t$, and 0 otherwise.

The reduced net $N_{red} = (P_{red}, T_{red}, F_{red})$ obtained by applying the implicit place rule on N is defined by
- $P_{red} = P \setminus \{p\}$
- $T_{red} = T$
- $F_{red} = F \cap ((P_{red} \times T_{red}) \cup (T_{red} \times P_{red}))$
- $M_0^{red} = M_0|_{P_{red}}$

■ 4

Fig. 1. Implicit Place Rule

Theorem 3. *Linear Programming Rules preserve LTL conditions*

Let $\Sigma_{\neg\varphi}$ be a synchronized net system and $\Sigma^{red}_{\neg\varphi}$ be the net system obtained by applying the dead transition rule and the implicit place rule. Conditions (i) and (ii) of Theorem 2 hold for $\Sigma_{\neg\varphi}$ if and only if they hold for $\Sigma^{red}_{\neg\varphi}$.

Proof: Applying the dead transition rule preserves the firing sequences of the net system. In [CS90] it has been proven that removing implicit places from the net system also preserves the firing sequences. Therefore the rules do not affect the conditions (i) and (ii) of Theorem 2.

4.2   Local Rules

In this section we present some local reduction rules. They differ from the linear programming rules in the sense that the linear programming rules globally investigate the whole net whereas the local rules only inspect small sub-nets. The local rules are taken from the literature, but we have improved some of them by weakening their conditions. This allows us to apply the rules more often and to obtain larger reductions.

Abstraction Rule. In this section we introduce a new abstraction rule that to the best of our knowledge has not been mentioned before in the literature. The abstraction rule is graphically described in Fig. 2. The main idea of the abstraction rule is that any occurrence sequence of the net can be reordered into an occurrence sequence where an occurrence of t has to be immediately preceded by an occurrence of an input transition of p. The reduction hides the occurrence of t by merging t with the input transitions of p. But we have to ensure that the input transitions of p are enabled in the reduced net if and only if the transition t is enabled in the original net. Therefore the presets of the input transitions of p are extended by the input places of t. Our rule differs from the abstraction rule in [DE95] in such way that in our rule the transition t may have more than one place in its preset. So our rule describes a generalization of [DE95].

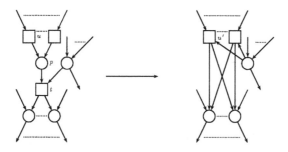

Fig. 2. Abstraction Rule

Definition 5. *Abstraction Rule*

Let $N = (P, T, F)$ be a net. A place $p \in P$ and a transition $t \in T$ satisfy the abstraction rule if and only if

- ${}^\bullet p \neq \emptyset$, $p^\bullet = \{t\}$
- $\forall u \in {}^\bullet p \colon u^\bullet = \{p\}$
- $t^\bullet \neq \emptyset$
- $M_0(p) = 0$

The reduced net $N_{red} = (P_{red}, T_{red}, F_{red})$ obtained by applying the abstraction rule on N is defined by:

- $P_{red} = P \setminus \{p\}$
- $T_{red} = T \setminus (\{t\} \cup \{u \in {}^\bullet p \mid {}^\bullet u \cap {}^\bullet t \neq \emptyset\})$
- $F_{red} = (F \cap ((P_{red} \times T_{red}) \cup (T_{red} \times P_{red}))) \cup (({}^\bullet p \cap T_{red}) \times t^\bullet) \cup (({}^\bullet t \cap P_{red}) \times ({}^\bullet p \cap T_{red}))$
- $M_0^{red} = M_0|_{P_{red}}$

∎ 5

Theorem 4. *Abstraction Rule preserves LTL conditions*

Let $\Sigma_{\neg\varphi}$ be a synchronized net system and $\Sigma_{\neg\varphi}^{red}$ be the net system obtained by applying the abstraction rule. Conditions (i) and (ii) of Theorem 2 hold for $\Sigma_{\neg\varphi}$ if and only if they hold for $\Sigma_{\neg\varphi}^{red}$.

Pre-Agglomeration Rule. The pre-agglomeration rule is graphically described in Fig. 3. The main idea of this rule is that any occurrence sequence of a net can be reordered into an occurrence sequence where the occurrence of transition t is immediately followed by an occurrence of an output transition of p. The reduction hides the occurrence of t by merging t with the output transitions of p. This structural condition implies that one can delay the firing of the transition t. The pre-agglomeration rule has been introduced by Berthelot [Ber85] and has been used also in [PPP00] but our rule differs from their rule in the following way: In [Ber85,PPP00] the restriction holds that $\forall q \in {}^\bullet t \colon q^\bullet = \{t\}$. This restriction is not neccessary in our approach.

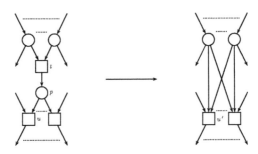

Fig. 3. Pre-Agglomeration Rule

Definition 6. *Pre-Agglomeration Rule*

Let $N = (P, T, F)$ be a net. A place $p \in P$ and a transition $t \in T$ satisfy the pre-agglomeration rule if and only if

- $^\bullet p = \{t\}$, $p^\bullet \neq \emptyset$
- $^\bullet t \neq \emptyset$, $t^\bullet = \{p\}$
- $M_0(p) = 0$

The reduced net $N_{red} = (P_{red}, T_{red}, F_{red})$ obtained by applying the pre-agglomeration rule on N is defined by:

- $P_{red} = P \setminus \{p\}$
- $T_{red} = T \setminus (\{t\} \cup \{u \in p^\bullet \mid {}^\bullet u \cap {}^\bullet t \neq \emptyset\})$
- $F_{red} = (F \cap ((P_{red} \times T_{red}) \cup (T_{red} \times P_{red}))) \cup ({}^\bullet t \times (p^\bullet \cap T_{red}))$
- $M_0^{red} = M_0|_{P_{red}}$

∎ 6

Theorem 5. *Pre-Agglomeration Rule preserves LTL conditions*

Let $\Sigma_{\neg\varphi}$ be a synchronized net system and $\Sigma_{\neg\varphi}^{red}$ be the net system obtained by applying the pre-agglomeration rule. Conditions (i) and (ii) of Theorem 2 hold for $\Sigma_{\neg\varphi}$ if and only if they hold for $\Sigma_{\neg\varphi}^{red}$.

Post-Agglomeration Rule. The post-agglomeration rule is graphically described in Fig. 4. Let us consider the sets $^\bullet p$ and $p^\bullet$ of transitions. The main idea of the post-agglomeration rule is that any occurrence sequence of the net containing a transition $h \in {}^\bullet p$ and a transition $u \in p^\bullet$ can be reordered into an occurrence sequence such that the occurrence of h is immediately followed by the occurrence of u. This structural condition implies that one can anticipate the firing of u. The post-agglomeration rule has been introduced in [Ber85] and has been used also in [PPP00].

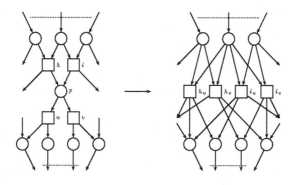

Fig. 4. Post-Agglomeration Rule

Definition 7. *Post-Agglomeration Rule*

Let $N = (P, T, F)$ be a net. A place $p \in P$ satisfies the post-agglomeration rule if and only if

- $^\bullet p \neq \emptyset$, $p^\bullet \neq \emptyset$
- $\forall t \in p^\bullet : {}^\bullet t = \{p\}$
- $M_0(p) = 0$

The reduced net $N_{red} = (P_{red}, T_{red}, F_{red})$ obtained by applying the post-agglomeration rule on N is defined by:

- $P_{red} = P \setminus \{p\}$
- $T_{red} = (T \setminus (^\bullet p \cup p^\bullet)) \cup (^\bullet p \times p^\bullet)$
- $\forall q \in P_{red}, \forall u \in T_{red} \setminus (^\bullet p \times p^\bullet):$
  $F_{red}(q, u) = F(q, u), F_{red}(u, q) = F(u, q)$
- $\forall q \in P_{red}, \forall t_1 t_2 \in (^\bullet p \times p^\bullet):$
  $F_{red}(q, t_1 t_2) = F(q, t_1), F_{red}(t_1 t_2, q) = F(t_1, q) + F(t_2, q)$
- $M_0^{red} = M_0|_{P_{red}}$

■ 7

Theorem 6. *Post-Agglomeration Rule preserves LTL conditions*

Let $\Sigma_{\neg\varphi}$ be a synchronized net system and $\Sigma_{\neg\varphi}^{red}$ be the net system obtained by applying the post-agglomeration rule. Conditions (i) and (ii) of Theorem 2 hold for $\Sigma_{\neg\varphi}$ if and only if they hold for $\Sigma_{\neg\varphi}^{red}$.

5   Implementation Issues

Special care has been taken to speed-up the net reductions. Therefore we mention some facets of our implementation in this section. First of all, we apply the dead transition rule because it investigates globally the whole net and yields sets of places and transitions which can be removed from the net. As mentioned in chapter 4.1 the dead transition rule detects places which never carry tokens. The solution is not unique, and therefore this solution might not give us all such places. We have to repeat the application of the dead transition rule to detect maybe more such places. $Y = 0$ is always a solution of the equation system. Therefore we use the objective function *Maximize $Y^T \cdot 1$*, $0 \leq Y \leq 1$, to get other possible solutions.

The following rules are part of a loop which will be repeated until no changes can be made. First, we detect redundant places and remove them from the net. A place p is called *redundant* if and only if there exists a place q such that p and q have the same initial markings and equal sets of input/output transitions. Redundant places would also be detected by applying the implicit place rule but it takes more time to solve a linear equation system than to pass through a list of places and to compare their pre- and postsets. We use efficient data structures for storing nets which allow fast comparisons of pre- and postsets of places and transitions. This universal data structure provides fast access to single nodes [Röm00]. Places, transitions and arcs are represented by nodes of doubly linked adjacent lists. Then we apply the abstraction, pre-agglomeration and post-agglomeration rules. These rules are based on simple operations conducted on our efficient data structures. They yield fast transformations compared to the linear programming techniques. We apply the implicit place rule at last for the following two reasons: The number and size of the equation systems to be solved depend on the number of places and transitions of the nets. Therefore it is reasonable to make them as small as possible before applying the implicit place rule. The creation of the equation system for each place can be done without significant loss of time because the objective function *Minimize $Y^T \cdot M_0 + \mu$* and the part $Y^T \cdot \mathbf{N}$ are equal for all places. We have to build them only once and can reuse them for all equation systems. We only have to make minor modifications on the right side, and have to add few rows for each place.

6   Experimental Results

In this section we will present our experimental results. Our aim is to show that we obtain good reduction ratios in practice and that the reduction times are very small compared with the verification times for the LTL model-checking task.

The systems we have used are as follows:

- buf(100): Buffer with capacity 100 generated by Römer.
- fifo(20): 1-bit-FIFO with depth 20 [Mar86,RCP95].
- plate(5): Production cell which handles 5 plates [LL95,HD95].
- dph(7): Variant of the dining philosophers with a butler [Cor94].
- furnace(3): Manages the temperature data for 3 furnaces [Cor94].
- key(4): Manages keyboard/screen interaction in a window manager for 4 customer tasks [Cor94].
- rw(3r1w), rw(1r2w): Address a scalable and bottleneck-free readers/writers synchronization algorithm for shared memory parallel machines [Hel93].

	$\Sigma_{\neg\varphi}$		Unf($\Sigma_{\neg\varphi}$)			$\Sigma_{\neg\varphi}^{red}$			Unf($\Sigma_{\neg\varphi}^{red}$)		
	$|S|$	$|T|$	$|B|$	$|E|$	t_{LTL}	$|S|$	$|T|$	t_{red}	$|B|$	$|E|$	t_{LTL}
buf(100)	205	107	10111	5054	1274.9	7	8	41.6	13	5	< 0.1
fifo(20)	171	132	64167	42107	47229.8	39	31	3.9	736	364	2.5
plate(5)	239	214	1803	810	35.8	117	128	2.9	998	416	8.4
dph(7)	71	127	72472	35021	11280.9	31	101	1.0	2971	1406	4.5
furnace(3)	58	105	33363	19322	1541.7	36	92	0.6	12819	7640	132.4
key(4)	169	180	138052	68585	49516.4	121	153	2.8	95416	57062	31196.4
rw(3r1w)	111	276	29717	15862	3828.5	80	202	0.9	25882	12078	1796.2
rw(1r2w)	214	1488	19874	9770	2384.8	159	981	8.2	17644	7550	1395.4
slotring(8)	85	86	24734	17195	5379.5	54	55	0.5	15283	7956	930.4
slotring(10)	105	106	67266	48145	44407.4	67	68	0.7	44562	23901	8842.9

Fig. 5. Experimental results

- slotring(n): Slotted ring protocol with n nodes [PRCB94].

The LTL properties we checked have the form:

- $\neg F(P_1 \wedge P_2)$ ($F \mathrel{\hat=} eventually$)
- $G((P_1 \wedge \neg P_2 \wedge \neg P_3) \vee (\neg P_1 \wedge P_2 \wedge \neg P_3) \vee (\neg P_1 \wedge \neg P_2 \wedge P_3))$ ($G \mathrel{\hat=} always$)

We have applied the first property on all systems except the production cell, and have checked the invariant (second property) for the production cell.

All experiments were performed on a SUN Ultra 60 with 1.5 GByte of RAM and a 295 MHz UltraSPARC-II CPU. The rules which are based on linear programming techniques use CPLEX^{TM} (version 6.5.1) as its underlying LP-solver. The LTL model-checker of [EH00] (*unfsmodels 0.9*) is implemented by Heljanko [Hel01].

Figure 5 shows the results. The columns $|S|$ ($|B|$) and $|T|$ ($|E|$) denote the numbers of places (conditions) and transitions (events) of the net (unfolding). The columns t_{LTL} and t_{red} denote the times for the LTL model-checking (*unfsmodels*) and for our reduction procedure. The times are measured in seconds.

The results show that we obtain very considerable reduction ratios for the systems buf(100), fifo(20), plate(5), dph(7), and also for the slotted ring protocols slotring(n). For these systems the unfoldings of the reduced nets are much smaller than the unfoldings of the original nets. This strongly affects the model-checking algorithm of [EH00] because it uses unfolding-based verification techniques. In fact, our practical results confirm this assumption. The LTL model-checking times for the reduced nets are much smaller than the verification times for the original systems. For instance, the model-checking time for the fifo(20) system can be decreased from 13 hours to only 3 seconds. Also the verification time for the slotring(10) protocol may be reduced from 12 hours to only 2 and a half by applying our reduction algorithm before the LTL verification. As one can see our optimized reduction algorithm takes only a few seconds, a negligible time compared to the actual verification times.

	Weakened Rules					Original Rules from literature				
	$\Sigma^{red}_{\neg\varphi}$		$\mathrm{Unf}(\Sigma^{red}_{\neg\varphi})$			$\Sigma^{red}_{\neg\varphi}$		$\mathrm{Unf}(\Sigma^{red}_{\neg\varphi})$		
	$|S|$	$|T|$	$|B|$	$|E|$	t_{LTL}	$|S|$	$|T|$	$|B|$	$|E|$	t_{LTL}
dph(7)	31	101	2971	1406	4.5	50	114	27922	13281	709.5
slotring(10)	67	68	44562	23901	8842.9	67	77	125351	62067	31674.9

Fig. 6. Experimental results with original rules

As mentioned in chapter 4.2 we have improved some of the local reduction rules compared to the rules known in the literature by weakening their conditions. This allows us to apply the rules more often and to obtain larger reductions. To confirm this assumption we have implemented the original rules just as taken from the literature and have conducted some experiments.

The results which are shown in Fig. 6 meet our expectations. Applying the original rules without weakening their conditions yields much larger unfolding sizes and LTL verification times. For instance, the dining philosophers system (dph(7)) can only be reduced to 50 places and 114 transitions (in contrast to 31 places and 101 transitions by applying our weakened rules). This leads to an unfolding which is ten times larger, and also the verification time for the LTL property grows from 5 seconds to 12 minutes. Applying the original and our weakened rules on the slotted ring protocol the reduced nets differ from 9 transitions. On the first look, this seems to be a negligible difference but indeed it has an considerable effect with respect to the unfolding size and the verification time. Preprocessing the net with our weakened rules causes that the unfolding is three times smaller, and that the verification time can be decreased from 9 hours to only 2 and a half. The unfolding of the reduced slotring(10) net obtained by applying the original rules is even larger than the unfolding of the original net. This is caused by the post-agglomeration rule which use can lead to larger unfoldings in some cases.

The reduction ratio for a system depends on the LTL property to be checked. All places corresponding to atomic propositions of the formula and all their input and output transitions remain untouched by our reduction rules. This entails that the number of places and transitions that must not be removed from the system depends on the number of places corresponding to the atomic propositions in the LTL property.

Altogether, our results have shown that our reduction algorithm yields a very efficient and suitable preprocessing technique for LTL model-checking.

7   Conclusions

We have presented techniques that allow to remove redundancies from systems modelled as 1-safe Petri nets. These techniques are of two kinds: (1) Linear programming techniques for detecting subinvariants and implicit places, and (2) local net reductions. We have shown that the conditions for some local net reductions known from the literature can be weakened if one is interested in model-checking LTL using the approach of Esparza and Heljanko [EH00,EH01]. Moreover, we have presented a number of experimental results. These results have confirmed that many redundancies may be detected and removed with our reduction techniques. Furthermore the results have shown that our reduction algorithm runs very fast, and that it has an considerable effect on the model-checking algorithm of [EH00,EH01]. Altogether, our reductions seem to be a good preprocessing technique for model-checking LTL with the approach of [EH00, EH01]. Due to space limitations we have omitted the proofs of the Theorems. The full version of this paper (including all proofs) is available in [ES01].

Acknowledgements. We would like to thank Keijo Heljanko for valuable comments and sending us an implementation of the LTL model-checker.

References

[Ber85]   G. Berthelot. Checking properties of nets using transformations. In *Advances in Petri Nets*, LNCS 222, pages 19 – 40. Springer-Verlag, 1985.

[Cor94]   J. C. Corbett. Evaluating Deadlock Detection Methods, 1994.

[CS90]   J. M. Colom and M. Silva. Improving the Linearly Based Characterization of P/T Nets. In *Advances in Petri Nets*, LNCS 483, pages 113 – 145. Springer-Verlag, 1990.

[DE95]   J. Desel and J. Esparza. Free Choice Petri Nets. Cambridge University Press, 1995.

[EH00]   J. Esparza and K. Heljanko. A new Unfolding Approach to LTL Model Checking. In *ICALP'00*, LNCS 1853, pages 475 – 486. Springer-Verlag, 2000.

[EH01]   J. Esparza and K. Heljanko. Implementing LTL Model Checking with Net Unfoldings. Accepted paper for SPIN'01, 2001.

[ES01]   J. Esparza and C. Schröter. Net Reductions for LTL Model-Checking. Full version, available at
           `ftp://131.159.22.8/pub/theory/schroete/reduct.ps.gz`, 2001.

[HD95]   M. Heiner and P. Deusen. Petri net based qualitative analysis - A case study. Technical report I-08/1995. Brandenburg Technische Universität Cottbus, 1995, 1995.

[Hel93]   H. Hellwagner. Scalable Readers/Writers Synchronization on Shared-Memory Machines. Esprit P5404 (GP MIMD), Working Paper, 1993.

[Hel01]   K. Heljanko. Unfsmodels 0.9. Available at
           `http://www.tcs.hut.fi/~kepa/experiments/spin2001/`, 2001.

[LL95]   C. Lewerentz and T. Lindner. Formal Development of Reactive Systems: Case Study Production Cell. LNCS 891. Springer-Verlag, 1995.

[Mar86]   A. J. Martin. Self-timed FIFO: An exercise in compiling programs into VLSI circuits. In *From HDL Descriptions to Guruanteed Correct Circuit Designs*, pages 133 – 153. Elsevier Science Publishers, 1986.

[McM92]   K. L. McMillan. Using Unfoldings to Avoid the State Explosion Problem in the Verification of Asynchronous Circuits. In *CAV'92*, LNCS 663, pages 164 – 174. Springer-Verlag, 1992.

[PPP00]   D. Poitrenaud and J. F. Pradat-Peyre. Pre- and Post-agglomerations for LTL Model Checking. In *ICATPN'00*, LNCS 1825, pages 387 – 408. Springer-Verlag, 2000.

[PRCB94]   E. Pastor, O. Roig, J. Cortadella, and R. M. Badia. Petri Net Analysis Using Boolean Manipulation. In *ATPN'94*, LNCS 815, pages 416 – 435. Springer-Verlag, 1994.

[RCP95]   O. Roig, J. Cortadella, and E. Pastor. Verification of Asynchronous Circuits by BDD-based Model Checking of Petri Nets. In *ATPN'95*, LNCS 935, pages 374 – 391. Springer-Verlag, 1995.

[Rei85]   W. Reisig. Petri Nets. Volume 4 of the EATCS Monographs on Theoretical Computer Science. Springer-Verlag, 1985.

[Röm00]   S. Römer. *Theorie und Praxis der Netzentfaltungen als Grundlage für die Verifikation nebenläufiger Systeme*. PhD thesis, Tech. Univ. München, 2000.

[Var96]   M. Y. Vardi. An automata theoretic approach to linear temporal logic. In *Logics for Concurrency: Structure versus Automata*, LNCS 1043, pages 238 – 265. Springer-Verlag, 1996.

[Wal98]   F. Wallner. Model checking LTL using net unfoldings. In *CAV'98*, LNCS 1427, pages 207 – 218. Springer-Verlag, 1998.

Formal Verification of the VAMP Floating Point Unit

Christoph Berg and Christian Jacobi

Saarland University
Computer Science Department
D-66123 Saarbrücken, Germany
Fax: +49/681/302-4290
{cb,cj}@cs.uni-sb.de

Abstract. We report on the formal verification of the floating point unit used in the VAMP processor. The FPU is fully IEEE compliant, and supports denormals and exceptions in hardware. The supported operations are addition, subtraction, multiplication, division, comparison, and conversions. The hardware is verified on the gate level against a formal description of the IEEE standard by means of the theorem prover PVS.

1 Introduction

Our institute at Saarland University is currently working on the formal verification of a complete microprocessor called VAMP. Part of this microprocessor is a fully IEEE compliant floating point unit (FPU). This paper describes the verification of the FPU in the theorem prover PVS [19].

The FPU we have verified is developed in the textbook on computer architecture by Müller and Paul [17]. The designs go down to the level of single gates. Along with the complete designs come paper proofs for the correctness of the circuits. These paper proofs served as guidelines for the formal proofs. We have specified and verified these designs on the gate level in PVS. Only small changes to the designs were necessary – some due to errors in [17], some to slightly simplify the proofs – with negligible impact on hardware cost and cycle time.

We have verified the designs with respect to a formalization of the IEEE standard 754 [10] (hereafter called "the standard"). We have partly used the formalization of the standard and the theory of rounding from [6,17], particularly the notion of factorings, round decomposition, and α-equivalence. Other parts of our IEEE formalization are influenced by Miner's formalization of the standard in PVS [14], particularly the definition of the rounding function.

The FPU we have verified supports both single and double precision. It can perform floating point addition, subtraction, multiplication, division, comparison, conversion between both floating point formats, and conversion between floating point numbers and integers. Denormal numbers are handled entirely in hardware. Exceptions and wrapped exponents are computed as mandated by the standard.

The verified VAMP processor will be implemented on a Xilinx FPGA.

Project Status. As mentioned above, the FPU we have verified is embedded in the VAMP microprocessor, which is currently being verified at our institute [12]. The VAMP is a

T. Margaria and T. Melham (Eds.): CHARME 2001, LNCS 2144, pp. 325–339, 2001.
© Springer-Verlag Berlin Heidelberg 2001

variant of the DLX [9,17], a 32 bit RISC processor based on the MIPS instruction set. The VAMP processor features a Tomasulo scheduler, delayed branch, a cache memory interface, precise interrupts, and the FPU described in this paper.

The verification of an in-order CPU core is complete, the verification of the Tomasulo out-of-order core will be completed in a few weeks [13]. The verification of the cache has just begun. The verification of the combinatorial floating point circuits and the FPU pipeline control is complete.

Our group has developed a translation tool to automatically convert the PVS specifications to Verilog HDL. This tool is already capable of translating the combinatorial floating point adder and rounding hardware to Verilog. We have used the Xilinx software to synthesize and simulate the Verilog code. In the end, we are going to implement the complete verified VAMP processor on a Xilinx FPGA Board.

All PVS specifications and proofs as well as the Verilog files are available at our web site.[1]

Paper Outline. In Sect. 2, we sketch the formalization of the IEEE standard. The implementation and verification of the combinatorial FPU is described in Sect. 3. We describe the errors that we have encountered during the verification at the end of Sect. 3. The pipelining of the combinatorial FPU is briefly discussed in Sect. 4. We conclude in Sect. 5.

Related Work. The verification of floating point algorithms and hardware using formal methods has received considerable attention over the last years.

As mentioned above, the formalization of the IEEE standard that we use is based on [6,14,17]. The notion of factorings, round decomposition, and α-equivalence is taken from [6,17]. We have formally verified this theory in [11]. Since the definition of the rounding function is informal in [6,17], we use a formal definition of rounding, which is based on Miners formalization of the standard [14].

Harrison has formalized the IEEE standard in the theorem prover HOL Light [8]. Both Miner and Harrison have no direct counterpart to the decomposition theorem and α-equivalence (cf. Sect. 2). They do not cover the actual implementation of operations or rounding.

Aagaard and Seger combine BDD based methods and theorem proving techniques to verify a floating point multiplier [1]. Chen and Bryant [3] use word-level SMV to verify a floating point adder. Exceptions and denormals are not handled in both verification projects.

Verkest et al. verify a binary non-restoring integer division algorithm [24]. Clarke et al. [5] and Ruess et al. [20] verify SRT division algorithms. Miner and Leathrum [15] verify a general class of subtractive division algorithms with respect to the IEEE standard.

Cornea-Hasegan describes the computation of division and square root by Newton-Raphson iteration in the Intel FPUs [4]. The verification is done using *Mathematica*.

---

[1] http://www-wjp.cs.uni-sb.de/projects/verification/

O'Leary et al. report on the verification of the gate level design of Intel's FPU using a combination of model-checking and theorem proving [18]. Denormals and exceptions are not covered in the paper. Their definition of rounding is not directly related to the IEEE standard.

Moore et al. have verified the AMD K5 division algorithm [16] with the theorem prover ACL2. Russinoff has verified the K5 square root algorithm as well as the Athlon multiplication, division, square root, and addition algorithms [21,22,23]. In all his verification projects, Russinoff proves the correctness of a register transfer level implementation against his formalization of the IEEE standard using ACL2. Russinoff does not handle exceptions and denormals in his publications; however, he states that he handles denormals in unpublished work (private communication). The definition of *sticky* in [16, 23] corresponds to our rounding of representatives.

2   IEEE Floating Point Arithmetic

To formally verify the correctness of a FPU, we need a formal notion of "correctness", i.e., a formalization of the IEEE standard which the FPU shall obey. In this section, we sketch the formalization of the IEEE standard used in our verification project. The formalization is primarily based on [6,14,17]. An extended version of this section is available as [11].

2.1   Factorings

We abstract IEEE numbers as defined in the standard to *factorings*. A factoring is a triple (s, e, f) with sign bit $s \in \{0, 1\}$, exponent $e \in \mathbb{Z}$, and significand $f \in \mathbb{R}_{\geq 0}$. Note that exponent range and significand precision are unbounded. The value of a factoring is

$$[\![s, e, f]\!] := (-1)^s \cdot 2^e \cdot f.$$

The standard introduces an exponent width N, from which constants $e_{min} := -2^{N-1} + 2$ and $e_{max} := 2^{N-1} - 1$ are derived. These constants are used to bound the exponent range.

We call a factoring (s, e, f) *normal* if $e \geq e_{min}$ and $1 \leq f < 2$. A factoring is called *denormal* if $e = e_{min}$ and $0 \leq f < 1$. We call a factoring an *IEEE factoring* if it is either normal or denormal.

Lemma 1. *Each* $x \in \mathbb{R}_{\neq 0}$, *has a unique factoring* $(\hat{s}, \hat{e}, \hat{f})$ *with* $1 \leq \hat{f} < 2$ *and* $[\![\hat{s}, \hat{e}, \hat{f}]\!] = x$. *Each* $x \in \mathbb{R}_{\neq 0}$ *has a unique IEEE factoring* (s, e, f) *with* $[\![s, e, f]\!] = x$. *Zero has two IEEE factorings* $(0, e_{min}, 0)$ *and* $(1, e_{min}, 0)$, *called* $+0$ *and* -0, *respectively.*

Let $\hat{\eta}$ and η be the functions that map (non-zero) reals x to their corresponding factorings $(\hat{s}, \hat{e}, \hat{f})$ and (s, e, f), respectively. We define $\eta(0) := (0, e_{min}, 0)$.

Lemma 2. *Let $x \in \mathbb{R}$ with $x \neq 0$ in the context of $\hat{\eta}$. It holds:*[2]

$$\hat{\eta}_e(x) = \lfloor \log_2 |x| \rfloor, \qquad \eta_e(x) = \begin{cases} \lfloor \log_2 |x| \rfloor & \text{if } x \neq 0 \text{ and } \lfloor \log_2 |x| \rfloor \geq e_{min} \\ e_{min} & \text{otherwise,} \end{cases}$$

$$\hat{\eta}_f(x) = |x| \cdot 2^{-\hat{\eta}_e(x)}, \qquad \eta_f(x) = |x| \cdot 2^{-\eta_e(x)}.$$

Let P be the significand precision as defined in the standard. A significand f is called *representable*, if f has at most $P-1$ digits behind the binary point, i.e., if $2^{P-1} \cdot f \in \mathbb{N}_0$. We call an IEEE-factoring (s, e, f) *semi-representable*, if f is representable. We call an IEEE-factoring *representable*, if it is semi-representable, and furthermore $e \leq e_{max}$ holds. We call a real x (semi-)representable, if $\eta(x)$ is (semi-) representable.

We will only investigate semi-representable factorings in the following (i.e., we allow e to exceed e_{max}). In order to "round" semi-representable factorings to representable ones, one has to decide whether to round to infinity or to the largest representable number in case $e > e_{max}$. This decision depends only on the sign and the rounding mode, and therefore is trivial.

Representable numbers exactly correspond to the representable numbers as defined in the standard. Common values for (N, P) are $(8, 24)$ and $(11, 53)$, called single and double precision, respectively. The standard defines an encoding of single and double precision IEEE factorings into bit strings of length 32 and 64, respectively. To enhance the readability of our formulas in the following, we consider factorings instead of their bit string encodings.

2.2 Rounding

We proceed with the definition of the rounding function. The IEEE standard defines four rounding modes: round to nearest, up, down, and to zero. We define a function $r_{int}(\cdot, \mathcal{M})$ for each rounding mode $\mathcal{M} \in \{near, up, down, zero\}$, which rounds reals x to integers [14]:

$$r_{int}(x, up) := \lceil x \rceil \qquad r_{int}(x, near) := \begin{cases} \lfloor x \rfloor & \text{if } x - \lfloor x \rfloor < \lceil x \rceil - x \\ \lceil x \rceil & \text{if } x - \lfloor x \rfloor > \lceil x \rceil - x \\ x & \text{if } \lfloor x \rfloor = \lceil x \rceil \\ 2\lfloor \lceil x \rceil / 2 \rfloor & \text{otherwise} \end{cases}$$

$$r_{int}(x, down) := \lfloor x \rfloor \qquad r_{int}(x, zero) := (-1)^{sign(x)} \cdot \lfloor |x| \rfloor$$

By scaling by 2^{P-1}, reals are rounded to rationals with $P - 1$ fractional bits:

$$r_{rat}(x, \mathcal{M}) := 2^{-(P-1)} \cdot r_{int}(x \cdot 2^{P-1}, \mathcal{M}).$$

Further scaling with 2^e, $e := \eta_e(x)$, yields the IEEE rounding function:

$$rd(x, \mathcal{M}) := 2^e \cdot r_{rat}(x \cdot 2^{-e}, \mathcal{M}).$$

---

[2] $\eta_e(x)$ denotes the e-component of the factoring $\eta(x) = (s, e, f)$; analogous for other components and $\hat{\eta}$.

It is not obvious that this definition conforms with the IEEE standard. We prove the theorems stating this conformance in [11].

The rounding of reals x can be decomposed into three steps: η-computation (sometimes called pre-normalization in the literature), significand rounding, and a post-normalization.

The η-computation step computes the IEEE factoring $\eta(x)$, where x is the number to be rounded. The significand round step then rounds the significand computed in the η-computation to $P - 1$ digits behind the binary point. This is formalized in the function $sigrd$:

$$sigrd(X, \mathcal{M}) := \left| r_{rat}\left((-1)^s \cdot f, \mathcal{M}\right) \right|,$$

where $X = (s, e, f)$ is an arbitrary IEEE factoring, and $\mathcal{M} \in \{near, up, down, zero\}$ is a rounding mode.

In the case that the significand round returns 0 or 2, the factoring has to be post-normalized. If the significand round returns 2, the exponent is incremented, and the significand is forced to 1; if the significand round returns 0, the sign bit is forced to 0 in order to yield $\eta(0)$. The post-normalization is defined as follows:

$$postnorm(X, \mathcal{M}) = \begin{cases} (s, e, sigrd(X, \mathcal{M})) & \text{if } 0 < sigrd(X, \mathcal{M}) < 2, \\ (s, e + 1, 1) & \text{if } sigrd(X, \mathcal{M}) = 2, \\ (0, e_{min}, 0) & \text{if } sigrd(X, \mathcal{M}) = 0. \end{cases}$$

Theorem 1 (Decomposition Theorem). *For any real x, and rounding mode $\mathcal{M} \in \{near, up, down, zero\}$, it holds*

$$postnorm\bigl(\eta(x), \mathcal{M}\bigr) = \eta\bigl(rd(x, \mathcal{M})\bigr).$$

The benefit of having the decomposition theorem is that it simplifies the design and verification of the rounder (cf. Sect. 3.3).

2.3 Exceptions

The IEEE standard defines five exceptions: invalid operation (INV), division by zero (DIVZ), overflow (OVF), underflow (UNF), and inexact result (INX). Our formalization of these exceptions is taken from [17], as the implementation in the actual hardware is. For example, the inexact result exception is formalized as

$$INX(x, \mathcal{M}) := \bigl(x \neq rd(x, \mathcal{M})\bigr),$$

where x is the infinitely precise result of a floating point operation. The definition of the other exceptions is similiar.

Lemma 3. *Let x be a real. It holds $INX(x, \mathcal{M}) \iff \eta_f(x) \neq sigrd(\eta(x), \mathcal{M})$.*

In case of underflow or overflow with the respective trap handler enabled, the standard mandates scaling the result into the representable range, and passing the scaled result to the trap handler. This is called *wrapped exponent*. The handling of wrapped exponents is as in [17].

2.4 α-Equivalence

We now define the concept of α-equivalence and α-representatives [17]. As we will see in theorem 3, this concept is a very concise way to speak about sticky-bit computations.

Let α be an integer. Two reals x and y are said to be α-*equivalent* ($x \equiv_\alpha y$), if $x = y$ or if there exists some $q \in \mathbb{Z}$ with $q \cdot 2^\alpha < x, y < (q+1) \cdot 2^\alpha$, i.e., if both x and y lie in the same singleton $\{2^\alpha\}$ or in the same open interval between two consecutive integral multiples of 2^α. Clearly, if such an q exists, it must be $q_\alpha(x) := \lfloor x \cdot 2^{-\alpha} \rfloor$. The α-representative of x is defined as

$$[x]_\alpha = \begin{cases} x & x = q_\alpha(x) \cdot 2^\alpha \\ (q_\alpha(x) + \frac{1}{2}) \cdot 2^\alpha & \text{otherwise,} \end{cases}$$

i.e., if x is an integral multiple of 2^α, the representative of x is x itself, and the midpoint of the interval between the surrounding multiples of 2^α otherwise. The following lemma summarizes some important facts:

Lemma 4. *Let* $x, y \in \mathbb{R}$, *and* $\alpha, k \in \mathbb{Z}$.

1. $\equiv_\alpha$ *is an equivalence relation,*
2. $x \equiv_\alpha [x]_\alpha$,
3. $x \equiv_\alpha y \iff [x]_\alpha = [y]_\alpha$,   *(representative equivalence)*
4. $x \equiv_\alpha y \iff -x \equiv_\alpha -y$, *and* $[-x]_\alpha = -[x]_\alpha$,   *(negative value)*
5. $x \equiv_\alpha y \iff 2^k \cdot x \equiv_{\alpha+k} 2^k \cdot y$, *and* $[2^k \cdot x]_{\alpha+k} = 2^k \cdot [x]_\alpha$,   *(scaling)*
6. $x \equiv_\alpha y \iff x + k \cdot 2^\alpha \equiv_\alpha y + k \cdot 2^\alpha$,   *(translation)*
7. $x \equiv_\alpha y \implies x \equiv_{\alpha+k} y$ *if* $k \geq 0$,   *(coarsening)*
8. $x = 0 \iff x \equiv_\alpha 0 \iff [x]_\alpha = 0$,   *(zero value)*
9. $0 < x < 2^\alpha \implies [x]_\alpha = 2^{\alpha-1}$.   *(small value)*

The following theorem describes the computation of IEEE-factorings corresponding to representatives:

Theorem 2. *Let* $x \in \mathbb{R}$, $(s, e, f) = \eta(x)$, *and* $p \in \mathbb{N}$. *The IEEE-factoring of* $[x]_{e-p}$ *can be computed by computing the representative* $[f]_{-p}$ *of* f:

$$\eta([x]_{e-p}) = (s, e, [f]_{-p}).$$

Next, we show that the representative of f can be computed by a *sticky-bit computation*. Let $f \geq 0$ be a real in binary format $f_k, \ldots, f_0, f_{-1}, \ldots, f_{-l} \in \{0,1\}^{k+l+1}$ such that $f = \sum_{i=-l}^{k} f_i \cdot 2^i$. Let $p \in \mathbb{Z}$, $k \geq -p > -l$. The $(-p)$-sticky-bit of f is the logical OR of all bits $f_{-p-1}, \ldots, f_{-l}$:

$$sticky_{-p}(f) := \bigvee_{i=-l}^{-p-1} f_i.$$

Theorem 3. *The representative* $[f]_{-p}$ *of* f *can be computed by replacing the less significant bits by the sticky bit:*

$$[f]_{-p} = \left(\sum_{i=-p}^{k} f_i \cdot 2^i \right) + 2^{-p-1} \cdot sticky_{-p}(f)$$

Theorems 2 and 3 together allow a very efficient computation of representatives (respectively their IEEE-factorings) by or-ing the less significant bits in an OR-tree, and replacing them by the sticky-bit. This technique is well known [7], but introducing the formalism with α-representatives allows for a very concise argumentation about these sticky-computations.

The valuable property of α-representatives is that rounding x and its representative $[x]_{e-P}$ yields the same result:

Theorem 4. *Let x be an arbitrary real, $(s, e, f) = \eta(x)$, and $\mathcal{M}$ be a rounding mode. It holds*

$$rd(x, \mathcal{M}) = rd([x]_{e-P}, \mathcal{M}).$$

The significand round can be performed on the representative $[f]_{-P}$ of f:

$$sigrd((s, e, f), \mathcal{M}) = sigrd((s, e, [f]_{-P}), \mathcal{M}).$$

Corollary 1. *By lemma 4.7, theorem 4 also holds for any $\alpha \leq e - P$:*

$$rd(x, \mathcal{M}) = rd([x]_\alpha, \mathcal{M}).$$

As a consequence, one can detect the OVF, UNF and INX exceptions by analysis of the representative of x:

Corollary 2. *Again, let $\alpha \leq e - P$. It holds*

$$INX(x, \mathcal{M}) \iff INX([x]_\alpha, \mathcal{M}),$$

and analogously for UNF and OVF.

Corollaries 1 and 2 facilitate the verification of the FPU in that they allow the decomposition of the FPU into computation units and a rounding unit. The computation unit performs the operation, e.g., a multiplication, and delivers a result to the rounder which is α-equivalent to the infinitely precise result of the operation (with the appropriate α). The rounder therefrom computes the correctly rounded result and the exception signals. During the verification of the computation units, the rounding algorithm and exceptions do not matter, and during the verification of the rounder, the operations do not matter. In fact, using the α-equivalence interface, the first author has verified the addition unit independently of the rounding unit, which was verified at the same time by the second author.

2.5   Correctness of the FPU

The standard requests that every floating point operation shall return a result obtained as if one first computed the exact result with infinite precision, and then rounded this exact result. We therefore call the FPU correct, if for each operation $\circ \in \{+, -, \times, \div\}$ on all representable numbers x and y, the FPU returns the IEEE bit string encoding of the factoring

$$\eta(rd(x \circ y, \mathcal{M})).$$

Furthermore, the FPU must compute the correct exception signals.

3   Verifying the VAMP FPU

Figure 1 shows the top-level schematic of the FPU. Floating point operands are passed into the floating point unpacker FPUNPACK, integer operands are passed into the fixed point unpacker FXUNPACK. Integer operands are used in conversion from integers to floating point numbers.

The floating point unpacker converts the operands from the IEEE format into a more convenient format. It translates the exponent from biased integer into two's complement format, and reveals the hidden significand bit. In case of multiplication and division, the unpacker normalizes denormal significands and adjusts the exponents accordingly. Single and double precision operands are embedded into the same internal format. Furthermore, the floating point unpacker handles special cases such as operations on $\pm\infty$, NaN, zeros etc.

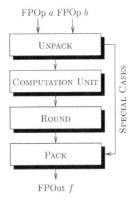

FPOp a FPOp b

UNPACK

COMPUTATION UNIT

ROUND

PACK

SPECIAL CASES

FPOut f

Fig. 1. VAMP FPU

From the floating point unpacker, non-special operands are fed into one of the computation units, namely addition and subtraction, multiplication and division, comparison, and the conversion unit.

Let $x = a \circ b$ with $\circ \in \{+, -, \times, \div\}$ be the exact result of an operation to be performed by the functional units. Instead of feeding x into the rounder, the functional units compute a factoring (s_i, e_i, f_i) which rounds to the same floating point number as x does (cf. corollary 1):

$$[\![s_i, e_i, f_i]\!] \equiv_\alpha x \text{ with } \alpha \leq \eta_e(x) - P.$$

This factoring needs not to be an IEEE-factoring. The rounder computes the floating point result and the exceptions from (s_i, e_i, f_i). After rounding, the circuit PACK transforms the rounded floating point result into the IEEE format.

Together with the conversion unit, the rounder is capable to convert between single and double precision floating point numbers, and to convert floating point numbers into the integer format.

The comparison unit outputs a flag indicating the result of the comparision performed. We have implemented and verified the comparison and conversion units, but do not discuss this further in this paper.

In the following sections, we describe the construction and verification of the computation units and the rounder. Exemplarily, we prove the correctness of the addition algorithm. The proof is a transcript of the actual PVS proof using standard mathematical notation instead of PVS notation for the sake of readability. The proof is similar to the proof given in [17] which, however, has larger gaps then the proof given here. *The significance of the proof presented here is that it is formally verified.*

We do not describe the proofs of the other components due to lack of space.

3.1   Adder

The floating point adder has IEEE-factorings (s_a, e_a, f_a) and (s_b, e_b, f_b) as inputs. The adder therefrom computes the sum (or difference in case of subtraction), (s_s, e_s, f_s), which is fed into the rounder. Let $a := [\![s_a, e_a, f_a]\!]$ and $b := [\![s_b, e_b, f_b]\!]$ be the values of the operands.

Since the unpacker embeds single and double precision inputs into the same internal format, we do not distinguish between single and double precision in the adder. The final rounding stage will round the result to the appropriate precision. We therefore fix $P = 53$ in this section.

To simplify the description, we assume that the adder shall perform an addition. If it shall perform a subtraction, b is replaced with $-b$ by inverting the sign bit s_b.

The exact sum is denoted by $S := a + b$. We assume that $a \neq 0, b \neq 0$, and $S \neq 0$, since these are special cases handled by the unpacker.

Addition Algorithm. The informal description of the addition algorithm is

1. The larger exponent of e_a and e_b is the result's exponent e_s.
2. Assume that $e_a \geq e_b$, otherwise swap a and b.
3. Align the significand f_b by shifting it $\delta := e_a - e_b$ to the right: $f_b' := 2^{-\delta} \cdot f_b$.
4. Add both significands with respect to the sign bits: $f_s' := (-1)^{s_a} \cdot f_a + (-1)^{s_b} \cdot f_b'$. The absolute value of f_s' is the result's significand, $f_s := |f_s'|$.
5. The result's sign s_s can be computed as $s_s := s_a \oplus (f_s' < 0)$.

As the alignment shift in step 3 would require a shifter of size $e_{max} - e_{min} \approx 2^{11}$, this is impractical. We therefore approximate the shifted significand by its $(-P - 1)$-representative:

$$f_b' := [2^{-\delta} \cdot f_b]_{-(P+1)}.$$

This does not change the result of the operation, since both values are rounded to the same value by the rounder:

$$rd(S, \mathcal{M}) = rd\big(2^{e_a} \cdot ((-1)^{s_a} \cdot f_a + (-1)^{s_b} \cdot f_b'), \mathcal{M}\big).$$

From corollary 1 we know that it suffices to supply a value to the rounder that is α-equivalent to the sum S, where $\alpha \leq \eta_e(x) - P$ must hold. From lemma 2 we know that $\hat{\eta}_e(x) \leq \eta_e(x)$. Therefore it suffices to prove the following theorem:

Theorem 5. *Let $\hat{e} := \hat{\eta}_e(S)$. It holds*

$$S \equiv_{\hat{e}-P} 2^{e_a} \cdot \big((-1)^{s_a} \cdot f_a + (-1)^{s_b} \cdot f_b'\big). \tag{1}$$

Proof. By definition, we have

$$S = [\![s_a, e_a, f_a]\!] + [\![s_b, e_b, f_b]\!] = (-1)^{s_a} \cdot 2^{e_a} \cdot f_a + (-1)^{s_b} \cdot 2^{e_b} \cdot f_b$$
$$= 2^{e_a} \cdot \big((-1)^{s_a} \cdot f_a + (-1)^{s_b} \cdot 2^{-\delta} \cdot f_b\big).$$

The claim (1) is therefore equivalent to

$$2^{e_a} \cdot \big((-1)^{s_a} \cdot f_a + (-1)^{s_b} \cdot 2^{-\delta} \cdot f_b\big) \equiv_{-(P-\hat{e})} 2^{e_a} \cdot \big((-1)^{s_a} \cdot f_a + (-1)^{s_b} \cdot f_b'\big). \tag{2}$$

Assume $\delta < 2$. Since f_b is a representable significand with at most $P - 1$ fractional digits, it holds

$$f_b' = [2^{-\delta} \cdot f_b]_{-(P+1)} = 2^{-\delta} \cdot f_b.$$

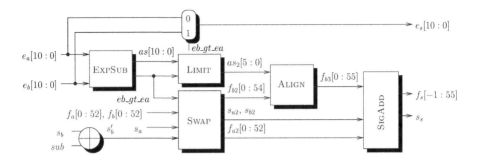

Fig. 2. The adder

This proves (2) for this case. Now let $\delta \geq 2$. By the definition of f_b', we have

$$2^{-\delta} \cdot f_b \equiv_{-(P+1)} f_b'.$$

Successively rewriting with lemma 4 (parts 4,5,6) yields

$$(-1)^{s_b} \cdot 2^{e_a - \delta} \cdot f_b \equiv_{e_a - (P+1)} (-1)^{s_b} \cdot 2^{e_a} \cdot f_b',$$
$$(-1)^{s_a} \cdot 2^{e_a} \cdot f_a + (-1)^{s_b} \cdot 2^{e_a - \delta} \cdot f_b \equiv_{e_a - (P+1)} (-1)^{s_a} \cdot 2^{e_a} \cdot f_a + (-1)^{s_b} \cdot 2^{e_a} \cdot f_b'.$$

Now assume $\hat{e} - P \geq e_a - (P+1)$. Lemma 4.7 then implies

$$(-1)^{s_a} \cdot 2^{e_a} \cdot f_a + (-1)^{s_b} \cdot 2^{e_a - \delta} \cdot f_b \equiv_{\hat{e}-P} (-1)^{s_a} \cdot 2^{e_a} \cdot f_a + (-1)^{s_b} \cdot 2^{e_a} \cdot f_b'.$$

This proves (2). It remains to show that $\hat{e} - P \geq e_a - (P+1)$. By lemma 2, this is equivalent to

$$\hat{e} = \hat{\eta}_e(S) = \left\lfloor \log_2 |(-1)^{s_a} \cdot 2^{e_a} \cdot f_a + (-1)^{s_b} \cdot 2^{e_b} \cdot f_b| \right\rfloor \geq e_a - 1. \qquad (3)$$

Since the operands are IEEE factorings, $f_b < 2$. Since $\delta \geq 2$, we have $2^{-\delta} \leq \frac{1}{4}$. Together, this yields

$$2^{-\delta} \cdot f_b < \frac{1}{2}.$$

Since $e_b \geq e_{min}$, and $\delta = e_a - e_b \geq 2$, we know that $e_a > e_{min}$, and hence $f_a \geq 1$. We now have

$$\left|(-1)^{s_a} \cdot f_a + (-1)^{s_b} \cdot 2^{-\delta} \cdot f_b\right| \geq \frac{1}{2}.$$

Multiplying with 2^{e_a} and taking logarithms yields (3). The floor brackets $\lfloor \cdot \rfloor$ may be dropped since $e_a - 1$ is integer.   □

The representative $[2^{-\delta} \cdot f_b]_{-(P+1)}$ can be computed with a shift distance limited to B (we later fix $B = 63$). This avoids the need for a very large shifter.

Lemma 5. *For $B > P$, let $\delta' = \min(\delta, B)$. It holds*

$$[2^{-\delta} \cdot f_b]_{-(P+1)} = [2^{-\delta'} \cdot f_b]_{-(P+1)}$$

Proof. The case $\delta \leq B$ is trivial. Let $\delta > B > P$. Then by lemma 4.9, it holds

$$[2^{-\delta'} \cdot f_b]_{-(P+1)} = 2^{-(P+2)} = [2^{-\delta} \cdot f_b]_{-(P+1)}.$$ □

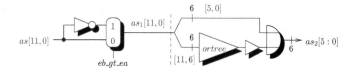

Fig. 3. Circuit LIMIT

Adder Sardware. The adder (Fig. 2) is a straightforward implementation of the described algorithm using basic components [2]. If a subtraction is to be performed, s_b is negated, yielding s'_b. Circuit EXPSUB computes the difference $as := e_a - e_b$ and the flag $eb\_gt\_ea := (e_b > e_a)$. The result's exponent e_s is selected by a multiplexer. Circuit SWAP swaps a and b in case $e_b > e_a$. The shift distance is limited in circuit LIMIT to $B := 63$. Circuit ALIGN performs the actual alignment shift. It primaly consists of a 64-bit shifter and a sticky computation, which collects the bits shifted out during the alignment. Circuit SIGADD performs the addition, i.e., steps 4 and 5 from our informal description.

The verification of the adder is straightforward: prove the correctness of the sub-circuits, and combine them using the above lemma and theorem.

Verifying the Gate Level. As an example for the detail level our proofs operate on, we present the LIMIT circuit (Fig. 3) that calculates the shift distance as_2 for circuit ALIGN. First, an approximation of the absolute value of the shift distance $as = e_a - e_b$ is computed.

$$as_1 = \begin{cases} as & \text{if } as \geq 0 \\ -as + 1 & \text{if } as < 0. \end{cases}$$

If one of the high order bits $as_1[10:6]$ is set, then $as_1 > 63$. In this case, the low order bits of as_1 are forced to one by the OR-gates. Otherwise, the shift distance as_1 is unchanged. It holds

$$as_2 = \min\{as_1, 63\}.$$

Both statements are easily verified in PVS.

In case $e_b > e_a$, the described computation introduces an error of 1 in the shift distance. This is compensated by pre-shifting the significand by one bit in circuit SWAP in this case. This detour is done to reduce the cycle time of the adder. The approximation of the absolute value can be computed with the delay of a single inverter. If one computed the exact absolute value of as, one would introduce the delay of an incrementer that would increase the length of the critical path of the adder.

3.2 Multiplier and Divider

The product of two floating point numbers can be computed by adding the exponents and multiplying the significands. The less significant bits of the significand's product are then compressed by means of a sticky bit computation. The so computed representative of the product is then passed to the rounder. Implementation and verification of this algorithm are straightforward.

In order to compute the quotient of two floating point numbers, one subtracts the exponents, and computes the quotient of the significands. The latter is the interesting part of the MULT/DIV unit.

Let f_a and f_b be the two significands. We may assume $1 \leq (f_a, f_b) < 2$, since the unpacker provides normalized significands. In our FPU, we use Newton-Raphson iteration to compute an approximation of f_b^{-1}. We start with an initial approximation x_0 with $0 < |x_0 - f_b^{-1}| < 2^{-8}$, which is loaded from a lookup table with 256 entries. In PVS, the lookup table is defined as a function mapping addresses $a \in \{0, \ldots, 255\}$ to bitvectors $b \in \{0, 1\}^8$. We have verified the content of the lookup table by automatically checking all 256 entries. The verification takes 5 minutes on a 500 MHz AMD Athlon.

The analysis of the actual Newton-Raphson iteration and the following computation of the representative $[f_a/f_b]_{-P}$ of the significand quotient is described very detailed in [17]. The translation of the proofs to PVS is therefore straightforward.

Before passing the result to the rounder, the significand is left-shifted by one bit to yield a significand in the range $[1, 4)$ as required by the rounder. The exponent is adjusted accordingly.

3.3 Rounder

Let x be the exact result of an operation, and let (s_i, e_i, f_i) be the input factoring to the rounder. This factoring is not necessarily an IEEE-factoring. Let $(s, e, f) = \eta(x)$ be the IEEE factoring of x. The rounder specification requires the input factoring to satisfy $[\![s_i, e_i, f_i]\!] \equiv_\alpha x$, where $\alpha \leq e - P$. Here, P is the precision of the operation's destination format.

The rounding unit is decomposed into the η-shifter, the representative computation, the significand round, and the post-normalization stage (Fig. 4). The η-shifter computes an IEEE-factoring (s_n, e_n, f_n) with $s_n = s, e_n = e, f_n \equiv_{-P} f$. Two cases have to be distinguished:

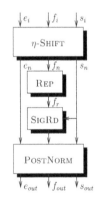

Fig. 4. Rounder

1. In case of an addition/subtraction, the exponent e_i satisfies $e_i \geq e_{min}$ by construction (Sect. 3.1). However, the significand f_i lies in the interval $(0, 4)$, and can – due to cancellation – be less than 1 even if $e_i > e_{min}$. In the latter case, the significand has to be shifted left.

2. In case of multiplication and division, the input significand f_i lies in the interval $[1, 4)$, since the inputs to the multiplier/divider were normalized by the unpacker. The exponent e_i, however, does not necessarily satisfy $e_i \geq e_{min}$.[3] In the case where $e_i < e_{min}$, the significand f_i has to be shifted right by $e_{min} - e_i$ digits. Since this shift could be very far, the shift distance is limited similarly to the adder alignment shift explained in Sect. 3.1.

The η-shifter outputs f_n with 128 binary digits. The circuit REP computes the representative $f_r := [f_n]_{-P}$. This is done using an OR-tree, as suggested by theorem 3. We then have $(s_n, e_n, f_r) = \eta([x]_{e-P})$ by theorem 2.

The next circuits SIGRD and POSTNORM exactly correspond to the functions $sigrd$ and $postnorm$ from Sect. 2.2. The significand round on f_r is performed by investigation

---

[3] For example, the multiplication $2^{e_{min}} \times 2^{e_{min}}$ generates inputs $(s_i, e_i, f_i) = (0, e_{min} + e_{min}, 1)$ to the rounder. Note that $e_{min} < 0$, and therefore $2 \cdot e_{min} < e_{min}$.

of the 3 least significant bits of f_r, and either chopping or incrementing the higher order bits [17]. If this effectively changes the significand, then the inexact result exception INX is signalled according to lemma 3.

The post-normalization increments the exponent and forces the significand to 1 if normalization is necessary.

Theorems 1 and 4 together imply the correctness of the rounder:

$$(s_{out}, e_{out}, f_{out}) = \eta\big(rd(x, \mathcal{M})\big).$$

After rounding, the circuit PACK outputs the IEEE bit string encoding of this factoring. In case of an untrapped overflow, however, the circuit PACK outputs either the format's maximal value, or infinity, depending on the sign and the rounding mode.

The correctness of the unpacker, the computation units, the rounder, and the packer together imply the correctness of the whole FPU.

3.4 Errors Encountered

We briefly describe some of the errors in [17] that we have encountered during the verification of the FPU in PVS:

The specification of the rounder interface (pg. 392) is wrong. There it is required that an overflow does not occur if a denormal significand f_i is fed into the rounder, i.e., $f_i < 1 \Rightarrow \neg OVF(x, \mathcal{M})$. This is necessary to detect overflows correctly (pg. 397). However, the requirement is not strong enough: it must hold that the input exponent e_i is less than e_{max} in case of a denormal input significand, i.e., $f_i < 1 \Rightarrow e_i \leq e_{max}$. Otherwise, the proof on page 397 fails.

The divider does not obey the rounder specification (neither the old nor the new one). A division of $1 \cdot 2^{e_{max}}$ by $(2 - 2^{-P+1}) \cdot 2^{e_{min}}$ overflows, but the input significand into the rounder $f_i \approx 1/2$ is denormal. This bug can be fixed by left-shifting f_i by 1 and appropriately adjusting the exponent e_i in case of divisions (cf. Sect. 3.2).

On page 400, a carry-in is fed into a compound adder, although compound adders do not feature a carry-in. A similar error was found in the exponent addition circuit in the multiplier (pg. 383).

In circuit SIGRD (pg. 406), chopping the significand in single precision mode leaves non-zero digits after the least significand bit. The claims in Sect. 8.4.5 are therefore wrong. This can be fixed by tying the bits after the least significand bit to zero.

In the significand round, the circuit for the decision whether to chop or to increment the significand is wrong (pg. 407). The XOR has to be replaced by an XNOR gate.

In the adder, the computation of the sign bit is wrong (pg. 371).

The proofs in [17] partly have large gaps. These gaps had to be filled during the verification in PVS. Most proof gaps could be filled without revealing errors in [17], but some proof gaps hid errors, e.g., the errors listed above. Having formally verified the proofs in PVS ultimately gives us the certainty that the design of the FPU is correct – under the assumption that PVS is sound.

4   FPU Control

So far we have verified combinatorial circuits. In order to implement the FPU in hardware with reasonable cycle time, one has to insert pipelining registers. Since multipliers are very expensive, one cannot fully pipeline the iterative Newton-Raphson algorithm. A loop has to be incorporated into the pipeline structure to reuse the multiplier in each iteration. This saves hardware costs, but considerably complicates control and the correctness proof.

In [17], the FPU is integrated into an in-order variant of the DLX-processor. In our verification project, the FPU will be integrated into a Tomasulo based out-of-order DLX-variant. It is therefore necessary to design a new control automaton for the FPU in order to exploit the benefits of the out-of-order scheduler.

After pipelining, the FPU has a variable latency, and operations are finished out-of-order. The latency of the FPU is 1 cycle for comparison and for operations involving special operands. It is 5 cycles for addition, subtraction, and multiplication. The division unit has latency 16 and 20 cycles in single and double precision, respectively. Two divisions can be performed interleaved without increased latency.

We have verified the new FPU control using a combination of PVS's modelchecking and theorem proving capabilities We omit the control implementation details here, since they are not specific to FPUs.

5   Summary and Future Work

We have formally verified a fully IEEE compliant floating point unit. The supported operations are addition, subtraction, multiplication, division, comparison, and conversions. The FPU handles denormals and exceptions as required by the IEEE standard. The hardware has been verified on the gate level with respect to a formal description of the IEEE standard using the theorem prover PVS.

The proofs in PVS used paper proofs from [17] as guidelines. However, some of the proofs in [17] were erroneous, and most proofs had gaps needed to be filled in PVS. Those gaps hid errors in the design in [17]. Having formally verified the proofs (and filled the proof gaps) in PVS gives us the certainty that now the hardware is correct with respect to its specification.

To the best of our knowledge, this is the first time that a floating point unit that supports addition/subtraction, multiplication/division, comparison, conversions, denormals, and exceptions in hardware has been formally verified on the gate level, and the designs and proofs scripts are made publicly available.

The amount of work needed to develop the PVS hardware description and proofs was roughly a year for each of the authors. Since theorem proving strongly profits from experience, we think we would succeed in at most half the time now on a comparable project.

We are currently working on the integration of the FPU into the VAMP processor and the translation of the VAMP to Verilog. The VAMP processor including the FPU will then be implemented on a Xilinx FPGA.

Acknowledgements. The authors would like to thank Sven Beyer, Daniel Kröning, Dirk Leinenbach, Wolfgang Paul, and Jochen Preiß for valuable discussions.

References

1. M. D. Aagaard and C.-J. H. Seger. The formal verification of a pipelined double-precision IEEE floating-point multiplier. In *ICCAD*, pages 7–10. IEEE, Nov. 1995.
2. C. Berg, C. Jacobi, and D. Kroening. Formal verification of a basic circuits library. In *IASTED International Conference on Applied Informatics*. ACTA Press, 2001.
3. Y.-A. Chen and R. E. Bryant. Verification of floating point adders. In *CAV'98*, volume 1427 of *LNCS*, 1998.
4. M. Cornea-Hasegan. Proving the IEEE correctness of iterative floating-point square root, divide, and remainder algorithms. *Intel Technology Journal*, Q2, 1998.
5. E. M. Clarke, S. M. German, and X. Zhao. Verifying the SRT division algorithm using theorem proving techniques. In *CAV'96*, volume 1102 of *LNCS*, 1996.
6. G. Even and W. Paul. On the design of IEEE compliant floating point units. In *Proceedings of the 13th Symposium on Computer Arithmetic*. IEEE Computer Society Press, 1997.
7. D. Goldberg. *Computer Arithmetic*. In [9], 1996.
8. J. Harrison. A machine checked theory of floating point arithmetic. In *TPHOL '99*, volume 1690 of *LNCS*. Springer, 1999.
9. J. L. Hennessy and D. A. Patterson. *Computer Architecture: A Quantitative Approach*. Morgan Kaufmann, San Mateo, CA, second edition, 1996.
10. Institute of Electrical and Electronics Engineers. *ANSI/IEEE standard 754–1985, IEEE Standard for Binary Floating-Point Arithmetic*, 1985.
11. C. Jacobi. A formally verified theory of IEEE rounding. Unpublished, available at www-wjp.cs.uni-sb.de/~cj/ieee-lib.ps, 2001.
12. C. Jacobi and D. Kroening. Proving the correctness of a complete microprocessor. In *GI Jahrestagung 2000*. Springer, 2000.
13. D. Kroening. *Formal Verification of Pipelined Microprocessors*. PhD thesis, Saarland University, Computer Science Department, 2001.
14. P. S. Miner. Defining the IEEE-854 floating-point standard in PVS. Technical Report TM-110167, NASA Langley Research Center, 1995.
15. P. S. Miner and J. F. Leathrum. Verification of IEEE compliant subtractive division algorithms. In *FMCAD-96*, volume 1166 of *LNCS*, pages 64–, 1996.
16. J Moore, T. Lynch, and M. Kaufmann. A mechanically checked proof of the AMD5K86 floating point division program. *IEEE Transactions on Computers*, 47(9):913–926, 1998.
17. S. M. Mueller and W. J. Paul. *Computer Architecture. Complexity and Correctness*. Springer, 2000.
18. J. O'Leary, X. Zhao, R. Gerth, and C.-J. H. Seger. IA-64 floating point operations and the IEEE standard for binary floating-point arithmetic. *Intel Technology Journal*, Q4, 1999.
19. S. Owre, N. Shankar, and J. M. Rushby. PVS: A prototype verification system. In *CADE 11*, volume 607 of *LNAI*, pages 748–752. Springer, 1992.
20. H. Ruess, N. Shankar, and M. K. Srivas. Modular verification of SRT division. In *CAV'96*, volume 1102 of *LNCS*, 1996.
21. D. M. Russinoff. A mechanically checked proof of IEEE compliance of the floating point multiplication, division and square root algorithms of the AMD-K7 processor. *LMS Journal of Computation and Mathematics*, 1:148–200, 1998.
22. D. M. Russinoff. A mechanically checked proof of correctness of the AMD K5 floating point square root microcode. *Formal Methods in System Design*, 14(1):75–125, Jan. 1999.
23. D. M. Russinoff. A case study in formal verification of register-transfer logic with ACL2: The floating point adder of the AMD Athlon processor. In *FMCAD-00*, volume 1954 of *LNCS*. Springer, 2000.
24. D. Verkest, L. Claesen, and H. De Man. A proof on the nonrestoring division algorithm and its implementation on an ALU. *Formal Methods in System Design*, 4, 1994.

A Specification Methodology by a Collection of Compact Properties as Applied to the Intel® Itanium™ Processor Bus Protocol

Kanna Shimizu[1], David L. Dill[1], and Ching-Tsun Chou[2]

[1] Computer Systems Laboratory, Stanford University, Stanford, CA 94305, USA
kannas@stanford.edu, dill@cs.stanford.edu
[2] Intel Corporation, 3600 Juliette Lane, SC12-401, Santa Clara, CA 95052, USA
ching-tsun.chou@intel.com

Abstract. In practice, formal specifications are often considered too costly for the benefits they promise. Specifically, interface specifications such as standard bus protocol descriptions are still documented informally, and although many admit formal versions would be useful, they are dissuaded by the time and effort needed for development.
We champion a formal specification methodology that attacks this cost-value problem from two angles. First, the framework allows formal specifications to be feasible for signal-level bus protocols with minimal effort, lowering costs. And second, a specification written in this style has many different uses, other than as a precise specification document, resulting in increased value over cost. This methodology allows the specification to be easily transformed into an executable checker or an simulation environment, for example.
In an earlier paper, we demonstrated the methodology on a widely-used bus protocol. Now, we show that the generalized methodology can be applied to more advanced bus protocols, in particular, the Intel® Itanium™ Processor bus protocol. In addition, the paper outlines how writing and checking such a specification revealed interesting issues, such as deadlock and missed data phases, during the development of the protocol.

1 Introduction

As digital circuits become larger, designs are broken up into more and more pieces, increasing the importance of design integration. The popularity of IP (Intellectual Property) cores, and the increased awareness that their interfaces must be clearly defined, is an affirmation of this trend. This development necessitates that functional interface specifications be correct and precise because they serve a pivotal role when integrating designs. However, specifications are still written in natural languages and not formal languages, forfeiting an opportunity for analysis, automated checks, and preciseness. In many cases, specifications widely in use are wordy, ambiguous, and contradictory; all problems that can be resolved by formal specification development. A subtle point missed in

T. Margaria and T. Melham (Eds.): CHARME 2001, LNCS 2144, pp. 340–354, 2001.

practice is that an informal specification may have inconsistencies that a human reader will not notice, or may be missing rules that a human reader may infer automatically, but, because it is inconsistent and incomplete, any correctness reasoning is impossible. A practical consequence of this is that a good protocol compliance checker can not be created from such a document.

Despite these arguments, in practice, formal specifications are rarely used. The reason seems to lie with the perceived high cost of formal development: mainly, the lengthy development time and the investment needed for formal verification training. For many, the value of a correct specification does not justify these costs. This paper is one step in a broader effort to reduce the cost by making the specification process more reasonable, and to increase the value of formal documents by developing direct applications for them. Methodology, as opposed to tool or language development, is the key to achieving this goal.

Until recently, formal specification research has focused mostly on developing tools or languages. Very little work has been done on how to develop complete specifications. Tools assume the existence of them, or are just used with an ad hoc list of properties. Our goal is to develop a methodology that produces a self-contained, complete, behavioral specification, while adhering to the cost-value goal. Currently, we are focusing on signal-level bus protocol descriptions, since they are both important and challenging to specify.

There has been a few other bus protocol specification projects. In 1999, Chauhan, Clarke, Lu and Wang [CCLW99] specified PCI (Peripheral Component Interconnect) protocol [SIG95] using CTL. Our specification has advantages that a CTL one does not have. In 1998, Mokkedem, Hosabettu, and Gopalakrishnan formalized higher-level properties of PCI involving communication over bus bridges [MHG98]. Their specification is almost unrelated to the one here, which focuses on the low-level behavior of individual signals. In fact, the methodology is most closely related to a 1998 paper by Kaufmann, Martin, and Pixley [KMP98], which proposed using logical constraints for environment modeling. However, they do not give guidance on how the constraints should be written, which is the focus of this project.

In [SDH00], we present a specification style that was developed using the core subset of the PCI protocol as an example. We also describe two debugging methods based on model checking, that were found to be effective with the style. As a second paper in this series, the primary contributions of this paper are the following. First, the work shows that the generalized methodology can be applied to more advanced bus protocols, in particular, the Intel® Itanium™ Processor bus protocol[Cor]. Second, it demonstrates how writing and checking such a specification reveals interesting issues during the development of the protocol. This point is illustrated with a few examples from the specification effort.

It has been found that the expressiveness of the style is not a problem despite the protocol's pipeline feature, that the increased complexity did not affect model checker performance, and the method's debugging strategy is still effective. The technique uncovered several issues in a development version of the protocol which allow deadlock of the bus and missed data phases. The protocol engineers have recognized the significance of these discoveries and subsequently made changes to the protocol to disallow such scenarios. Furthermore, with this method, the total time to write and check the description was 2 man-months.

The Specification Style

The specification style addresses the cost problem in two ways. First, formal verification expertise is not required to write the specification. In contrast, many specification frameworks require knowledge of LTL (linear time temporal logic) or CTL (computation tree logic)[CE81] as evidenced by numerous projects such as CMU's PCI specification [CCLW99] and IBM Haifa's FoCs software [ABG+00]. Because it does not require the complex constructs of these languages, the formal specification can be and has been written in a hardware description language such as Verilog, a language familiar to many engineers. This feature partly counters the training cost argument against formal specifications. More generally, the style is language-independent (indeed, the specification can still be written in LTL or CTL if desired). The methodology can be applied to any of the existing tool frameworks and languages such as SMV[McM], FoCs[ABG+00] or LUSTRE [HCRP91].

Second, because the style is based on writing many small protocol rules and purposely avoids writing large state machines, the specifications have been found to be easier to write and maintain. Many descriptions comprise of one large state machine with actions specified for each state. Designing such a state machine correctly is a complex, error-prone task. This method instead relies on many small state machines, but the bulk of the specification is done using compact rules.

To maximize its value, the style allows the specification to have multiple functions. For a formal verification framework using a model checker, *with no extra work,* the specification can be used to constrain the inputs to the design under test, and at the same time, it can be used as the list of properties to check for. For example, this was done by Govindaraju and Dill; using the PCI specification written in this style, they were immediately able to formally verify a PCI driver implementation [GD00]. The specification constrained the state space, and also gave them properties to check for. This assume-guarantee approach is not new, but the style easily allows it while an *ad hoc* specification probably would not. For designs that are too complex to be formally verified and can only be simulated, the specification can be directly used as a conventional checker monitor during simulation runs. This is possible because the specification is executable and can be written in a language such as Verilog. For example, at Intel®, the specification that was developed for the Itanium™ processor bus has been translated into a C++ simulation checker.

Specification Writing. At the foundation of the specification style is the concept of monitors. Monitors observe the output signals of interacting agents, flag illegal behavior, and assign blame to the erring agent. The monitor concept allows executable specifications to be written for non-deterministic behavior, and allows specifications to be written as a conjunction of simple properties (in these respects, monitors have many of the advantages of temporal logic). The concept is not new; numerous specification frameworks are based on monitors.

Fig. 1. A Monitor

Within this monitor framework, the style is based on using multiple constraints to collectively define the correctness of an agent's signalling behavior. The constraints are simple propositional formulas with a time construct. The basic form is,

$$prev(signal_0 \wedge ... \wedge signal_1 \vee (statemachine = value)) \rightarrow signal_i \vee ... \wedge signal_n$$

where "$\rightarrow$" is the logical symbol for "implies". The antecedent is an expression containing signal values and state machine values, and the consequent is an expression of signal values. The *prev* construct allows the state of the signal a cycle before to be expressed, and it can be nested although in the examples attempted, only up to a doubly nested *prev* was needed. For readability and to facilitate debugging and implementation, the constraints are written as an implication with the past condition as the antecedent and the current condition as the consequent. For example, the protocol constraint,

$$prev(trdy \wedge stop) \rightarrow stop$$

means "if the signals *trdy* and *stop* were true in the previous cycle, then *stop* must be true in the current cycle" where a "true" signal is asserted and a "false" signal is deasserted.

Due to various desired characteristics on the specification, such as guaranteeing the existence of an implementation, there is one important restriction placed on the constraints. *Each constraint can only constrain the behavior of one agent.* This leads to the ability to blame just one agent when a specification constraint is violated during agent interaction. If signals o_a and q_a are outputs of the same agent and r_b is an output of another agent, the first constraint obeys this restriction, while the second doesn't.

$$correct : previous(r_b) \rightarrow o_a \wedge q_a$$

$$incorrect : previous(r_b) \rightarrow o_a \wedge r_b$$

The PCI protocol was specified with this restriction, and the targeted subset of the Itanium[TM] processor bus protocol was also successfully specified with this restriction.

The *prev* construct is not enough to retain state information. State machines are needed for this. But instead of relying on large state machines, the style relies on many, small, standard state machines which each track one thread of information. An example is a 2-state, set-and-reset machine which becomes set when a certain event happens and stays set until it is no longer needed. It is used to record certain information such as whether the transaction is a read or a write. Another example is a counter which counts the number of cycles from a certain event, or counts the number of occurrences of a special event. Only these two types of state machines were needed. A key point is that a protocol can be written with a few standard state machines, even if it is as complex as a pipelined protocol.

In addition to the benefits already described, there are numerous advantages to writing a specification as a "collection of small constraints and state machines." Since most existing natural-language specifications, such as the PCI specification, are already written as a list of rules, the translation to this type of specification requires less manipulation and results in fewer opportunities for error. Furthermore, not only is such a specification easier to write, it is easier to read and understand than a complicated, large state machine description. And finally, on a theoretical note, this specification style together with a simple-to-check property guarantees the existence of a implementation (and more precisely, a stronger property called *receptiveness*). Although many of these qualities are hard to quantify, is it noted that our PCI specification has been used by others working on related projects such as Clarke *et al.* [CGY+00] and Aloul *et al.* [AS00].

Specification Checking. Two debugging methods were found to be effective for this style when developing the PCI specification. Their main purpose is to check if the specification is overly restrictive, or under-restrictive, and whether it agrees with the protocol designer's intent.

The *dead state check* finds contradictions in the specification. It searches for a state where all the specification constraints have been true so far, but there is no next state where all of the constraints are true. This indicates that the specification places contradictory requirements on the system at some point in its execution. To determine this, it is sufficient to model check the specification with the following property,

$$AG(all\_correct \rightarrow EX(all\_correct))$$

where *all_correct* true indicates that none of the constraints have been violated so far. This property guarantees that the system can execute infinitely obeying the protocol no matter which path it chooses, as long as at each step, an *all_correct* next state is chosen.

The *characteristic check* checks whether the specification guarantees certain properties. These characteristics are logical statements about agent events and are expressed in CTL.[1] They are checked against the specification using a CTL

---

[1] It must be emphasized that the *specification* constraints are simple, bounded, linear time properties and these *checking* properties are more complex, unbounded, CTL formulas.

model checker and by the counterexamples provided by the model checker, the check finds bad scenarios allowed by the specification. This technique requires knowledge of the protocol because the characteristics cannot be deduced automatically from the specification.

2   Specification

A core subset of the Itanium^{TM} Processor bus protocol was specified; the request phase, the snoop phase, the response phase, the data phase, and the deferred phase are all covered, but, for example, the arbitration phase is not. Higher level properties of deferred transactions, such as the assurance of completion, are not specified or checked because they are probably better treated in a different specification.

In this section, it is illustrated how the specification style specifies advanced features that were not part of the simpler PCI protocol. There is a description of pipeline specification, the treatment of the protocol's time-unbounded rules, and an explanation of a reaction timing issue.

Definitions Here are some terms that are used throughout this paper.

constraints	The small, propositional formulas in the *formal* protocol specification describing agent behavior
rules	Specification properties in the *informal*, English specification provided by the protocol designer. There is no one-to-one relationship between constraints and rules. In most cases, one rule corresponds to multiple constraints.
history variables	Variables that retain information from the past. This can be as simple as the value of a single bus signal from one cycle ago, or as complicated as a counter.
agents	The entities communicating with one another using the interface, or in this case, the bus.

2.1   Pipelining

First, an un-pipelined version of an example constraint will be described. The protocol has six phases. Transactions must go through the arbitration phase, request phase, error phase, snoop phase, response phase, and data phase. The type of transaction, whether it is a write or a read, is determined during the request phase. An important signaling event, the assertion of *trdy*, happens during the response phase. Thus, the example rule, "*trdy* must be asserted for a write transaction," requires the transaction type (write) to be stored from the request phase until the response phase. Also, the end of the response phase is defined by the assertion of *rs* and so the rule requires that *trdy* be asserted before *rs*.

Consequently, to specify the constraint, two auxiliary variables are sufficient. First, there is a history variable *write* which becomes true during the request phase if the transaction is a write, and stays so until the transaction is completed.

The second history variable, *trdy_happened*, becomes true when *trdy* is asserted and stays so for the duration of the transaction.

Using these variables, the constraint becomes, "if a transaction is a write, then if *trdy* assertion hasn't happened yet, *rs* cannot be asserted (i.e. the response phase can't complete)."

$$prev(write) \rightarrow (prev(\neg trdy\_happened) \rightarrow \neg rs)$$

For a pipelined version, while transaction i is in the response phase for example, transaction $i - 1$ can be processed concurrently, say in its request phase. Thus, there needs to be multiple *write* variables and *trdy_happened* variables for each outstanding transaction. Assume that there is a mechanism to tag each transaction with an ID number. This same number can be used as a subscript on the history variables to create separate variables for each transaction. Thus, the history variables become $write_i$ instead of *write*, and $trdy\_happened_i$ instead of *trdy_happened*. Also, the constraints are activated only when a particular transaction reaches a particular pipeline phase. Consequently, each constraint developed in the un-pipelined version is (indirectly) indexed by a transaction ID in the pipelined version. The example constraint now becomes,

$$prev(write_i \wedge (response\_phase = i)) \rightarrow (prev(\neg trdy\_happened_i) \rightarrow \neg rs)$$

which is, "if transaction i is a write and is undergoing the response phase, then if *trdy* assertion hasn't happened yet, *rs* cannot be asserted." Thus, to create a pipelined version from an un-pipelined specification, the constraints and history variables are replicated and indexed by transaction IDs.

The transaction ID assigning process is implemented by counters. In this scheme, the transaction ID corresponds to the order in which the transaction started. The first transaction that undergoes the arbitration phase (which is the first phase in the pipeline) is assigned the ID of 0. The ith transaction that undergoes the arbitration phase is assigned the ID of $i - 1$. Since there can only be eight outstanding transactions, the IDs are assigned modulo 8. This ID scheme works because most phases for a transaction happen in-order;[2] the ith transaction to undergo the request phase is the the ith transaction to go through the response phase.

For each phase, there is a signal expression which indicates that the phase has completed. For a particular phase, if the number of occurrences of these "complete" signaling is known, the number of transactions that have undergone this phase so far is also known. In this way, the specification can keep track of which transaction is being processed in each phase. For example, the request phase's completion event is the assertion of *ads*. The request phase counter increments at every occurrence of this, and the counter value indicates the ID of the transaction currently in the request phase. As this transaction moves onto subsequent phases, each phase counter will have this same value.

---

[2] Data phases can be deferred and completed later so they do not necessarily happen in order.

Thus, the general form for constraints is, "if transaction i is a write (or read) and it is currently undergoing the response (or request or snoop) phase, then p must hold."

$$prev(transaction\_type_i \land (some\_phase = i)) \to p$$

(where for example, $transaction\_type_i$ is a "write" and $some\_phase$ is a "response phase" counter)

Thus, using these counters and appropriate history variables, a pipelined protocol can be easily specified with small constraints. The one drawback of this methodology is the linear increase in the number of state variables, which may be a problem when model checking. However, this is only a problem with verification and not the specification. The specification scales well; the constraints only have to be duplicated with the counter values and variable subscripts changed. In fact, the methodology specifies a certain intricate Itanium^{TM} bus feature effectively and simply. Although there are only five main pipeline stages, the protocol supports eight outstanding transactions at any time. The extra transactions are buffered at each stage. With the monitor-style constraints, *the buffers do not have to be explicitly modeled* to specify the agent-bus behavior. There are simply eight copies of the constraints for the eight outstanding transactions.

2.2   Time-Unbounded Rules

Compared to the tightly timed PCI protocol, the Itanium^{TM} Processor bus has a less constrained timing relationship among the different signal events. With PCI, most rules fall into the category, "exactly n cycles after event a happens, event b *must* happen." The PCI protocol is a *time-bounded* protocol where most events are guaranteed to happen within a certain time span.

In comparison, there are no such guarantees with the Itanium^{TM} Processor bus protocol. Most of the rules follow the form, "any time after n cycles have elapsed since event a, event b *may* happen." The environment must expect that the event can happen at n, or $n + 1$, or $n + 2$, and so on, and be designed accordingly. An example from the protocol is "*trdy* may be deasserted a minimum of 3 cycles after the deassertion of the previous *trdy*".

Furthermore, there are no rules stating that an event must eventually happen (the so-called *liveness* property). In essence, the protocol is a *time-unbounded* protocol. Hence, the protocol allows the bus agents to have more freedom in ordering events, and optimizing bus performance. However, this extra degree of freedom leads to more corner cases in the specification that need to be checked.

Constraint Style. In the prior section, it is explained that the constraints for this protocol are written in the form,

$$prev(transaction\_type_i \land (some\_phase = i)) \to p$$

The form of p will now be described. The most natural form, considering the "time-unbounded" characteristics of the protocol, is

$$p : \neg prev(q) \to \neg r$$

"If q is true in the previous cycle, then r *may* be true in the current cycle."
The expression for q is a *trigger* condition which enables a certain exchange
or a change of state. A trigger condition which signals that an agent is ready
to receive data is one example. Another example is a trigger condition which
indicates the completion of an event so that a bus signal can be deasserted.

One consequence of this time-unbounded feature, is that the dead state check,
in its original form, does not catch contradictions. This is due to the fact that
*because most actions are not required to happen at any given time (the time-
unbounded characteristic), the bus can always choose to loop in the current
state and "do nothing."* If the check searches for a legal (all the specification
constraints are true) current state which has no legal next state, with a time-
unbounded protocol, the current state is always a legal next state and so the
check is vacuous. Therefore, the dead state definition needs to be expanded so
that the test check for the following desired property,

$$AG(all\_correct \rightarrow EX(all\_correct \wedge \neg same))$$

where *same* is true if all the state variables, except the timer-like variables which
increment at each clock, have the same value as in the previous state. Thus, this
check ensures that, at every legal state, there is at least one possible legal next
state where some change happens and the bus is not forced to stay in the current
state. And so, a check that was effective for PCI is modified for the Itanium[TM]
protocol so that a wider class of anomalies are detected.

2.3   2-Clocks or 1-Clock Reaction Time

Unlike the PCI protocol, the Itanium[TM] process bus is a latched protocol where
there is a 1 cycle delay from when the bus agent asserts (or deasserts) a signal to
when the action appears on the bus. Thus, when observing events on the bus, a
reaction to a trigger event happens (at earliest) in two cycles instead of one. On
cycle n, a trigger condition becomes true on the bus; on cycle $n + 1$, the agent
asserts a signal in response; and on cycle $n+2$, the assertion appears on the bus.
And so most time-bounded constraints are in the form, $prev(prev(input)) \rightarrow$
output where *input* and *output* are bus signal expressions.

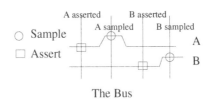

The Bus

Fig. 2. The Latched Bus Protocol - 2 Cycle Response Time

However, the response to a trigger *from the same agent* may happen in one
cycle. There is no one-cycle delay through the latched bus in this case. For

example, there is a rule, "*ids* cannot be asserted in cycle N if *trdy* is sampled asserted on clock N." As a constraint, this becomes $prev(trdy) \rightarrow \neg ids$. This requirement is possible only because both *trdy* and *ids* are controlled by the same agent at all times. The agent does not need to wait for the trigger $(trdy)$ to appear on the bus to react $(\neg ids)$ to it.

The difficulty arises when the trigger condition of a rule is a *mix* of external and internal signals. An example is "*input* $\wedge$ *output*$_0$ true requires *output*$_1$ to become true." If an external signal causes the trigger expression to become true (*output*$_0$ was already true and *input* just became true), then a two cycle reaction time is needed. However, if the agent's own signal causes the trigger condition to become true (*input* was already true and *output*$_0$ just became true), it can react in one cycle. Thus, the rule needs to be separated into two constraints depending on the situation. However, the English specification fails to distinguish between the two cases, and states a blanket requirement allowing a one cycle reaction timing at all times. This is problematic because if a particular agent asserts a reaction earlier (one cycle) than expected by other agents (two cycles), it may lead to an undefined state. In fact, the modified dead state check detected a scenario where this misstated rule led to a contradiction.

3 Debugging

Several issues were found with the development version of the Itanium™ processor bus protocol using the methodology's debugging procedures. Some are omission of rules that are arguably implicit in the official specification, but violate the completeness concept. Others are serious enough to cause data phases to be missed unnoticed, or cause a deadlock. These were resolved by revising the Intel® protocol specification.

3.1 Found by Dead State Check

The dead state check mainly found cases of missing rules. The informal specification tends to state the sufficiency conditions of an action, while leaving necessary conditions implicit. For example, a sufficient condition for the assertion of *trdy*, "if the transaction is a write, then a *trdy* assertion must happen," is stated in the specification. Logically, it specifies $write\_transaction \rightarrow trdy\_happens$. However, a necessary condition that *trdy* can only be asserted at certain times, is missing. The specification should state that "*trdy* can be asserted *only if* the transaction is a write or has a snoop-initiated data transfer." Else, the system will reach an undefined state because the agents do not expect *trdy* to be asserted during a read, for example. By adding such a rule, the specification is made more complete so that a simulation checker can catch erroneous behavior at the earliest time. Overall, there were five cases of such omissions where the specification does not state that a particular event can happen *only if* certain conditions are true.

3.2   Found by Characteristic Check

Missing Trigger Condition and Resulting Deadlock. There is a pair of communicating agents: a data sending agent, the *Sender*, and a receiving agent, the *Receiver*. When the bus signal *trdy* is true, the Receiver is signaling that it is ready to receive data. When *dbsy* is true, the Sender is sending data. So when ¬*dbsy* is true, the Sender is idle, and is ready to start the next data transfer. Thus, when *trdy* ∧ ¬*dbsy* is true, both agents are ready; consequently, *trdy* ∧ ¬*dbsy* is a trigger condition that allows a new data phase to start.

The protocol is designed so that the Receiver keeps *trdy* true until the data sending agent is idle and *dbsy* is deasserted (¬*dbsy*). Thus, the normal sequence of events is as shown in Fig. 3. Note the one cycle delay between a signal change and its appearance on the bus because of the latched property.

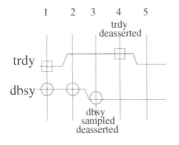

Time	Receiver	Sender
1	asserts *trdy* and observes that *dbsy* is active	driving *dbsy*
2	keeps *trdy* asserted	deasserts *dbsy*
3	observes that *dbsy* is low	idle
4	in response, deasserts *trdy*	idle
5	idle	idle

Fig. 3. *trdy* and *dbsy* relation

What makes the protocol tricky is that if *dbsy* is already deasserted, as an optimization, *trdy* can assert and deassert right away (Fig. 4). Note that, unlike the normal sequence, the trigger (*trdy* ∧ ¬*dbsy*) is true for only one cycle in this case. Thus, this scenario lets the idle state, ¬*trdy* ∧ ¬*dbsy*, happen a cycle earlier (that is why it's an optimization).

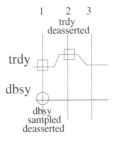

Time	Sender	Receiver
1	asserts *trdy* and observes that *dbsy* is inactive	idle
2	deasserts *trdy* as an optimization	idle
3	idle	idle

Fig. 4. When Optimized : *trdy* and *dbsy* relation

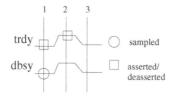

Fig. 5. Missed Trigger Condition : $trdy \wedge \neg dbsy$ is missed

It is important that *trdy* and *dbsy* never become true on the same clock because of the following possibility (Fig. 5). On cycle 1, the Receiver samples *dbsy* low so it does the optimization, but because in cycle 2, *dbsy* is high, *trdy* is inadvertently deasserted before the trigger condition becomes true. So the awaited trigger condition never happens and a data phase cannot start.

In most cases, the protocol is designed so that *trdy* and *dbsy* cannot become true on the same clock, but, in a special case, a loophole in the protocol allows this. Normally, because *trdy* and *dbsy* are handshake signals between two communicating agents, they are not asserted at the same time. However, there is a case where a third agent can assert *dbsy* and thereby break this rule. This special case is when an agent which previously deferred to complete a data phase (the *Deferrer*), takes advantage of the apparently idle bus to complete the deferred data phase. Meanwhile, there is a separate ongoing transaction where an Receiver agent is about to assert *trdy* to communicate to a Sender agent. The Deferrer asserts *dbsy* to start the data transfer at the same time the Receiver asserts *trdy*, and, as a result, *trdy* and *dbsy* are asserted at the same time. Note that, in this scenario, the *trdy* and *dbsy* assertions are not for the same data transfer. This overlooked case causes the normal data transfer to wait forever for the (missed) trigger condition of $trdy \wedge \neg dbsy$.

In a similar case,[3] a deadlock occurs because a transaction cannot proceed unless the data phase completes.[4] But the data phase can not happen because the trigger condition did not become true.

Since this trigger condition is crucial for data phases, and optimizations often lead to unexpected scenarios, the existence of the sequence shown in Fig. 5 was one of the first checked properties. Although the model checking properties cannot be automatically deduced, after the specification process, properties testing for suspicious sequences can be developed with little difficulty.

---

[3] For the protocol experts: this happens when 1. the responding agent and the requesting agent are the same and 2. it is a write with a snoop-initiated data transfer, where the second *trdy*, which is for the snoop-initiated transfer, happens exactly when the data phase (which was allowed to be indefinitely delayed because of 1.) for the first *trdy* starts.

[4] For the protocol experts: this requirement is because the second data phase is snoop-initiated, and it must happen together with the response.

Dropped Data Phase. Under certain circumstances, a write data phase can be delayed indefinitely.[5] The immediate danger of this is that the data phase never happens, and the system proceeds without any trace of the phantom data phase. The basic signalling mechanism of a write transaction is,

1. The Receiver indicates a "ready" state by asserting *trdy* true.
2. The Receiver may deassert *trdy* false before the Sender starts the data phase.
3. The Sender, acknowledging the "ready" signal, starts the data transfer by asserting *dbsy*.

Thus, the normal sequence of events is, "*trdy* is asserted and then deasserted, data is transmitted, *trdy* is asserted and then deasserted, data is transmitted, ..." Now, consider the sequence "*trdy* is asserted and then deasserted, *trdy* is asserted again, data is transmitted, ... (Fig. 6)." The one-to-one correspondence between a *trdy* assertion and a data transfer breaks down. The second *trdy* assertion should not have happened before the start of the first data transfer. Consequently, the data phase for the first *trdy* misses its window to start the transfer. This happens only in the case where a data phase can be delayed indefinitely.

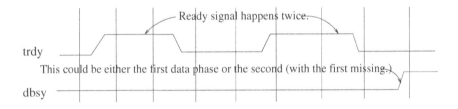

Fig. 6. Early second *trdy* assertion

This was found by model checking whether the specification allows a second *trdy* assertion before the start of a pending data phase. Coincidentally, this problem was also discovered using simulation by the testing team at Intel, but since our methodology does not require an implementation, it found the problem in a shorter time with less effort.

4   Conclusion

The formal specification for the core subset of the Itanium Processor bus protocol consists of 46 independent constraints which can be replicated eight times for the pipeline depth of eight. To minimize model checking complications, a pipeline depth of two was used for debugging the specification. Again, this is not a limitation of the specification methodology, which scales well. The description

---

[5] For the protocol experts: this happens when the responding agent and the requesting agent are the same.

file for a pipeline depth of two, written in Cadence SMV [McM], is 650 lines long, excluding the variable declarations. The formal specification was debugged using the techniques described; the current specification has no dead states and all characteristics checked for, hold. Model checking was done using Cadence SMV [McM] and all characteristics checks can be completed within two minutes on a Pentium Pro system with 128Mb of memory.

Acknowledgement. This research was supported by GSR contract SA2206-23106PG-2. Many thanks to Mani Azimi and Sridhar Lakshmanamurthy at Intel®, and Chris Wilson at Stanford University.

References

[ABG+00]   Y. Abarbanel, I. Beer, L. Gluhovsky, S. Keidar, and Y. Wolfsthal. FoCs - Automatic Generation of Simulation Checkers from Formal Specification. In *International Conference on Computer-Aided Verification*, volume 1855 of *Lecture Notes in Computer Science*. Springer-Verlag, 2000.

[AS00]   F. Aloul and K. Sakallah. Efficient Verification of the PCI Local Bus using Boolean Satisfiability. In *International Workshop on Logic Synthesis (IWLS)*, 2000.

[CCLW99]   P. Chauhan, E. M. Clarke, Y. Lu, and D. Wang. Verifying IP-Core based System-On-Chip Designs. In *Proceedings of the IEEE ASIC conference*, September 1999.

[CE81]   E.M. Clarke and E.A. Emerson. Synthesis of synchronization skeletons for branching time temporal logic. In *Logic of Programs: Workshop*, volume 131 of *Lecture Notes in Computer Science*, May 1981.

[CGY+00]   E. Clarke, S. German, Y. Lu, H. Veith, and D. Wang. Executable Protocol Specification in ESL. In *Proceedings of the Third International Conference of Formal Methods in Computer-Aided Design*, November 2000.

[Cor]   Intel Corporation. Itanium Processor Bus Protocol Specification. Internal document.

[GD00]   Shankar G. Govindaraju and David L. Dill. Counterexample-guided choice of projections in approximate symbolic model checking. In *Proceedings of International Conference on Computer-Aided Design*, November 2000. San Jose, CA.

[HCRP91]   N. Halbwachs, P. Caspi, P. Raymond, and D. Pilaud. The synchronous dataflow programming language lustre. *Proceedings of the IEEE*, 79(9):1305–1320, September 1991.

[KMP98]   M. Kaufmann, A. Martin, and C. Pixley. Design Constraints in Symbolic Model Checking. In *International Conference on Computer-Aided Verification*, 1998.

[McM]   Kenneth McMillan. http://www-cad.eecs.berkeley.edu/~kenmcmil/smv/.

[MHG98]   A. Mokkedem, R. Hosabettu, and G. Gopalakrishnan. Formalization and Proof of a Solution to the PCI 2.1 Bus Transaction Ordering Problem. In *Proceedings of the Second International Conference, Formal Methods in Computer-Aided Design*, volume 1522 of *Lecture Notes in Computer Science*. Springer-Verlag, 1998.

[SDH00]   Kanna Shimizu, David L. Dill, and Alan J. Hu. Monitor-Based Formal
           Specification of PCI. In *Proceedings of the Third International Conference
           of Formal Methods in Computer-Aided Design*, November 2000.
[SIG95]   PCI SIG. PCI Local Bus Specification, Revision 2.2, 12 1995.

The Design and Verification of a Sorter Core

Koen Claessen[1], Mary Sheeran[1], and Satnam Singh[2]

[1] Chalmers University of Technology
[2] Xilinx, Inc.

Abstract. We show how the Lava system is used to design and analyse fast sorting circuits for implementation on Field Programmable Gate Arrays (FPGAs). We present both recursive and periodic sorting networks, based on recursive merging networks such as Batcher's bitonic and odd-even mergers. We show how a design style that concentrates on capturing *connection patterns* gives elegant generic circuit descriptions. This style aids circuit analysis and also gives the user fine control of the final layout on the FPGA. We demonstrate this by analysing and implementing four sorters on a Xilinx Virtex-II[TM] FPGA. Performance figures are presented.

1 Introduction

This paper describes the application of various formal and informal techniques to the design, implementation, optimisation and verification of a high speed sorter core realised on a large field programmable gate array (FPGA). Customers who buy FPGA cores expect them to have high performance and to have been carefully verified. Examples of the application of high speed sorting are graphics algorithms for rendering and ray tracing.

The design of the sorter core is based on recursively described butterfly networks which are composed to realise periodic sorters. There are a large number of different but similar sorting network designs, resulting in circuits of varying performance (depending on the lengths of intermediate wires). Having a design language that is well-suited to describing these networks has helped us to explore the design space far more effectively than is possible using conventional hardware description languages. We present four designs and instrument their performance. The user of the sorter core specifies the required speed performance, area requirements, latency and pipelining behaviour. Based on these requirements the sorter core selects one of the four sorter implementations presented here.

To produce an efficient sorter network on an FPGA, one must carefully manage the intermediate wire lengths and also make effective use of the available silicon resource. We demonstrate how a layout combinator based style of description allows us to generate a compact layout without the tedious calculations that are necessary in a conventional HDL.

The design environment used to describe and verify our circuit cores was developed at Xilinx and at Chalmers University, and is called Lava [3]. Circuit descriptions in Lava are written in the functional language Haskell [6].

T. Margaria and T. Melham (Eds.): CHARME 2001, LNCS 2144, pp. 355–368, 2001.

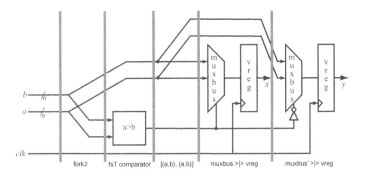

Fig. 1. Architecture of the 2-sorter

2   Design and Verification of a 2-Sorter

The basic element of the sorters and mergers that we present in the following sections is a two-sorter circuit. This circuit takes a pair of n-bit numbers and returns the pair sorted into ascending order. The butterfly networks implemented on FP-GAs are pipelined. This is achieved by registering the output of each 2-sorter.

Figure 1 shows the implementation architecture for the 2-sorter. This implementation uses a comparator to determine which of a or b is greater. The result of the comparator is used as the select signal to two bus-multiplexors. The outputs of the bus-multiplexors are registered to allow the construction of pipelined sorters.

The top-level Lava description of the two sorter is:

```
twoSorter clk = fork2 >-> fsT comparator >-> condSwap clk
```

This Lava description describes the circuit shown in Fig. 1 by composing in series several sub-circuits. The serial composition infix combinator is written as >->. This connects the output of the circuit on the left to the input of the circuit on the right. Furthermore, the circuits are laid out horizontally with a left to right information flow. Note how combinators compose *behaviour* and *layout*. Lava also provides combinators for right to left serial composition (<-<), top to bottom serial composition (\/), bottom to top (/\) and overlaid layout (>|>). There are also combinators for four sided tiles and many kinds of elaborate layout and wiring for real circuits have been successfully described using convenient combinators.

The `fork2` circuit simply duplicates its input, as shown in Fig. 1. The `fsT` combinator takes a circuit as a parameter and applies it to the first element of a pair leaving the second element unchanged. The comparator is implemented by using a subtracter laid out vertically (its definition is not given in this paper). The next stage uses `fsT` with `comparator` to perform a comparison on the first element of the forked value, namely (a,b). The next three stages implement a conditional swap circuit which will swap the values (a,b) if a is larger than b.

The source for the conditional swap circuit is shown below:

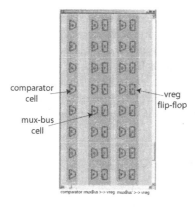

Fig. 2. FPGA floorplan of an 8-bit 2-sorter

```
condSwap clk =
   fork2List >-> hpar [muxBus >|> vreg clk, muxBus' >|> vreg clk]
```

This circuit is build using the **hpar** combinator which lays out a list of circuits horizontally. In this case, the two circuits laid out next to each other are bus-multiplexors that use a select signal to determine which input to transfer to the output. The **fork2List** combinator duplicates its input into a two element list. Each bus-multiplexor has its output connected to a register bus and the two circuits are overlaid so that they occupy the same locations on the FPGA (there is enough space in each cell to perform the multiplexing and registering). To avoid having to explicitly realise the inverter required for the second multiplexor a variant **muxBus'** is used which switches in the opposite sense from **muxBus**. This optimisation allows for a very compact floorplan as shown in Fig. 2 for a Xilinx Virtex ᵀᴹ FPGA.

The verification of the two-sorter was conducted using an in-house system developed for the formal verification of intellectual property cores [8]. A technology specific equivalence checker has been produced which takes as input two EDIF netlists and tries to establish their equivalence using sequential equivalence checking. Using this tool we have shown that the Lava implementation of the 2-sorter has the same behaviour as a golden VHDL behavioural description.

3 Describing Networks

Many circuits have a clear recursive or iterative pattern of construction. Typical examples are multipliers, mergers, sorters, interconnection networks and transforms (notably the Fast Fourier Transform). These are important functions, and indeed it turns out that many of the functions that we would like to implement as cores for Digital Signal Processing (DSP) display this kind of regularity.

For this reason, we have developed design, verification and implementation methods that are particularly suited to these regular circuits. The overall goal is to make a fast route to efficient FPGA implementations. Here, we consider how to describe recursive and iterative networks, taking mergers and sorters as examples. This means that we are particularly interested in *butterfly networks* like that shown on the right of Fig. 3, and in variants on them. As a first step, we introduce some useful connection patterns.

Often, we want to feed each of the pairs in a list of even length to different copies of a two-input two-output component, producing a list of the same length. We define:

```
evens f []      = []
evens f (a:b:cs) = f [a,b] ++ evens f cs
```

Here, `f` is the component in question, given as a parameter to the circuit. `[]` is the empty list, giving a base case. In the step, `a` and `b` are the first two elements of the input list, and `f [a,b]` is the output of the `f` component. That list is concatenated (using `++`) to the result of applying `evens f` to the rest of the input list, `cs`.

A circuit in which `g` works on the bottom half of the input list and `h` on the top half is written `parl g h`. The special case in which the two components are the same arises often and so gets its own abbreviation.

```
two g = parl g g
```

We can simulate this circuit on the Lava prompt. As an example, we simulate `two reverse`, where `reverse` is the wiring pattern that reverses its inputs.

```
Lava> simulate (two reverse) [1..16]
[8,7,6,5,4,3,2,1,16,15,14,13,12,11,10,9]
```

We also introduce the pattern `ilv`, for *interleave*. Whereas `two f` applies `f` to the top and bottom halves of a list, `ilv f` applies `f` to the odd and even elements (see Fig. 3). We define it in terms of the wiring pattern *riffle*, which performs the perfect shuffle on a list. Think of taking a pack of cards, halving it, and then interleaving the two half packs (as you do before you deal out the cards). If you now *unriffle* the pack, you reverse the process, returning the pack to its original condition. (This is somewhat more difficult to accomplish with aplomb at the poker table.)

```
Lava> simulate riffle [1..16]
[1,9,2,10,3,11,4,12,5,13,6,14,7,15,8,16]
```

```
Lava> simulate unriffle [1..16]
[1,3,5,7,9,11,13,15,2,4,6,8,10,12,14,16]
```

Note that unriffling the sequence from 1 to n divides into its odd and its even elements. We use this fact to define `ilv` in terms of `two`.

```
ilv f = unriffle >-> two f >-> riffle
```

Now we are in a position to define a connection pattern for butterfly circuits.

```
bfly 1 f = f
bfly n f = ilv (bfly (n-1) f) >-> evens f
```

The smallest butterfly is just a single `f` component with two inputs and two outputs. A butterfly of size n, for n greater than zero, consists of two interleaved

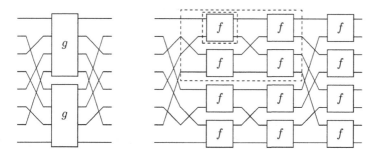

Fig. 3. ilv g and bfly 3 f

butterflies of size $n-1$, the output of which is fed into a stack of f components, which is made using **evens**.

This pattern corresponds to a divide and conquer algorithm: there are two recursive calls on sub-parts of the input, and then a final phase (the call of **evens**) in which the results are combined. This butterfly-shaped connection pattern is shown on the right in Fig. 3. The figure indicates one of the smaller butterflies using dotted lines. This butterfly is interleaved with another one. (Compare with the diagram on the left of the figure.) The call of **evens** can be seen on the right of the figure.

3.1 Batcher's Bitonic Merger and Sorter

One of the best known uses of the butterfly network is in the building of mergers and sorters based on a two-input two-output comparator, that is a two-sorter. It turns out that **bfly n cmp** sorts some input patterns, including those whose first half is sorted and second half is sorted into reverse order (provided that **cmp** is a two-sorter). This allows us to build a recursive sorter.

For lists of length 1, the base case, sorting is just the identity function (**id**). Otherwise, we make two recursive calls to smaller sorters, and reverse the output of the second, feeding the result into a merger. The merger (**bfly n cmp**) is known as *Batcher's bitonic merger* [1].

```
sortB 0 cmp = id
sortB n cmp = parl (sortB (n-1) cmp) (sortB (n-1) cmp >-> reverse) >->
              bfly n cmp
```

The recursive structure of Batcher's Bitonic Sorter is illustrated in Fig. 4. In part (a) of the figure a sorter is designed by recursively sorting the two halves of the input and then merging the result (after reversing one of the sub-sorts). Part (b) shows that a merger can be made from a butterfly of two sorters. The two sorter component is shown as a box with 2S written inside it. The merger shown here is Batcher's bitonic merger. Part (c) shows that one of the smaller sorters can be decomposed using exactly the same strategy i.e. two sub-sorters and a merger. Part (d) shows that the merger inside this sub-sorter is just a butterfly of size 2. The resulting sorting network is shown in Fig. 5.

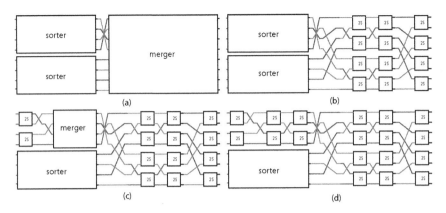

Fig. 4. Recursive structure of Batcher's Bitonic Sorter

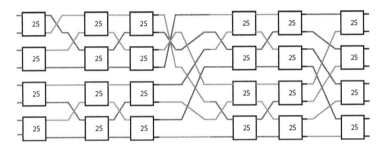

Fig. 5. Batcher's Bitonic Sorter `sortB`

Note that our sorter is parameterised on the comparator component. So, for instance, if cmp is a two-sorter on integers, then sortB n cmp is sorter on integer lists of length 2^n, but later we will plug other circuit-level comparator components into the same sortB function. So we have really designed the connection pattern that must be used to connect comparators. We have not in any way tied ourselves down to comparators of a particular type. We will later use this fact more than once, when visualising merging networks and when verifying sorters.

3.2   Batcher's Odd Even Merger and Sorter

The other well-known merging network is Batcher's *odd even merger*. To describe it, we need a another connection pattern, mid. The circuit mid f passes the first and last inputs of its input list through unchanged, and applies f to the remaining elements. For example, simulating mid reverse [0..7] gives [0,6,5,4,3,2,1,7].

Now, the connection pattern for the odd-even merge is defined as

```
mergeOE 1 cmp = cmp
mergeOE n cmp = ilv (mergeOE (n-1) cmp) >-> mid (evens cmp)
```

The subcircuit `mid (evens cmp)` on the right places components not on even pairs, but on *odd* pairs. For comparison, look back at the definition of `bfly`, the connection pattern for the bitonic merger. The pattern is almost identical! The big difference is that the butterfly has `evens cmp` as its last column, while the odd even merger has `mid (evens cmp)`. The reader is encouraged to sketch an odd even merger for 8 inputs, along the lines of our picture of the butterfly in Fig. 3.

The odd even merger sorts inputs whose top and bottom halves are sorted, so the definition of the corresponding recursive sorter is short and sweet.

```
sortOE 0 cmp = id
sortOE n cmp = two (sortOE (n-1) cmp) >-> mergeOE n cmp
```

4   Visualising Networks

We have introduced some merging and sorting networks, and made claims about their behaviour. How do we convince ourselves that we have got our mergers and sorters right?

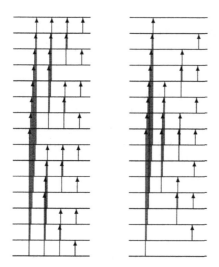

Fig. 6. The topology of the butterfly and odd even merge networks

A good way to check that one has described these standard networks correctly is to draw pictures of them and compare with reliable presentations in the literature. A standard way to visualise sorting or merging networks is shown in Fig. 6. Values in the network flow from left to right. The arrows indicate comparators, which operate on values flowing on the wires at each end of the arrow (possibly swapping them). These figures show Batcher's bitonic and odd even mergers for 16 inputs.

We have generated these pictures using a simple non-standard interpretation. We replace the standard comparator by one where each "wire" carries both a number (as before) and a list of pairs of numbers, the comparisons done so far in the circuit. The comparator records which two numbers it compares, appending the pair to the comparisons list on one of its outputs. In addition, we add depth indicators to the circuit. A depth counter is incremented each time it flows through such a marker. Then, the new circuit is simulated on a sorted input list. We simply write some Haskell code to naively generate pictures from the resulting output. We exploit the fact that we have a full programming language available to code up lightweight analyses that help with the design problem at hand. Here, we exploit the fact that we have written generic descriptions of the connection patterns for the mergers. This enables us to plug in non-standard components for use in circuit analysis.

5   Verifying Mergers and Sorters

The recursive structure of the mergers and sorters encourages us to reason about them by induction. At present, we perform such induction proofs by hand (although it is our ambition to provide automatic assistance). For examples of hand proofs about sorting and permutation networks, see for example references [5,9, 10]. What we have available in the current Lava system is support for reasoning about *fixed size circuits*. What we tend to do in practice is to use this facility to check instances of more general conjectures – a sanity check that can be a great time-saver.

A network sorts correctly if (1) it always produces sorted output, and (2) if the output is a permutation of the input. In performing automatic proofs about fixed-size networks of comparators, we are fortunate in being able to use the zero-one principle [7]: A sorting network built only of wiring patterns and two-input two-output comparators sorts arbitrary numbers if it sorts numbers in $\{0,1\}$. This remarkable fact means that we only need to check that our networks, when built from combinational two-bit comparators (twoBitSort, which consists of just one *or* and one *and* gate) sorts lists of bits. Here, we again see the advantage of having generic connection patterns, into which we can plug many components.

A list of bits is sorted if implication holds between adjacent bits. A circuit permutes an input list of bits, if there are equally many ones in the input as there are in the outputs. If these two checks return true for all inputs, then we have a sorter. (An alternative would be to compare our sorters with a known correct sorter. We often perform such equivalence checking, but have chosen not to do so in this paper.)

To check the first sorting network property, we first define in what case a list of bits is sorted.

```
sorted []       = high
sorted [x]      = high
sorted (x1:x2:xs) = (x1 ==> x2) <&> sorted (x2:xs)
```

Then, we define the property, which is parameterised by a sorting network sort, and a natural number n.

```
prop_Sorts sort n =
  forAll (list (2^n)) $ \ inp ->
    sorted (sort n twoBitSort inp)
```

Read this property definition as: "For all lists of size 2^n called inp, the output of the sorting network sort is sorted, when given inp as input".

We can check these kinds of properties using external tools such as SAT-solvers and BDD-based tools. In order to do so, we have to specify at what size we want to verify the property. The logical level at which these tools work is not powerful enough to perform the verification for all sizes at once. We use symbolic evaluation to generate an input file to the external tool, which contains the unfolded definition of the circuit and the property. As an example, we present the verification of the sortedness property for size 4 (that is 16 inputs), using Prover Technology's propositional prover.

```
Lava> verify (prop_Sorts sortB 4)
Proving: ... (t=0.5) Valid.
```

Given a suitable circuit count, which outputs a binary number indicating how many of its inputs are ones, the permutation property can be easily formulated as below. We show the verification for size 5 (64 inputs) using the BDD-based tool VIS.

```
prop_Permutes sort n =
  forAll (list (2^n)) $ \ inp ->
    count inp <==> count (sort n twoBitSort inp)
```

```
Lava> vis (prop_Permutes sortOE 5)
Vis: ... (t=3.0) Valid.
```

Unfortunately, the verification of the two properties presented above runs out of steam at around 64 inputs, for both SAT-solvers and BDD-based tools. The problem is that the verification problem simply becomes too hard for current-day technology.

However, all is not lost. If we did a proof of the sorters by hand, we would use inductive reasoning. An inductive proof of the correctness of for example Batcher's odd even sorter clearly relies on the fact that the merger sorts two appended sorted lists of equal length. We can actually check this conjecture for fixed sizes of the merger!

The property that we then want to check has a list of bits as input, and a single bit as output. That bit should be 1 if the property holds of the given merger (which is the parameter merge). So, if out is the output of the merger when it has a bit-sorter as its component, we check that out is sorted if the two halves of inp are.

```
sortsTwoSorted merge n inp = ok
  where
    (inpL, inpR) = halveList inp
    out          = merge n twoBitSort inp
    ok           = (sorted inpL <&> sorted inpR) ==> sorted out
```

To make this into a property that can be checked by a verification tool dealing with fixed sized circuits, we must set the size of the list:

```
prop_SortsTwoSorted merge n =
  forAll (list (2^n)) $ \ inp ->
    sortsTwoSorted merge n inp
```

It is read as "for all lists of size 2^n called inp, the merger sorts two appended sorted lists". This property is rather easy to check for for example Prover Technology's propositional prover. For example, size 7 (that is 128 inputs) takes only about 3 seconds.

These results seem to indicate that monolithic proofs of sorting networks are hard for both SAT-solving and BDD-based methods, and that splitting the proofs up in smaller lemmas can help. Concluding, we were able to verify the two sorting properties of our sorting networks for small sizes (up to 64 inputs). The lemmas about the mergers have been verified for more than 256 inputs.

6   Periodic Networks

We have seen how to give elegant descriptions of two well-known recursive sorters. Now, we turn our attention to the so-called periodic sorters. These are sorters that can be made by composing a number of identical circuits. The best known of these sorters is odd even transposition sort. For $2N$ inputs, and cmp a comparator, $N/2$ copies in series of the circuit

```
evens cmp >-> mid (evens cmp)
```

makes a sorter. But this is a rather large sorter! Interestingly, there are mergers that can be composed in series to give sorters that are only about twice as big as the recursive sorters that we have seen. We are attracted by the prospect of building very regular, very fast sorters out of a single merger component. We can then concentrate optimisation efforts on that merger, aiming for a small footprint on the FPGA. Also, having a single merger makes it possible to experiment with space/time trade-offs, by reusing a single component repeatedly over time.

We first check to see whether or not the bitonic merger can be composed in sequence to make a sorter. (The combinator hrep puts n copies of circuit in serial composition, horizontally.)

```
bflyCompose n cmp = hrep n (bfly n cmp)
```

Verifying prop_Sorts bflyCompose n fails already when n is 2, indicating that for example [low,high,low,high] is a counterexample. (This list passes through each butterfly unchanged.) So we need to look further.

Sheeran has earlier studied the periodic balanced merger of Dowd et al [4,9]. The Ruby descriptions can be directly translated into Lava. We introduce a new and somewhat mysterious connection pattern vee related to ilv, and build a connection pattern similar to the butterfly, but replacing ilv by vee.

The wiring pattern alt swaps every second of the even pairs of a list. So, for example, the list [1..16] is permuted to

```
[1,2,4,3,5,6,8,7,9,10,12,11,13,14,16,15]
```

by the application of alt.

```
vee f = alt >-> ilv f >-> alt
```

```
vfly 1 f = f
vfly n f = vee (vfly k f) >-> evens f
```

Here, we need a generated picture to help understanding, see Fig. 7, which illustrates `vfly 4 cmp`. It becomes clear why this is called the balanced merger. This rather beautiful and symmetrical merger can be composed to make a sorter.

```
sortV n cmp = hrep n (vfly n cmp)
```

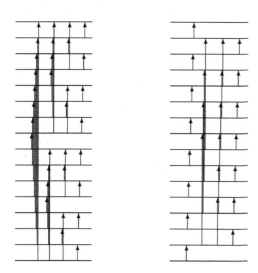

Fig. 7. The balanced merger and a periodic variant of the odd even merger

The same `prop_Sorts` that we used before confirms that `sortV 4`, that is four copies of `vfly 4` in series, does indeed construct a sorter for 16 inputs (using Prover Technology's basic propositional prover). The proof for 32 inputs takes too long. On the other hand, we can quickly confirm that the periodic balanced merger sorts two interleaved sorted lists, even for 128 inputs.

By experimenting with variants on the `ilv` connection pattern in the odd even merge, we have found another appealing periodic merger. You can think of `ilv` as marking the odd and even elements of the input and output lists with different colours, and then operating separately on each colour. What if we first divide the list up into pairs, and then mark alternate pairs with different colours? Then, one of the components operates on elements 0,1,4,5,8,9 etc. while the other operates on the remaining elements. We define

```
que f = ilv unriffle >-> two f >-> ilv riffle
```

You are encouraged to sketch this connection pattern.

Now, we replace each `ilv` in the odd even merge pattern by `que`.

```
qfly 1 f = f
qfly n f = que (qfly (n-1) f) >-> mid (evens f)
```

The resulting comparator network is shown on the right in Fig. 7. It is, we believe, a periodic network proposed by Canfield and Williamson [2]. Note that the lengths of the stretched comparators are shorter than in any of the networks that we have seen. This is the best periodic merger that we have found so far. We call the resulting sorter sortQ. Again, this network sorts two interleaved sorted lists, and this can be checked using a SAT-solver.

Why, though, does this merger compose to form a sorter? The key insight is that the merger not only transforms lists satisfying ilv sorted to lists satisfying sorted, but also lists satisfying ilv (ilv sorted) to lists satisfying ilv sorted. In general, the merger increases sortedness by "removing one ilv". So a sequence of n mergers takes an unsorted list (of length 2^n), that is a list that repeatedly interleaves 2^n sorted singleton lists (n calls of ilv) to a sorted list (no calls of ilv). A useful lemma in this proof is the fact that

que (ilv f) = ilv (ilv f)

Our recursive descriptions of networks are very different from the standard ways of describing networks in the extensive literature on sorting networks. We have a language with which to describe networks! We have found that this connection pattern oriented style of description has enabled both informal and formal proofs about our networks. In addition, this style of description gives us an elegant way to control the final layout when implementing circuits on FPGAs. This is the topic of the following section.

7   Implementation

Four of the sorters presented in the previous section have been implemented on a Xilinx Virtex-II XC2V3000 FPGA to allow us to evaluate area and speed performance. We can measure the size of designs in terms of look-up table (LUT) and register pairs. The XC2V3000 FPGA has 256 LUT rows and 112 LUT columns. The 2-sorter is 3 LUTs wide and n LUTs tall when used to compare two n-bit numbers.

The layout produced by each of the sorters results in a solid rectangular area which is the result of tightly tiled 2-sorters. No gaps are left between columns and the routing software manages to find enough wiring resources to connect up the stages of the butterfly network. The FPGA floorplan of the sorters sortB and sortQ is shown in Fig. 8. The sorter sortV is like sortB but wider, and the sorter sortOE is like the sorter sortQ but narrower.

Note that we have to add extra delays on some of the wires of sortOE and sortQ for the pipelined version of the sorter by changing the mid combinator.

The results of placing and routing the four pipelined sorters are shown in Table 1. Each of these sorters sorts 32 16-bit numbers. All the sorters have a footprint which is 256 LUTs high. The width (in LUTs) of the sorters is shown in the table. The footprint of the design is 256 times the width and we use this area metric rather than the gate count.

Based on the information in the table a sorter core can use any of these sorters (or variants) to satisfy speed, area, pipelining and latency requirements. The sortB sorter offers the most compact pipelined sorter. If a faster pipelined

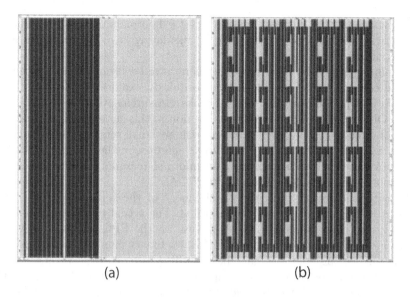

(a)   (b)

Fig. 8. FPGA floorplan of the sorters (a) sortB (b) sortQ

Table 1. Performance results

Sorter	Speed	Latency (ticks)	Gate Count	Area of XC2V3000	Width
sortB	127 MHz	14	153,350	43%	45
sortOE	147 MHz	14	136,886	41%	55
sortV	121 MHz	20	248,070	70%	75
sortQ	152 MHz	20	222,870	66%	99

sorter is required then sortOE operates 20MHz faster but requires 10 extra columns. The periodic sortV sorter has little merit as shown since it is slow and wide. However, by using just one stage (instead of five) and feeding the output back into the input we get a very compact sorter (15 columns) but the penalty is that this sorter can not be pipelined. The fastest pipelined sorter is sortQ but this sorter also has the largest footprint. If a high speed non-pipelined sorter is needed then just one stage of sortQ can be used with the output fed back into the input which reduces the width to 18 columns. When minimising latency is an issue the sorter core will select sortB when saving area is more important than speed and sortOE when speed is more important than area.

We have shown in Lava that prop_Sorts is a valid safety property for all the pipelined sorters, for input sizes up to 128 bits.

8   Conclusion

The butterfly descriptions presented here are concise, and they result in highly optimised layouts. It is virtually impossible to produce such netlists from conventional hardware description languages. This is either because a synthesis based approach usually has little or no support for floorplanning or if a structural approach is used then the complexity of calculating (x, y) co-ordinates makes non-trivial layouts infeasible.

The butterfly-based sorters presented here have been built from a rich set of reusable combinators. This allowed us to quickly explore the design space of recursive and periodic sorters, and to make measurements of area and performance. Our earlier work on circuit description in this style indicated that large classes of circuits can be covered by surprisingly small sets of combinators.

What we have presented here is a *whole solution*, in the sense that it allows elegant circuit descriptions, formal and informal reasoning about the circuits, and a route to fast implementations on FPGAs.

The actual verifications in the development of the sorting networks which are described in this paper are the following. The actual two-sorter component (working on binary numbers) which we use on the FPGAs has been verified against a behavioural description in VHDL, for binary number sizes up to 8 bits. The various sorting networks have been verified to be correct sorting networks using the zero-one principle for sizes up to 64 inputs. Lastly, we have shown that the pipelined sorters always produce sorted outputs, for sizes up to 128 inputs.

"Virtex" and "Virtex-II" are trademarks of Xilinx Inc.

References

1. K.E. Batcher. Sorting networks and their applications. In *AFIPS Spring Joint Computing Conference*, volume 32, 1969.
2. E.R. Canfield and S.G. Williamson. A sequential sorting network analagous to the Batcher merge. In *Linear and Multilinear Algebra, 29*, pages 43–51, 1991.
3. Koen Claessen and Mary Sheeran. A tutorial on Lava: A hardware description and verification system. Available from http://www.cs.chalmers.se/~koen/Lava, 2000.
4. M. Dowd, Y. Perl, L. Rudolph, and M. Saks. The periodic balanced sorting network. *JACM*, 36:738–757, 1989.
5. Geraint Jones and Mary Sheeran. The study of butterflies. In Graham Birtwistle, editor, *Proc. 4th Banff Workshop on Higher Order*. Springer Workshops in Computing, 1991.
6. Simon Peyton Jones, John Hughes, (editors), Lennart Augustsson, Dave Barton, Brian Boutel, Warren Burton, Joseph Fasel, Kevin Hammond, Ralf Hinze, Paul Hudak, Thomas Johnsson, Mark Jones, John Launchbury, Erik Meijer, John Peterson, Alastair Reid, Colin Runciman, and Philip Wadler. Report on the Programming Language Haskell 98, a Non-strict, Purely Functional Language. Available from http://haskell.org, February 1999.
7. D.E. Knuth. *Sorting and Searching, vol. 3 of The Art of Computer Programming*. Addison-Wesley, 1973.
8. Carl Johan Lillieroth and Satnam Singh. Formal verification of FPGA cores. *Nordic Journal of Computing*, 6:299–319, 1999.
9. Mary Sheeran. Sorts of butterflies. In Graham Birtwistle, editor, *Proc. 4th Banff Workshop on Higher Order*. Springer Workshops in Computing, 1991.
10. Mary Sheeran. Puzzling permutations. In *Proc. Glasgow Functional Programming Workshop*, 1996.

Refinement-Based Formal Verification of Asynchronous Wrappers for Independently Clocked Domains in Systems on Chip

Xiaohua Kong, Radu Negulescu, and Larry Weidong Ying

Microelectronics And Computer Systems Laboratory
Dept. of ECE, McGill University
Montreal, Quebec, Canada H3A 2A7
Tel: (514) 398-2194, Fax: (514) 398-4470
{kong,radu,larry}@macs.ece.mcgill.ca

Abstract. In this paper we propose a novel refinement-based technique to formally verify data transfer in an asynchronous timing framework. Novel data transfer models are proposed to represent data communication between two locally independent clock domains. As a case study, we apply our technique to verify data transfer in a previously published architecture for globally asynchronous locally synchronous on-chip systems. In this case study, we find several race conditions, hazards, and other dangers that were not mentioned in the original publication, and we find additional delay constraints that avoid some of the detected dangers.

1 Introduction

With smaller feature sizes and larger die areas, integrating systems on a chip has become a feasible solution to satisfy unrelenting demands to provide more functionality at higher clock rates. Future on-chip systems designs are expected to involve the integration of multiple subsystems with independently clocked domains, including memory cores and processor cores. An advantageous scheme for realizing communication among clock domains is by means of glue logic circuits that operate in a self-timed manner, i.e., without a clock. Robust glue logic circuits are critical to ensure safe and stable data transfer at high operating speeds.

Synchronizing inputs with a local clock requires arbitration, which has a high probability of metastability at high operating speeds. Pausible clock schemes, proposed in [20], handle metastability by extending the passive phase of the clock long enough for metastability to resolve.

Globally-asynchronous locally-synchronous architectures (GALS), proposed by [33], [8] and [9] among others, show that a system can be partitioned into several independently clocked domains (subsystems) that communicate in a self-timed manner. To isolate each locally-synchronous domain from its globally-asynchronous environment, [1], [8] and [9] introduced an elegant design, called asynchronous wrapper, used to equip each locally-synchronous domain. Asynchronous wrappers serve as

T. Margaria and T. Melham (Eds.): CHARME 2001, LNCS 2144, pp. 370-385, 2001.
© Springer-Verlag Berlin Heidelberg 2001

controllers for data transfer between individual domains, and deliver a locally generated pausible clock for the synchronous part of circuitry [8].

The interface-based design methodology of [18] proposes to define a system on chip architecture by a high-level block partition on one hand, and by multi-level communication policies on the other hand. Communication refinement was promoted in [18] as a technique for rapid development of system on chip architectures by successive refinement applied to communication policies. In this paper, we show how to formally verify a key step of communication refinement for a GALS architecture, by verifying that a channel configuration with asynchronous wrappers meets a higher-level data-transfer specification that abstracts away the communication protocol used.

Verification of communication refinement is both important and non-trivial. We believe that the adoption of GALS architectures in future on-chip systems design will depend on the extent to which these novel architectures can guarantee safe and stable data transfer, while preserving robustness of the system designs. These concerns can be generally attributed to data and system safety issues. Although the simulations and tests reported in [8] and [9] show that their design works properly, such simulations and tests can only cover a small fraction of all possible configurations of relative delays in asynchronous communication, while we find that various malfunctions may still emerge under different relative delays. Applying formal verification, we find the potential pitfalls and provide delay constraints that permit to size the circuits so that the pitfalls are avoided.

The main challenges in our verification tasks are as follows: we have to combine in uniform models two essentially different synchronization paradigms (edge-triggered synchronous and handshake asynchronous); our data transfer specifications must be kept simple in the presence of non-synchronized clock cycles, which entails, for instance, that data can not be updated on every clock cycle on the sender and receiver sides; and, our specification and verification must be kept simple in the presence of interleavings of datapath and control signals.

In this paper, we: 1) propose new data transition model to represent the implicit relationship between clock and data validity events; 2) construct comprehensive implementation models for the asynchronous wrapper and the asynchronous communication scheme; 3) report several design pitfalls, including hazards in a design, obtained from 3D synthesizing tool, which claimed by [9] to be hazard-free; 4) provide relative timing constraints that were not mentioned by [8] and [9], along with fault diagnosis which indicates that the disregard of these constraints can cause system deadlock or erroneous data transfers.

Formal methods for verifying protocol conversion and refinement have been proposed for instance in [2], [3], [15] and [19]; [2] applies some of these ideas to the design of asynchronous circuits. However, our method is based on transforming high-level specifications that use data validity events into low-level specifications that use signal transition events, whereas the cited previous work only interfaces protocols by means of physical circuits which add latency and buffering. The result of our specification is similar to a CFSM (communication finite state machines) as in [15]. However, [15] does not include specifications in terms of data validity events.

Following [11], our verification method uses metric-free models for systems that rely on relative timing. Compared to timed methods, such as [5] and [10], metric-free verification is less computationally-intensive and integrates more easily within a hier-archical verification framework that permits non-determinism. Metric-free verification methods for relative timing are also used in [17].

2 Preliminaries

In subsection 2.1, we briefly overview the necessary notions from our verification framework, called *process spaces*; for more details, we refer the reader to [11] and [12]. In subsection 2.2., we briefly overview the *active-edge* specifications introduced in [7] and [13], which are used here to simplify the specification of control signals. In subsection 2.3, we briefly overview the asynchronous data communication schemes from [9] that are used in our verification case study.

2.1 Process Spaces

Process spaces are a general theory of concurrency, parameterized by the execution type. Systems are represented in terms of their possible executions, which can be taken to be sequences of events, functions of time, etc., depending on the level of detail desired in the analysis. This permits to trade tractability against precision of analysis by choosing an execution type. Modular and hierarchical verification are supported. If executions are taken to be finite traces (finite sequences of events), close relationships exist to several previous treatments of concurrency, such as [5] and [6].

Let E be the set of all possible executions. A *process* p is a pair (X, Y) of subsets of E such that $X \cup Y = E$. A process represents a contract between a device and its environment, from the device viewpoint. Executions in $X \cap Y$, called *goals*, denoted by **g** p, are legal for both the device and the environment. Executions from outside X, called *escapes*, denoted by **e** p, represent bad behavior on the part of the device. Finally, executions from outside Y, called *rejects*, denoted by **r** p, represent bad behavior on the part of the environment. We also use **as** p (accessible) and **at** p (acceptable) to denote X and Y respectively.

Process spaces can be used to build models of circuit behavior in a manner similar to conventional state machines. For an example of the models used in this paper, consider the C-element in Fig.1 (a). If the inputs a and b have the same logical value, the C-element copies that value at the output c; otherwise, the output value remains unchanged. Waveforms are represented by finite sequences of actions corresponding to signal transitions, such as *abcbac* for the waveform in Fig. 1 (b). In this paper, we use the term *trace* to refer to such a sequence of actions. We sometimes indicate that a certain action represents a rising or falling transition, as in $a+ b+ c+ b- a- c-$.

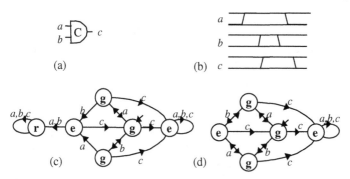

Fig. 1. Example processes: (a) The C-element symbol we use; (b) Waveform; (c) Hazard-intolerant model; (d) Inertial model.

If all signals start low, the C-element can be represented by the process in Fig. 1 (c), where **r**, **g**, and **e** stand for reject, goal, and escape. Illegal output events lead to an escape state with self loops on all subsequent events, call it a *permanent escape*, and illegal input events lead to a reject state that cannot be left either, call it a *permanent reject*. The state where *ab* leads is also marked **e**, making it illegal for the device to complete its operation by stopping there.

The model in Fig. 1 (c) is a *hazard-intolerant* model. There are variations of the CMOS cell models, because, in the presence of hazards, the behavior of a CMOS cell is not fully standardized. A hazard is a situation where an output transition is enabled and then disabled without being completed. For example, execution *abb* is a hazard for the C-element in Fig. 1. Hazard-intolerant models simply require the environment to avoid hazards, by stating that each execution that includes a hazard will lead to a permanent reject. The model in Fig. 1 (d) is an *inertial* model. Inertial models ignore hazards by filtering out input pulses that are shorter than the delay of the gate.

Our processes can be used to model not only gates or cells, but also relative timing assumptions of the following form:

$$D (b_1 b_2 ... b_n) > D (a_1 a_2 ... a_m)$$

where a_1, ..., a_m, b_1, ..., b_n are events such that a_1 is the same as b_1, and the Ds are the durations of the chains of events. Such a constraint, called a *chain constraint* [14], enforces that the *b* chain of events will not be completed before the *a* chain (unless one of the *a* or *b* actions involved occurs out of order).

Treating constraints as processes rather than linear inequalities permits us to deal with cases of deadlock and non-determinism, where the inequalities might not apply. Chain constraints can be implemented by transistor sizing. The absence of numerical information in the process models for chain constraints leads to more efficient verification using existing tools for non-timed analysis. Metric-free verification under relative timing constraints was first presented in [14] and [11].

In this paper, we only use the following operations and conditions on processes:

- *Refinement* is a binary relation, written $p \sqsubseteq q$, meaning "process q is a satisfactory substitute for process p".
- *Product* is a binary operation, written $p \times q$, yielding a process for a system of two devices operating "jointly". Product is defined by **as** $(p \times q) = $ **as** $p \cap$ **as** q and **g** $(p \times q) = $ **g** $p \cap$ **g** q.
- *Robustness* is a unary predicate on processes, written R_E, defined by **r** $p = \varnothing$, which represents a notion of absolute correctness: the device is "fool-proof" and can operate in any environment.
- *Reflection*, written $- p$, defined by **as** $(- p) = $ **at** p and **at** $(- p) = $ **as** p, represents a swap of roles between environment and device.

Refinement is reflexive, transitive, and antisymmetric. Product is commutative, associative, and idempotent. Furthermore, for processes p, q, and r,

$$p \sqsubseteq q \ ! \ \ p \times r \sqsubseteq q \times r.$$

These properties suffice to break a verification problem into several layers of partial specifications, and each layer into several modules, and to verify only one module at a time instead of verifying the overall problem in one piece.

Manipulations of finite-word processes are implemented by a tool called FIREMAPS [11] (for finitary and regular manipulation of processes and systems). This tool uses a BDD library [5] which offers basic routines for manipulating large Boolean functions. FIREMAPS implements the process space operations and conditions mentioned above, and has built-in constructors for hazard-intolerant, and inertial models, and for chain constraints. In addition, if refinement does not hold, FIREMAPS can produce a *witness execution* that pinpoints the failure. Such witness executions are used for fault diagnosis.

2.2 Active-Edge Specifications

Most digital circuit components are designed to synchronize on edges of their synchronization signals. Often, only one set of edges (rising or falling) of a synchronization signal are taken into account; such edges are called *active*. The other set of edges, called *passive*, are not used for synchronization. Correspondingly, simpler state-machine representations can be obtained by only referring to the active edges for control signals.

We use the term *active-edge specification* to refer to processes that represent circuit components by ignoring passive edges. For example, as illustrated in [13], the role of the clock signal in an edge-triggered flip-flop can be described by an active-edge specification. We refer to [13] on how to build active-edge specification by a "semi-hiding" operation, and to both [7] and [13] on how to reconstruct full specifications from active-edge specifications.

We use the term *transition-event specification* to denote a process in which both rising edges and falling edges of a signal are specified in the executions. Examples of transition-event models including their specifications presented in subsection 2.1, where each signal transition is represented by an occurrence of the corresponding action.

2.3 Asynchronous Communication

An alternative to clocking for digital circuits is to use handshake protocols to ensure coherent data communication between modules. In such protocols, data communication is synchronized by request and acknowledge signals. This alternative is particularly convenient for communication in heterogeneous designs with modules that have different timing schemes.

The design in [9] uses an early single-rail four-phase protocol [16], illustrated in Fig. 2. In this protocol, data should be guaranteed to be valid between the active edges of the request and acknowledge signals (bundled data). As an example, Fig. 2 shows an early four-phase handshake protocol for a push data channel [16], which is used in the asynchronous wrapper configuration presented in [9].

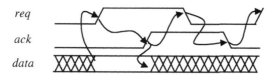

Fig. 2. Four-phase handshake protocol with early data valid scheme

Asynchronous wrappers in GALS architectures provides an interface between locally-synchronous domains and a globally-asynchronous environment. Data communication among independently clocked domains uses a four phases handshake protocol as illustrated in Fig. 2. Arbitration is required to synchronize the handshake signals and associated data with one of the clocked domains. The asynchronous wrapper circuits proposed in [1] and [9] attempted to realize failure-free communication in presence of metastability by performing arbitration between local clocks and handshaking control signals.

An asynchronous wrapper is composed of one pausible clock generator for each independently clocked domain, one input port and one output controller for each data port, and several other control ports. Two types of control ports proposed by [12], *Demand-type* (*D*-port) and *Poll-type* (*P*-port), are employed to control data transfer between two clock domains in a self-timed manner. Fig. 3 illustrates our specifications for a *D*-input port, *P*-input port and *D*-output port from [9]. These specifications were built by us on the basis of informal explanations, extended-burst-mode specifications [21], and waveforms provided by [9]. The only point where we depart from [9] in these specifications is by allowing retraction of the *Pen* signal in Fig. 3 (b), because this retraction actually occurs in the channel configuration shown in [9] if the clock domain that generates the *Pen* signal has a significantly higher frequency than the other clock domain involved in the data transfer. No such retraction is needed for the

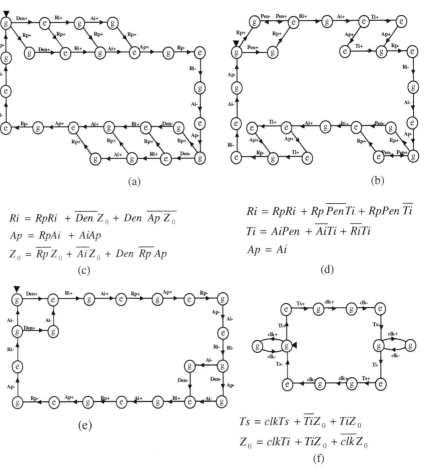

$$Ri = RpRi + \overline{Den}\, Z_0 + Den\, \overline{Ap}\, \overline{Z_0}$$
$$Ap = RpAi + AiAp$$
$$Z_0 = \overline{Rp}\, Z_0 + \overline{Ai}\, Z_0 + Den\, \overline{Rp}\, Ap$$

(c)

$$Ri = RpRi + Rp\, \overline{Pen}\, Ti + RpPen\, \overline{Ti}$$
$$Ti = AiPen + \overline{Ai}\, Ti + \overline{Ri}\, Ti$$
$$Ap = Ai$$

(d)

(e)

$$Ts = clkTs + \overline{Ti}\, Z_0 + TiZ_0$$
$$Z_0 = clkTi + TiZ_0 + \overline{clk}\, Z_0$$

(f)

Fig. 3. Asynchronous wrapper control ports: (a) *D*-input port specification; (b) *P*-input port specification; (c) Boolean functions of *D*-input port; (d) Boolean functions of *P*-input; (e) *D*-output port specification (f) Translator specification and its implementation

Den signal, because D-type ports block their local clocks. The Boolean functions shown in Fig. 3 are the synthesis results reported in [9].

The primary difference between *D*-ports and *P*-ports is that after an enable signal event, a *D*-port will fire a request to pause the clock before next rising clock edge, whereas a *P*-port does not pause the clock after an enable signal event until it receives corresponding events from handshake line.

Local clocks are generated by ring oscillators with arbitration [20]. A clock pause request will shift the start of the next clock cycle and stretch only the low phase of the clock. Competition between clock pause request *Ri* (see Fig. 4) and clock restart request is arbitrated by a four-phase mutual exclusion element. If both requests arrive simultaneously, then mutual exclusion element will make a non-deterministic selection to pick up one, while guaranteeing that the grant outputs are mutually excluded at all

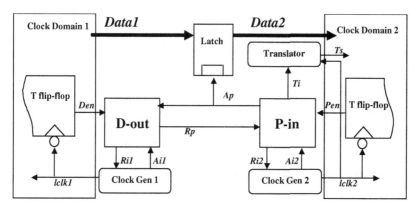

Fig. 4. Data channel between two independently clocked domains

times. Since the P-port doesn't have a predetermined clock cycle to transfer data, an extra control unit called translator is required to synchronize data transfer from the asynchronous wrapper to the receiver side of a local clock domain and provide the receiver correct timing to sample the available data at its data input.

Based on different data transfer schemes, several channel configurations can be obtained from various combinations of D-type and P-type control ports [9]. Following [9], we select D-type output with P-type input configuration as our verification case study. Following [8] and [9], Fig. 4 shows this control ports configuration. A mutual exclusion element is present in each clock generator.

A data transfer cycle is started by $Den+$. With the constraint reported by [9], a D-output clock pause request $Ri1+$ should reach the clock generator within half of its clock cycle. After $Ai1+$, clock1 ($lclk1$) remains low and $Rp+$ is issued. Clock2 ($lclk2$) will be stopped after both $Rp+$ and $Pen+$ are fired. $Rp+$ is acknowledged by $Ap+$. Clock2 can restart anytime after $Rp-$, whereas clock1 is only allowed to restart after $Ap-$, which indicates the completion of the handshake operation cycle. A latch is inserted in the data path in order to decouple send and receive operations, and ensures availability of stable data at its output when the latch is in opaque state. Two T flip-flops are added as next-state logic in each clock domain, to obtain hazard-free enable signals at their outputs. The incoming signals of the T flip-flops are not necessarily hazard-free.

3 Data Transfer Modeling

To verify correctness of data exchange, we build a high-level specification of data transfer, we convert the high-level specification to a low-level specification by automated means, and check refinement of the low-level specification by the implementation models of the asynchronous wrappers. Essentially, communication among locally-synchronous blocks is implemented by data lines bundled with control signals. We expect the specification should:

- Include features of both control path and data path.
- Be sufficiently precise so that it can support automated verification.
- Be as simple as possible, so that it can be easily developed and understood by designers.
- Be sufficiently general so that it can be easily mapped to communication refinement.

We introduce a new technique to build data transfer specifications independently of the synchronization scheme, and we apply this technique to our verification case study. Although the examples we use in this paper focus on data transfer between two independently clocked domains, the same data transfer specification modeling technique can be applied in a more general asynchronous context.

3.1 Data Transfer Specifications

A robust data transfer scheme requires that, under any circumstance, data issued by the sender should be received correctly by the receiver after a certain delay. The notion of correctness used here can be decomposed into the following aspects:

- Data integrity: received data preserves its original sending value.
- Stream integrity: no data items are lost or duplicated during data transmission, and the order of data items is preserved through the transfer.

As modeling stream integrity would require numerous states to represent data and control signal interleavings in a transition-event representation, we use instead a fictitious data event called *validity event* in our high-level specifications. Our technique abstracts away any irrelevant transition events and only considers data events triggered at active edges of clocks. This model fits nicely into a communication refinement paradigm, by permitting to isolate data validity events from the particulars of the synchronization signals used.

A validity-event data-transfer specification incorporates the following assumptions:

- After a control signal is fired by the sender, there is a data validity event at the input port of the data channel.
- After data is properly sampled by the receiver, there is a data validity event at the input port of the receiver.
- Data integrity is preserved in the data transmission, for instance, for each valid input "0" (or "1") there is a valid output "0" ("1"), and *vice versa*.
- Stream integrity is preserved in the data transmission.

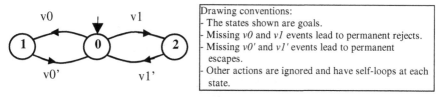

Fig. 5. Propagation of validity events

For the asynchronous wrapper, the high-level data-validity specification is the process shown in Fig. 5. The "0" data validity event *v0* is propagated as *v0'* to the end of data channel. Same applies for the "1" data validity events.

3.2 Fusion Processes

In order to transform the high-level specification of Fig. 5 into a full-blown   transition-event specifications, we introduce *fusion* processes to "glue" validity events, transition events, and control events (e.g. clock) which trigger the validity events. Such processes effectively force glued events to occur simultaneously by forbidding other events from occurring in between. For instance, Fig. 6 (a) illustrates a fusion process where the data transition event *Data1* is fused with its validity events *v0* and *v1* by control event *lclk1*, which is the active edge of a local clock. Note that signal *Data1* can toggle arbitrary in a clock cycle, while the validity events are related to the logical level that signal *Data1* has right before active-edge of *lclk1*: If *lclk1* comes when *Data1* is low, validity event *v0* will be issued and there will be no another validity event until the next active edge of *lclk1*, though *Data1* might keep changing between two active clock events. Same as Fig. 6 (a), Fig. 6(b) fuses the transition events of *Data2*, with the active edges of control signal event *Ts* and with corresponding

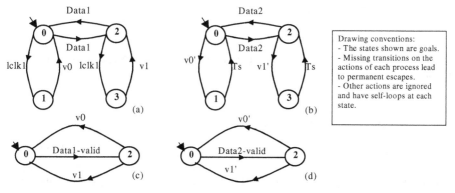

Fig. 6. Example of Fusion Processes (a). Fusion of Data1; (b) Fusion of Data2; (c) Invariant fusion of Data1; (d) Invariant fusion of Data2

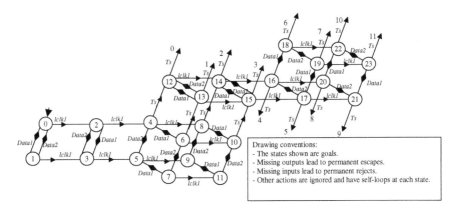

Fig. 7. Specification constructed from fusions

validity events. Fig. 6(c) and (d) are fusion processes to glue the validity events with a higher-level data-invariant validity event, for the case that where one data-invariant validity event is required to represent either of the lower-level validity events in analysis.

We present in Fig. 7 a state diagram for the process obtained by applying fusion process to "glue" both *data1* and *data2*. To obtain the full-blown transition-event specification, we first compute the product of the high-level specification in Fig. 5 with the fusion events in Fig. 6 (a) and (b), then by hiding the high-level data-validity events. In Fig. 6, only the active edges of the clock signals are represented. Further, we transform the active-edge signals into transition-event signals using the procedures from [7] and [13]. The result of these transformations is the full specification of signal transitions shown in Fig. 7. These transformations are implemented in FIREMAPS.

4   Case Study

In subsection 4.1, we present the construction of asynchronous wrapper models. In subsection 4.2, we discuss the refinement-based verification procedures. Finally, we report our verification results and analysis in subsection 4.3.

4.1   Models Construction

We assume the initial states for each component with all signals are low. The process space specifications for each asynchronous wrapper port type are as shown in Fig. 3. To obtain hazard-free data (stable data) in the design from [8] and [9], the incoming data for asynchronous wrapper are registered before being put onto data channel. Accordingly, a D flip-flop model is included to the data path in our asynchronous wrapper model and validity event data can be obtained at the output of D flip-flop.

To describe the edge-triggered components (T, D), as well as the latch component, we use the active-edge specification models of [7] and [13]. For the lowest-level equations in Fig. 3, we built hazard-intolerant models in FIREMAPS, assuming each equation is implemented by a complex gate (a single standard Boolean gate). The data transfer model of the asynchronous wrapper is constructed based on validity events (see section 3). Fig. 7 shows the resulting model. The translator specification is based on the state machine shown in [8] for that component. The four-phase mutex used to implement clock generation as shown in [8] and [9] was modeled by its standard transition-event process. D-output port timing constraint reported at [9] is included in our model, and is built in the form of chain constraint (see subsection 2.1) to represent this timing restriction in design.

4.2   Refinement-Based Verification Cycle

Our first refinement-based verification cycle is started by adding the D-output port delay constraints reported in [9] to our asynchronous wrapper's intermediate specification model, and checking the refinement with its high-level specification. FIREMAPS checks refinement and diagnoses the dangers detected in the implementations by providing a set of witness executions that constitute counter-examples to refinement.

We repeat the verification cycle by adding new relative timing constraints to avoid the witness executions detected at previous verification cycles. Relative timing constraints are found through a detailed analysis of asynchronous wrapper design with the help of witness executions obtained from FIREMAPS. These cycles repeat until FIRMAPS eventually reports that the refinement condition is satisfied.

4.3   Verification Results

To check the claim in [9] that hazard-free implementation of port controller could be obtained by synthesizing its extended-burst-mode specification by the 3-D synthesis tool, we verify the individual control components D-input control port and P-input control port against the hazard-intolerant models of their equations from [9].

Subsequently, we present the most significant of our detected relative timing constraints. We explain the problems indicated by the counter-examples to refinement that are obtained by FIREMAPS when one of the respective constraints is not included, but all the other constraints are included. This shows how ignorance of each constraint in the asynchronous wrappers design may lead to system failures and improper data manipulation under particular timing conditions.

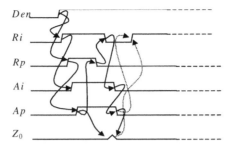

Fig. 8. Illustration waveforms (drawn) of D-input port hazard

Control Ports Verification. Here we verify the refinement relationship between specification and implementation for D-input ports and P-input ports (Fig. 3), respectively. Both D-type and P-type control ports are used to control data transfer between two independently clocked domains in asynchronous wrappers. To ensure robust data transfer under fewer delay constraints, hazard-free control port circuitry is preferred. Our verification results show that the refinement relationship does hold for P-input, but does not hold for D-input. In other words, the implementation of the P-input is hazard-free, while a hazard exits in the implementation of the D-input.

Assuming low initial states of all voltage signals, FIREMAPS reports a witness execution: $Den+$ $Ri+$ $Rp+$ $Ai+$ $Ap+$ $Rp-$ $Ri-$ $Ai-$ $Ap-$ $Ri+$ for D-input (Fig. 3(c)), showing that a race condition exists at Z_0. Notice that: starting initially with $Den+$, after $Rp-$, with certain propagation delay, Z_0 should rise (Fig. 8). However, if the delay of $Ri-$ to $Ai-$ (clock pause releasing acknowledge) plus the propagation delay of $Ai-$ to $Ap-$ is less than the propagation delay of $Ap+$ (previous Ap event) to Z_0+, then the Z_0+ event will be cancelled, causing Ri to rise again. Fig. 8 depicts the corresponding trace waveform (drawn). Further, since there are no clock events to trigger the T flip-flop, no Den events will be generated. Therefore, the corresponding local clock will be stopped indefinitely, and this stopping effect may ripple through other clock domain since no data acknowledge will be fired subsequently by the local domain, thus causing system stall.

We further find that the implementation of the translator block does not refine its specification. Please refer to Fig. 3 (f) for detailed specification and implementation description of the translator (after Fig. 5 in [8]). From an initial state where all signals are low, a $Ts+$ event cannot occur after $Ti+$ is fired. This consequently halts events that are supposed to be fired after $Ts+$. If there are no Ts events to trigger the sampling action at the receiver side, no data transfers will occur. This flaw can be fixed by re-synthesizing the state-graph specification in Fig. 3 (f).

Overall Data Transfer Verification. Here we verify a step of communication refinement by checking whether our data-transfer specification in Fig. 7 is satisfied by the channel configuration in [9]. At the present stage, this part of our verification is based on safety models, which only detect the presence of invalid events. Further investigation will be needed to verify liveness properties by using stronger specifications. By applying our verification techniques, we detect several relative timing constraints that were not reported by the designers. To simplify our presentation of results, we only refer to the validity events data1-valid and data2-valid below, instead of any data event. Relative timing constraints are represented as chain constraints. Signal labels refer to Fig. 4.

1. $D(Ti+ Ts+ lclk2+ lclk2- Ts-) < D(Ti+ Ti-)$. The $Ts+$ event, which should indicate to the receiver block that data has been received, should be fired and become stable within two consecutive Ti events. In other words, the delay from $Ti+$ to $Ts+$ should be less than half the period of clock 2. Failing to satisfy this constraint might lead to data being sampled twice at the receiver side, which leads to erroneous duplication of data items. The worst-case scenario is where every data item is duplicated at the receiver side. In [12], there are no mentions of this danger for duplication. Even though the duplication of items might be fixed inside the receiver block by another level of the communication protocol, the duplication would still undesirable because the computation tasks for receiver would be doubled.

2. $D(Ai2+ Ti+ Ts+) < D(Ai2+ Ap+ Rp- Ai2- lclk2+)$. The delay from $Ai2+$ (which is to acknowledge the pausing of clock2) to $Ts+$ (which is to indicate the receiver the arrival of data) should be less than the delay from $Ai2+$ to restart the clock2. Otherwise, the receiver will not sample the available data at its data input, due to absence of a triggering event $Ai2+$; moreover, the data which was supposed to be sampled by the receiver will be flushed away by the next incoming data by restarting of clock2. In this situation, the data loss will be permanent and unrecoverable.

3. $D(Den+ Rp+ Ap+) < D(Den+ data\text{-}valid)$. Data should be put at the input port of the latch before the latch switches from transparent to opaque; otherwise, before data getting stable, improper data states will propagate through latch and be sampled by receiver. This constraint sets a delay time bound for the latch to switch its state.

4. $D(Rp+ Ap-) < D(Rp+ Ts+ lclk2+)$. A $Ts+$ event should be issued later than the $Ap-$ event to ensure a stable and valid data at the output of the latch, which was triggered by $Ap-$ and switched to opaque state already; otherwise, the $Ts+$ event will trigger the receiver to sample a data item which is not guaranteed to be correct.

5. $D(Pen+ Ai2+) < D(Pen+ lclk2+)$. The delay path from the P-input enabling signal $Pen+$, to the clock pausing acknowledge signal, $Ai2$, should take less time than the issuing of the next $lclk2$ event; otherwise, the next P-input enabling signal can be ignored as the result of a race condition.

6. $D(Pen+ Rp- Ri2+ lclk2+ lclk2- Pen-) < D(Pen+ Rp+ Ri2+ lclk2+ lclk2- Ai2+ Ti+ Ri2-)$. The relative timing interval between $Rp+$ and $rclk2+$ is arbitrary. If $Rp+$ is issued closely enough to $lclk2+$, then, $Ri2+$, which was supposed to be triggered by both $Rp+$ and a Pen event ($Pen+/-$) can not win the arbitration over $lclk2+$. Therefore, $lclk2$ will not be paused immediately after the arrival of $Rp+$, the next clock event $lclk2-$ will be fired, and, further, Pen will be reset. Thus, a Ti event would be canceled, before the *Ts* is set to high. Meanwhile, the acknowledge *Ai2* will still be sent to the D-port, though no data is sampled by the receiver. Moreover, if the implementation of the translator is not totally hazard-free, *Ts* will quickly return to low if *Pen-* is issued right after a *Ts+* event. If the clock2 pause request *(Ri2)* cannot withdraw before *Pen-* comes, *Ti* will be reset again and still might affect the state of *Ts*. We add the above constraint, which implies the delay from *lclk2-* to *Pen-* is longer than the delay from *Ai2+* to *Ri2-*, so that the data item will not be lost during transfer.

5   Conclusion

In [9], the authors introduced an elegant design, called asynchronous wrapper, used to interface locally-synchronous domains to a global asynchronous handshake environment. This design was reported in [9] to be synthesized by an asynchronous synthesis tool, called 3D, and was considered in [9] to be hazard free. In addition, [9] reported several delay constraints to avoid race conditions in their circuits.

We introduce a novel technique to construct and verify specifications for data communication between independent clock domains. We apply our technique to the asynchronous wrapper design in [9], and we find several hazards, additional race conditions, and other faults. We indicate the corresponding observable failures.

Hazards and race conditions often escape detection by non-exhaustive methods, such as simulation, prototyping, and even testing of a few fabricated circuits; given the manufacturing variations and parameter fluctuations present in large systems on chip, only a minor percentage of the fabricated circuits can have the same delay configuration as the tested or simulated ones. We believe that the availability of formal verification techniques for communication refinement is necessary for the creation of robust GALS architectures and for the subsequent acceptance of such architectures by the industry.

Acknowledgments. We thank Mark De Clercq, Mark Greenstreet, and the anonymous CHARME reviewers for suggestions of improvement of the presented material. We are grateful for financial support from NSERC, FCAR, CFI, MEQ, and McGill University.

References

1.   Bormann, D., Cheung, P.: "Asynchronous Wrapper for Heterogeneous Systems." In Proc. Int. Conf. Computer Design (ICCD), Oct. 1996.
2.   Brown, G., Luk, W., O'Leary, J.: "Retargeting a Hardware Compiler Proof Using Protocol Converters." In Proc. Int. Symp. on Advanced Research in Asynchronous Circuits and Systems (ASYNC), pp: 54-64, 1994.
3.   Calvert, K.L.; Lam, S.S.: "Formal methods for protocol conversion." In IEEE Journal on Selected Areas in Communications, Jan. 1990, pp: 127 –142.
4.   Chapiro, D. M.: Globally-Asynchronous Locally-Synchronous Circuits. PhD thesis, Stanford University, U.S.A., Oct. 1984.

5. Dill, D. L.: Trace theory for Automatic Hierarchical Verification of Speed-Independent Circuits. (An ACM Distinguished Dissertation.) MIT press, 1989.
6. Hoare, C. A. R.: Communicating Sequential Processes. Prentice Hall, 1985.
7. Kong, X., Negulescu, R.: "Formal Verification of Pulse-Mode Asynchronous Circuits." In Proc. Asia and South Pacific Design Automation Conference (ASP-DAC), Jan. 2001.
8. Muttersbach, J,. Villiger, T,. Fichtner, W.: "Practical Design of Globally-Asynchronous Locally-Synchronous Systems." In Proc. Int. Symp. on Advanced Research in Asynchronous Circuits and Systems (ASYNC), Mar. 2000.
9. Muttersbach, J,. Villiger, T,. Kaeslin, H,. Felber, N,.Fichtner, W.: "Globally-Asynchronous Locally-Synchronous Architectures to Simplify the Design of On-Chip Systems." In Proceedings of ASIC/SOC Conference, pp. 317 -321, 1999.
10. Myers, C.J.; Rokicki, T.G.; Meng, T.H.-Y: "POSET timing and its application to the synthesis and verification of gate-level timed circuits." IEEE Transactions on Computer-Aided Design of Integrated Circuits and Systems, Volume:20, pp: 769 -786, June 1999.
11. Negulescu, R. : Process Spaces and Formal Verification of Asynchronous Circuits. PhD thesis, University of Waterloo, Canada, 1998.
12. Negulescu, R.: "Process spaces." In Proc. Int. Conf. on Concurrency Theory (CONCUR), pp. 196-210, Aug. 2000.
13. Negulescu, R., Kong X.: "Semi-hiding Operators and the Analysis of Active-Edge Specifications for Digital Circuits." In Proc. Int. Conf. on Application of Concurrency to System Design (ICACSD-2001), Apr. 2001.
14. Negulescu, R., Peeters, A.: "Verification of speed-dependences in single-rail handshake circuits." In Proc. Int. Symp. on Advanced Research in Asynchronous Circuits and Systems, pp. 159-170, 1998.
15. Okumura, K.: "Generation of proper adapters and converters from a formal service specification." In Proc. of Ninth Annual Joint Conference of the IEEE Computer and Communication Societies. The Multiple Facets of Integration, pp: 564 –571, 1990.
16. Peeters, A. M. G.: Single-Rail Handshake Circuits. PhD thesis, Eindhoven Univ. of Tech nology, Eindhoven, The Netherlands, Jun. 1996.
17. Peña, M. A., Cortadella, J., Kondratyev, A. and Pasor, E.: "Formal verification of safety properties in timed circuits." In Proc. Int. Symp. on Advanced Research in Asynchronous Circuits and Systems (ASYNC), pp 2-12, 2000.
18. Rowson, J.A., Sangiovanni-Vincentelli, A.: "Interface-Based Design." In Proceedings of the 34th Design Automation Conference (DAC), pp.: 178 -183, 1997.
19. Tao, Z.P., v. Bochmann G., Dssouli, R.: "An efficient method for protocol conversion." In Proc. Int. Conf. on Computer Communications and Networks, pp. 40-47, 1995.
20. Yun, K. Y., Donohue, R. P.: "Pausible clocking: a first step toward heterogeneous systems." In Proc. Int. Conf. Computer Design (ICCD), pp. 118 –123, 1996.
21. Yun, K. Y., Dill, D. L.: "Automatic synthesis of extended burst-mode circuits: Part I and Part II." IEEE Transactions on Computer-Aided Design, vol. 20, no. 2, pp. 101-132, 1999.

Using Abstract Specifications to Verify PowerPC[TM][*] Custom Memories by Symbolic Trajectory Evaluation

Jayanta Bhadra[1,2], Andrew Martin[1], Jacob Abraham[2], and Magdy Abadir[1]

[1] Motorola Inc.
[2] The University of Texas at Austin
Jayanta.Bhadra@Motorola.Com

Abstract. We present a methodology in which the behavior of a switch level device is specified using abstract parameterized regular expressions. These specifications are used to generate a finite automaton representing an abstraction of the behavior of a block of memory comprised of a set of such switch level devices. The automaton, in conjunction with an Efficient Memory Model [1], [2] for the devices, forms a symbolic simulation model representing an abstraction of the array core embedded in a larger design under analysis. Using Symbolic Trajectory Evaluation, we check the equivalence between a register transfer level description and a schematic description augmented with abstract specifications for one of the custom memories embedded in the MPC7450 PowerPC processor.

1 Introduction

At Somerset, Symbolic Trajectory Evaluation (STE) is routinely used to check equivalence between Register Transfer Level (RTL) and switch level views of embedded custom memories [3]. The assertions are generated automatically from the RTL description of the memory under verification using the technique described by Wang [4]. The assertions are verified against a switch level model of the circuit using STE. The switch level model is obtained from transistor netlists using Anamos [5], which partitions the design into channel connected subcomponents, and then analyzes each component as a set of simultaneous switch equations. Although Anamos is quite sophisticated as far as switch level analyzers are concerned, there are still many analog circuit effects that it ignores. That is why, in spite of its sophistication, an Anamos switch level analysis, which views the circuit as a system of switches, is unsuited to a custom static RAM core circuit, which is an inherently analog design. It is quite easy to demonstrate input sequences for which the resulting switch level model predicts one result, while a more sophisticated analog simulation would predict another.

One obvious approach to address these problems would be to increase the sophistication of the switch level model. Given that the circuitry in a custom

---

[*] PowerPC is a trademark of the International Business Machines Corporation, used under license therefrom.

T. Margaria and T. Melham (Eds.): CHARME 2001, LNCS 2144, pp. 386–402, 2001.
© Springer-Verlag Berlin Heidelberg 2001

static RAM core solves a fundamentally analog problem: how to drive a very large load with a very small bitcell, we feel that this approach is unlikely to prove satisfactory in the long term. This is because, an improved switch level model would continue to suffer from the same inaccuracy problem – representing analog devices by switches.

In reality, the array core is designed to operate over a very limited range of input stimuli. Each such "pattern" is validated by extensive analog simulation over a variety of process corners and operating conditions using a circuit simulator such as Spice. These "certified" patterns are known to work. Any other pattern that has not been simulated with Spice, is assumed to fail. This paper presents a notation, based on regular expressions, for describing such "certified" patterns, and the effect that each has on the internal state and outputs of the RAM core. From this notation, we build an automata based simulation model representing an abstraction of the array cores embedded in the larger design under analysis. This abstract model updates its internal state – representing the state of the RAM – as specified, provided that the input conforms to at least one "certified" input pattern. If, however, an "uncertified" input sequence is presented, the RAM internal state and its outputs are set to the "unknown" value "X", where they will remain until altered by a new "certified" input pattern.

In addition to being more conservative than the Anamos generated model, our abstraction of the array cores is also more efficient. The verification methodology, in which the abstraction is used, is based upon symbolic simulation. In the course of a verification, symbolic values are "written" to symbolic addresses within the memory core. Our BDD based implementation (as reported by Krishnamurthy et al in [3]), using an Anamos generated model, must necessarily maintain several unique BDDs for each bitcell in the array. This leads to a BDD table whose size is at best a linear function of the size of the array being verified. In contrast, the abstract approach is able to make use of the well known Efficient Memory Model (EMM) due to Velev et al [1,2], which represents the state of the memory using an association list. Each symbolic write into an array need only add one element to the head of this list. As a result, it is possible to represent the result of a symbolic write using only as many BDD nodes as are required to represent the address and data – typically a logarithmic function of the array size.

The approach presented here represents a fine balance between implementation efficiency and expressive power. The semantics of the regular expressions have been designed to admit a tractable automata based implementation. Moreover, in several places the result is weakened to provide a tractable implementation at the expense of possible false negative verification results. In practice, we do not believe that these false negative results will be problematic.

An interesting facet of the combination of regular expressions representing an implementation, with STE, is the marriage of two opposing forms of nondeterminism. The traditional partially ordered state space of STE represents a form of demonic non-determinism. Values that are lower in the partial order represent an undetermined choice between the values which lie above them. A specification is satisfied if and only if the consequent portion is satisfied by the weakest *trajectory* satisfying the antecedent. That is, every non-deterministic choice that the implementation can make must satisfy the specification. Increasing the amount of non-deterministic choice reduces the number of specifications a given model will satisfy. In contrast, non-determinism that results from alternation in a regular expression describing legal input sequences for a RAM, is angelic. It suffices to find one satisfactory production for any given input sequence. Increasing the amount of non-deterministic choice increases the number of specifications a given model will satisfy.

In this paper we present a methodology by which the user is allowed to represent or *abstract* a block of switch level devices by a set of regular expression specifications. A state machine model is obtained from the specifications and an EMM is used to store values written into the devices. The state machine controls the read/write operations performed on the EMM and provides the outputs from the abstracted block. Our goal is to show that the assertions generated from the RTL description of the circuit corresponds to the switch level description of the circuit augmented with the composition of the state machine model and the EMM. As a future work, we propose to show that the composition of the state machine model and the EMM is a conservative approximation of the abstracted switch level device block. This would let us claim that the switch level model *implies* the RTL model.

2   STE Background

Symbolic Trajectory Evaluation [6] requires a system expressed as a model of the form $\langle S, \preceq, y \rangle$ where S is a set of states, $\preceq$ is a partial order on the states ($\preceq \subseteq S \times S$) and $y : S \to S$ is a state transition function. S must form a complete lattice and y must be monotonic under $\preceq$, i.e., if $s \preceq t$ then $y(s) \preceq y(t)$.

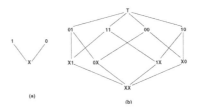

Fig. 1. Lattice Structures

Traditionally, for switch level verification, STE operates over the ternary logic domain $\mathcal{T} = \{0, 1, X\}$, where X denotes an "unknown" value. In order to formalize the concept of an "unknown" value, define a partial order $\preceq$ on $\mathcal{T}$ as illustrated in Fig. 1(a), so that $X \preceq 0$, and $X \preceq 1$. States of an STE model are vectors of elements taken from $\mathcal{T}$. The partial order over $\mathcal{T}$ is extended pointwise to yield a partial order on the space $\mathcal{T}^n$. Unfortunately, $\langle \mathcal{T}^n, \preceq \rangle$ is not a complete lattice, since the least upper bound does not exist for every pair of vectors in $\mathcal{T}^n$. Introduction of a new *top* element, $\top$, solves this problem. Intuitively, $\top$ can be viewed as an "overconstrained" state, in which some node is both 0 and 1 at the same time. This makes the state space to be $S = \mathcal{T}^n \cup \{\top\}$. For example, Fig. 1(b) shows S for $n = 2$.

In STE, the state of the circuit model includes values on the input, internal and output nodes of the circuit. So n is the collective number of all circuit nodes. The state transition function expresses constraints on the values the nodes can take one time unit later, given the node values at the current time, under some discrete notion of time. Since the value of an input is controlled by the external environment, the circuit does not constrain its value; hence the transition function sets it to X. On the other hand, the value of any other node is determined by the functionality of the circuit and the circuit state. Figure 2 shows the next state function of a unit delay inverter.

Fig. 2. Transition function of a unit delay inverter (in.out)

A sequence of states, or *behavior*, $\sigma = s_0 s_1 s_2 \ldots$ defines a trajectory if it has at least as much information as given by the application of the next-state function to successive elements,

$$\forall i \geq 0 : (y(s_i) \preceq s_{i+1})$$

The set of all trajectories $\{\sigma\}$ is called the language of the system, L. Specifications are *trajectory assertions* of the form $A_\phi \rightarrow C_\phi$. A_ϕ and C_ϕ are functions (called *trajectory formulas*) from valuations, ϕ, of a set of symbolic variables to predicates over state sequences. STE gives a procedure to determine the set of assignments of values to the variables such that every trajectory of the system that satisfies A_ϕ also satisfies C_ϕ.

The complete lattice, S, represents a form of non-determinism that fundamentally affects the validity of a specification in S. Consider two distinct trajectories $\sigma_1, \sigma_2 \in L$. If $\sigma_1 \preceq \sigma_2$, where $\preceq$ is naturally extended over state sequences, then the states in σ_1 are *not higher* than those in σ_2 in the partial order over S. Each state in σ_1 contains less information, or, more Xs or unconstrained node values than the corresponding state in σ_2. Any specification that is satisfied by σ_1 has to be satisfied by all the trajectories that are obtained by *all* possible choices on the unconstrained node values in the states of σ_1. Constraining nodes to values stronger than those in the states of σ_1 means obtaining trajectories that are necessarily *higher* than σ_1. Since σ_2 is one of the non-deterministic choices that can be made by constraining the node values in the states of σ_1, the specifications that are satisfied by σ_1 constitute a subset of those satisfied by σ_2. So, as one moves *down* the lattice, thus weakening the states in the corresponding trajectories, one decreases the number of specifications that are satisfied by a given circuit model. Hence, the partially ordered state space S represents a demonic non-determinism.

Observation 1 *If Θ is a specification and functions y_1 and y_2 are such that $\forall s \in S, y_1(s) \preceq y_2(s)$, then if the STE model $\langle S, \preceq, y_1 \rangle$ satisfies Θ, then the STE model $\langle S, \preceq, y_2 \rangle$ also satisfies Θ.*

3   Preliminaries of Array Abstraction

3.1   The Broader Picture

We aim at establishing that the state machine model produced from the regular expression definitions, when combined with the surrounding switch level model of the circuit, satisfies the RTL specification. Let M be the block of an array being abstracted, E be its environment circuitry and $\hat{M}$ be the abstraction. Let V_{RTL} be the set of variables in P_{RTL}, the RTL specification, and $\phi \in \mathcal{B}^{V_{RTL}}$ be some assignments to the variables, where $\mathcal{B} = \{0,1\}$. The verification obligation is to show that the composition of the environment and the abstraction models the property: $\forall \phi \in \mathcal{B}^{V_{RTL}} \{(E \parallel \hat{M}) \models P_{RTL}(\phi)\}$. As a future work, we intend to show that the state machine model is a conservative approximation of the switch level model, thus establishing $(M \rightarrow \hat{M})$, which in turn would imply $(E \parallel M) \rightarrow (E \parallel \hat{M})$. This, in conjunction with the current work, would prove that the switch level model satisfies the RTL specification.

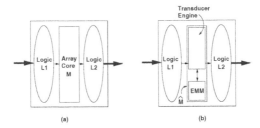

Fig. 3. Abstracting an Array Core

The array core (M) is a piece of logic containing all the bitcells of the array, that accepts certain inputs from some logic module ($L1$ in Fig. 3(a)), stores the values of the bitcells and produces outputs to another logic module ($L2$). The environment E is modeled collectively by $L1$ and $L2$. Figure 3(b) shows the abstract module ($\hat{M}$) that replaces the array core. The sub-module labeled "Transducer Engine" is a state machine model generated from abstract specifications of the array core. It accepts the inputs from $L1$, updates its internal state(s), *writes* or *reads* data to and from the sub-module labeled as "EMM" and provides its outputs to $L2$.

3.2   A Simple Example

Bitcells represent an integral part of an array core in custom-designed memories. They are read from or written to by the application of complex sequences of values to specific control signals (like precharge, word selects etc). In order to abstract an entire column of bitcells forming a single channel connected subcomponent, one needs to look at the inputs of the column and enumerate the set of legal sequences of control signal assignments for reads and writes on the column. One way to represent such a set is by using what we call *Parameterized Regular Expressions* (PREs). PREs extend regular expressions to include variables and outputs. Variables provide a way in which the description becomes succinct and outputs help define an automaton that produces output strings when input strings are provided. Before we present the PRE syntax details in Sect. 4, let us give a simple example.

In our example the array core is controlled by the signals $wl, precharge, wr\_en$ and din which are the set of word select lines, the precharge line, the write enable and the set of data input lines respectively. We define a predicate "$safe\_state$" to represent the condition in which none of the w word select lines (wl) are high.

$$safe\_state() = \bigwedge_w (\neg wl[i]), i = 0, 1, \ldots, w - 1$$

The predicate $safe\_state$ is special since both read and write condition sequences start from and terminate in the $safe\_state$. Also $safe\_state$ represents the state of the circuit when nothing is happening as far as read or write is concerned.

In the following discussion, high and low mean logical one and zero respectively. Consider a write operation where data d is written into location i. Initially, the $safe\_state$ condition holds. After that, the $safe\_state$ condition should keep holding and at the same time the precharge should be high for at least 10 time steps. Following that the $safe\_state$ condition should keep holding even if the precharge goes low. Then exactly one word select line (line i) should go high and the write enable should be low. Then the same word select line should remain high in conjunction with the write enable being high, precharge being low and some data (d) being on the data lines. This is maintained for at least 6 time steps and then the write happens in the addressed bitcell. After that the condition is weakened to keep the same word select line high and if the write enable is high then the data lines are also maintained at d.

This family of conditions is expressed by the following PRE in which one can specify sequences of conditions by using semi-colons; provide annotations for the time for which a particular condition should hold; and specify when the writes to the bitcells actually occur. In the following PRE, the predicate $one\_hot(wl)$ represents the condition that exactly one word select line is high and all others are low.

$var\ i : 0..3; ;$ (* address width *)
$var\ d : 0..31; ;$ (* data width *)
$write(i, d) = safe\_state();$
$\qquad (safe\_state() \land precharge)^{10};$
$\qquad safe\_state()^*;$
$\qquad (one\_hot(wl) \land wl[i] \land \neg wr\_en)^* ;$
$\qquad (one\_hot(wl) \land wl[i] \land wr\_en \land (din = d) \land \neg precharge)^5;$

$$(one\_hot(wl) \wedge wl[i] \wedge wr\_en \wedge (din = d) \wedge \neg precharge)^*;$$
$$one\_hot(wl) \wedge wl[i] \wedge wr\_en \wedge (din = d) \wedge \neg precharge :: WR(i, d);$$
$$(one\_hot(wl) \wedge wl[i] \wedge (wr\_en \Rightarrow (din = d)))^*;;$$

The annotation WR signifies that the write (updating the bitcell contents) completes at this point of execution. Similar to the write operation, we can specify the read operation using condition sequences over input signals of the bitcell column. A bitcell column can thus be abstracted by a set of PREs representing all its transactions (reads and writes).

4   Parameterized Regular Expressions and STE

In this section, we will define a syntax for Parameterized Regular Expressions (PREs) and show how a finite automaton can be generated from a PRE definition. We will then prove that the finite automaton is suitable for use in a symbolic simulator employing STE.

4.1   PRE Definition and Finite Automaton

A PRE expresses a family of legal behaviors. Each behavior represents (a) a sequence of assignments to the inputs of the abstraction, $\hat{M}$, (b) a sequence of assignments to the outputs of $\hat{M}$ and (c) a set of writes to the memory in $\hat{M}$. All of these are associated with precise timing information detailing durations and relative times. The alternations in the PRE definition result in an angelic form of non-determinism. A sequence of conditions on the input signals that matches any of the alternative behaviors defined by a PRE, is accepted as legal. Thus by increasing the alternation in a PRE definition one can increase the number of legal behaviors. PREs are defined with respect to a set of boolean valued variables, A, a finite input alphabet Σ_i, and a finite output alphabet Σ_o which is partially ordered by the relation $\preceq_{\Sigma_o}$. Σ_o forms a complete lattice with respect to $\preceq_{\Sigma_o}$ with $\perp_{\Sigma_o}$ and $\top_{\Sigma_o}$ as the bottom and top elements respectively.

Definition 1 PRE:

$$
\begin{aligned}
R \ ::= \ & P \ ; \ Q \\
| \ & P \ + \ Q \\
| \ & P^* \\
| \ & P^n \\
| \ & a :: o
\end{aligned}
$$

Here P and Q are PREs, and the operators ; $+$ and $*$ are sequence, choice, and Kleene Star respectively. P^n represents a short hand for the sequence $P; P; P; \ldots$ (n times). In the terminal "$a :: o$", a is a predicate over variable assignments in $\mathcal{B}^A$ and input symbols in Σ_i; o is a function mapping variable assignments in $\mathcal{B}^A$ to output symbols in Σ_o, where $\mathcal{B} = \{0, 1\}$. Given a PRE R, the Non-Deterministic Finite Automaton (NDFA) with ϵ is a tuple $N_\epsilon(R) = \langle s, t, S, T, W, A, \Sigma_i, \Sigma_o \rangle$ where A, Σ_i, Σ_o are the set of variables, the input alphabet and the output alphabet of the PRE respectively, S is a finite set of states, $s \in S$ is the start state, $t \in S$ is the end state, T is a set of transitions, $T \subseteq S \times \mathcal{B}^A \times (\Sigma_i \cup \{\epsilon\}) \times S$ and W is an output function, $W : S \times \mathcal{B}^A \to \Sigma_o \cup \{\epsilon\}$. A transition in T is of the form $s_1 \xrightarrow{I,v} s_2$, where $s_1, s_2 \in S$ are the source and destination states respectively, v is an assignment of values from $\mathcal{B}$ to the variables, and $I \in \Sigma_i \cup \{\epsilon\}$ is either an input symbol, or the special symbol ϵ. A *run* of N_ϵ is a sequence $vs_0, vs_1, vs_2, \ldots$ where $v \in \mathcal{B}^A$ is a variable assignment, $\forall j \geq 0 : s_j \xrightarrow{I_j,v} s_{j+1} \in T$, where $\forall j \geq 0 : I_j \in \Sigma_i \cup \{\epsilon\}$, $\forall j \geq 0 : s_j \in S$ and $s_0 = s$, the starting state of N_ϵ. The corresponding output of N_ϵ is the sequence $\sigma_0\sigma_1\sigma_2\ldots \in \Sigma_o^*$, such that $\forall j \geq 0 : \sigma_j = W(s_j, v)$. The role of v can be understood by considering a different run $v's_0, v's_1, v's_2, \ldots$, such that $v' \neq v$. Although the state sequence involved in this run is same as that of the former, the outputs obtained from the two runs are different as $v' \neq v$.

Hopcroft and Ullman give a description of regular expressions and the corresponding construction of automata from them [7]. The regular expressions described by Hopcroft result in automata that are language acceptors. PREs, in contrast, are transducers – automata that output a string when provided with an input string. Moreover, PREs have explicit variables which are absent in the classical automata theory. Nonetheless, a few simple modifications enable us to construct an NDFA with ϵ transitions (N_ϵ) in a fashion that resembles the classical automata theory. For a PRE R, $N_\epsilon(R)$ can be constructed inductively using the definition of PREs. For the PRE of the form "$a :: o$", we construct an NDFA containing only two states s and t and a set of transitions $T = \{s \xrightarrow{I,v} t | a(I,v)\}$. The resultant NDFA is $\langle s, t, \{s,t\}, T, W, A, \Sigma_i, \Sigma_o \rangle$, where A is a set of variable names, $W(s, v) = \perp_{\Sigma_o}$ and $W(t, v) = o(v)$.

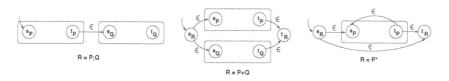

Fig. 4. Inductive Construction of N_ϵ

The method for inductively constructing NDFAs from PREs of the form $P;Q$, $P + Q$ and P^*, given the NDFAs of P and Q, can be derived by simple modifications of the techniques described in Hopcroft [7] as illustrated in Fig. 4. Let PREs P and Q generate ϵ-NDFAs $N_\epsilon(P) = \langle s_P, t_P, S_P, T_P, W_P, A, \Sigma_i, \Sigma_o \rangle$ and $N_\epsilon(Q) = \langle s_Q, t_Q, S_Q, T_Q, W_Q, A, \Sigma_i, \Sigma_o \rangle$ where the respective state sets are disjoint, i.e., $S_P \cap S_Q = \phi$. The general inductive construction procedure can be characterized by one that adds a few new transitions to *merge* the NDFAs and unites the state spaces and the rest of the transitions. In the following discussion, we assume that $s_R, t_R \notin S_P, S_Q$. If $R = P;Q$ then $N_\epsilon(R) = \langle s_P, t_Q, S_P \cup S_Q, T_P \cup T_Q \cup \{t_P \xrightarrow{\epsilon} s_Q\}, W_R, A, \Sigma_i, \Sigma_o \rangle$ where $W_R(x, v) = \epsilon$ if $x = s_Q$, $W_P(x, v)$ if $x \in S_P$, and $W_Q(x, v)$ otherwise. If $R = P + Q$ then $N_\epsilon(R) = \langle s_R, t_R, S_P \cup$

$S_Q \cup \{s_R, t_R\}, T_P \cup T_Q \cup \{s_R \overset{\epsilon}{\to} s_P, s_R \overset{\epsilon}{\to} s_Q, t_P \overset{\epsilon}{\to} t_R, t_Q \overset{\epsilon}{\to} t_R\}, W_R, A, \Sigma_i, \Sigma_o \rangle$
where $W_R(x, v) = \epsilon$ if $x \in \{s_P, s_Q, t_R\}$, $\perp_{\Sigma_o}$ if $x = s_R$, $W_P(x, v)$ if $x \in S_P$, and
$W_Q(x, v)$ otherwise. If $R = P^*$, then $N_\epsilon(R) = \langle s_R, t_R, S_P \cup \{s_R, t_R\}, T_P \cup \{s_R \overset{\epsilon}{\to} s_P, s_R \overset{\epsilon}{\to} t_R, t_P \overset{\epsilon}{\to} s_P, t_P \overset{\epsilon}{\to} t_R\}, W_R, A, \Sigma_i, \Sigma_o \rangle$ where $W_R(x, v) = \perp_{\Sigma_o}$ if
$x = s_R$, ϵ if $x \in \{t_R, s_P\}$, and $W_P(x, v)$ otherwise.

4.2   Obtaining a Deterministic Automaton from N_ϵ

Given a PRE, we aim at constructing an STE-usable Deterministic Finite Automaton (DFA)[1] . Given a regular expression described by Hopcroft [7], we can construct an NDFA with ϵ transitions (state space say S) and then convert it to an NDFA without ϵ transitions (state space S) and then determinize it using standard subset construction methods to obtain a DFA (state space 2^S) that acts as an acceptor of the language of the regular expression. However, since we are dealing with variables and outputs and our final automaton is going to be a transducer over infinite strings, the conversion from a PRE definition to an STE-usable DFA involves a series of automata constructions. Parallels can be drawn between our procedure and the standard procedure.

We begin by removing the so-called ϵ transitions to yield an automaton that consumes one input symbol and produces one output symbol during each state transition. At this "ϵ-removal" step we also introduce some additional non-determinism to enable non-deterministic *restarting* of runs in the automaton [2]. We then perform a determinization step, yielding a deterministic automaton, D. The standard determinization procedure would result in an exponential blowup in the size of the state space, which would be impractical. Instead, we use a sparse state space representation and construct a determinized conservative approximation to the original automaton N. Finally, we introduce the standard STE lattice structure over the input, state and output spaces to yield a determinized transition system suitable for use as a model for STE. The construction thus produced is a conservative approximation of the precise intended semantics of PREs, and thus in theory may give false negative verification results. Our admittedly limited experience to date suggests that this does not become a problem in practice.

Given a state q in N_ϵ, let ϵ-*closure*(q) denote the set of all states that are reachable from q via only ϵ transitions. Let ϵ-*closure*(Q), where Q is a set of states, be $\bigcup_{q \in Q} \epsilon$-*closure*$(q)$, that is, the union of the ϵ-closures of all the elements of Q.

Given a PRE R and $N_\epsilon(R) = \langle s, t, S, T, W, A, \Sigma_i, \Sigma_o \rangle$, define an NDFA without ϵ transitions, N with state space $S \times \mathcal{B}^A$. States of N consist of two components – a control component represented by S and a data component represented

---

[1] A DFA that has i) a state space that forms a complete lattice under a certain partial order and ii) a monotonic next state function.

[2] The non-determinism is added at this stage rather than being included in the original automaton N_ϵ purely as a technical convenience to facilitate the inductive definition of N_ϵ.

by assignments of boolean values to the variables in A. Define the starting state $s_N = \{s_0\} \times \mathcal{B}^A$ as the set of starting states of N, signifying that the variables can be non-deterministically assigned values when the execution of N starts. The transitions of N are derived from two sources. The first, T_{N_1}, results from the non-ϵ transitions of N in the classical way: $(s, v) \xrightarrow{I} (t, v) \in T_{N_1}$ if and only if $\exists r : r \in \epsilon\text{-}closure(s) \wedge r \xrightarrow{I, v} t \in T$. The second source, T_{N_2}, is introduced to allow the automaton to "restart" at any point, assigning new values to the variables, $(s, v_1) \xrightarrow{I} (t, v_2) \in T_{N_2}$ if and only if there is a starting state $(s_0, v_2) \in s_N$ such that $(s_0, v_2) \xrightarrow{I} (t, v_2) \in T_{N_1}$.

Definition 2 NDFA:
Given a PRE R and $N_\epsilon = \langle s_0, t, S, T, W, A, \Sigma_i, \Sigma_o \rangle$, define an NDFA N obtained from N_ϵ to be the tuple $N = \langle s_N, S_N, T_N, W_N, A, \Sigma_i, \Sigma_o \rangle$ where A remain unchanged from N_ϵ, $S_N = S \times \mathcal{B}^A$ is the set of states and $s_N = \{s_0\} \times \mathcal{B}^A$ is the set of starting states of N. Define the transition relation $T_N = T_{N_1} \cup T_{N_2}$, where T_{N_1} and T_{N_2} are as already defined. Define the output function $W_N : S_N \to \Sigma_o$ to be $W_N(\langle s, v \rangle) = \perp_{\Sigma_o}$ if $W(s, v) = \epsilon$, else, $W(s, v)$ for all $s \in S, v \in \mathcal{B}^A$.

The standard subset construction method for computing a deterministic automaton, say $\hat{D}$, from N, results in a state space $S_{\hat{D}} = 2^{S \times \mathcal{B}^A}$, which is doubly exponential in A. To arrive at a more efficient solution, we define a DFA D with a state space $S_D = 2^S \times \mathcal{Q}^A$, where $\mathcal{Q} = \{0, 1, X, \top\}$. Thus, a state in S_D consists of a control component from S and a data component consisting of an assignment from $\mathcal{Q}$ to every variable in A. S_D encodes the state information of $S_{\hat{D}}$ in a conservative but concise manner. The subset construction method represents a set of states (P) from N by a single state in $\hat{D}$. The state in D that maps to P is obtained by computing the greatest lower bound of the states in P in the partially ordered state space S_D (the partial order defined later) and hence is a conservative approximation of the entire set P. Every state in the state space S_D can be associated with a set of states in $S_{\hat{D}}$, but not vice versa; thus making S_D more sparsely populated than $S_{\hat{D}}$.

Define a partial order $\preceq_D$ over S_D as $\forall \langle s_1, \phi_1 \rangle, \langle s_2, \phi_2 \rangle \in S_D$, $\langle s_1, \phi_1 \rangle \preceq_D \langle s_2, \phi_2 \rangle$ if and only if $s_1 \subseteq s_2$ and $\phi_1 \preceq \phi_2$. For example, if $S = \{s, t\}$ and $A = \{v_0, v_1\}$, and states $s_0 = \langle \{s\}, 00 \rangle$, $s_1 = \langle \{s\}, 01 \rangle$ and $s_2 = \langle \{s\}, 0\top \rangle$ are members of S_D, then, $s_0 \preceq_D s_2$ and $s_1 \preceq_D s_2$. The partially ordered state space S_D represents a form of angelic non-determinism. The state s_2, that is *higher* in the partial order, represents a non-deterministic choice between related *lower* states s_0 and s_1. A specification that is satisfied by a trajectory σ containing either s_0 or s_1, is also satisfied by one that can be obtained from σ by replacing occurrence of s_0 or s_1 by s_2. As one goes higher up in the partial order the number of specifications that are validated by a given circuit model increases.

The automaton D has a starting state which assigns $\top$ to all the variables in A, representing an angelic non-deterministic choice between all possible values. An execution along a path in D *weakens* the value of each variable as it is assigned in order to satisfy a predicate along a particular arc. A state D represents a set of states in N. A transition in N takes place when its predicate is satisfied by the values of the inputs as well those of the variables associated with the source and destination states. A transition in D assigns the greatest lower bound of the values that are necessary to take the *corresponding set* of transitions in N. Thus the assignment of variables in D is a conservative approximation. Also, the output generated from a state in D is the greatest lower bound of the outputs generated by the corresponding set of states in N, thus making it weaker than the weakest output produced by any state in the state set in N. In practice, being conservative in this construction does not affect the verification.

Definition 3 DFA:
Let R be a PRE with ϵ-NFA and NDFA respectively defined as
$N_\epsilon = \langle s_0, t, S, T, W, A, \Sigma_i, \Sigma_o \rangle$ *and* $N = \langle s_N, S_N, T_N, W_N, A, \Sigma_i, \Sigma_o \rangle$ *where the symbols are as described earlier. Define a DFA obtained from R as*
$D = \langle \langle \{s_0\}, \top^A \rangle, S_D, T_D, W_D, A, \Sigma_i, \Sigma_o \rangle$ *where $\top^A$ is the assignment of $\top$ to all the variables in A and the state space $S_D = 2^S \times \mathcal{Q}^A$. The transition function $T_D : S_D \times \Sigma_i \to S_D$ is defined using a function ψ*

$$\psi(\alpha, \phi, i) = \{(t, a) | \exists s \in \alpha : a \preceq \phi \wedge (s, a) \xrightarrow{i} (t, a) \in T_N \}$$

Define the state transition function, T_D, as

$$\forall \alpha, \phi : \langle \alpha, \phi \rangle \in S_D, \alpha \neq \{\} : T_D(\langle \alpha, \phi \rangle, i) = \langle \alpha', \phi' \rangle$$

where, $\alpha' = \{t | \exists a : (t, a) \in \psi(\alpha, \phi, i)\}$, is the set of projections onto the states obtained from ψ, and, $\phi' = glb\{a | \exists t : (t, a) \in \psi(\alpha, \phi, i)\}$, is the glb of the set of projections onto the assignments to variables obtained from ψ. Define the output function, $W_D : S_D \to \Sigma_o$ as $W_D(\langle \alpha, \phi \rangle) = glb\{W_N(\langle t, \phi \rangle) | t \in \alpha\}$, where W_N is extended from its previous definition by simple monotonic quarternary extension of its domain, making it $W_N : S \times \mathcal{Q}^A \to \Sigma_o$.

The state $\langle \{\}, \top^A \rangle$ can be reached only by accepting an input symbol that neither starts any fresh legal input string nor satisfies any of the outgoing transitions from any state of D. This in turn means that $\langle \{\}, \top^A \rangle$ is the state which acts as a recognizer of an illegal input symbol. Also the output generated from $\langle \{\}, \top^A \rangle$ is $\perp_{\Sigma_o}$. However, we can encounter certain strings that might have a legal prefix of length zero or more followed by substrings that do not start any legal transaction. These illegal substrings need to have an execution on the automaton. This is served by *adding* some *new edges* to D.

Whenever the starting state of D_A is reached, the automaton is starting afresh and the EMM is initialized to the empty list. This is accomplished by the $\perp_{\Sigma_o}$ output at the starting state. Also all the variables are reset to $\top$. D_A can reach the sink state $\langle \{\}, \top^A \rangle$ via transitions inherited from D, where the EMM is reset to the empty list by producing the output $\perp_{\Sigma_o}$. All the variables are also reset to $\top$. The new edges in D_A enables it to continue to remain in $\langle \{\}, \top^A \rangle$ and maintain the variables and the EMM at $\top$ and empty list respectively until the start of a fresh legal transaction is recognized. This mechanism enables D_A to recognize illegal input substrings and produce the required output strings.

Definition 4 Augmented DFA:
Let R be a PRE with ϵ-NFA, NDFA and DFA respectively defined as
$N_\epsilon = \langle s_0, t, S, T, W, A, \Sigma_i, \Sigma_o \rangle$, $N = \langle s_N, S_N, T_N, W_N, A, \Sigma_i, \Sigma_o \rangle$, and
$D = \langle \langle \{s_0\}, \top^A \rangle, S_D, T_D, W_D, A, \Sigma_i, \Sigma_o \rangle$ *where the symbols are defined as ear-*
lier. Define the augmented DFA $D_A = \langle \langle \{s_0\}, \top^A \rangle, S_D, T_A, W_D, A, \Sigma_i, \Sigma_o \rangle$,
where $T_A : S_D \times \Sigma_i \to S_D$ *is defined as:* $T_A(\langle \{\}, \top^A \rangle, i) = T_D(\langle \{s_0\}, \top^A \rangle, i)$
and for all other states $s \in S_D : T_A(s, i) = T_D(s, i)$.

4.3   Obtaining an Automaton for STE

Until now the input variables were assumed to be assigned boolean values. In
order to function in a ternary simulation environment we have to extend the
input alphabet to the ternary domain – one that is partially ordered. This results
into a further extension of the state space. As the input *weakens*, one encounters
values that are lower in the partial order of the input space. Weaker values
represent an undetermined choice between stronger input values that lie above.
A state transition that is *certain* for a given input might become *uncertain*
as the input weakens. Hence the state space extends to $S_F = (\mathcal{T}^S \times \mathcal{Q}^A) \cup$
$\{\top\}$, where $\mathcal{T} = \{0, 1, X\}$ and $\mathcal{Q} = \{0, 1, X, \top\}$, in order to incorporate this
non-determinism. Weaker state assignments represent an undetermined choice
between stronger state assignments. For instance, being *"uncertainly* present"
in a state is weaker than either being "definitely present" or being "definitely
not present" in the state. Also, being "uncertainly present" represents the non-
deterministic choice between being "definitely present" and being "definitely not
present" in the state. A specification is satisfied if and only if the consequent
portion is satisfied by the weakest trajectory (sequence of states) satisfying the
antecedent. As the trajectory weakens (via the weakening of the constituent
states), every nondeterministic choice that is possible in the state space must
satisfy the specification. So increasing the amount of nondeterministic choice
reduces the number of specifications a given model will satisfy. Thus the state
space S_F represents a form of *demonic* non-determinism, wherein, increasing
non-determinism results into non-acceptance of input strings. Parallels can be
drawn between the non-determinism exhibited by S_F and the state space of an
STE circuit model, as discussed in Sect. 2. Later in this section, we will show
that S_F is suitable for STE. An automaton having S_F as its state space, when
viewed as a transducer, has a weaker state (that is lower in the partial order)
producing outputs that are weaker than the stronger states lying above it.

We extend Σ_i to another finite alphabet Σ_i' such that Σ_i' is partially ordered
by the relation $\preceq_{\Sigma_i'}$, $\Sigma_i \subseteq \Sigma_i'$ and Σ_i' is a complete lattice with $\top_{\Sigma_i'}$ and $\perp_{\Sigma_i'}$ as
the top and bottom elements respectively. Also, two unequal symbols $a, b \in \Sigma_i$
are incomparable in $\preceq_{\Sigma_i'}$. Next, we extend the definition of D_A into one which
is obtained by simple monotonic ternary extension of the inputs of D_A.

Definition 5 Final DFA:
Let R be a PRE with ϵ-NFA, NDFA and augmented DFA respectively defined as
$N_\epsilon = \langle s_0, t, S, T, W, A, \Sigma_i, \Sigma_o \rangle$, $N = \langle s_N, S_N, T_N, W_N, A, \Sigma_i, \Sigma_o \rangle$, *and*

$D_A = \langle \langle \{s_0\}, \top^A \rangle, S_D, T_A, W_D, A, \Sigma_i, \Sigma_o \rangle$. Define the final DFA $D_F = \langle s_f, S_F, y, W_F, A, \Sigma_i', \Sigma_o \rangle$ where A, Σ_i and Σ_o are as defined before. The state space is $S_F = \mathcal{T}^S \times \mathcal{Q}^A$, and the next state function is $y : S_F \times \Sigma_i' \to S_F$, which is defined using the functions φ and ξ. In the following definitions $\alpha \in \mathcal{T}^S$, $\alpha' \in 2^S$ and $\phi \in \mathcal{Q}^A$. Let $\varphi(\alpha, \phi, i) =$

$$\{\langle \alpha_D, \phi_D \rangle | \exists \alpha' \in 2^S, \exists \phi' \in \mathcal{Q}^A, \exists i' \in \Sigma_i : \alpha \preceq \xi(\alpha') \wedge \phi \preceq \phi'$$

$$\wedge\, i \preceq_{\Sigma_i'} i' \,\wedge\, T_A(\langle \alpha', \phi' \rangle, i') = \langle \alpha_D, \phi_D \rangle \}$$

where if the states in S be arbitrarily one-to-one mapped to the set $\{0, 1, 2, \ldots, |S| - 1\}$ by the function λ then,

$\xi(\alpha') = \alpha_0' \alpha_1' \ldots \alpha_{|S|-1}'$ such that $\alpha_{\lambda(i)}' = 1$ if $i \in \alpha'$, $\alpha_{\lambda(i)}' = 0$ otherwise.

Define the starting state as $s_f = \langle \xi(\{s_0\}), \top^A \rangle$ and the next-state function as $y(\langle \alpha, \phi \rangle, i) = \langle \hat{\alpha}, \hat{\phi} \rangle$ where

$$\hat{\alpha} = glb\{\xi(\alpha_D) | \exists \phi_D : \langle \alpha_D, \phi_D \rangle \in \varphi(\alpha, \phi, i)\}$$

$$\hat{\phi} = glb\{\phi_D | \exists \alpha_D : \langle \alpha_D, \phi_D \rangle \in \varphi(\alpha, \phi, i)\}$$

Let $\xi(\alpha) = \alpha_0 \alpha_1 \ldots \alpha_{|S|-1}$. Define a function ρ as $\rho(\alpha) = 1$ if and only if $\exists k \in \{0, 1, \ldots, |S| - 1\}$ such that $\alpha_k = 1$; otherwise 0. Finally, the output function, $W_F : S_F \to \Sigma_o$, is defined as

$$W_F(\langle \alpha, \phi \rangle) = \; if \; \rho(\alpha) = 1 \; then \; glb \; \{W_N(\langle k, \phi \rangle) | \alpha_{\lambda(k)} \preceq 1\}, \; otherwise, \; \perp_{\Sigma_o}$$

where α_i is the i-th bit of $\xi(\alpha)$.

Intuitively, the automaton D_F does not output the bottom element, $\perp_{\Sigma_o}$, if and only if D_F is definitely in at least one of the states of N. In such a case, D_F outputs the glb of the outputs of all the states in which it is definitely or possibly present in. D_F has a partially ordered state space. States that are weaker lie below the ones that are stronger. In situations where D_F reaches a weak state where it is not certainly present in at least one of the states of N then it outputs $\perp_{\Sigma_o}$. This situation results from inputs that are not strong enough to push D_F to be present in stronger states. The output $\perp_{\Sigma_o}$ is also produced in cases where an illegal input sequence has been detected by D_F. In such cases because of the output $\perp_{\Sigma_o}$, the EMM is re-initialized to an empty list, where it is maintained until the start of a fresh legal string is recognized. Unlike D_F, the output of N is never supposed to be $\perp_{\Sigma_o}$ in any situation other than when N is present in its start state. The fundamental reason why the outputs generated by D_F is weakened under the certain situations arises because of a fundamental difference between D_F and N. R and hence N only refer to exact, legal and finite input strings which can definitely take N to some states in N whereas D_F has the capability of recognizing weak, or illegal, or infinite input strings as well.

4.4   Using the Final DFA in an STE Engine

In order to establish that the final DFA is suitable for symbolic simulation purposes in an STE environment, we need to demonstrate that its state space forms a complete lattice under a partial order relation and that its next-state function is monotonic under that relation. We define a partial order relation on the state space S_F.

Definition 6 *Define* $\sqsubseteq$, *a partial order over* $S_F \cup \{\top\}$. *Let* $\langle a, \phi_a \rangle, \langle b, \phi_b \rangle \in S_F$. *Define* $\langle a, \phi_a \rangle \sqsubseteq \langle b, \phi_b \rangle$ *if and only if* $a \preceq b \wedge \phi_a \preceq \phi_b$ *and* $\forall \langle a, \phi_a \rangle \in S_F$: $\langle a, \phi_a \rangle \sqsubseteq \top$.

Lemma 1 $\langle S_F \cup \{\top\}, \sqsubseteq \rangle$ *forms a complete lattice.*

Proof Outline. The proof follows from a) both $\langle \mathcal{T}^S, \preceq \rangle$ and $\langle \mathcal{Q}^A, \preceq \rangle$ are partial orders, b) $\sqsubseteq$ is defined using $\preceq$ and c) $\forall \langle a, \phi_a \rangle \in S_F : \langle a, \phi_a \rangle \sqsubseteq \top$.   □

Lemma 2 *The function* y *is monotonic under the* $\sqsubseteq$ *partial order.*

Proof Outline. Let us take two elements of S_F, say, $\langle \alpha, \phi_\alpha \rangle$ and $\langle \beta, \phi_\beta \rangle$ such that $\langle \alpha, \phi_\alpha \rangle \sqsubseteq \langle \beta, \phi_\beta \rangle$, then we are required to prove that $\langle \hat{\alpha}, \hat{\phi_\alpha} \rangle \sqsubseteq \langle \hat{\beta}, \hat{\phi_\beta} \rangle$ where, for an input $i \in \Sigma'_i$, $y(\langle \alpha, \phi_\alpha \rangle, i) = \langle \hat{\alpha}, \hat{\phi_\alpha} \rangle$ and $y(\langle \beta, \phi_\beta \rangle, i) = \langle \hat{\beta}, \hat{\phi_\beta} \rangle$. The proof can be divided into two parts: showing $\hat{\alpha} \preceq \hat{\beta}$ and $\hat{\phi_\alpha} \preceq \hat{\phi_\beta}$.

Using $\langle \alpha, \phi_\alpha \rangle \sqsubseteq \langle \beta, \phi_\beta \rangle$, one can show $\varphi(\beta, \phi_\beta, i) \subseteq \varphi(\alpha, \phi_\alpha, i)$. That, in turn can be used to show that $\hat{\phi_\alpha} \preceq \hat{\phi_\beta}$ and $\hat{\alpha} \preceq \hat{\beta}$ and hence the proof.   □

4.5   Outputs from PREs

Until this point, we have been able to define PREs, give an operational account of their semantics based on a sequence of automata constructions without being specific about the precise nature of the output alphabet. In practice, however, PRE outputs are used to provide inputs to the EMMs, and to represent the outputs of the circuit elements that they are intended to abstract. The same generic format $write(enable, addr, data)$ can be used to represent both cases. In this format $enable$ is a ternary value, while $addr$ and $data$ are both vectors of ternaries. The entire PRE output is a vector of such values. The tuples have the obvious meaning when writing to an EMM – the $enable$ bit determines whether to write to the EMM or not, while the vectors $addr$ and $data$ supply the address and data for the write respectively. In case of primary outputs of the block being abstracted, the tuple is a degenerate case in which the address is of length 0, that is, an output is represented as an EMM with only a single address. If the enable bit is on, new data is output, if it is off, the most recent output is repeated. In both the general case of the EMM output and the special case of a primary output the output symbols are partially ordered by bitwise extension of the standard ternary partial order.

As defined in Sect. 4.1, the terminal in a PRE definition is of the form
"$a :: o$", where a is a predicate on the input symbols and assignments to variables
of the abstraction $\hat{M}$ and o is the output defined as functions mapping from
assignments to variables to output symbols of $\hat{M}$.

$$o ::= \; / * empty * /$$
$$| \; WR(i, d)$$
$$| \; output[output\_pin\_name := val, \ldots]$$

The annotations WR and $output$ enable the user to specify write operations
to the EMM in $\hat{M}$ and the signal values to be assigned at the output pins of
$\hat{M}$. The write annotation "$WR(i, d)$" stands for a memory write that happens
with variable i as the address and variable d as the data and produces a "write
tuple" $write(1, i, d)$ where the 1 stands for the enabled bit being high signifying
a *definite* write. The $output\_pin\_name$ and val are the name and the specified
value of an output pin respectively. val can be any of $\{0, 1, X, RD(j), \neg RD(j)\}$
where $RD(j)$ represents a read of the value of the array entry symbolically
indexed by a variable j. When o produces the *empty* terminal, the predicate a
can be viewed as augmented with the write tuple $write(0, \top, \top)$ to represent *no
change* to the memory contents.

The EMM is modeled using an associative list mapping symbolic data to
symbolic addresses. The state of the EMM is a finite lattice formed by ternary
partial order $\preceq$. The *write* command, $write(enable, address, data)$, is monotonic
over the EMM state space with respect to the ternary partial order. The "*no
change*" command, which is the identity function for the state space of the EMM,
is $write(0, \top, \top)$. The output $\perp_{\Sigma_o}$ from the PRE meant for an EMM translates
to the write tuple $write(X, X, X)$, which resets the EMM to an empty list.

The outputs can be treated as an EMM with one bit of memory. When an
output becomes $0, 1, X$, or some value a (result of an array read) the corresponding EMM commands would be $write(1, , 0)$, $write(1, , 1)$, $write(1, , X)$ and
$write(1, , a)$ respectively. The address in this case does not matter and hence is
omitted.

If in a particular terminal in a PRE nothing is specified about a memory write
or value assignments to an output, we assume that the corresponding write tuple
generated for the corresponding EMM is the identity $write(0, \top, \top)$, preserving
its contents.

5   Experimental Results

We prototyped a tool building on that reported by Krishnamurthy et al [3] and
augmented it with the technique presented in this paper. We conducted our experiments on a 360MHz Sun UltraSparc-II with 512MB of memory. Our example
circuit was a segment array from the MPC7450 PowerPC microprocessor. The
array had 512 bitcells, a read port and a write port. At first the PRE definitions
of the read and the write operations were written. Then the RTL description
was verified against the switch level model augmented with the abstract PRE

descriptions in about 4 minutes time. The method reported by Krishnamurthy et al [3] takes about a minute to do the same correspondence check without the PREs. The time difference is due to some BDD reordering taking place and we speculate that this is because of the way we handle variables in the implementation. This issue is currently being addressed.

The real effectiveness of the methodology was demonstrated by a fault injection experiment. We injected a fault by tying the precharge to ground, so that the bitcells are never precharged before a read or a write operation takes place. The earlier method [3] failed to discover the bug whereas our prototype was able to discover it and generate a witness execution sequence exposing it. The bug that we found is an instance of an "uncertified" input pattern that is not allowed as an input to the memory but is permitted by the switch level model. Thus, the bug is a member of a class of circuit level bugs that are abstracted away by the switch level model but are exposed by the current methodology. This experiment points out the importance of our methodology as it can serve as a "stricter check-point" when the custom memories have passed the earlier verification flow [3]. Although we believe that more time is consumed by our prototype because of implementation details, it hardly seems fair to compare a method that checks a schematic against an RTL with another that checks a composition of a schematic and an abstraction of the rest of the schematic against an RTL, since they are doing two fundamentally different things.

6   Related Work, Conclusions, and Future Directions

Checking the correspondence between the switch and the gate level views of the memory has been addressed by many researchers [5], [10], [11], [12], [4], [3]. While others focussed on using STE to verify equivalence by proving functional properties on both the RTL and the schematic view of the circuit, Krishnamurthy et al reported work on generating the assertions automatically from the RTL and to cross-check them against the schematic [3]. This removes the onus off the user to come up with a so-called "complete" set of assertions. Our work advances this approach by enabling the schematic view to be "weaker" in order for us to discover more buggy behavior.

We have provided a way in which memory can be modeled using regular expressions which represent a family of conditions that are visible to the portion of the memory that is being abstracted out. We have shown that the state transition model obtained from such a specification can be used in an STE framework. We have conducted experiments on an industrial strength circuit and demonstrated the applicability of our approach.

Future work will address the issue of checking whether the state machine model produced is a conservative approximation of the actual bitcell behavior. This could be done by verifying the specifications against a switch level bit-cell model, or even better, by verifying against a suite of Spice simulations. One way to get behavioral specifications is to obtain them automatically from the Spice simulations performed on these switch level designs. Work in this area is another candidate for future work.

References

1. M. N. Velev, R. E. Bryant, A. Jain. Efficient "Modeling of Memory Arrays in Symbolic Simulation". *CAV, 1997*, Proceedings. LNCS, Vol. 1254, Springer, 1997, pp. 388-399.
2. M. N. Velev, R. E. Bryant. "Efficient Modeling of Memory Arrays in Symbolic Ternary Simulation". *TACAS, 1998*.
3. N. Krishnamurthy, A. K. Martin, M. S. Abadir, J. A. Abraham. "Validating PowerPC Microprocessor Custom Memories" *IEEE Design and Test of Computers*, Vol. 17, No. 4, Oct-Dec 2000, pp. 61-76.
4. L.-C. Wang, M. S. Abadir, N. Krishnamurthy. "Automatic Generation of Assertions for Formal Verification of PowerPC Microprocessor Arrays Using Symbolic Trajectory Evaluation". *35th ACM/IEEE DAC*, June, 1998.
5. R. E. Bryant. "Algorithmic Aspects of Symbolic Switch Network Analysis". *IEEE Transactions on Computer-Aided Design of Integrated Circuits and Systems*, 6(4), July 1987.
6. C.-J. H. Seger and R. E. Bryant. "Formal verification by symbolic evaluation of partially-ordered trajectories". *Formal Methods in System Design*, 6(2):147-189, March, 1995.
7. J. E. Hopcroft and J. D. Ullman. *Introduction to Automata Theory, Languages, and Computation*. Addison-Wesley Publishing Company, 1979, pp. 1-45.
8. R. E. Bryant. "Graph-Based Algorithms for Boolean Function Manipulation". *IEEE Transactions on Computers*, 35(8), August 1986.
9. R. E. Bryant. "Verifying a Static RAM Design by Logic Simulation", *Fifth MIT Conference on Advanced Research in VLSI*, 1988, pp. 335-349.
10. N. Ganguly, M. S. Abadir, M. Pandey. "PowerPC array verification methodology using formal techniques". *International Test Conference 1996*, pp.857-864.
11. M. Pandey, R. Raimi, D. L. Beatty, R. E. Bryant. "Formal Verification of PowerPC arrays using Symbolic Trajectory Evaluation". *33rd ACM/IEEE DAC*, June 1996, pp.649-654.
12. M. Pandey, R. Raimi, R. E. Bryant, M. S. Abadir. "Formal Verification of Content Addressable Memories using Symbolic Trajectory Evaluation". *34th ACM/IEEE DAC*, June 1997.

Formal Verification of Conflict Detection Algorithms

Ricky Butler[1], Víctor Carreño[1], Gilles Dowek[2], and César Muñoz[3]

[1] Assessment Technology Branch, Mail Stop 130, NASA Langley Research Center
Hampton, VA 23681-2199
{r.w.butler,v.a.carreno}@larc.nasa.gov
[2] INRIA, Domaine de Voluceau - Rocquencourt - B.P. 105
78153 Le Chesnay Cedex, France
gilles.dowek@inria.fr
[3] ICASE, Mail Stop 132C, 3 West Reid Street, NASA Langley Research Center
Hampton VA 23681-2199
munoz@icase.edu

Abstract. Safety assessment of new air traffic management systems is a main issue for civil aviation authorities. Standard techniques such as testing and simulation have serious limitations in new systems that are significantly more autonomous than the older ones. In this paper, we present an innovative approach for establishing the correctness of conflict detection systems. Fundamental to our approach is the concept of *trajectory*, which is described by a continuous path in the x-y plane constrained by physical laws and operational requirements. From the model of trajectories, we extract, and formally prove, high level properties that can serve as a framework to analyze conflict scenarios. We use the AILS (Airborne Information for Lateral Spacing) alerting algorithm as a case study of our approach.

1 Introduction

Due to rapid growth in air travel, air traffic congestion has become an international problem and it is expected to worsen in the next two decades. Many concepts have been proposed to alleviate this problem. In many of these concepts – such as free-flight [9] –, the responsibility for aircraft separation is partially or completely moved from a centralized air traffic controller to a distributed aircraft system. The consensus in the avionics community is that the distribution will increase the efficiency of the air space system and terminal areas. However, a major concern of engineers, scientists, and civil aviation authorities is that the implementation of new approaches may compromise the overall system safety.

In the avionics community, testing and simulation are the standard methods for assuring the safety of digital systems. For instance the *Airborne Information for Lateral Spacing (AILS)* algorithm [10] - a conflict alerting algorithm used for parallel landing without control of the traffic controller authorities - has been extensively simulated and tested. So far, no major flaws have been detected.

T. Margaria and T. Melham (Eds.): CHARME 2001, LNCS 2144, pp. 403–417, 2001.
© Springer-Verlag Berlin Heidelberg 2001

However, neither testing nor simulation can give a definitive answer to questions such as: *"Does there exist a trajectory leading to a conflict without an alarm being issued?"* or *"What is the safety time between a conflict and a prior conflict detection?"* Given the critical nature of the problem, we believe that such analysis should be mechanically checked via a theorem proving system, such as PVS, or other automated proving techniques, e.g., model-checking.

Programs that are used for the monitoring and control of physical devices are constantly interacting with the physical world. The specification of such programs cannot usually be expressed as a mere input/output relation, but also involves concepts related to the physical environment where they are deployed. Hence, proving the correctness of such a system requires reasoning not only about the system itself, but also about its physical environment. For instance, to prove that a conflict detection algorithm does not allow conflicts without first issuing an alarm, we need to reason not only about the properties of the algorithm but also about the properties of the trajectories of the aircraft.

An air traffic control system, such as a conflict detection algorithm, is a hybrid system. It consists of simultaneous discrete and continuous behaviors. The discrete behavior is inherent to the algorithmic implementation on an embedded digital computer. Whereas, the continuous behavior arises from the kinematics of the aircraft. It is, of course, possible to discretize the continuous trajectory of an aircraft, approximating it by discrete segments of trajectories. Discretization of continuous trajectories allows the definition of trajectories by a transition relation. This way, the property we want to prove can be rephrased as a finite state automaton reachability property and it can be established by model checking and/or formal theorem proving. More or less complex extensions to this approach have been often used in the literature to model air traffic control problems (for a list of works, see [11]). These techniques have been shown to be effective for modeling systems where control logic modes trigger continuous and dynamic changes of the state. For instance, the TCAS alerting system for preventing midair collision was modeled in [6] using a hybrid automata approach.

From our experience, the formal verification task can considerably profit from standard mathematical analysis when a continuous model, rather than a discrete one, is used. Indeed, we have used semi-discrete trajectories in a previous effort to verify the AILS algorithm [2]. In that effort, it was not possible to take advantage of many useful properties of elementary calculus to prove the desired properties. For example, minimums and maximums of a differentiable function can easily be obtained by computing the derivative of the function, making it equal to zero, and solving for time. To find minimums and maximums of a function in a semi-discrete model requires an extraordinary effort. Moreover, introducing such a discretization of time and space may complicate the reasoning (it is more complicated, for instance, to reason about floats than about real numbers, because properties such as associativity of addition do not hold for floats).

In this paper, we propose an innovative approach, based on the formalization of continuous mathematics within a higher-order logic framework, for establishing the correctness of conflict detection systems. In this approach, a trajectory is given by differentiable functions in $\mathbb{R} \to \mathbb{R}$, which map *time* into each one of the components of a state aircraft, e.g., *heading, position, bank angle*, and *ground speed*. Using standard calculus, geometry, and kinematics, we formally asses safety properties of conflict detection algorithms.

In [5], Kuchar and Yang characterize three kinds of trajectory models: *nominal, worst-case*, and *probabilistic*. In the nominal approach, the future aircraft state, i.e., position, speed, heading, bank angle, is projected from the current state according to physics laws. In the worst-case approach, the future state is projected by following a policy of extreme values for specific state variables. In a probabilistic model, uncertainties such as weather conditions or extrapolation errors are taken into account to calculate the most probable aircraft trajectories. In our higher-order logic formalization, we can quantify over all (nominal) trajectories and thus worst-case trajectories are just particular cases of nominal trajectories.

Using our approach, we have been able to prove that the AILS algorithm is *correct*, i.e., (1) there is no possible conflict between two aircraft without a prior alarm and (2) the time between an alarm and a potential conflict is at least 10 seconds. We have also proven that the AILS alerting algorithm is not *certain*, i.e., an alarm can be issued when no potential conflict exists. Although AILS will be our running example, we believe that the continuous model of aircraft trajectories we build in this paper is rather general and may be used to prove properties of other algorithms and concepts.

The remainder of this paper is organized as follows. In Sect. 2, we present a model of continuous trajectories, which is the core of the framework for the verification of conflict detection algorithms. We study the correctness and certainty properties of the AILS Alerting algorithm in Sect. 3. The last section summarizes our work and contains concluding remarks. Along this paper, and for readability reasons, we have used standard mathematics and traditional logic reasoning. Nevertheless, our development has been formally checked in the general verification system PVS [8]. See [7] for an extended version of this paper. All the PVS theories and proofs are available at http://shemesh.larc.nasa.gov/fm/ftp/ails/.

2   Conflict Detection Framework

2.1   Scope

Our framework consists of three elements:

1. a continuous model of aircraft trajectories,
2. the concept of intruder and evader aircraft, and
3. the correctness and certainty properties.

Aircraft Trajectories. At the basis of our conflict detection framework is the concept of *aircraft trajectory*. A *trajectory* describes a continuous path in the

x-y (horizontal) plane subject to constraints imposed by the aircraft dynamics.[1]
It consists of *differentiable* functions in $\mathbb{R} \to \mathbb{R}$, which map *time* into *heading*,
position, bank angle, and *ground speed* of the aircraft.

We assume two things about the velocity vector of an aircraft. First, an
aircraft is moving at constant ground speed v, hence the velocity vector is completely determined by its polar angle θ called the heading of the aircraft. Second,
the variation of the heading is proportional to the tangent of the bank angle of
the aircraft. Formally, this can be expressed by the equations

$$x'(t) = v \cos(\theta(t)) \tag{1}$$
$$y'(t) = v \sin(\theta(t)) \tag{2}$$
$$\theta'(t) = (g/v) \tan(\phi(t)) \tag{3}$$

where x, y, θ, ϕ are differentiable functions mapping time to location coordinates,
heading, and bank angle, respectively. Equations 1 and 2 state that the derivative
of the position functions gives the velocity vector of the aircraft. Equation 3
relates the bank angle with the heading of the aircraft. That equation states
that the rate of direction change of an aircraft is proportional to the tangent of
the bank angle by a factor of g/v, where g is the gravitational force.

The constant ground speed assumption is imposed by the AILS system designers. It is justified by the fact that during AILS operations, aircraft are on
final approach and their velocities are restricted.

In addition to the above physical constraints, we impose a maximum bank
angle operational constraint for commercial aircraft to be 35°, i.e.,

$$|\phi(t)| \leq 35\pi/180. \tag{4}$$

Henceforth, we use the constant `MaxBank` $= 35\pi/180$.

Evader and Intruder Aircraft. We assume a pair of aircraft, one labeled
evader and the other *intruder*. In the AILS system, the evader represents an
aircraft flying on normal conditions while the intruder represents a blundering
aircraft. The AILS algorithm runs twice on each airplane: the first execution
treats the local aircraft as the evader an the foreign aircraft as the intruder; the
second execution interchanges the roles of intruder and evader aircraft.

Multiple aircraft scenarios can be modeled as sequential composition of pairwise aircraft conflict detection algorithms. Notice that, in contrast to conflict
detection algorithms, conflict *resolution* algorithms for multiple aircraft system
are usually not compositional as a pair of aircraft could create new conflicts in
previously solved aircraft. Conflict resolution, however, is beyond the scope of
this paper.

Equations 1-4 apply to both evader and intruder aircraft. State functions
representing the state of the evader and intruder aircraft are subscripted with

---

[1] The vertical separation is typically handled separately. This will be studied in future
work.

lowercase letters e and i, respectively. The AILS system assumes that only one aircraft is diverting from its intended landing path. This assumption is also a system designer assumption. It is based on a probabilistic failure assessment. Hence, an additional restriction is imposed on evader trajectories by constraining the bank angle $\phi_e(t) = 0$. This constraint makes the heading of the evader constant and its trajectories straight lines. Without loss of generality, we can chose a coordinate system where the x-axis coincides with the evader trajectory making the heading angle of the evader aircraft always 0. Thus, the equations for the evader can be rephrased and integrated

$$x_e(t) = X_e + v_e t \qquad (5)$$
$$y_e(t) = Y_e \qquad (6)$$
$$\theta_e(t) = 0 \qquad (7)$$
$$\phi_e(t) = 0 \qquad (8)$$

where X_e and Y_e are the coordinates of the initial evader position.

Correctness and Certainty. Conflict detection is based on the ability to predict future aircraft locations for a given lookahead time $T > 0$. Two aircraft have a *(potential) conflict* at time T, if there exists a trajectory leading to a distance between the aircraft less than a given value ConflictRange at time T. Assuming that the aircraft have reliable access to accurate data flight information, two key properties that must be established for a conflict detection algorithm are: (1) any future conflicts within the lookahead time will be detected, and (2) a conflict detection reflects a potential conflict within the lookahead time. The first property is called *correctness* and the latter one is *certainty*. Notice that correctness means that conflicts will not go undetected and certainty means that the algorithm will not detect conflicts that do not exist, possibly leading to false alarms.

Since possible conflicts that are not detected may lead to mid-air collisions, correctness is a much more critical feature, from a safety point of view, than certainty. However, false alarms will have negative effects both on safety and in the overall performance of the airspace system [4].

Predictions of aircraft trajectories are made to determine if a conflict exists in a given lookahead time. Two types of information can be used for prediction: (1) intent information for medium to long lookahead times; and (2) state information for short to medium lookahead times. Intent information refers to information in flight plans, destination, in route way points, etc. State information uses the airplane heading, speed and location to predict future aircraft states. In this paper, we are only concerned with trajectory prediction based on state information.

2.2   Main Lemmas

We have modeled the motion of aircraft by differentiable functions from $\mathbb{R}$ to $\mathbb{R}$. To establish the basic properties of trajectories, we shall need several lemmas that concern differentiable functions, coordinate systems, and objects in motion in general. We present in this section, the main lemmas we have had to prove on such topics.

Elementary Differential Calculus. Like most theorem provers, PVS has little automated support for non-linear arithmetic and real analysis. We have extended the pre-defined theory of real numbers and the theory of differential functions developed in [1] with theories dealing with trigonometric and other transcendental functions.

Non-effective real functions are declared in PVS as uninterpreted functions. Their behavior is given axiomatically. For example, cos and sin are functions from reals to the real interval $[-1 \dots 1]$ satisfying, among other properties, $\sin(a)^2 + \cos(b)^2 = 1$. In a similar way, $\sqrt{.}$ is a function from non-negative reals to non-negative reals such that $\sqrt{a}^2 = a$ for $a \geq 0$. From this axiom, we can prove, for instance, that $\sqrt{a^2} = a$ for $a \geq 0$.

The concept of differentiability and derivative is treated the same way. We use an uninterpreted predicate `Differentiable` over functions mapping real numbers to real numbers and an uninterpreted higher-order function `D` (for derivative) mapping `Differentiable` functions to functions from real numbers to real numbers. Typical axioms assert that sin is differentiable and that its derivative is cos. For our proofs, we have managed to avoid the concept of differentiability over an interval. For instance, the square root function is not assumed to be differentiable, but we have an axiom stating that for all differentiable functions f mapping real numbers to *positive* real numbers, the function $x \mapsto \sqrt{f(x)}$ is differentiable. We do not know if this trick is sufficient in general, or if a general theory of differentiability over an interval is needed in other examples.

Another important axiom defining the concept of derivative is the following theorem of calculus (that can be seen as a consequence of Rolle's theorem).

Theorem 1 (`monotonic_anti_deriv`).

$$\forall f, g : \mathcal{R} \to \mathcal{R}. \ \forall a, b : \mathcal{R}. \ a \leq b \quad \supset$$
$$(\forall c : \mathcal{R}. \ a \leq c \leq b \ \supset \ f'(c) \leq g'(c)) \quad \supset$$
$$f(b) - f(a) \leq g(b) - g(a).$$

In the verification process it is sometimes necessary to perform calculations on expressions containing non-effective functions such as the trigonometric functions. It is tempting to use approximation series to define, for instance, sin and cos. However, mixing approximation series and axiomatic definitions of trigonometric functions may be a source of paradoxes. Say for example that sin and cos compute approximate values of the real ones. It will be very unlikely that $\sin(a)^2 + \cos(a)^2$ evaluates to 1 for any value of a. In order to avoid that kind of inconsistencies, we mix approximations and uninterpreted functions in a very rigorous way. Assume we want to prove that $e_1[\sin(a)]^+ \leq e_2[\cos(b)]^+$, where $e[s]^+$ stands for a context expression e containing a distinguished positive occurrence of s. Then, we find a computable upper bound of $\sin(a)$, say

$\text{sin}_{ub}(a)$, and a computable lower bound of $\cos(b)$, say $\cos_{lb}(b)$. Finally, we prove $e_1[\sin(a)]^+ \leq e_2[\cos(b)]^+$ as follows

$$e_1[\sin(a)]^+ \leq e_1[\sin_{ub}(a)]^+ \tag{9}$$
$$e_1[\sin_{ub}(a)]^+ \leq e_2[\cos_{lb}(b)]^+ \tag{10}$$
$$e_2[\cos_{lb}(b)]^+ \leq e_2[\cos(b)]^+ \tag{11}$$

Most of the times, Formulas 9 and 11 are simple to discharge. If $e_1[\sin(a)_{ub}]^+$ and $e_2[\cos(b)_{lb}]^+$ are computable then we prove Formula 10 by evaluating the expressions. Otherwise, we use the same technique to remove other non-computable values. Eventually, we will get two expressions that we can evaluate. This technique is so used and simple that we have developed PVS strategies to automate the work. As for computable definitions of $\sin_{ub}, \sin_{lb}, \cos_{ub}, \cos_{lb}$, we have used partial approximation by series:

$$\sin_{lb}(a) = \sum_{i=1}^{4}(-1)^{i-1}\frac{a^{2i-1}}{(2i-1)!} \qquad \sin_{ub}(a) = \sum_{i=1}^{5}(-1)^{i-1}\frac{a^{2i-1}}{(2i-1)!}$$

$$\cos_{lb}(a) = 1 + \sum_{i=1}^{3}(-1)^{i}\frac{a^{2i}}{(2i)!} \qquad \cos_{ub}(a) = 1 + \sum_{i=1}^{4}(-1)^{i}\frac{a^{2i}}{(2i)!}$$

and the axioms

Axiom 1 (PI)

$$3.14 \leq \pi \leq 3.15.$$

Axiom 2 (SIN)

$$0 \leq a \leq \pi \supset \sin_{lb}(a) \leq \sin(a) \leq \sin_{ub}(a).$$

Axiom 3 (COS)

$$-\pi/2 \leq a \leq \pi/2 \supset \cos_{lb}(a) \leq \cos(a) \leq \cos_{ub}(a).$$

The fact that we have stated many axioms in this section reflect the fact that we have focused on developing proofs of properties of our framework for analysis of conflict detection algorithms using well-known results of calculus, not on developing a calculus library. On a longer term research effort, these uninterpreted constants should be replaced by definitions and these axioms would become theorems.

Geometry. An important technique used in our formal development is to take as reference a new system of coordinates where the origin is the position of the evader aircraft at a given time T, i.e., $(x_e(T), y_e(T))$, and the x-y plane has been rotated by $\theta_i(0)$ degrees. We recall that $\theta_i(0)$ is the heading of the intruder aircraft at time 0. The new $\hat{x}$-$\hat{y}$ plane, is defined as follows

$$\hat{x}(t) = \cos(\theta_i(0))[x(t) - x_e(T)] + \sin(\theta_i(0))[y(t) - y_e(T)] \tag{12}$$
$$\hat{y}(t) = \cos(\theta_i(0))[y(t) - y_e(T)] - \sin(\theta_i(0))[x(t) - x_e(T)] \tag{13}$$

We have formally proven several properties related to changes of coordinate systems. For instance, lemma `isometric` states that distances are invariant under rotation and translation of the coordinate system.

Lemma 1 (`isometric`).

$$(x_1(t) - x_2(t))^2 + (y_1(t) - y_2(t))^2 = (\hat{x}_1(t) - \hat{x}_2(t))^2 + (\hat{y}_1(t) - \hat{y}_2(t))^2.$$

Kinematics. Let us now turn to the lemmas that involve moving objects. These lemmas were needed to prove safety properties of the AILS algorithm, but they are still rather general and some of them have been reused to prove properties over other air traffic control algorithms.

The first example concerns the ability to determine whether the straight-line trajectories of two aircraft are diverging or converging and to find the point of closest separation of the projected trajectories. This amounts to finding the minimum of the distance between two straight-line trajectories. If the evader aircraft is assumed to have heading 0 and the intruder aircraft has heading θ, then the equations defining the *projected trajectories* are

$$x_e(t) = x_e(0) + v_e t$$
$$y_e(t) = y_e(0)$$
$$x_i(t) = x_i(0) + v_i t \cos(\theta)$$
$$y_i(t) = y_i(0) + v_i t \sin(\theta)$$

and the distance between the *projected trajectories* at time t, $R(t)$, can be computed as follows:

$$\Delta_x(t) = x_i(t) - x_e(t)$$
$$\Delta_y(t) = y_i(t) - y_e(t)$$

$$R(t) = \sqrt{\Delta_x(t)^2 + \Delta_y(t)^2} \tag{14}$$

To find the minimum of $R(t)$, first the derivative of $R(t)$ is computed:

$$R'(t) = \frac{\Delta_x(t)\Delta'_x + \Delta_y(t)\Delta'_y}{R(t)}$$

where

$$\Delta'_x = v_i \cos(\theta) - v_e$$
$$\Delta'_y = v_i \sin(\theta)$$

We have formally verified that when $R'(t + \tau) = 0$, the time τ, relative to t, is the time of closest separation between the aircraft. The solution to this equation is:

$$\tau(t) = -\frac{\Delta_x(t)\Delta'_x + \Delta_y(t)\Delta'_y}{\Delta'^2_x + \Delta'^2_y} \tag{15}$$

It is important to note that τ is undefined, i.e., denominator is zero, when the aircraft are parallel and the ground speeds are equal.

For any time t, if $\tau(t)$ is negative or zero, the tracks are diverging or parallel, respectively. If $\tau(t)$ is greater than zero, the tracks are converging and $\tau(t)$ is the time of closest separation relative to t. In PVS:

Lemma 2 (derivative_eq_zero_min).

$$R(t_1 + \tau(t_1)) \leq R(t_1 + t_2).$$

Lemma 3 (asymptotic_decrease_tau).

$$t_1 \leq t_2 \leq \tau(t) \supset R(t + t_1) \geq R(t + t_2).$$

Lemma 4 (asymptotic_increase_tau).

$$\tau(t) \leq t_1 \leq t_2 \supset R(t + t_1) \leq R(t + t_2).$$

A second example concerns the maximum and minimum distance traveled by an aircraft in a given time whose speed is constant and bank angle bounded. In particular, vt is the farthest distance, i.e., via straight line, that can be reached by an aircraft moving at constant speed v in t seconds. That property is called YCNGFTYS, which stands for *You Cannot Go Faster Than Your Speed*, and it has been formally verified in PVS.

Theorem 2 (YCNGFTYS).

$$t \geq 0 \supset \sqrt{(x(t) - x(0))^2 + (y(t) - y(0))^2} \leq vt.$$

For an aircraft moving at constant speed v and with a constant bank angle ϕ, Fig. 1, the distance from the position at time 0 to the position at time t is given by the formula

$$m(v, \phi, t) = 2r(v, \phi) \sin(vt/2r(v, \phi)) \tag{16}$$

where $r(v, \phi)$ is the turn radius of the aircraft.

The turn radius $r(v, \phi)$ can be calculated as follows.

$$vt/r(\phi, v) = (g/v)\tan(\phi)t \text{ (From Equation 3)}$$
$$v/r(v, \phi) = (g/v)\tan(\phi) \text{ (Simplifying } t).$$

Thus,

$$r(v, \phi) = v^2/(g\tan(\phi)). \tag{17}$$

According to Formula 4, the maximum change of heading per second of an aircraft moving at constant speed v is given by

$$\rho(v) = (g/v)\tan(\texttt{MaxBank}). \tag{18}$$

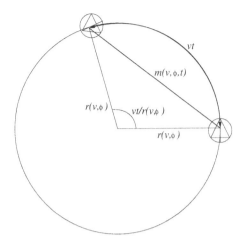

Fig. 1. Distance traveled in curved trajectory

From Equation 17 and Equation 18:

$$r(v, \texttt{MaxBank}) = v/\rho(v) \tag{19}$$

and from Equation 16 and Equation 19:

$$m(v, \texttt{MaxBank}, t) = 2r(v, \texttt{MaxBank}) \sin(\rho(v)t/2). \tag{20}$$

When $0 \leq \rho(v)t \leq 2$, we have formally verified in PVS that $m(v, \texttt{MaxBank}, t)$ is the minimum distance traveled by an aircraft moving at constant speed v in t seconds[2]. The property is called $\texttt{YCNGSTYS}$, which stands for *You Cannot Go Slower Than Your Speed.*

Theorem 3 ($\texttt{YCNGSTYS}$).

$$0 \leq \rho(v)t \leq 2 \supset m(v, \texttt{MaxBank}, t) \leq \sqrt{(x(t) - x(0))^2 + (y(t) - y(0))^2}.$$

According to theorems $\texttt{YCNGFTYS}$ (Theorem 2) and $\texttt{YCNGSTYS}$ (Theorem 3), for an aircraft moving at constant ground speed v, the inner circle of radius $m(v, \texttt{MaxBank}, t)$ and the outer circle of radius vt, both centered at the current position of the aircraft, delimit the area that could be reached by the aircraft at time t.

3   Verification of the AILS Alerting Algorithm

The AILS system is intended to enable independent parallel landings to closely spaced parallel runways during reduced visibility conditions. An integral part of

---

[2] We conjecture that the property still holds for $0 \leq \rho(v)t \leq 2\pi$; but, we could not find a formal proof of this proposition.

the system is the alerting algorithm which provides a sequence of alarms when the aircraft are diverting from their intended landing paths.

The original AILS algorithm was written in FORTRAN at Langley Research Center. It has been revised several times and the latest version, flown in the Boeing 757 experimental aircraft, was written by Honeywell. Since the AILS alerting system is a multilevel alarm system, it is possible to prove properties of the different levels of alarms. The alarm we are most concerned with is the alarm the evader aircraft should receive when the intruder aircraft is threatening its airspace. This is the alarm we will be discussing in the remaining of this section.

We have specified that algorithm in PVS by a predicate `ails_alert` that takes the initial measured states of an intruder aircraft and an evader aircraft, and returns `true` or `false` depending on whether the alarm is issued or not. The specification of the algorithm is the translation of the original FORTRAN code to a higher-order logic description. The translation was almost one-to-one with the exception of replacing DO-LOOPS statements with recursive definitions.

3.1   Correctness

The verification of the alerting algorithm consists of showing that for all trajectories that could lead to a conflict within a given time T, the alerting algorithm will issue an alarm. We called this property the correctness of the AILS Alerting Algorithm. Using the trajectory framework described in previous sections and the model of the algorithm, we formulated the properties necessary for the verification.

We started the verification process by trying to show correctness for a value of T equals 19 seconds which is the lookahead time for the algorithm. Soon, we discovered that the correctness property was not provable for $T = 19s$. Indeed, a counter example for $T = 10.5$ seconds was found. The counter example says that there exists a trajectory which will bring the two aircraft within 10.5 seconds of a conflict without an alarm being issued. This counter example is in some sense minimal, as we have been able to prove correctness for $T \leq 10$ seconds and the time step between two consecutive calls of the algorithm is 0.5 seconds.

To prove the correctness property for $T \leq 10$, it is sufficient to prove that if a conflict is going to occur in a time ranging between $9.5s$ and $10s$ then an alarm is raised. Note that if a conflict is going to occur in less that $9.5s$ an alarm has already been raised[3]. Hence, we have the following correctness theorem.

Theorem 4 (`ails_correctness`).

$$\forall i, e. \ 9.5 \leq T \leq 10 \ \wedge \ conflict_{ie}(T) \ \supset$$
$$ails\_alert(measure2state(i, 0), measure2state(e, 0))$$

where `measure2state`(a, t) is the state of the aircraft a at time t and the proposition $conflict_{ie}(T)$ means that there exists a potential conflict between the aircraft at time T.

---

[3] Due to operational constraints, when the AILS system is first engaged during a final approach, there is a safe window of at least 9.5 seconds where no conflict can occur.

The proof of this theorem uses the notion of state of an aircraft consisting of its location, heading, and bank angle. The states of the aircraft are part of the geometrical description of the problem, but also are the data processed by the AILS algorithm. Thus, they constitute the interface between the algorithm and its physical environment. In this paper, we have assumed that these measurements are made without error. Although with ADS-B exchange of information the errors can be made very small, it should be included in future work.

The proof proceeds by proving independently that (1) if a conflict is to occur in a time ranging between 9.5s and 10s, then the pair of aircraft is in a certain region $\mathcal{G}$ of the state space; and (2) if it is in this region $\mathcal{G}$, then an alarm is issued. The proof of (1) involves only the geometry and the kinematics of the physical environment, while the second proof concerns only the specification of the algorithm. The region $\mathcal{G}$ constitutes the interface between the two proofs, in the same way as the measured state is the interface between the algorithm and its physical environment.

More precisely, the first lemma proves that if a conflict is to occur, then

1. the distance of the initial position of the intruder to the final position of the evader ranges between a distance `MinDistance` and `MaxDistance` that are the minimum and the maximum distance the intruder may run in a time ranging between 9.5s and 10s (as determined by theorems 2 and 3),
2. the initial heading of the intruder also ranges between two bounds,
3. $\tau(0) > 0$, i.e., the aircraft are converging at current time.

In this last case, we have to prove that that if the aircraft are diverging at current time, although the intruder can change heading, it cannot do it sufficiently fast to create a conflict. The second lemma proves that the AILS algorithm issues an alarm if all these conditions hold.

It is important to note that what we have shown for $9.5 \leq T \leq 10$ is a worst case scenario. In simulations with actively flying airline pilots, it was found that the pilot reaction time was in average approximately 2 seconds. Also, according to the logic of the AILS alerting system, the intruder aircraft should receive two path deviation indications and one traffic caution indication before the evader aircraft is alerted.

3.2   Uncertainty

We have also proven that although the AILS algorithm is correct, it is not certain, i.e., it may issue false alarms.

Theorem 5 (ails_uncertainty).

$$\exists s_i, s_e : \textbf{\textit{State}}. \ \forall i, e. \ 0 < T \leq 10$$
$$s_i = measure2state(i, 0) \ \wedge \ s_e = measure2state(e, 0) \qquad \supset$$
$$ails\_alert(s_i, s_e) \ \wedge \ \neg conflict_{ie}(T).$$

To prove this theorem, we simply find states of intruder and evader aircraft that issue an alarm, but where the aircraft cannot conflict within 10 seconds.

4   Conclusion

In this paper, we have presented the foundation for a new approach to verifying the safety of conflict detection algorithms that may one day be deployed in the national airspace. Such algorithms are an enabling technology for free flight, where pilots are allowed to fly their own preferred trajectories. The introduction of these algorithms in a free-flight context raises significant safety issues. Historically the trajectories of aircraft have been managed by ground controllers through use of aircraft position data obtained from radar. Under this approach, the primary responsibility for maintaining aircraft separation has been borne by the air traffic controller. But under a free-flight approach, much of the responsibility for maintaining separation will be transferred to the pilots *and the software which provides them aircraft positions*, i.e., via Cockpit Display of Traffic Information (CDTI), and warnings of potential conflicts. We believe that current methods for gaining assurance about the safety of ground-based decision-aid software are inadequate for many of the software systems that will be deployed in the future in support of free flight. The current approach is based upon human-factors experimentation using high fidelity simulations. In the current approach, where the responsibility for safety resides in the human controller, this is clearly the right approach. The primary question to be answered is whether the software provides the controllers with useful information that aids them in their decision making. But as software takes on more and more of the responsibility for generating aircraft trajectories and detecting potential conflicts and perhaps even producing (and executing?) the evasive maneuvers, we will need additional tools to guarantee safety. It is our view that it is essential that the correctness of the algorithm be established for *all possible* situations. Simulation and testing cannot accomplish this. Although simulation and controlled experimentation are clearly necessary, they are not sufficient to guarantee safety. This can only be done by analytical means, i.e., formal verification. We should also note that it will also be necessary to demonstrate that the implementation of these algorithms in software is correct. This refinement verification, in our view, must also be accomplished using formal methods. We hope to explore this issue with our colleagues in future work.

The trajectory model used in this paper is the result of investigating different approaches. Earlier work looked at more discrete versions with the expectation that this would lead to a more tractable verification task. Unfortunately the discretization of the trajectories led to significant (and accumulating) modeling error that led to erroneous conclusions. In the end, we have settled on modeling trajectories as differentiable functions over real numbers. These trajectories are constrained by the dynamics of an aircraft. These constraints enable one to establish high level properties that delineate when a conflict is possible. In this paper we have developed a formal theory about trajectories that can serve as the basis for the formal analysis of conflict detection and resolution (CD&R) algorithms. There are several limitations to this formal theory that will be addressed in future work. These include: (1) the theory only deals with 2 aircraft,

(2) the vertical dimension is not modeled, and (3) aircraft data measurement errors are not modeled.

Because the trajectories of the aircraft are modeled by differentiable functions over real numbers and the discrete algorithms are periodically executed on a digital computer, this problem domain falls into the domain of hybrid models. The hybrid nature of this domain makes the verification problem especially difficult. Automatic methods such as model checking cannot directly handle the continuous trajectories, i.e., infinite state space, and discretization leads to unacceptable errors. We are forced to reason about such systems in the context of a fully general theorem prover designed to handle a rich logic such as higher-order logic, type theory, or ZFC set theory. We have used the PVS theorem prover in our work and found this tool to be sufficient to handle this problem domain, but our work was often impeded by PVS's baroque method for dealing with nonlinear arithmetic. Although PVS provides a suite of decision procedures that automate much of the tedium of theorem prover, in this arena, one must wrestle with the prover in order to make progress. Adding capability for reasoning about formulas containing non-linear arithmetic in theorem provers is a current area of research.

Future work will concentrate on applying this modeling framework to specific CD&R algorithms and perhaps to self-spacing and merging algorithms designed to increase capacity in the terminal area. We would also like to develop formal methods for analyzing conflict resolution schemes and the safety of algorithmically-generated evasive maneuvers [3]. The CD&R methods must be generalized to cover sets of aircraft constrained by formally specified notions of aircraft density (static or dynamic). We would also like to generalize the methods to encompass measurement error and data errors. This is a necessary step toward developing formal methods useful for the design and implementation phases of realistic avionics.

Acknowledgment. The authors would like to thank Alfons Geser, Michael Holloway, and the anonymous referees for their helpful comments on preliminary versions of this paper.

References

1. B. Dutertre. Elements of mathematical analysis in PVS. In J. Von Wright, J. Grundy, and J. Harrison, editors, *Ninth international Conference on Theorem Proving in Higher Order Logics TPHOL*, volume 1125 of *Lecture Notes in Computer Science*, pages 141–156, Turku, Finland, August 1996. Springer Verlag.

2. V. Carreño and C. Muñoz. Aircraft trajectory modeling and alerting algorithm verification. In J. Harrison and M. Aagaard, editors, *Theorem Proving in Higher Order Logics: 13th International Conference, TPHOLs 2000*, volume 1869 of *Lecture Notes in Computer Science*, pages 90–105. Springer-Verlag, 2000. An earlier version appears as report NASA/CR-2000-210097 ICASE No. 2000-16.

3. G. Dowek, C. Muñoz, and A. Geser. Tactical conflict detection and resolution in a 3-D airspace. Technical Report NASA/CR-2001-210853 ICASE Report No. 2001-7, ICASE-NASA Langley, ICASE Mail Stop 132C, NASA Langley Research Center, Hampton VA 23681-2199, USA, April 2001.
4. J. Kuchar and Jr. R. Hansman. A unified methodology for the evaluation of hazard alerting systems. Technical Report ASL-95-1, ASL MIT Aeronautical System Laboratory, January 1995.
5. J. Kuchar and L. Yang. Survey of conflict detection and resolution modeling methods. In *AIAA Guidance, Navigation, and Control Conference*, volume AIAA-97-3732, pages 1388–1397, New Orleans, LA, August 1997.
6. J. Lygeros and N. Lynch. On the formal verification of the TCAS conflict resolution algorithms. In *Proceedings 36th IEEE Conference on Decision and Control*, San Diego, CA, pages 1829–1834, December 1997. Extended abstract.
7. C. Muñoz, R.W. Butler, V. Carreño, and G. Dowek. On the verification of conflict detection algorithms. Technical Report NASA/TM-2001-210864, NASA Langley Research Center, NASA LaRC Hampton VA 23681-2199, USA, May 2001.
8. S. Owre, J. M. Rushby, and N. Shankar. PVS: A prototype verification system. In Deepak Kapur, editor, *11th International Conference on Automated Deduction (CADE)*, volume 607 of *Lecture Notes in Artificial Intelligence*, pages 748–752, Saratoga, NY, June 1992. Springer-Verlag.
9. Radio Technical Commission for Aeronautics. Final report of the RTCA board of directors' select committee on free flight. Technical Report Issued 1-18-95, RTCA, Washington, DC, 1995.
10. L. Rine, T. Abbott, G. Lohr, D. Elliott, M. Waller, and R. Perry. The flight deck perspective of the NASA Langley AILS concept. Technical Report NASA/TM-2000-209841, NASA, January 2000.
11. C. Tomlin, G. Pappas, and S. Sastry. Conflict resolution for air traffic management: A study in multi-agent hybrid systems. *IEEE Transactions on Automatic Control*, 43(4), April 1998.

Induction-Oriented Formal Verification in Symmetric Interconnection Networks

Eric Gascard and Laurence Pierre

LIM - CMI / Université de Provence - 39, rue Joliot-Curie
13453 Marseille Cedex 13 - FRANCE
{Eric.Gascard,Laurence.Pierre}@lim.univ-mrs.fr

Abstract. The framework of this paper is the formal specification and proof of *applications distributed on symmetric interconnection networks*, e.g. the torus or the hypercube. The algorithms are distributed over the nodes of the networks and use well-identified communication primitives. Using the notion of *Cayley graph*, we model the networks and their communications in the inductive theorem prover *Nqthm*. Within this environment, we mechanically perform correctness verifications with a specific invariant oriented method. We illustrate our approach with the verification of two distributed algorithms implemented on the hypercube.

1 Introduction

We propose a methodology for the specification and the formal verification of distributed algorithms designed to be implemented on symmetric interconnection networks such as the ring, the torus or the hypercube [14]. In areas like signal processing or computer vision, the necessity of improving the performance of the applications is increasing inexorably. The parallelization of the algorithms in view of an implementation on specific interconnection networks[1] provides a solution to this problem [15,16]. The program is distributed over the nodes of the network, and the processes can exchange data by means of various types of communication procedures.

Our ultimate goal is to develop a specialized environment for the design and validation of such hardware/software architectures. To that aim, various aspects have to be taken into account: *(i)* the formal specification of the *hardware structures* and the description and validation of the usual *communication operations* over these architectures, *(ii)* the development of a methodology for the specification and the formal proof of *application programs* that make use of these communication functions, *(iii)* to make the approach accessible to application programmers, it is necessary to provide a way of mechanizing the transformation from their *source code to the formal model*.

---

[1] For instance, the ring of the HP/Convex SPP-1200, the tori of the Cray T3E and of the Intel Paragon XP, or the hypercube of the nCUBE 3.

T. Margaria and T. Melham (Eds.): CHARME 2001, LNCS 2144, pp. 418–432, 2001.

All these aspects depend on the formal proof tool that we choose. In the framework of regular structures like symmetric interconnection networks, the induction-based prover *Nqthm* [4] provides valuable assistance. Its mechanisms allow to reason at a high-level of abstraction and to deal with parameterization (here the size of the network is a parameter). We model interconnection networks in Nqthm as *Cayley graphs* [1], which give a representation of the mathematical groups defined by generators: the vertices are the elements of the group and the arcs are the actions of the generators. Cayley graphs have been extensively used to solve problems of routing [2] or information dissemination [8,3] in interconnection networks. To our knowledge, this model has never been applied to formal verification. We will show that one of the advantages of this approach is that it allows to reason independently on the structures on the one hand and on the algorithms on the other hand. Moreover, most of the specifications and verifications are performed at the processor level i.e., the statements of most theorems only involve one (universally quantified) processor, as Sects. 5 and 6 will demonstrate.

After a brief presentation of the networks and their communication primitives in Sects. 2 and 3, we describe our models and proof methodology in Sects. 4 and 5. We concentrate on the hypercube, and we make the hypothesis that the behaviour of the processes that constitute the application is synchronous (i.e., either they run on a synchronous SIMD computer, or they are designed for a MIMD architecture and include resynchronization phases). Verifying the correctness of such a distributed program is understood here as proving its equivalence with its sequential counterpart. Section 6 provides two illustrative examples, Sect. 7 recalls some related works, and we conclude in Sect. 8.

2 Cayley Graphs and Interconnection Networks

Interconnection networks can be modelled by finite graphs: the vertices represent the nodes and the edges are the communication lines. A special class of networks, called *symmetric interconnection networks*, has the property that the network viewed from any vertex looks the same (*Vertex Symmetry*). The *Cayley graph* model is proposed in [1] for designing and analyzing symmetric interconnection networks, such as the cycle, the torus and the hypercube presented here.

2.1 Cayley Graphs

Definition 1. *Let G and S be a group and a subset of this group, the Cayley digraph of G and S, denoted $Cay(G, S)$, is such that its vertices are the elements of G and its arcs are all ordered pairs $(g, g \otimes s)$ where $g \in G$, $s \in S$ and $\otimes$ is the law of the group.*

The elements of a subset S of a group G are called *generators* of G, and S is said to be a *generating set*, if every element of G can be expressed as a finite product of their powers. We say that G is generated by S.

If S is a generating set of G, then the digraph $Cay(G, S)$ is strongly connected. If S is unit free (does not contain the identity) and closed under inverses (if $s \in S$, then $s^{-1} \in S$) then $Cay(G, S)$ is a graph.

2.2   Cycle and Torus

The **cycle/ring** of length n, C_n, is the graph whose nodes are labelled by integers ranging from 0 to $n - 1$ and whose edges connect i ($0 \leq i < n$) to $(i + 1)$ mod n, its degree is 2. It is the Cayley graph of the additive groupe $\mathbb{Z}_n$ generated by $\{-1, +1\}$. The actions of the generators g_1 and g_2 of C_n on a vertex s are expressed by $s \otimes g_1$ and $s \otimes g_2$, that we model by $g_1(s, n)$ and $g_2(s, n)$:

$$\begin{cases} g_1(s, n) = (s + (n - 1)) \text{ mod } n \text{ if } s \in [0, n - 1] \\ g_2(s, n) = (s + 1) \text{ mod } n \text{ if } s \in [0, n - 1] \end{cases}$$

The **torus** $T_{n,m}$ is the cartesian product of two cycles of lengths n and m. $T_{n,m}$ has $n \times m$ vertices, its degree is 4. It is the Cayley graph of the group $\mathbb{Z}_n \times \mathbb{Z}_m$ generated by $S = \{(0, 1), (0, -1), (1, 0), (-1, 0)\}$. See the torus $T_{3,3}$ on Fig. 1.

2.3   Hypercube

The n-dimensional **hypercube** H_n consists of 2^n processors, its degree is n. Its nodes can be labelled by binary strings $(x_1 x_2 \ldots x_n)$; there is an edge between two nodes if their labels differ in exactly one position (the dimension). H_n is also the Cayley graph of the permutation group G generated by the n transpositions $g_i = \langle 2i - 1, 2i \rangle$, $1 \leq i \leq n$ on the set $X = \{1 \ldots 2n\}$. With this view, each vertex is a permutation $(a_1, a_2, \ldots, a_{2n})$ such that $(a_{2i-1}, a_{2i}) = (2i - 1, 2i)$ if $x_i = 0$ and $(a_{2i-1}, a_{2i}) = (2i, 2i - 1)$ if $x_i = 1$, see Fig. 2.

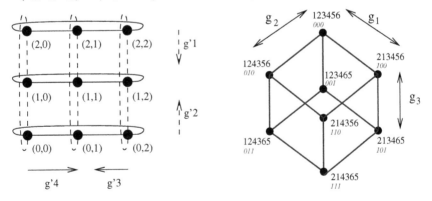

Fig. 1. Cayley graph for the torus $T_{3,3}$ **Fig. 2.** Cayley graph for the hypercube H_3

3   Communication Operations

We have developed Nqthm libraries for some widespread communication primitives on these interconnection networks. Let us briefly recall them, more details can be found in [14,9].

- *Broadcast:* a node (processor) sends the same message to every other node.
- *Scatter:* a node distributes portions of a message among all the other nodes.
- *Reduction:* assume that each processor P_i, $1 \leq i \leq N$, holds a value v_i. Let $\oplus$ be a user-specified commutative, associative operation on the elements v_i. After the reduction computation, each processor P_i holds the value $\oplus_{k=1}^{N} v_k$.
- *Reduction-to-one:* the reduction-to-one operation is similar but deposits the result $\oplus_{k=1}^{N} v_k$ on a specified node.
- *Prefix Like Computation:* it is also similar to the reduction, but each processor P_i ultimately holds the value $\oplus_{k=1}^{i} v_k$.
- *Fold:* assume that each processor P_i holds a vector z_i of length l. After the fold operation, each processor P_i holds a part of length l/N of $\sum_{k=1}^{N} z_k$.
- *Expand:* conversely, after the expand operation, each processor P_i holds the vector of length $l * N$ which is the concatenation of all the vectors z_i.

4   Modelling Methodology

Our purpose is to model in the Boyer-Moore logic the networks presented in Sect. 2 and their ad hoc realizations of the communication operations of Sect. 3. Every concept is given a purely functional representation.

4.1   Modelling the Networks

Every interconnection network is characterized by generators. The function that computes s', the label of the node that can be reached by applying the generator g_i to the node s ($s' = s \otimes g_i$) in a graph of size n will be referred to as g_i, i.e. $s' = g_i(s, n)$. We define a function $Cayley$ for each graph, it returns its set of vertices. In the case where the number of generators is fixed whatever the size is (e.g. the cycle or the torus) the Cayley graph can be defined as follows: starting from $I(n)$, the identity element of σ_n (the group of all permutations over n elements), we iteratively compute the vertices. At each step, new vertices are computed, by application of the generators g_i to the vertices obtained in the previous iterations. The construction stops when no new vertex can be obtained. The necessary number of steps is equal to the diameter of the graph. In the case where the number of generators is a function of the size of the Cayley graph i.e., the graph is hierarchical (e.g. the hypercube), an alternative approach described below can be taken.

Definition 2. *The graph $G(n)$, of dimension n, is hierarchical if there exists a non null function $\mathcal{K}$ such that for $n > 1$, $G(n)$ is decomposable into $\mathcal{K}(n)$ disjoint hierarchical subgraphs that are isomorphic to $G(n-1)$.*

We call $Iso(i, G(n))$, $1 \leq i \leq \mathcal{K}(n)$, the i-th subgraph isomorphic to $G(n)$. $G(n)$ can be associated with $Cayley(n, n)$, where $Cayley$ is defined as follows (k is the level of decomposition):

<u>**funct**</u> $Cayley(k, n) \equiv$
<u>**if**</u> $k = 0$ <u>**then**</u> $\{I(n)\}$ <u>**else**</u> $\bigcup_{1 \leq i \leq \mathcal{K}(k)} Iso(i, Cayley(k-1, n))$ <u>**fi**</u>

This methodology has been used to encode the interconnection networks in Nqthm. For the hypercube, the function above is instanciated with $\mathcal{K}$ as the constant function 2. We have also verified basic properties for each graph, such as its finiteness and its vertex symmetry. Table 1 gives the corresponding numbers of Nqthm events (an *event* is a function definition or a theorem).

Table 1. Number of events to model Cayley graphs in Nqthm

	# definitions	# intermediate lemmas	# theorems
Cycle	14	60	16
Torus	26	108	20
Hypercube	14	32	3

4.2   Modelling the Neighbourhood of Communicating Processors

Our proof methodology makes use of *invariants* that call on the notion of *neighbourhood* of a vertex. $Neighb(s,t,n)$ represents the set of vertices that have communicated, directly or not, with the vertex s at step t of the communication algorithm running on a network of size n. Depending on the communication operation, one or several direct neighbour(s), reachable by one or several generators, are added to this set at each step; this is characterized by the set $C_t \equiv \{\, i \mid g_i(s,n) \text{ communicates with } s \text{ at step } t \,\}$.

$\underline{\text{funct}}\ Neighb(s,t,n)\ \equiv$
$\underline{\text{if}}\ t = 0\ \underline{\text{then}}\ \{s\}$
        $\underline{\text{else}}\ Neighb(s,t-1,n) \cup \bigcup_{i \in C_t} Neighb(g_i(s,n),t-1,n)$
$\underline{\text{fi}}$

Example. Neighbourhood for the broadcast in the hypercube. At every step t, each processor communicates with its direct neighbour in the tth dimension, see Fig. 3 where each number represents both the dimension and the step number.

$\underline{\text{funct}}\ Neighb_{bh}(s,t,n)\ \equiv$
$\underline{\text{if}}\ t = 0\ \underline{\text{then}}\ \{s\}$
        $\underline{\text{else}}\ Neighb_{bh}(s,t-1,n)$
            $\cup\ Neighb_{bh}(g_t(s,n),t-1,n)$
$\underline{\text{fi}}$

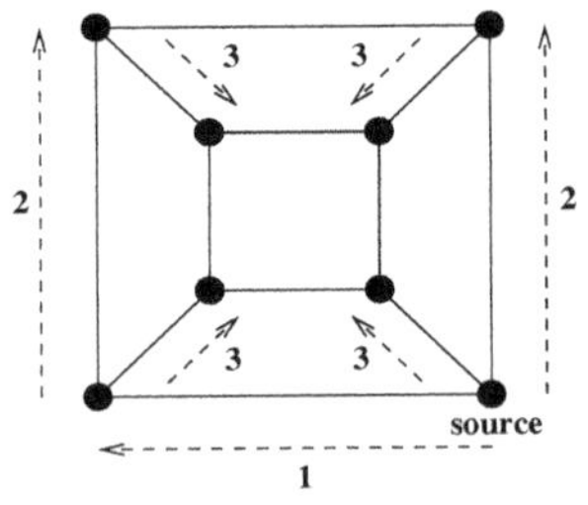

Fig. 3. Broadcast on H_3

4.3   Modelling the Communications

It remains to formalize, for every communication operation and every architecture, the code executed by each processor. For each operation, we define a function that takes the following arguments:

– s: the processor where the algorithm is running;
– k: the number of iterations;
– ini: initial state of the system (associative list of pairs of the form (s, val_{init}) where s is a vertex and val_{init} is the initial state of its local register(s)). It is associated with a function $Take\_value$ such that $Take\_value(s, ini) = val_{init}$;
– n: the size of the network.

The function returns the value(s) stored in the register(s) of the processor after k iterations. For instance, a function $Broadcast_h(s, k, ini, n)$ is defined to model the broadcast operation on a n-dimensional hypercube, see its Nqthm definition in the next section.

4.4   Link with an MPI-Based Implementation

In this section we consider the link between our formal model and a source code written in a usual programming language. This aspect is fundamental w.r.t. a possible integration of our approach in a development environment. The overall method described in this paper is adapted to the validation of applications designed to run on distributed memory SIMD or MIMD target architectures, provided that synchronization takes place between the elementary point-to-point send/receive operations. MPI (Message Passing Interface) [20] is a standard for the realization of message passing procedures; MPICH is one of its implementations that allows to develop applications on a network of workstations. In view of designing a methodology of transformation from a C program, using MPI point-to-point primitives, to our formalism, we study the correspondance between these two representations in the case of the collective operations presented in Sect. 3.

Here is our principle for implementing every procedure: since each processor cannot know in which iteration it will receive (and then send) the message which is broadcasted, scattered and the like, it executes "send" and "receive" operations at every step. It receives both a message and a "flag" that indicates if this message is relevant. In that case, it updates its register(s) and sets its own flag to true. The C version of the $broadcast$ function for H_n is given below.

```
int broadcast_h (char *s, int n, void *buffer, int count,
                 MPI_Datatype datatype, int *flag, void *mybuffer,
                 int *myflag, MPI_Comm comm)
{ char *neighb; /* neighbour in the k-th dimension */
  int k; MPI_Status st;
  MPI_Aint l; MPI_Type_extent(datatype,&l);
  for (k=0; k<n; k++)
  { /* transfer of data with the neighbour: */
    neighb = gen_hyper(k, s, n); /* permutation of the neighbour */
    MPI_Send(mybuffer,count,datatype,PermutationToInt(neighb),k,comm);
    MPI_Send(myflag,1,MPI_INT,PermutationToInt(neighb),k,comm);
    MPI_Barrier(comm);       /* synchronization */
```

```
MPI_Recv(buffer,count,datatype,PermutationToInt(neighb),k,comm,&st);
MPI_Recv(flag,1,MPI_INT,PermutationToInt(neighb),k,comm,&st);
if (Hyper_sendp (*myflag)) /* if my flag is 1, data are unchanged */
  { }
else if (Hyper_sendp (*flag))
     /* if the neighbour's flag is 1, I update my data */
     { memcpy(mybuffer, buffer, count*l);
       *myflag = 1;
     }
     /* otherwise, data are unchanged */
}
return 1;
}
```

```
int Hyper_sendp (int flag) { return (flag==1); }
```

The Nqthm function broadcast_h (see below) is its faithful translation; it is defined as described in the previous section. Its definition is recursive, its parameter k corresponds to the iteration variable in the C function. It includes standard hypotheses on the types and constraints related to the parameters, systematically introduced in each function. Its second let gives the values of the variables that characterize the data to be sent (though the "send" operation is not explicit in our functional representation). The third let corresponds to the variables associated with the data that are received from the neighbour(s). In both cases, the Lisp expressions can be generated in a systematic way from the C source, knowing the names of the Nqthm *accessor* functions (buf-rcv, buf-snd, . . .). A test is included to stop recursion; its condition corresponds to the initialisation of the iteration variable in the C program, and the returned value is given by the expression that computes the initial value of the set of buffers for processor s (in ini). In the rest of the function body, there are as many nested "if" statements as in the C function:

- in the case(s) where data are updated in the C function, the Lisp function returns the new status of the set of buffers for processor s, using the expression (cbuffer buffer (message t buffer)) which means that the new value of mybuffer, resp. myflag, is buffer, resp. 1 (value t). In the broadcast operation, the message to be sent is exactly the one that has been received.
- in the case(s) where data are unchanged in the C function, the value of the set of buffers remains the same and the Lisp function recurses with k decreased by 1. In the broadcast example, the corresponding expression is (broadcast_h s (sub1 k) ini n).

```
(defn broadcast_h (s k ini n)
  (if (and (not (zerop n)) (numberp k) (leq k n)
           ...)   ; + other hypotheses
      (let ((neighb (gen_hyper k s n)))   ; neighbour
        (let (; data to be sent:
              (mybuffer (buf-rcv (broadcast_h s (sub1 k) ini n)))
```

```
              (myflag (get-flag (buf-snd (broadcast_h s (sub1 k) ini n)))))
         (let (; data received:
              (buffer (get-msg (buf-snd
                                  (broadcast_h neighb (sub1 k) ini n))))
              (flag (get-flag (buf-snd
                                  (broadcast_h neighb (sub1 k) ini n)))))
           (if (zerop k)   ; to stop recursion
               (take_value s ini)
             (if (hyper_sendp myflag) ; data are unchanged
                 (broadcast_h s (sub1 k) ini n)
               (if (hyper_sendp flag)
                   ; the new buffer contains ''buffer'' and ''true'':
                   (cbuffer buffer (message t buffer))
                 ; otherwise data are unchanged
                 (broadcast_h s (sub1 k) ini n)))))))
    (empty)))

(defn hyper_sendp (flag) (equal flag t))
```

5   Proof Methodology

5.1   Description of the Method

Our proof method is the same for every communication primitive and every network. First we exhibit and prove an invariant $\mathcal{I}$ and then, using a proof obligation $\mathcal{O}$ that is related to the function $Neighb$, we complete the proof of the main theorem $\mathcal{T}$. The invariant as well as the characteristics of $Neighb$ depend on the problem, but we will show that $\mathcal{I}$ can easily be deduced from $\mathcal{T}$. The form of $\mathcal{T}$, $\mathcal{I}$ and $\mathcal{O}$ clearly demonstrates that the verifications are performed at the processor level; they involve one (universally quantified) processor s. In the following, we assume that the graph is built after t_c iterations, $Comm$ denotes the communication function which needs t_f iterations to complete, and S represents the specification function. The form of $\mathcal{T}$ slightly differs according to whether the operation involves a source processor or not. In the case where there is no specific source, the equality $\mathcal{T}$ to be proven is of the form:

$$\forall s \in Cayley(t_c, n).\, Comm(s, t_f, ini, n) = S(s, Cayley(t_c, n), ini, n)$$

The statement of $\mathcal{I}$ generalizes the one of $\mathcal{T}$ to k iterations, considering the neighbourhood of s. After k iterations, s holds the value given by the specification function S applied to the values of the processors in its neighbourhood:

$$\forall s \in Cayley(t_c, n), \forall k \leq t_f.\, Comm(s, k, ini, n) = S(s, Neighb(s, k, n), ini, n)$$

The proof obligation $\mathcal{O}$ necessary to get $\mathcal{T}$ states that, at the end of the operation, every processor has taken part in the communication:

$$\forall s \in Cayley(t_c, n).\, Neighb(s, t_f, n) = Cayley(t_c, n)$$

When the operation involves a source (e.g. broadcast), the idea is similar but the theorems make use of the source, and they assume its uniqueness.

Table 2 gives the numbers of Nqthm events for the communication operations that we have validated using this methodology (the first column is for the specification of the "neighbourhoods", "-" means that the proofs are under development).

Table 2. Communication operations in Nqthm (numbers of events)

	Neighb.	Broadcast	Reduction	Scatter	Gather	Prefix	Fold	Expand
Cycle	74	53	11	-	-	-	-	-
Torus	108	31	12	-	-	-	-	-
Hypercube	75	21	3	12	19	63	103	51

5.2   Example: Broadcast on the Hypercube

To validate the broadcast operation on H_n we prove that, after n steps, every processor has received the message sent by the source.

$\forall s \in Cayley_h(n,n).\ Uniquenessp(Source(ini)) \implies$
$Broadcast_h(s,n,ini,n) =\ Take\_Value(Source(ini),ini)$

The invariant $\mathcal{I}$ generalizes this statement:

$\forall s \in Cayley_h(n,n), \forall k \leq n.\ Uniquenessp(Source(ini)) \implies$
$Broadcast_h(s,k,ini,n) =$
if $s \in Neighb_{bh}(Source(ini),k,n)$ **then** $Take\_Value(Source(ini),ini)$
else $Take\_Value(s,ini)$

The lemma $\mathcal{O}$ states that, after n steps, every processor is in the neighbourhood of the source.

$\forall s \in Cayley_h(n,n).\ s \in Neighb_{bh}(Source(ini),n,n)$

6   Applicative Examples

The libraries we have developed can be used for the verification of application programs (we verify that they are equivalent to their sequential counterparts).

6.1   Computational Geometry Algorithm

This algorithm [19] solves a problem of computational geometry: given a query point z, determine whether it lies in a region R. Here, z is a planar point and R is a polygon. For a N-polygon, the program can be implemented on a hypercube with $N = 2^n$ processors. Initially, a source processor s holds the query point z and the set E of the N edges e_i that defines the polygon. The algorithm is:

1. the source s *scatters* the edges over the network, one edge per processor
2. it *broadcasts* the query point to all other processors
3. each processor P_i calls on a function *Intersectp* that returns $l_i = 1$ (otherwise 0) if its edge intersects the horizontal line containing z to the left of z
4. every processor computes $\sum_{i=1}^{N} l_i$; it is a *reduction* procedure with the addition as operator. If this sum is odd then z is internal to the polygon.

Each processor has two buffers, their values are enclosed in brackets in the function definitions below and are accessible by the functions acc_1 and acc_2. We model each step of the algorithm by a function $Step_i$. The expressions $Build\_ini_i(\ldots)$ give the global state of the network after the step number i.

1. The source processor s sends to each processor P_i the edge e_i:

$Step1(P_i, s, E, z, n) \equiv$
   $[\ Scatter_h(P_i, n, Build\_ini_0(s, E), n)\ ;\ \textbf{if}\ P_i = s\ \textbf{then}\ z\ \textbf{else}\ \perp]$

Correctness lemma: each P_i has received the edge e_i.

$\forall P_i, s \in Cayley(n, n).\ Step1(P_i, s, E, z, n) = [e_i\ ;\ \textbf{if}\ P_i = s\ \textbf{then}\ z\ \textbf{else}\ \perp]$

2. The processor s broadcasts z to every other processor:

$Step2(P_i, s, E, z, n) \equiv$
   $[\ acc_1(Step1(P_i, s, E, z, n))\ ;\ contents\ of\ the\ 1st\ buffer\ unchanged$
   $Broadcast_h(P_i, n, Build\_ini_1(\ldots), n)\ ]$

Correctness lemma: each P_i has received z.

$\forall P_i, s \in Cayley(n, n).\ Step2(P_i, s, E, z, n) = [\ e_i\ ;\ z\ ]$

3. Each processor P_i computes l_i:

$Step3(P_i, s, E, z, n) \equiv$
   $[\ Intersectp(acc_1(Step2(P_i, s, E, z, n)), acc_2(Step2(P_i, s, E, z, n)))\ ;$
   $acc_2(Step2(P_i, s, E, z, n)))\ ]\ ;\ contents\ of\ the\ 2nd\ buffer\ unchanged$

Correctness lemma: each P_i has correctly computed its l_i.

$\forall P_i, s \in Cayley(n, n).\ Step3(P_i, s, E, z, n) = [\ Intersectp(e_i, z)\ ;\ z\ ]$

4. Each processor computes $\sum_{k=1}^{2^n} l_k$:

$Step4(P_i, s, E, z, n) \equiv$
   $[\ Reduction\_Sum_h(P_i, n, Build\_ini_3(\ldots), n)\ ;$
   $acc_2(Step3(P_i, s, E, z, n)))\ ]\ ;\ contents\ of\ the\ 2nd\ buffer\ unchanged$

Correctness of the complete algorithm: each processor knows if z is internal or external to the polygon.

$\forall P_i, s \in Cayley(n, n).\ Step4(P_i, s, E, z, n) = [\ \sum_{k=1}^{2^n} Intersectp(e_k, z)\ ;\ z\ ]$

6.2   Matrix-Vector Product

Let us consider the algorithm proposed in [9] to perform a matrix-vector multiplication on the hypercube. Consider the product $y = Ax$ where A is an $n \times n$ matrix and x and y are vectors of length n. The algorithm computes y on H_d (there are $p = 2^d$ processors, n is evenly divisible by p, and d is even i.e., $\sqrt{p} = 2^{\frac{d}{2}}$ is a natural number).

Decomposition and Assignment of Data. The matrix A is decomposed into square blocks of size $(n/\sqrt{p}) \times (n/\sqrt{p})$. Each block is denoted $\mathcal{A}[\alpha, \beta]$, α, β running from 0 to $\sqrt{p}-1$, see Fig. 4. The input vector x and product vector y are also divided into $\sqrt{p}$ pieces. The subvector number β of x, resp. the subvector number α of y, will be denoted $\mathcal{X}[\beta]$, resp. $\mathcal{Y}[\alpha]$. Each pair of subscripts (α, β) can be associated with one of the p processors of H_d, which will be in charge of computing its contribution to $\mathcal{Y}[\alpha]$ in terms of $\mathcal{A}[\alpha, \beta]$ and $\mathcal{X}[\beta]$. This processor will be referred to as $P_{\alpha\beta}$; its registers initially holds the data $\mathcal{A}[\alpha, \beta]$ and $\mathcal{X}[\beta]$.

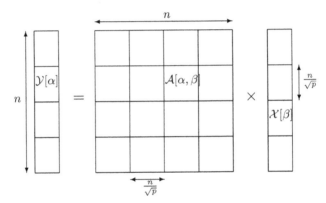

Fig. 4. Distribution of data

Simplified Matrix–Vector Product. The method is the following (see Fig. 5 for the algorithms of the *fold* and *expand* operations):

- $P_{\alpha\beta}$ computes its contribution to $\mathcal{Y}[\alpha]$. This is a vector of length $n/\sqrt{p}$ which we denote by $\mathcal{Z}[\alpha, \beta]$; thus $\mathcal{Z}[\alpha, \beta] = \mathcal{A}[\alpha, \beta] \times \mathcal{X}[\beta]$ and $\mathcal{Y}[\alpha] = \Sigma_\beta \mathcal{Z}[\alpha, \beta]$

- The second step uses the *fold* operation to sum vectors $\mathcal{Z}[\alpha, \beta]$. This operation is executed between a group of $\sqrt{p}$ processors; it requires $log_2(\sqrt{p}) = \frac{d}{2}$ steps, halving the length of the vectors involved at each step. Within each step, a processor first divides its vector z into two equal size subvectors, z_1 and z_2 (notation $(z_1|z_2)$). One of these subvectors is sent to another processor P, and a subvector w is received from P. It is summed element-by-element with the retained subvector. As a result, each processor $P_{\alpha\beta}$ has a unique, n/p-length portion of the fully summed vector denoted $\mathcal{Y}[\alpha]_{[\beta]}$.

– The last step uses the *expand* operation. Each processor in the row has a subvector $z = \mathcal{Y}[\alpha]_{[\beta]}$. At each step, it sends z to another processor P and receives w from P. These two subvectors are concatenated in the correct order, as indicated by the notation '|', to form the updated value of z. This operation requires $log_2(\sqrt{p})$ steps. This primitive combines the subvectors $\mathcal{Y}[\alpha]_{[\beta]}$ and places the complete vector $\mathcal{Y}[\alpha]$ in each processor of row α.

Fold operation for $P_{\alpha\beta}$	Expand operation for $P_{\alpha\beta}$
$z := \mathcal{Z}[\alpha, \beta]$ **for** $i := 0$ **to** $log_2(\sqrt{p}) - 1$ $\quad (z_1\|z_2) = z$ $\quad P := P_{\alpha\beta}$ with i^{th} bit of β flipped $\quad$ **if** bit i of β is 1 **then** $\quad\quad$ **Send** z_1 to processor P $\quad\quad$ **Receive** w from processor P $\quad\quad z := z_2 + w$ $\quad$ **else** $\quad\quad$ **Send** z_2 to processor P $\quad\quad$ **Receive** w from processor P $\quad\quad z := z_1 + w$ $\mathcal{Y}[\alpha]_{[\beta]} := z$	$z := \mathcal{Y}[\alpha]_{[\beta]}$ **for** $i := log_2(\sqrt{p}) - 1$ **to** 0 $\quad P := P_{\alpha\beta}$ with i^{th} bit of β flipped $\quad$ **Send** z to processor P $\quad$ **Receive** w from processor P $\quad$ **if** bit i of β is 1 **then** $\quad\quad z := w\|z$ $\quad$ **else** $\quad\quad z := z\|w$ $\mathcal{Y}[\alpha] := z$

Fig. 5. Communication primitives for processor $P_{\alpha\beta}$

As in the previous example, we model each step of the algorithm by a function $Step_i$, which is associated with an intermediate correctness lemma.

1. Each processor $P_{\alpha\beta}$ of a d-dimensional hypercube computes locally a matrix-vector product. The expression $acc_1(Take\_value(P_{\alpha\beta}, Build\_ini_0(\ldots)))$ returns $\mathcal{A}[\alpha, \beta]$ and $acc_2(Take\_value(P_{\alpha\beta}, Build\_ini_0(\ldots)))$ returns $\mathcal{X}[\beta]$.

$$Step1(P_{\alpha\beta}, A, X, d) \equiv$$
$$Matrix\_Vector\_Product(acc_1(Take\_value(P_{\alpha\beta}, Build\_ini_0(\ldots)))$$
$$acc_2(Take\_value(P_{\alpha\beta}, Build\_ini_0(\ldots))))$$

2. $P_{\alpha\beta}$ computes $\mathcal{Y}[\alpha]_{[\beta]}$ by applying the fold operation. The function $Build\_ini_1$ returns the global state of the network after the first step.

$$Step2(P_{\alpha\beta}, A, X, d) \equiv Fold(P_{\alpha\beta}, \tfrac{d}{2}, Build\_ini_1(\ldots), d)$$

Correctness lemma: each $P_{\alpha\beta}$ has received $\mathcal{Y}[\alpha]_{[\beta]}$.

$$\forall P_{\alpha\beta} \in Cayley(d, d). \, Step2(P_{\alpha\beta}, A, X, d) =$$
$$\left(\sum\nolimits_{e \in Neighb(P_{\alpha\beta}, \frac{d}{2}, d)} Matrix\_Vector\_Product(\mathcal{A}[\alpha_e, \beta_e], \mathcal{X}[\beta_e])\right)_{[\beta]}$$

3. The processor $P_{\alpha\beta}$ computes $\mathcal{Y}[\alpha]$ by applying the expand operation. The function $Build\_ini_2$ returns the global state after the second step.

$$Step3(P_{\alpha\beta}, A, X, d) \equiv Expand(P_{\alpha\beta}, \tfrac{d}{2}, Build\_ini_2(\ldots), d)$$

Correctness lemma: Processor $P_{\alpha\beta}$ has computed $\mathcal{Y}[\alpha]$.

$$\forall P_{\alpha\beta} \in Cayley(d,d).\, Step3(P_{\alpha\beta}, A, X, d) =$$
$$\sum_{e \in Neighb(P_{\alpha\beta}, \frac{d}{2}, d)} Matrix\_Vector\_Product(\mathcal{A}[\alpha_e, \beta_e], \mathcal{X}[\beta_e])$$

Correctness of the complete algorithm:

$$\forall P_{\alpha\beta} \in Cayley(d,d).\, Step3(P_{\alpha\beta}, A, X, d) =$$
$$\sum_{j=0}^{\sqrt{p}-1} Matrix\_Vector\_Product(\mathcal{A}[\alpha, j], \mathcal{X}[j])$$

7 Related Works

The validation of applications involving distributed processes has been widely studied. Many approaches are related to the verification of communication protocols and make use of *model checking* techniques; some others put the emphasis on the correct *parallelization* of sequential code, like [5]. Here we focus only on works which are close to ours i.e., that are concerned with the verification of the equivalence between a parallel algorithm and its sequential counterpart, by means of *theorem proving* techniques. They can be classified into two main categories.

In the first one, the algorithm is modelled as a function that takes as argument the set of data distributed in the system (its *state*) and returns the state obtained after one computation step. The proof consists in showing that the state after a number N of iterations satisfies the *specification*. The function and the data structures give an explicit view of the network. In [7], we prove the correctness of a parallel algorithm for finding the maximum of set of values on a $n \times n$ 4-neighbour torus, using Nqthm. The method uses two recursive functions to model the network; the main data structure is a list of list of natural numbers. Point-to-point communications are expressed by updating the elements of this list according to the values of their neighbours. In [11], RRL is used in conjunction with *powerlists* data structures [17]. The authors verify the *Batcher's Merge Sorting Network* and prove correct the mapping of multi-dimensional arrays expressed as powerlists into hypercube networks. The verification of the FFT with ACL2 is described in [6]; the underlying representation is also based on powerlists. This notation is well suited to the specification and verification of certain types of parallel algorithms running on regular structures. Kornerup discusses in [13] the modelling of hypercube algorithms with this notation.

The second approach makes explicit the *send* and *receive* point-to-point primitives and the network structure. The algorithm is expressed by a function which describes the updating of the *global state* of the system, which consists of the private states of the processes together with the bag of messages that are in transit (sent but not yet accepted by the destination process). This model allows the specification of asynchronous communications, which could introduce nondeterminism. To model nondeterminacy, a free variable *oracle* gives a symbolic view of the list of messages that are in transit. Hesselink explains this modelling methodology in [10]. The function that expresses the distributed algorithm takes as parameters the global state s, the oracle ora, the iteration number n and the description of the network (a graph) g. The proof methodology is based on invariants describing properties of the global state, and proofs are performed with Nqthm. Arbitrary interconnection networks can be considered, the only hypothesis is that the corresponding graph is connected.

There are common aspects between these approaches and ours. Like in the first one, we consider synchronized algorithms and the point-to-point communications are implicit in the formal model. A similarity with both approaches is that we use an explicit representation of the interconnection networks. However, a major advantage of modelling these structures by means of Cayley graphs is that our proof methodology works at the processor level instead of having to reason on the whole network. The graph is not encoded as a static structure, but the processor neighbours are dynamically computed using the generator functions. In many cases, the correspondance between this processor view and the global system view is straightforward. In a near future, we plan to consider more complex applications and to improve our method with techniques like the one described in [18] to map a "uniprocessor" view to a "multiprocessor" one.

8 Conclusion

We have proposed a new approach for the validation of parallel algorithms running on symmetric interconnection networks. The proof methodology exploits their topological properties. We have built reusable libraries for reasoning about widespread networks and their collective communication primitives. Two examples have illustrated the usefulness of these libraries to perform correctness verifications for distributed programs; other applications are being developed. Table 3 gives the corresponding numbers of Nqthm events. In each case, about 70 % of events have simply been imported from the libraries (without reproving them). A part of the remaining 30 % could be generated automatically, from the contents of the source code and from the specification. However, human intervention is needed for the other ones. Since there is a common basis in the underlying reasoning related to most of them, we plan to design a development environment with an interactive user-interface where the programmer could input his source code and the specification, ask for the generation of the corresponding events by means of a mechanized translator, and then guide the proof tool when needed. To that goal, we are currently re-implementing our libraries in the up-to-date prover Acl2 [12].

Table 3. Numbers of Nqthm events for the verification of the examples

	# Events (def. + lemmas)	# Events reused from our libraries	# Events specific to the algorithm
Computational Geometry	142	111 (78 %)	31 (22 %)
Matrix-vector product	417	286 (68 %)	131 (32 %)

References

[1] S. B. Akers and B. Krishnamurthy. A Group Theoretic Model for Symmetric Interconnection Networks. *IEEE Transactions on Computers*, 38(4), 1989.

[2] B.W. Arden and K.W. Tang. Representations and Routing of Cayley Graphs. *IEEE Transactions on Communications*, 39(11), November 1991.

[3] J-C. Bermond, T. Kodate, and S. Perennes. Gossiping in Cayley Graphs by Packets. In *Proceedings of Franco-Japanese conference Brest July 95*, volume 1120 of *Lectures Notes in Computer Science*. Springer Verlag, 1996.

[4] R. Boyer and J Moore. *A Computational Logic Hand-book*. Perspectives in Computing, Vol. 23. Academic Press, Inc., 1988.

[5] R. Couturier and D. Méry. An experiment in parallelizing an application using formal methods. In *CAV'98*. LNCS 1427, Springer Verlag, 1998.

[6] R. A. Gamboa. Mechanically Verifying the Correctness of the Fast Fourier Transform in ACL2. In *Third International Workshop on Formal Methods for Parallel Programming: Theory and Applications (FMPPTA)*, 1998.

[7] E. Gascard and L. Pierre. Two Approaches to the Formal Proof of Replicated Hardware Systems using the Boyer-Moore Theorem Prover. In *Proc. International Workshop on First Order Theorem Provers (FTP'97)*, October 1997.

[8] C. GowriSankaran. Broadcasting on Recursively Decomposable Cayley Graphs. *Discrete Applied Mathematics*, 53:171–182, 1994.

[9] B. Hendrickson, R. Leland, and S. Plimpton. An Efficient Parallel Algorithm for Matrix–Vector Multiplication. *Int. J. High Speed Computing*, 7(1):73–88, 1995.

[10] W. H. Hesselink. A mechanical proof of Segall's PIF algorithm. *Formal Aspects of Computing*, 9:208–226, 1997.

[11] D. Kapur and M. Subramaniam. Automated Reasoning about Parallel Algorithms using Powerlists. In *Proc. Algebraic Methodology and Software Technology, AMAST'95*, volume 936 of *LNCS*, pages 416–430. Springer-Verlag, 1995.

[12] M. Kaufmann and J S. Moore. An Industrial Strength Theorem prover for a Logic Based on Common Lisp. *IEEE Trans. on Software Engineering*, 23(4), April 1997.

[13] J. Kornerup. Mapping Powerlists onto Hypercubes. Technical Report TR-94-04, Department of Computer Sciences, University of Texas at Austin, 1994.

[14] V. Kumar, A. Grama, A. Gupta, and G. Karypis. *Intoduction to Parallel Computing: Design and Analysis of Algorithms*. Benjamin Cummings Pub., 1994.

[15] T.M. Kurç, C. Aykanet, and B. Özguç. A parallel scaled conjugate-gradient algorithm for the solution phase of gathering radiosity on hypercubes. *The Visual Computer*, 13(1):1–19, 1997.

[16] Y. Lee, S. Horng, T. Kao, and Y. Chen. Parallel computation of the Euclidean distance transform on the mesh of trees and the hypercube computer. *Computer Vision and Image Understanding*, 68(1):109–119, October 1997.

[17] J. Misra. Powerlist: A structure for parallel recursion. *ACM Transactions on Programming Languages and Systems*, 16(6):1737–1767, November 1994.

[18] J S. Moore. A mechanically checked proof of a multiprocessor result via a uniprocessor view. *Formal Methods in System Design*, 14(2):213–228, 1999.

[19] K. Qiu and S.G. Akl. Novel Data Communication Algorithms on Hypercubes and Related Interconnection Networks and Their Applications in Computational Geometry. Technical Report 97-415, Department of Computing and Information Science, Queen's University, Kingston, Ontario, December 1997.

[20] M. Snir, S. Otto, S. Huss-Lederman, D. Walker, and J. Dongarra. *MPI: The Complete Reference*. The MIT Press, 1996.

A Framework for Microprocessor Correctness Statements

Mark D. Aagaard[1], Byron Cook[2], Nancy A. Day[3], and Robert B. Jones[4]

[1] Electrical and Computer Engr., University of Waterloo, Waterloo, ON, Canada
markaa@swen.uwaterloo.ca
[2] Prover Technology, Portland, OR, USA
byron@prover.com
[3] Computer Science, University of Waterloo
nday@cs.uwaterloo.ca
[4] Strategic CAD Labs, Intel Corporation, Hillsboro, OR, USA
rjones@ichips.intel.com

Abstract Most verifications of out-of-order microprocessors compare state-machine-based implementations and specifications, where the specification is based on the instruction-set architecture. The different efforts use a variety of correctness statements, implementations, and verification approaches. We present a framework for classifying correctness statements about safety that is independent of implementation representation and verification approach. We characterize the relationships between the different statements and illustrate how existing and classical approaches fit within this framework.

1 Introduction

The increased parallelism provided by *out-of-order execution* in microprocessors has made correctness statements for verification complicated, varied, and even controversial. We studied published verifications of out-of-order microprocessors and discovered a wide variety of correctness statements, verification techniques, and processor implementations. Some correctness statements initially appear to be similar, such as the ones based on Burch-Dill style flushing [BD94], but differences emerge after close examination. Other statements are difficult to compare at first, but later reveal similarities. The goal of this work is to provide a foundation for clarifying the meaning of individual correctness statements; precisely comparing different statements; and analyzing the interaction between processor features, verification strategy, and correctness statements.

Most recent verification efforts verify a state-machine-based microarchitectural implementation against a state-machine-based instruction-set architecture. The verification efforts focus on safety; liveness is usually dealt with as a secondary concern. In keeping with these trends, we focus on the verification of safety between microarchitectural implementations and instruction-set architectures. We include deterministic and non-deterministic state machines with finite or infinite state spaces. We do not yet include specifications that are collections of properties, e.g. [BB94,McM98,PJB99].

The result of our investigation and analysis is a framework that precisely describes and classifies correctness statements about safety between state machines. It allows cor-

T. Margaria and T. Melham (Eds.): CHARME 2001, LNCS 2144, pp. 433–448, 2001.

rectness statements to be analyzed independent of verification techniques and microarchitectural features. In this paper, we introduce the framework, present its mathematical basis, and describe how existing out-of-order microprocessor correctness statements fit within the framework.

2 Modeling with State Machines

We assume that both the specification and implementation have program memories as part of their state. Therefore, our state machines do not take instructions as inputs. Approaches that take instructions as inputs in their correctness statements (e.g. [BD94,JSD98,BBCZ98]) can be augmented with program memories that produce the input trace. Interrupts can also be treated as part of the state space by adding appropriate control circuitry to read the interrupt input trace from a store. We assume that state machines generate infinite traces, where "termination" of a program is denoted by repeating the final state of the program. Definition 1 shows the formalism we use to describe state machines.

Definition 1 (State machines). A state machine M is a triple (Q, Q°, N) where:
- Q is the set of possible state values and is a Cartesian product of internal (hidden) state components and externally-visible state components.
  Q_e is the set of possible external state values.
  $\Pi_e : Q \to Q_e$ is the corresponding projection function.
  $q_1 \stackrel{\Pi}{=} q_2$ says that q_1 and q_2 have equivalent external state: $\Pi_e(q_1) = \Pi_e(q_2)$.
- $Q^\circ \subseteq Q$ is the set of initial states.
- $N \subseteq Q \times Q$ is the next-state relation.
  $N^k(q, q')$ means q' is reachable from q in k steps of N.
  When N is a function, we write it as n.

The components of a state machine M will be subscripted with "s" for specification and "i" for implementation. We assume a machine can always make a transition, i.e. $\forall q \in Q. \exists q' \in Q. N(q, q')$. We allow machines to *self-loop*, that is transition from a given state back to itself.

In verification, the state space of the implementation often needs to be limited to reachable states, or an over-approximation of reachable states. This challenging task is done by finding and proving invariants. Invariants are treated with varying degrees of emphasis in the literature. In our framework we consider the invariants to be encoded in Q, the set of states for the machine.

3 Correctness Statements

A well-established definition of correctness is that of *trace containment*: every trace of external observations generated by the implementation can also be generated by the specification. A disadvantage of trace containment is that verifying it can require information about an entire trace. Another traditional definition of correctness is *simulation*: if an implementation state is externally equal to a specification state, then executing one instruction in both the implementation and specification results in states that are externally equal. Simulation is usually easier to verify than trace containment, because simulation refers to individual transitions, rather than entire traces. Formal verification of sequential microprocessors has generally been done using simulation-style correctness statements. Similar correctness statements are also used in other domains such as cache-coherence protocols (e.g. [PD96,SA97,NG98]).

Pipelining and other optimizations increase the gap between the behavior of the implementation and the specification, thus making it more difficult to consider only one step within the implementation and specification traces. Pipelined machines begin executing new instructions before previous ones retire. Machines with out-of-order retirement retire instructions in a different order than the specification. A superscalar machine may externally appear to do nothing for a number of steps and then, in a single step, update the register file with the results of several instructions.

To describe how out-of-order verifications use simulation-style correctness statements, we separate the notions of 1) how to align the implementation trace against the specification trace to determine which states should match, and 2) what it means for an implementation state to successfully match a specification state.

When verifying non-pipelined machines, the traces can be aligned at every transition and two states match if the externally-visible state components are equal. To verify pipelined machines, the alignment often needs to be at looser intervals than every transition, or external equivalence needs to be replaced by a looser relationship. With out-of-order microprocessors, the notions of alignment and matching are necessarily even more complicated. A common alignment technique is to check the implementation when it is in a *flushed* state (i.e. no in-flight instructions). A common matching relationship is Burch-Dill style flushing [BD94], which uses an abstraction function to retire all in-flight instructions in the implementation and project the externally-visible state, and then checks for equality with the specification state.

Our framework uses four parameters to characterize a correctness statement: alignment, match, implementation execution, and specification execution. *Alignment* is the method used to align the executions of the implementation and specification (Sect. 3.1). *Match* is the relation established between the aligned implementation and specification states (Sect. 3.2). *Implementation execution* and *specification execution* describe the type of state machines used (Sect. 3.3). Section 4 shows how the correctness statements of existing work fit within our framework.

3.1 Alignment

Alignment describes which states in the execution traces are tested for matching. Figure 1 illustrates the four kinds of alignment that we have found used in out-of-order microprocessor verification. *Pointwise alignment* (P) is the classic commuting diagram, which compares every transition. *Stuttering alignment* (S) allows the specification to stutter, i.e. two or more consecutive implementation states can match the same specification state. In *must-issue alignment* (M), the specification takes a single step, and the implementation takes as many steps as are necessary to reach an unstalled state, and then issue an instruction. A predicate *isStalled* indicates when the implementation cannot take a "productive" step, and is generally defined to be true when the implementation cannot issue an instruction. Finally, *flush-point alignment* (F) says that if there is a trace between flushed implementation states, then there must exist a trace in the specification between any pair of states that match the flushed implementation states. A predicate *isFlushed* indicates flushed implementation states. Instruction-set architectures execute one instruction per step; therefore all of their states are flushed.

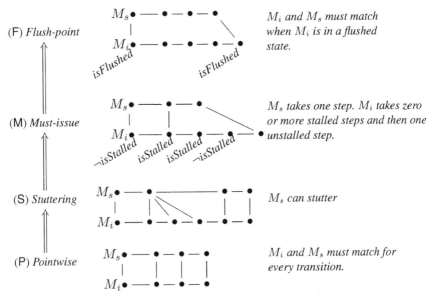

(F) *Flush-point* — M_i and M_s must match when M_i is in a *flushed* state.

(M) *Must-issue* — M_s takes one step. M_i takes zero or more stalled steps and then one unstalled step.

(S) *Stuttering* — M_s can stutter

(P) *Pointwise* — M_i and M_s must match for every transition.

In each diagram, the horizontal lines between states are the specification and implementation traces. The vertical lines between states show the where the implementation state must match the specification state.

Fig. 1. Options and total order for the alignment parameter

We place the four kinds of alignment in a total order as illustrated by the arrows in Fig. 1. This order is based on generality where alignments higher in the order are weaker. For example, stuttering correctness implies flush-point for any instance of the predicate *isFlushed*.

3.2 Match

Instantiations for the match parameter are relations $\mathcal{R}$ over an implementation state q_i and specification state q_s that mean "q_i is a correct representation of q_s". Figure 2 shows the matches that we found used in out-of-order microprocessor verification. The arrows show the total order, where definitions lower in the order are instances of higher values.

A *general match* (G) is any relation between implementation and specification states. The *abstraction match* (A) uses a function (*abs*) to map an implementation state to a point that is externally equivalent to the specification state. The *equality match* (E) requires that the implementation and specification states be externally equivalent. The tightest match is the *refinement map* (R), which is an abstraction function that preserves the externally-visible part of the implementation state. Refinement differs from equality because the refinement map is a function, so each implementation state matches exactly one specification state. The literature overloads words such as "refinement" and "abstraction". The mathematics in Fig. 2 give the precise definitions that we use.

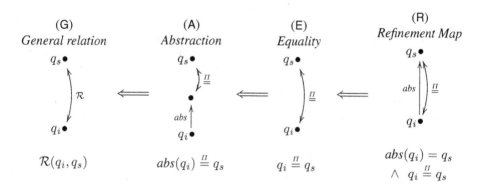

Fig. 2. Options and total order for the match parameter

If the specification does not have any internal state (i.e. all of the state components are externally visible), then equality and refinement both reduce to $\Pi_e(q_i) = q_s$. Because refinement is a tighter match than equality, we will call such cases refinement.

In addition to the instances listed here, other options are possible. For example, Pnueli *et al.* [DP97,PA98,AP99,AP00] have a matching relation that uses both concretization and abstraction functions. Two states match if concretizing the specification state produces the same result as abstracting the implementation state. In their examples, their concretization functions are identity or projection, so their match specializes to abstraction in our framework.

3.3 Execution

The third and fourth parameters of the framework are the methods for describing the traces of the implementation and specification. In the literature we find both *deterministic* (D) and *non-deterministic* (N) implementations and specifications. In a deterministic machine, the transition relation is instantiated as: $N(q, q') \equiv q' = n(q)$. Implementations are often modeled with non-determinism because of scheduling circuitry. On the other hand, most instruction-set architectures are deterministic, so most specification machines are deterministic. Exceptions include specifications with imprecise exceptions or external interrupts. For our purposes, we consider deterministic machines as instances of non-deterministic machines in the total order for the execution parameters.

3.4 Correctness Space

By choosing different values of the parameters, we arrive at a variety of correctness statements. We use four-letter acronyms to describe the values of the parameters:

$<alignment> <match> <impl. execution> <spec. execution>$

For example, "PADD" denotes point-wise alignment (P), abstraction match (A), and deterministic implementation (D) and specification (D).

Each parameter has a total order on its instantiations. Together, these total orders induce a partial order over correctness statements, which serves to map out the space of correctness statements for microprocessor implementations (Fig. 3). The partial order is based on the generality of the correctness statements. For example, FGNN (at the top of the partial order) is more general than PADD because pointwise alignment implies flush-point alignment; an abstraction function is an instance of a general relation; and deterministic machines are instances of non-deterministic ones. Correctness statements lower in the order are less general in that they apply to fewer systems. We do not advocate any points in the correctness space over others. The classification serves to highlight the differences and similarities among approaches.

3.5 Mathematical Formulation

In this section we describe the mathematical formulations of correctness statements in the framework. We use $M_i \preccurlyeq_\mathcal{R} M_s$ to mean "M_i is correct with respect to M_s via the relation $\mathcal{R}$". All of the correctness statements have the general form of Definition 2. The base clauses remain largely unchanged from the one shown in Definition 2, so in the remainder of the paper, we will discuss only the induction clauses.

Definition 2 (General form of correctness statement).

$$(Q_i, Q_i^\circ, N_i) \preccurlyeq_\mathcal{R} (Q_s, Q_s^\circ, N_s) \quad \equiv \quad \left[\wedge \begin{array}{l} \forall\, q_i^\circ \in Q_i^\circ.\ \exists\, q_s^\circ \in Q_s^\circ.\ \mathcal{R}(q_i^\circ, q_s^\circ) \\ \langle inductive\ clause \rangle \end{array} \right]$$

The alignment parameter determines the form of the correctness statement. We show the correctness statements for the various values of the alignment parameter together with the most-general values for the other parameters (i.e. non-deterministic machines with a general relation match).

The most general combination in the correctness space is flush-point alignment with a general match and non-deterministic machines (FGNN, Definition 3). It says that if the implementation is in a flushed state q_i and can transition through some number of steps k to another flushed state q_i', then all specification states q_s that $\mathcal{R}$ says match q_i must transition through some number of steps j to some state q_s' that matches q_i'.

Definition 3 (Flush-point induction clause: FGNN).

$$\forall\, q_i, q_i' \in Q_i.\ \forall\, q_s \in Q_s.\ \exists\, q_s' \in Q_s.$$

$$\left[\wedge \begin{array}{l} isFlushed(q_i)\ \wedge\ \exists\, k.\ N_i^k(q_i, q_i')\ \wedge\ isFlushed(q_i') \\ \mathcal{R}(q_i, q_s) \end{array} \right] \implies \left[\wedge \begin{array}{l} \exists\, j.\ N_s^j(q_s, q_s') \\ \mathcal{R}(q_i', q_s') \end{array} \right]$$

The most general case of must-issue alignment (Definition 4) is MGNN. In must-issue alignment, the specification takes one step and the implementation takes as many steps k as are necessary to become unstalled; it then takes one unstalled transition from q_i^k to q_i^{k+1}.

Match

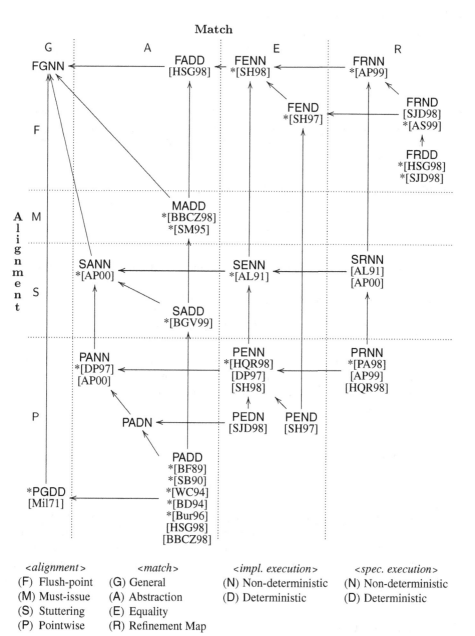

<alignment>   <match>   <impl. execution>   <spec. execution>
(F) Flush-point   (G) General   (N) Non-deterministic   (N) Non-deterministic
(M) Must-issue   (A) Abstraction   (D) Deterministic   (D) Deterministic
(S) Stuttering   (E) Equality
(P) Pointwise   (R) Refinement Map

Each point is annotated with citations for the works that use the particular correctness statements. Citations prefixed with * denote top-level correctness statements; others are used as intermediate correctness statements during the proofs. Section 4 provides further explanation.

Fig. 3. Space of correctness statements

Definition 4 (Must-issue induction clause: MGNN).
$$\forall q_i^0, q_i^1, \dots, q_i^{k+1} \in Q_i. \ \forall q_s \in Q_s. \ \exists q_s' \in Q_s.$$
$$\left[\begin{array}{l} \wedge \ \forall j < k. \ N_i(q_i^j, q_i^{j+1}) \ \wedge \ isStalled(q_i^j) \\ \wedge \ N_i(q_i^k, q_i^{k+1}) \ \wedge \ \neg isStalled(q_i^k) \\ \mathcal{R}(q_i^0, q_s) \end{array} \right] \implies \left[\wedge \begin{array}{l} N_s(q_s, q_s') \\ \mathcal{R}(q_i^{k+1}, q_s') \end{array} \right]$$

Stuttering alignment results in a simpler correctness statement (Definition 5) because it considers only one step of execution and requires no special predicates. The specification is allowed to *stutter*, e.g. consecutive implementation states may align with the same specification state.

Definition 5 (Stuttering induction clause: SGNN).
$$\forall q_i, q_i' \in Q_i. \ \forall q_s \in Q_s. \ \exists q_s' \in Q_s.$$
$$\left[\wedge \begin{array}{l} N_i(q_i, q_i') \\ \mathcal{R}(q_i, q_s) \end{array} \right] \implies \left[\wedge \begin{array}{l} (N_s(q_s, q_s') \ \vee \ (q_s' = q_s)) \\ \mathcal{R}(q_i', q_s') \end{array} \right]$$

Pointwise alignment (Definition 6) drops the stuttering disjunct from Definition 5.

Definition 6 (Pointwise Induction Clause: PGNN).
$$\forall q_i, q_i' \in Q_i. \ \forall q_s \in Q_s. \ \exists q_s' \in Q_s. \quad \left[\wedge \begin{array}{l} N_i(q_i, q_i') \\ \mathcal{R}(q_i, q_s) \end{array} \right] \implies \left[\wedge \begin{array}{l} N_s(q_s, q_s') \\ \mathcal{R}(q_i', q_s') \end{array} \right]$$

Section 4 describes other correctness statements as points in the correctness space by optionally instantiating $\mathcal{R}$ and the next-state relations in each of the above four definitions. Instantiating next-state relations with functions removes the need for some quantified variables.

3.6 Limitations

As with most formal verification approaches, the correctness framework presented in this paper does not exclude certain "pathologically bad" matching relations. A matching relation that includes all implementation and specification state pairings will result in a vacuously true correctness statement, as any pair of consecutive implementation states can be related to any pair of consecutive specification states. Another vacuous statement results from using stuttering alignment with an abstraction function that maps all implementation states to the same specification state.

4   Literature Survey

In this section, we show how a variety of correctness statements for out-of-order microprocessors are points in the correctness space created by the framework. While the phrase "we use the correctness statement of Burch and Dill [BD94]" appears in many papers, detailed examinations reveal that this is something of an approximation. For conciseness, we discuss only the inductive clauses of the correctness statements.

4.1 Historical Perspective

Milner's [Mil71] work in software verification led to a definition of simulation that is *pointwise alignment* of a general relation between a deterministic implementation and a deterministic specification (PGDD). Milner does not include a base clause. PGDD is derived from Definition 6 (pointwise alignment) by substituting next-state functions (n_s and n_i) for the next-state relations (N_s and N_i).

Definition 7 (Milner's *simulation*: PGDD).
$$\forall q_i \in Q_i. \ \forall q_s \in Q_s. \ \mathcal{R}(q_i, q) \implies \mathcal{R}(n_i(q_i), \ n_s(q))$$

Abadi and Lamport [AL91] define *refinement maps*, which in our parlance is *stuttering refinement* between non-deterministic machines (SRNN). They use refinement as a verification strategy to prove SENN (stuttering equivalence), which they call *implements*. Both SRNN and SENN are derived from Definition 5 (SGNN). For SRNN, refinement $(abs(q_i) = q_s \wedge q_i \stackrel{\Pi}{=} q_s)$ is substituted for the match $\mathcal{R}$, which results in Definition 8.

Definition 8 (Abadi and Lamport's *refines*: SRNN).
$$\forall q_i, q_i' \in Q_i.$$
$$\left[\wedge \begin{array}{c} N_i(q_i, q_i') \\ q_i \stackrel{\Pi}{=} abs(q_i) \end{array} \right] \implies \left[\wedge \begin{array}{c} N_s(abs(q_i'), abs(q_i)) \vee abs(q_i') = abs(q_i) \\ q_i' \stackrel{\Pi}{=} abs(q_i') \end{array} \right]$$

Their main result is that if SENN holds, then it is possible to construct an intermediate model from an implementation using history and prophecy variables such that the intermediate model will satisfy SRNN with the specification.

Several verifications of scalar pipelines use correctness statements that are relevant to this paper. Bose and Fischer [BF89] used PADD in the verification of a pipelined stack. In the first published verification of pipelined microprocessor, Srivas and Bickford [SB90], and Windley and Coe [WC94] verified PADD between their implementations and specifications. Srivas and Miller [SM95] proved MADD between a pipelined microprocessor at the instruction and micro-instruction levels of abstraction. All of the abstraction functions in these efforts were manually constructed. The complexity of superscalar microprocessors has led to efforts to automatically (Sect. 4.2) or systematically (Sect. 4.4) create abstraction functions.

4.2 Flushing

Burch and Dill [BD94] use PADD (pointwise abstraction) as their correctness statement. They observed that for in-order pipelines, it is possible to construct an abstraction function by forcing the implementation to issue a stream of *bubbles* that *flush* all of the in-flight instructions out to retirement. In this case, $abs \equiv flush$. Notions similar to flushing are also used for correctness criteria in other domains, such as cache-coherence protocol verification (e.g. [PD96,SA97,NG98]).

To flush a microprocessor, the implementation must be forced to issue a bubble. An implementation $\hat{n}_i$ must be constructed that has the necessary input and control circuitry for flushing. The function *flush* and the next-state function n_i used in the correctness statement are defined as shown below, where k is the number of bubbles used to flush the pipeline, i.e. the maximum latency of the pipeline:

$$flush(q_i) \equiv \hat{n}_i{}^k(q_i, \texttt{True})$$
$$n_i(q_i) \equiv \hat{n}_i(q_i, \texttt{False})$$

Stalls complicate the flushing abstraction, because the implementation cannot fetch an instruction to take a "productive" step. The framework provides us with several options for handling this. The first two are instances of PADD. The other three options; which use PADN, SADD, and MADD; are slightly more general than PADD.

Explicit Stalling of Specification. The first option is to construct a revised specification ($\hat{n}_s$) that takes a stall input s to indicate whether to self-loop or not:
$$\hat{n}_s(q_s, s) \equiv \texttt{if } s \texttt{ then } q_s \texttt{ else } n_s(q_s)$$
The correctness statement provides this stall information to the specification (Definition 9) [BD94,Bur96,HSG98,JSD98].

Definition 9 (Burch and Dill flushing: an instance of PADD).
$$\forall q_i \in Q_i. \; \hat{n}_s(flush(q_i), isStalled(q_i)) = flush(n_i(q_i))$$

This definition is derived from Definition 6 by replacing $\mathcal{R}$ with a match abstraction that uses *flush* as the abstraction function and using next-state functions rather than relations. Burch generalized this idea to superscalar microprocessors [Bur96], by generalizing the *isStalled* predicate to be a function that indicates to the specification how many instructions the implementation issued.

Modified Pointwise. The second option is to use the *isStalled* predicate directly in the PADD correctness statement rather than modifying the specification:
$$\wedge \begin{array}{ll} isStalled(q_i) & \Longrightarrow \quad flush(q_i) = flush(n_i(q_i)) \\ \neg isStalled(q_i) & \Longrightarrow \quad n_s(flush(q_i)) = flush(n_i(q_i)) \end{array}$$

Non-deterministic Specification. The third option is to use a non-deterministic specification that includes a state element that says whether the specification self-loops (PADN). The abstraction function maps the value of the implementation predicate *isStalled* to the new specification state element.

Stuttering. The fourth option is to use stuttering abstraction with flushing as the abstraction function. This method avoids the need to alter the specification. However, this permits the specification to stutter at anytime, not just when the implementation stalls. This is the approach taken by Bryant *et al.* [BGV99] and by Arvind and Shen [AS99]. Bryant *et al.* use SADD to verify a superscalar machine. Stuttering alignment for this kind of implementation allows the specification to issue $0-k$ instructions, where k is the maximum number of instructions the implementation can issue in one cycle. Arvind and Shen prove FRND for an out-of-order implementation with in-order retirement. Both the specification and implementation are represented as term rewriting systems. They omit the details of their proof, but it appears to rely on stuttering abstraction (SAND) between the implementation and specification.

Must-Issue. The fifth option is to use must-issue alignment with flushing as an abstraction function (Definition 10). Berezin *et al.* [BBCZ98] prove must-issue abstraction (MADD) for a processor with out-of-order retirement. The model is deterministic but some of the scheduling is left underspecified. They introduce intermediate models of the implementation and specification that are optimized for model-checking efficiency. They prove MADD between the intermediate implementation and intermediate specification. They relate this result to the real specification and implementation by proving pointwise abstraction (PADD) between each of the intermediate models and its respective concrete counterpart.

Definition 10 (Must-issue abstraction with flushing: an instance of MADD).

$$\forall q_i \in Q_i. \ \forall\, k. \ \forall\, j < k. \ isStalled(n_i^j(q_i)) \ \wedge \ \neg isStalled(n_i^k(q_i))$$
$$\implies \ n_s(flush(q_i)) \stackrel{\Pi}{=} flush(n_i^{k+1}(q_i))$$

When the abstraction function used in MADD is *flush*, then an implementation step from a stalled state should result in a state that matches the same specification state, i.e. $isStalled(q_i) \implies abs(q_i) = abs(n_i(q_i))$ which is equivalent to SADD.

4.3 Trace Tables

Sawada and Hunt [SH97] verified that a non-deterministic processor with out-of-order retirement satisfies *flush-point equality* with a deterministic specification (FEND, Definition 11). FEND results from substituting the equality match ($q_i \stackrel{\Pi}{=} q_s$) for $\mathcal{R}$ and a next-state function n_s for the next-state relation in Definition 3 (FGNN).

Definition 11 (Flush-point equality: FEND).

$$\forall\, q_i, q_i' \in Q_i. \ \forall\, q_s \in Q_s.$$
$$\left[\begin{array}{l} isFlushed(q_i) \ \wedge \ \exists\, k. \ N_i^k(q_i, q_i') \ \wedge \ isFlushed(q_i') \\ \wedge \\ q_i \stackrel{\Pi}{=} q_s \end{array} \right] \implies \left[\exists\, j. \ q_i' \stackrel{\Pi}{=} n_s^j(q_s) \right]$$

In later work, they enhanced their implementation to support in-order retirement, external interrupts, and precise exceptions [SH98,SH99]. The inclusion of interrupts led them to add non-determinism to their specification, to account for the problem of predicting how many instructions the implementation will have completed when an interrupt occurs. They kept flush-point equality as their alignment and match criteria, making their correctness statement FENN.

Throughout this work, their verification strategy was to build an intermediate model with history variables. The intermediate model contains an unbounded table, called a MAETT, with one entry for each issued instruction. In their first work [SH97], they prove pointwise equality (PEND) between the implementation and intermediate model and FEND (flush-point equality) between the intermediate model and specification, which together imply FEND. Similarly for their second model, they prove PENN and FENN respectively to conclude FENN.

4.4 Completion Functions

Hosabettu, Srivas, and Gopalakrishnan [HSG98,HSG99] prove that a deterministic out-of-order implementation satisfies *flush-point refinement* with a deterministic specification where the match is projection (Definition 12).

Definition 12 (Flush-point refinement with projection: an instance of FRDD**).**

$$\forall q_i \in Q_i. \; \forall k. \; isFlushed(q_i) \wedge isFlushed(n_i^k(q_i)) \implies \exists j. \; n_s^j(\Pi_e(q_i)) = \Pi_e(n_i^k(q_i))$$

Because their verification is completely within a theorem prover, they are able to use underspecified next-state functions (rather than relations) for their scheduler. They prove FRDD in three steps. They prove *pointwise abstraction* (PADD) and then apply induction to prove *flush-point abstraction* (FADD). They go from FADD to FRDD by proving that the abstraction of a flushed state is equivalent to projection. The abstraction is constructed with *completion functions*. Completion functions describe the effect of the completion of each in-flight instruction on the observable state, and are composed in program order. Hosabettu *et al.* [HGS00] also use the same correctness statement to verify an implementation with speculative execution and precise exceptions.

4.5 Incremental Flushing

Skakkebæk *et al.* [SJD98] verify that a deterministic implementation with in-order retirement satisfies *flush-point refinement* with a deterministic specification (FRDD). They build a non-deterministic intermediate model that computes the result of each instruction when it enters the machine and queues the result for later retirement. This intermediate model has hidden state relative to the implementation. The verification of the implementation against the intermediate model shows PEDN (pointwise equality). The verification of the intermediate model against the specification establishes FRND (flush-point refinement) by incrementally decomposing the monolithic flushing abstraction function into a set of simpler flushing steps. In [JSD98], they use a non-deterministic intermediate model with an abstracted scheduler that provides fine-grained control over instruction progress. This reduces the amount of manual abstraction required by strengthening the simpler flushing steps.

4.6 Variations on Refinement

The four works by the authors Damm, Pnueli, and Arons use a wide range of correctness statements and implementations. Damm and Pnueli [DP97] prove PANN (pointwise abstraction) for an implementation with out-of-order retirement. Their non-deterministic specification (NONDET) generates all possible traces of a program that obey data-dependencies, which allows them to use pointwise alignment. They introduce an intermediate model with auxiliary variables (TOMASULO) and prove PENN (pointwise equality) between the implementation and the intermediate model, and PANN between the intermediate model and the specification. For PANN their abstraction projects the current implementation state if all instructions have retired and otherwise returns the initial implementation state.

Arons and Pnueli [AP99] prove FRNN (flush-point refinement) for an implementation with out-of-order retirement. The specification can self-loop at every state, but is otherwise deterministic. They use an intermediate model with history variables and prove that whenever the implementation is flushed, the history variables match the implementation (FRNN). They verify that the intermediate model satisfies pointwise refinement

(PRNN) with the specification. Subsequently, Pnueli and Arons change their synchronization point from instruction issue to instruction retirement, which allows them to tighten their top-level correctness statement to be PRNN (pointwise refinement) for an implementation with in-order retirement [PA98].

Arons and Pnueli [AP00] verify SANN (stuttering abstraction) for a machine with speculative execution, precise exceptions, and in-order retirement. Their abstraction computes the abstract program counter based on the contents of the reorder buffer. They perform two different verifications, one based on induction over the size of the reorder buffer and one using abstraction functions. In the inductive proof, they use three intermediate models, and prove SRNN (stuttering refinement, relying on the result of [AL91]), PANN (pointwise abstraction), and SANN to conclude SANN overall.

4.7 Assume-Guarantee

Henzinger *et al.* [HQR98,Qad99] use a top-level correctness statement of pointwise equality (PENN), which they call *trace containment*, to prove the correctness of an out-of-order retirement processor where both the specification and implementation may have internal state. Their specification includes a non-deterministic stall signal and the scheduling in their implementation is non-deterministic. They construct abstraction and witness modules to bridge the gap between the specification and implementation. Using assume-guarantee reasoning, they reduce the problem to smaller proof obligations where the specification has no internal state. In these cases (which they call *projection refinement*), they prove PRNN (pointwise refinement).

4.8 Related Correctness Statements

Manolios [Man00] defines correctness based on *well-founded bisimulation*. He allows both the specification and implementation to be non-deterministic and to stutter, but also includes a liveness property that guarantees that they will stutter for only finitely many steps. This approach has not yet been applied to out-of-order implementations. If we excise the liveness criteria from his correctness statement, his work can be characterized as verifying that the implementation satisfies *stuttering equivalence* (SENN) against the specification and that the specification satisfies SENN against the implementation.

Fox and Harman [FH98] define a correctness statement that uses explicit time and temporal abstraction. Their theory supports arbitrary alignments based on temporal abstraction, that they call *retimings*, but the only supported matching instantiation is abstraction. They have used this statement in the verification of a superscalar machine with in-order retirement. For superscalar implementations, they align both the implementation and specification to an intermediate clock.

5 Discussion

We have presented a framework for describing microprocessor correctness statements that helps us to compare existing correctness statements and to highlight the differences among them. Our classification is meant as a stepping stone towards understanding the links between an implementation's features, the desired "strength" of correctness statement, and the verification techniques. Indeed, the framework has led us to a number of observations that we now discuss.

Machines with out-of-order retirement are problematic, because they can reach states that are not possible when executing instructions sequentially. One possibility is to use equality match, a deterministic specification, and flush-point alignment. Two other approaches support point-wise alignment: a non-deterministic specification that allows different retirement orderings [DP97] or an abstraction function that retires all in-flight instructions (e.g. flushing [BD94], or completion functions [HSG98]).

Sawada and Hunt [SH97] have verified the same implementation using flush-point equality (Definition 11) and Burch-Dill style pointwise abstraction (Definition 9). They found flush-point equality to be significantly easier. We speculate that flush-point equality is a verification convenience, i.e. realistic machines that satisfy flush-point equality will also satisfy pointwise abstraction or stuttering abstraction. In the case of machines without external interrupts, a flushing-style abstraction function should suffice, while a machine with interrupts would require a more sophisticated abstraction function to keep the interrupt trace aligned between the specification and implementation.

Stalls complicate the alignment of the implementation and specification. Pnueli and Arons [PA98] use PRNN (pointwise refinement) with a specification that self-loops when no instruction retires. Many others use pointwise abstraction where the abstraction function flushes the implementation and the specification self-loops when no instruction is issued. An emerging trend is to use flush-point equality or flush-point refinement, where the implementation and specification are compared only when the implementation is in a flushed state.

Verifying machines with exceptions complicates the instantiation of the match parameter. Most approaches in the literature synchronize the implementation and specification machines at instruction issue. However, Damm and Pnueli [DP97] and Pnueli and Arons [AP00] synchronize at retirement, an approach that makes it easier to handle exceptions. The synchronization point is encapsulated in the definition of the match parameter and is not distinguished by our framework.

In Fig. 3 almost all of the intermediate correctness statements lead to the top-level correctness statements by tracing along the edges in the graph. The two exceptions are incremental flushing [SJD98] and completion functions [HSG98], whose use of mechanized theorem proving enables these more complicated verification strategies.

We are formalizing the framework in a theorem prover and mechanically verifying the partial order between correctness statements. For a general matching relation, we have verified that pointwise implies stuttering, and stuttering implies flushing. We are investigating the logical relationships between must-issue and the other three.

Our framework is not an end in itself. Rather, it should be used as a foundation for further investigations and a deeper understanding of developments in the formal verification of microprocessors. There are values for the framework's parameters that we have not enumerated, and we anticipate that some of these will find useful application. For example, as other approaches besides Sawada and Hunt [SH98] begin to include external interrupts, we anticipate that additional points in the correctness space will be explored. It remains to be determined what the framework indicates about the relative "quality" of correctness criteria. It would also be fruitful to explore the potential of using the framework to predict the difficulty of different verification approaches.

Acknowledgments. We thank Andrew Martin of Motorola and the anonymous reviewers for their helpful comments on this work. The first and third authors are supported by the Natural Sciences and Engineering Research Council of Canada (NSERC).

References

[AL91]   M. Abadi and L. Lamport. The existence of refinement mappings. *Theoretical Computer Science*, 2(82):253–284, 1991.

[AP99]   T. Arons and A. Pnueli. Verifying Tomasulo's algorithm by refinement. In *Int'l Conference on VLSI Design*, pp 92–99, 1999.

[AP00]   T. Arons and A. Pnueli. A comparison of two verification methods for speculative instruction execution with exceptions. In *TACAS*, vol 1785 of *LNCS*, pp 487–502. Springer, 2000.

[AS99]   Arvind and X. Shen. Using term rewriting systems to design and verify processors. *IEEE Micro*, 19(3):36–46, 1999.

[BB94]   D. Beatty and R. Bryant. Formally verifying a microprocessor using a simulation methodology. In *DAC*, pp 596–602, 1994.

[BBCZ98] S. Berezin, A. Biere, E. Clarke, and Y. Zhu. Combining symbolic model checking with uninterpreted functions for out-of-order processor verification. In *FMCAD*, vol 1522 of *LNCS*, pp 369–386. Springer, 1998.

[BD94]   J. Burch and D. Dill. Automatic verification of pipelined microprocessor control. In *CAV*, vol 818 of *LNCS*, pp 68–80. Springer, 1994.

[BF89]   S. Bose and A. Fisher. Verifying pipelined hardware using symbolic logic simulation. In *ICCD*, pp 217–221, 1989.

[BGV99]   R. Bryant, S. German, and M. Velev. Processor verification using efficient decision procedures for a logic of uninterpreted functions. In *TABLEAUX*, vol 1617 of *LNAI*, pp 1–13. Springer, June 1999.

[Bur96]   J. Burch. Techniques for verifying superscalar microprocessors. In *DAC*, pp 552–557, 1996.

[DP97]   W. Damm and A. Pnueli. Verifying out-of-order executions. In *CHARME*, pp 23–47. Chapman and Hall, 1997.

[FH98]   A. Fox and N. Harman. An algebraic model of correctness for superscaler microprocessors. In *Prospects for Hardware Foundations*, vol 1546 of *LNCS*, pp 138–183. Springer, 1998.

[HGS00]   R. Hosabettu, G. Gopalakrishnan, and M. Srivas. Verifying advanced microarchitectures that support speculation and exceptions. In *CAV*, vol 1855 of *LNCS*, pp 521–537. Springer, 2000.

[HQR98]   T. Henzinger, S. Qadeer, and S. Rajamani. You assume, we guarantee: Methodology and case studies. In *CAV*, vol 1427 of *LNCS*, pp 440–451. Springer, 1998.

[HSG98]   R. Hosabettu, M. Srivas, and G. Gopalakrishnan. Decomposing the proof of correctness of pipelined microprocessors. In *CAV*, vol 1427 of *LNCS*, pp 122–134. Springer, 1998.

[HSG99]   R. Hosabettu, M. Srivas, and G. Gopalakrishnan. Proof of correctness of a processor with reorder buffer using the completion functions approach. In *CAV*, vol 1633 of *LNCS*, pp 47–59. Springer, 1999.

[JSD98]   R. Jones, J. Skakkebæk, and D. Dill. Reducing manual abstraction in formal verifi-
          cation of out-of-order execution. In *FMCAD*, vol 1522 of *LNCS*, pp 2–17. Springer,
          1998.

[Man00]   P. Manolios. Correctness of pipelined machines. In *FMCAD*, vol 1954 of *LNCS*, pp
          161–178. Springer, 2000.

[McM98]   K. McMillan. Verification of an implementation of Tomasulo's algorithm by com-
          positional model checking. In *CAV*, vol 1427 of *LNCS*, pp 110–121. Springer, 1998.

[Mil71]   R. Milner. An algebraic definition of simulation between programs. In *Proc. of
          2nd Int'l Joint Conf. on Artificial Intelligence*, pp 481–489. The British Comp. Soc.,
          1971.

[NG98]   R. Nalumasu and G. Gopalakrishnan. Deriving efficient cache coherence protocols
          through refinement. In *Formal Methods for Parallel Programming: Theory and
          Applications (FMPPTA'98)*, 1998.

[PA98]   A. Pnueli and T. Arons. Verification of data-insensitive circuits: An in-order-
          retirement case study. In *FMCAD*, vol 1522 of *LNCS*, pp 351–368. Springer, 1998.

[PD96]   S. Park and D. Dill. Protocol verification by aggregation of distributed transactions.
          In *CAV*, vol 1102 of *LNCS*, pp 300–310. Springer, 1996.

[PJB99]   V. Patankar, A. Jain, and R. E. Bryant. Formal verification of an ARM processor. In
          Int'l Conf. on VLSI Design, pp 282–287. IEEE; New York, NY, January 1999.

[Qad99]   S. Qadeer. *Algorithms and Methodology for Scalable Model Checking*. PhD thesis,
          Elec. Eng. and Comp. Sci., University of California at Berkeley, 1999.

[SA97]   X. Shen and Arvind. A methodology for designing correct cache coherence protocols
          for DSM systems. Technical Report CSG Memo 398 (A), MIT, June 1997.

[SB90]   M. Srivas and M. Bickford. Formal verification of a pipelined microprocessor. *IEEE
          Trans. on Software Engineering*, pp 52–64, September 1990.

[SH97]   J. Sawada and W. Hunt. Trace table based approach for pipelined microprocessor
          verification. In *CAV*, vol 1254 of *LNCS*, pp 364–375. Springer, 1997.

[SH98]   J. Sawada and W. Hunt. Processor verification with precise exceptions and specula-
          tive execution. In *CAV*, vol 1427 of *LNCS*, pp 135–146. Springer, 1998.

[SH99]   J. Sawada and W. Hunt. Results of the verification of a complex pipelined machine
          model. In *CHARME*, vol 1703 of *LNCS*, pp 313–316. Springer, 1999.

[SJD98]   J. Skakkebæk, R. Jones, and D. Dill. Formal verification of out-of-order execution
          using incremental flushing. In *CAV*, vol 1427 of *LNCS*, pp 98–109. Springer, 1998.

[SM95]   M. K. Srivas and S. P. Miller. Applying formal verification to a commercial micro-
          processor. In *CHDL*, pp 493–502, August 1995.

[WC94]   P. Windley and M. Coe. A correctness model for pipelined microprocessors. In
          Theorem Provers in Circuit Design, pp 32–51. Springer, 1994.

From Operational Semantics to Denotational Semantics for Verilog

Zhu Huibiao[1], Jonathan P. Bowen[1], and He Jifeng[2]

[1] Centre for Applied Formal Methods
South Bank University, SCISM, Borough Road, London SE1 0AA, UK
{huibiaz,bowenjp}@sbu.ac.uk http://www.cafm.sbu.ac.uk/
[2] United Nations University, UNU/IIST, P.O. Box 3058, Macau, China
jifeng@iist.unu.edu http://www.iist.unu.edu/

Abstract. This paper presents the derivation of a denotational semantics from an operational semantics for a subset of the widely used hardware description language Verilog. Our aim is to build an equivalence between the operational and denotational semantics. We propose a discrete time semantic model for Verilog. Algebraic laws are also investigated in this paper, with the ultimate aim of providing a unified set of semantic views for Verilog.

1 Introduction

Modern hardware design typically uses a hardware description language (HDL) to express designs at various levels of abstraction. An HDL is a high level programming language, with usual programming constructs such as assignments, conditionals and iterations and appropriate extensions for real-time, concurrency and data structures suitable for modelling hardware. Verilog is an HDL that has been standardized and widely used in industry [6]. Verilog programs can exhibit a rich variety of behaviours, including event-driven computation and shared-variable concurrency.

The semantics for Verilog is very important. At UNU/IIST, the operational semantics has been explored in [1,3,4,7]. Verilog's denotational semantics [9] has also been explored based on the operational semantics using Duration Calculus [8]. The two semantics can be considered equivalent informally. The question is how the two semantics can be proved equivalent formally.

The aim of this paper is to derive the denotational semantics for Verilog from its operational semantics. This ensures the consistency of the two semantics, making it possible to demonstrate their equivalence formally. The similar problem was also investigated in [5] for Dijkstra's sequential language. In our paper we define a transitional condition and the phase semantics for each type of transition. The denotational semantics can be treated as the sequential composition of those phase semantics.

This paper is organized as follows. Section 2 introduces the language and presents a discrete denotational semantic model. We also design a refinement calculus for the discrete model. Section 3 is devoted to deriving the denotational semantics from its operational semantics. We introduce the operational semantics, and define a function that maps any program text to a logic formula representing its denotational semantics.

T. Margaria and T. Melham (Eds.): CHARME 2001, LNCS 2144, pp. 449–464, 2001.

We derive the denotational semantics for each statement from the function by a formal proof in Sect. 4. We also discuss the algebraic laws that are well suited for symbolic calculation. The three semantics form a unifying model, proving different views useful for varying purposes when reasoning about Verilog.

2 The Discrete Denotational Model

2.1 The Syntax for Verilog

The language discussed in this paper is a subset of Verilog. It contains the following categories of syntactic elements introduced in [2].

1. Sequential Process (Thread):
   $S ::= PC \mid S \; ; \; S \mid \textbf{if } b \textbf{ then } S \textbf{ else } S \mid c\, S$
   where PC ranges over primitive commands.
   $PC ::= (x := e) \mid \textbf{skip} \mid \textbf{chaos}$
   and $c\, S$ denotes timing controlled statement, and c is a time control used for scheduling.
   $c ::= \#(\Delta) \mid @(\eta)$, where $\eta ::= v \mid\uparrow v \mid\downarrow v$
   Time delay $\#\Delta$ suspends the execution for exactly Δ time units. Δ is treated as the integer in this paper. Event guard $@(\uparrow v)$ is fired by the increase of the value of v, whereas $@(\downarrow v)$ is triggered by a decrease in v. Any change of v awakes the guard $@(v)$.

2. Parallel Process (Module):
   $P ::= S \mid P \parallel P$
   To accommodate the expansion laws of parallel construct, the language is equipped with a hybrid control event hc:
   $hc ::= @(x := e) \mid @(g) \mid \#(\Delta)$
   $g ::= \eta \mid g \textit{ or } g \mid g \textit{ and } g \mid g \textit{ and } \neg g$
   and the guarded choice $(hc_1\, P_1) \| \ldots \| (hc_n\, P_n)$

2.2 Denotational Semantic Model

Verilog processes are allowed to share program variables. In order to deal with this shared-variable feature, we describe the behaviour of a process in terms of a trace of *snapshots*, which records the sequence of atomic actions in which that process has engaged to some moment in time. Our semantic model contains a variable tr to denote that trace.

If a trace tr is not empty, the function "*last*" yields its last snapshot. Let tr_1, tr_2 be two traces. The notation $tr_1 \frown tr_2$ denotes the concatenation of tr_1 and tr_2. $tr_1 \preceq tr_2$ indicates that tr_1 is a prefix of tr_2. Suppose $tr_1 \preceq tr_2$, the notation $tr_2 - tr_1$ denotes the result of subtracting those snapshots in tr_1 from tr_2. We use the notation $tr_1 \textbf{ in } tr_2$ to indicate that tr_1 is contained in tr_2, i.e., there are sequences s and t such that $tr_2 = s \frown tr_1 \frown t$.

A snapshot is used to specify the behaviour of an atomic action, and expressed by a triple (t, σ, μ) where:
(1) t indicates the time when the atomic action happens;

(2) σ denotes the final values of program variables at the termination of an atomic action; (3) μ is the control flag indicating which process is in control: $\mu = 1$ states the atomic action is engaged by the process, whereas $\mu = 0$ implies it is performed by the environment.

We select the components of a snapshot using the projections:

$$\pi_1((t, \sigma, \mu)) =_{df} t \qquad \pi_2((t, \sigma, \mu)) =_{df} \sigma \qquad \pi_3((t, \sigma, \mu)) =_{df} \mu$$

Once a Verilog process is activated, it continues its execution until the completion of an atomic action; namely either it encounters a timing controlled statement, or it terminates successfully. An atomic action usually consists of a sequence of assignments as shown below.

Example 2.1: Consider the parallel program $P\|Q$ where $P =_{df} (x := 1; y := x + 1; z := x + 2)$ and $Q =_{df} x := 2$. Three assignments of P form an atomic action, and their execution is uninterrupted. The process Q can only be started at the beginning or at the end of the execution of P.   □

The execution of an atomic action is represented by a single snapshot. In order to describe the behaviour of individual assignment, we introduce a variable ttr to model the accumulated change made by the statements of the atomic action. On the completion of an atomic action, the corresponding snapshot is attached to the end of the trace to record its behaviour.

Example 2.2: Let $P =_{df} x := x + 1 ; y := y - 1 ; @(g)$. Assume that program variables x and y are 0 and 1 respectively when P is activated, and the activated time of P is at 0. The execution of $x := x + 1$ produces $ttr = \{x \mapsto 1, y \mapsto 1\}$ on its termination that specifies the change made by the assignment to variable x. The statement $y := y - 1$ in turn yields $ttr = \{x \mapsto 1, y \mapsto 0\}$ as the final value of ttr, which reflects the change incurred by the atomic action $(x := x+1; y := y-1)$. The snapshot $(0, \{x \mapsto 1, y \mapsto 0\}, 1)$ will be added to the end of the trace variable tr when $@(g)$ is encountered. After this adding, ttr will be assigned an empty value $null$.   □

Example 2.3: Let $P =_{df} x := 1 ; @(x := 2) ; x := 3$. The contribution of $(x := 1)$ is added to the end of the trace when assignment guard $@(x := 2)$ is encountered. This means $x := 1$ in this particular case is an atomic action. Although $@(x := 2)$ is an atomic action, it also stores its result in ttr. In order to distinguish assignment guard from assignment, we assign a control $flag$ with 0 to identify this case. The result of the assignment guard will be added when its sequential statement is encountered (not only guards).   □

We are now ready to represent the observation by a tuple

$$(\overleftarrow{time}, \overrightarrow{time}, \overleftarrow{tr}, \overrightarrow{tr}, ttr, ttr', flag, flag')$$

where
- $\overleftarrow{time}$ and $\overrightarrow{time}$ are the start point and the end point of a time interval over which the observation is recorded. We use $\delta(time)$ to represent the length of the time interval.

$$\delta(time) =_{df} (\overrightarrow{time} - \overleftarrow{time})$$

- $\overleftarrow{tr}$ stands for the initial trace of a program over the interval which is passed by its predecessor. $\overrightarrow{tr}$ stands for the final trace of a program over the interval.

$\overrightarrow{tr} - \overleftarrow{tr}$ stands for the sequence of snapshots contributed by the program itself and its environment during the interval.

• ttr and ttr' stand for the initial and final value of the variable ttr which are used to store the contribution of an atomic action over the interval.

• $flag$ and $flag'$ stand for the initial and final value of the control flag. There are two cases to indicate the end of its prior atomic action("$ttr = null$" or "$ttr \neq null \wedge flag = 0$").

Example 2.4: Let $P =_{df} x := 1 ; \#1,$ $Q =_{df} \#1 ; x := 2,$ $R =_{df} x := 3.$
Consider the trace of program $(P \parallel Q) ; R$.
The trace of P is $< (\overleftarrow{time}, \sigma_1, 1) , (\overleftarrow{time} + 1, \sigma_2, 0) >$.
The trace of Q is $< (\overleftarrow{time}, \sigma_1, 0) , (\overleftarrow{time} + 1, \sigma_2, 1) >$.
Hence, the trace of $P \parallel Q$ is $< (\overleftarrow{time}, \sigma_1, 1) , (\overleftarrow{time} + 1, \sigma_2, 1) >$.
R's trace is $< (\overleftarrow{time}, \sigma_3, 1) >$. Then the trace of $(P \parallel Q) ; R$ is
$$< (\overleftarrow{time}, \sigma_1, 1) , (\overleftarrow{time} + 1, \sigma_2, 1) , (\overleftarrow{time} + 1, \sigma_3, 1) >.$$
where $\sigma_1 = \{x \mapsto 1\},$ $\sigma_2 = \{x \mapsto 2\},$ $\sigma_3 = \{x \mapsto 3\}.$ □

We use the following diagram to indicate the trace behaviour of a process (and its environment). Here, "•" stands for the process's atomic action. "∘" stands for the environment's atomic action. The numbers on the vertical line stand for the snapshot sequences in the process's trace, whereas the number on the horizontal line represents the time when the atomic actions take place.

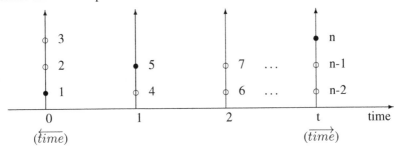

As in Temporal Logic, we introduce a binary "chop" operator to describe the composite behaviour of sequential composition.

Definition 2.5
$$P ^\frown Q =_{df} \exists t, s, tt, f \bullet \quad P[s/\overrightarrow{tr}, t/\overrightarrow{time}, tt/ttr', f/flag'] \\ \wedge Q[s/\overleftarrow{tr}, t/\overleftarrow{time}, tt/ttr, f/flag] \qquad □$$

The "chop" operator is associative, and distributes over disjunction. It has I has its unit and **false** as its zero, where

$$I =_{df} \delta(time) = 0 \wedge \overrightarrow{tr} = \overleftarrow{tr} \wedge ttr' = ttr \wedge flag' = flag.$$

A Verilog process may perform an infinite computation and enter a *divergent state*. To distinguish its chaotic behaviour from the stable ones we introduce the variables $ok, ok' : Bool$ into the semantic model, where $ok = true$ indicates the process has been started, and $ok' = true$ states the process has become *stable*.

A timing controlled statement can not start its execution before its guard is triggered. To distinguish its waiting behaviour from terminating one, we introduce another pair of variables $wait, wait' : Bool$. $wait = true$ indicates that the process starts in an intermediate state, and $wait' = true$ means the process is waiting. The introduction of intermediate waiting state has implications for sequential composition "$P;Q$": if Q is asked to start in a waiting state of P, it leaves the state unchanged, i.e., it satisfies the healthiness condition.

$(H)\,Q \;=\; \mathit{II} \lhd wait \rhd Q,$

where $\mathit{II} =_{df}$ **true** $\vdash (\delta(time) = 0) \wedge (\overrightarrow{tr} = \overleftarrow{tr}) \wedge (\bigwedge_{s \in \{ok,wait,ttr,flag\}} s' = s)$

$P \lhd Q \rhd R =_{df} (P \wedge Q) \vee (\neg Q \wedge R)$

$P \vdash R =_{df} (ok \wedge P) \Rightarrow (ok' \wedge R)$

Definition 2.6: Let P and Q be formulae. Define

$$P \;;\; Q =_{df} \exists w,o \bullet (P[w/wait',o/ok']^\frown Q[w/wait,o/ok])$$

Definition 2.7: A formula is called a *healthy formula* if it has the following form.

$$\mathbf{H}(Q \vdash W \lhd wait' \rhd T)$$

where, $\mathbf{H}(X) \;=\; \mathit{II} \lhd wait \rhd X$

Theorem 2.8: $\mathbf{H}(P)$ satisfies healthiness condition (H).

Theorem 2.9: If D_1, D_2 are healthy formulae, so are $D_1 \vee D_2$, $D_1 \lhd b \rhd D_2$ and $D_1 \;;\; D_2$, where

$$\mathbf{H}(Q_1 \vdash W_1 \lhd wait' \rhd T_1) \;;\; \mathbf{H}(Q_2 \vdash W_2 \lhd wait' \rhd T_2)$$
$$= \mathbf{H}(\neg(\neg Q_1 ^\frown \mathbf{true}) \wedge \neg(T_1 ^\frown \neg Q_2) \vdash (W_1 \vee (T_1 ^\frown W_2)) \lhd wait' \rhd (T_1 ^\frown T_2))$$

Corollary 2.10: If P is a healthy formula then

(1) $\mathit{II} \;;\; P \;=\; P$ (2) $\bot \;;\; P \;=\; \bot$ □

The union and intersection of arbitrary healthy formulae set are also healthy formulae. This implies that healthy formulae form a complete lattice under the implication order, which has a bottom element $\bot =_{df} \mathbf{H}(\mathbf{false} \vdash \mathbf{true})$ and a top element $\top =_{df} \mathbf{H}(\mathbf{true} \vdash \mathbf{false})$.

3 From Operational Semantics to Denotational Semantics

3.1 Transitional Condition and Phase Semantics

In order to derive Verilog's denotational semantics from its operational semantics we define a transitional condition and the phase semantics for each type of transition. The standard way to give an operational semantics is to define a set of transition rules based on configurations, such that any computation of a program can be generated from the transition rules. A configuration usually consists of four components (or five components in some cases):

(1) a program text P representing the rest of the program that remains to be executed;
(2) a data state σ (the second element of a configuration) denoting the initial data state of an atomic action;
(3) another data state σ' (the third element) representing the current data state during the execution of an atomic action ($\sigma' = \emptyset$ represents the previous atomic action ends and the new atomic action has not been scheduled);
(4) a control flag k (the fourth element) indicating which process is selected to execute:

$k = 0$ states the program P is waiting to be executed and its environment may perform triggering action or let time make advance, whereas $k = 1$ implies that P is being executed and neither time advance step nor triggering action can take place;
(5) a thread number i (in some configurations) denoting the i-th thread of process P is being executed (i.e., this thread obtains the control flag).

The relationship between a transition and the variables in the denotational model can be described by the following diagram of an example transition.

$$\pi_2(last(\overleftarrow{tr})) \quad ttr \quad flag \qquad\qquad \pi_2(last(\overrightarrow{tr})) \quad ttr' \quad flag'$$

$$\uparrow \qquad \uparrow \qquad \uparrow \qquad\qquad\qquad\qquad \uparrow \qquad \uparrow \qquad \uparrow$$

$$< P, \quad \sigma, \quad \sigma', \quad 0 > \quad \overset{<\sigma,\sigma'>}{\longrightarrow}_c \quad < P', \quad \sigma', \quad \emptyset, \quad 0 >$$

Let $O(\alpha_1, \alpha_2, \alpha_3, \alpha_4)$ stands for the observation of ttr and $flag$.

$$O(\alpha_1, \alpha_2, \alpha_3, \alpha_4) =_{df} ttr = \alpha_1 \wedge ttr' = \alpha_2 \wedge flag = \alpha_3 \wedge flag' = \alpha_4$$

We use "$ttr = notnull$" to indicate "$ttr \neq null$".

The transition rules can be grouped into the following types [7]. We define a transitional condition $\mathbf{Cond}_{i,j}$ and its corresponding phase semantics for each type of transition. Our map from operational semantics to denotational semantics is based on the phase semantics. Here, $\mathbf{Cond}_{i,j}$ stands for the transitional condition of the $\mathbf{j}$-th transition of type $\mathbf{T}_i$.

• Instantaneous transition

$\mathbf{T}_1$: The i-th thread of process P can perform an instantaneous action, and P enters the instantaneous section by its i-th thread being activated.

$$< P, \sigma, \emptyset, 0 > \quad \longrightarrow \quad < P, \sigma, \sigma, 1, i >, \qquad i \in \{1, 2\}$$
$$\mathbf{Cond}_{1,1} =_{df} \overrightarrow{tr} = \overleftarrow{tr} \wedge O(null, \pi_2(last(\overleftarrow{tr})), 0, 1)$$

$$< P, \sigma, \sigma', 1 > \quad \longrightarrow \quad < P, \sigma, \sigma', 1, i >, \qquad i \in \{1, 2\}$$
$$\mathbf{Cond}_{1,2} =_{df} \overrightarrow{tr} = \overleftarrow{tr} \wedge O(notnull, ttr, 1, 1)$$

$\mathbf{T}_2$: Within the instantaneous section, the i-th thread of the process P performs a transition, and remains in the section or terminates. This transition assigns the successor of P an active status.

$$< P, \sigma_0, \sigma, 1, i > \quad \longrightarrow \quad < P', \sigma_0, \sigma', 1, i >, \qquad i \in \{1, 2\}$$
$$< P, \sigma_0, \sigma, 1, i > \quad \longrightarrow \quad < P', \sigma_0, \sigma', 1 >, \qquad i \in \{1, 2\}$$

For a specific program P, σ' should be of the form $f(\sigma)$. The two transitional conditions are the same.

$$\mathbf{Cond}_{2,1} =_{df} \overrightarrow{tr} = \overleftarrow{tr} \wedge O(notnull, f(ttr), 1, 1)$$

$\mathbf{T}_3$: Within the instantaneous section, the i-th thread of a process may leave the instantaneous section. If the process is breakable, it can also leave the instantaneous section.

$$< P, \sigma_0, \sigma', 1, i > \quad \longrightarrow \quad < P, \sigma_0, \sigma', 0 >, \qquad i \in \{1, 2\}$$
$$< P, \sigma_0, \sigma', 1 > \quad \longrightarrow \quad < P, \sigma_0, \sigma', 0 >$$

The two transitional conditions are the same.

$$\text{Cond}_{3,1} =_{df} \overrightarrow{tr} = \overleftarrow{tr} \wedge O(notnull, ttr, 1, 0)$$

T_4: A transition represents that the program executes an assignment guard (i.e., assignment guard is regarded as an atomic action).

$$< P, \sigma, \emptyset, 0 > \;\longrightarrow\; < P', \sigma, \sigma', 0 >$$

For a specific process P, σ' should be of the form $f(\sigma)$.

$$\text{Cond}_{4,1} =_{df} \overrightarrow{tr} = \overleftarrow{tr} \wedge O(null, f(\pi_2(last(\overleftarrow{tr}))), 0, 0)$$

The above four types of transitions have the instantaneous feature. The corresponding phase semantics of each transition can be expressed as $Inst(\text{Cond}_{i,j})$.

where, $\text{Cond}_{i,j}$ can be the above seven transitional conditions.

$$Inst(\mathbf{X}) =_{df} \mathbf{H}(\mathbf{true} \vdash \neg wait' \wedge \delta(time) = 0 \wedge \mathbf{X})$$

"$\delta(time) = 0$" indicates those transitions consume zero time.

- Triggered transition

T_5: (1) A transition can be triggered by its sequential predecessor. This kind of transition is called the self-triggered transition.

$$< P, \sigma, \sigma', 0 > \;\xrightarrow{<\sigma,\sigma'>}_c\; < P', \sigma', \emptyset, 0 >$$

Here, c in notation $\xrightarrow{<\sigma,\sigma'>}_c$ represents the condition which triggers the transition. It has the form $c(\sigma, \sigma')$ based on a pair of states $< \sigma, \sigma' >$. If there is no this kind of condition, it can be understood as $\mathbf{true}$. If σ and σ' (i.e., $\pi_2(last(\overleftarrow{tr}))$ and ttr) are the same, σ' will not be attached to the end of the trace.

$$\text{Cond}_{5,1} =_{df} c(\pi_2(last(\overleftarrow{tr})), ttr) \wedge O(notnull, null, 0, 0)$$
$$\wedge\, (\overrightarrow{tr} = \overleftarrow{tr} \lhd \pi_2(last(\overleftarrow{tr})) = ttr \rhd \overrightarrow{tr} = \overleftarrow{tr}^\frown < (\overleftarrow{time}, ttr, 1) >)$$

This transition also lasts zero time. Its phase semantics is also $Inst(\text{Cond}_{5,1})$.

(2) A transition can be triggered by its parallel partner.

$$< P, \sigma, \emptyset, 0 > \;\xrightarrow{<\sigma,\sigma'>}_c\; < P', \sigma', \emptyset, 0 >$$

A process can also record the contribution of its environment's atomic action. But the control flag μ in the snapshot is 0. If σ and σ' are the same, the environment will not attach σ' to the end of the trace. Therefore, the process's trace remains unchanged (i.e., $\overrightarrow{tr} = \overleftarrow{tr}$) in this case.

$$\text{Cond}_{5,2} =_{df} O(null, null, 0, 0) \wedge c(\pi_2(last(\overleftarrow{tr})), \pi_2(last(\overrightarrow{tr})))$$
$$\wedge \left(\overrightarrow{tr} = \overleftarrow{tr} \vee \left(\begin{array}{c} \pi_1(\overrightarrow{tr} - \overleftarrow{tr}) = \overleftarrow{time} \wedge \\ \pi_3(\overrightarrow{tr} - \overleftarrow{tr}) = 0 \end{array} \right) \right)$$

Its phase semantics is also $Inst(\text{Cond}_{5,2})$. It means its environment's corresponding atomic action also lasts zero time.

- Time advancing transition

$\mathbf{T}_6$: $< P, \sigma, \emptyset, 0 > \overset{1}{\longrightarrow} < P', \sigma, \emptyset, 0 >$

$\mathbf{Cond}_{6,1} =_{df} \overrightarrow{tr} = \overleftarrow{tr} \wedge O(null, null, 0, 0)$

If process P can not do any other transitions at the moment, time will advance. We regard the unit of time advancing is 1. During this period, there are no atomic actions contributed by the process P itself and its environment. Hence, time advancing keeps the trace unchanged. Its phase semantics is:

$\mathbf{H}(\mathbf{true} \vdash \mathbf{Cond}_{6,1} \wedge (\delta(time) < 1 \lhd wait' \rhd \delta(time) = 1))$

3.2   Map from Operational Semantics to Denotational Semantics

Definition 3.1: A configuration $< P, \sigma, \sigma', 1 >$ (or $< P, \sigma, \emptyset, 0 >$) is called *a divergent state* if P can perform an infinite sequence of instantaneous transitions or self-triggered transitions; i.e., there exists an infinite set $\{D_i \mid i \in Nat\}$ of configurations such that $D_0 =< P, \sigma, \sigma', 1 >$ (or $< P, \sigma, \emptyset, 0 >$), and for all i,

- either $D_i \longrightarrow D_{i+1}$
- or $D_i =< P_i, \sigma_i, \sigma'_i, 0 >$, $\sigma'_i \neq \emptyset$, $D_i \overset{<\sigma_i, \sigma'_i>}{\longrightarrow}_{c_i} D_{i+1}$

where, Nat is the set containing all non-negative integers.   □

Definition 3.2: A computational sequence of program P is an empty sequence or any finite sequence leading P to the other state, that is:

$$D_0 \overset{\delta_1}{\longrightarrow} D_1 \ \ldots\ldots\ \overset{\delta_n}{\longrightarrow} D_n$$

where $D_0 =< P, \sigma_0, \emptyset, 0 >$ or $D_0 =< P, \sigma_0, \sigma'_0, 1 >$ or $D_0 =< P, \sigma_0, \sigma'_0, 0 >$ and $D_i =< P_i, \sigma_i, \emptyset, 0 >$ or $D_i =< P_i, \sigma_i, \sigma'_i, 1 >$ or $D_i =< P_i, \sigma_i, \sigma'_i, 0 >$ or $D_i =< P_i, \sigma_i, \sigma'_i, 1, j > (i = 1, \ldots, n$ and $j \in \{1, 2\})$

and $\overset{\delta_i}{\longrightarrow} (i = 1, \ldots, n)$ can be an instantaneous transition ($\longrightarrow$), a triggered transition ($\overset{<\sigma, \sigma'>}{\longrightarrow}_c$), or a time advancing transition ($\overset{1}{\longrightarrow}$).   □

If computational sequence seq is not empty, $seq[i]$ is the i-th transition ($D_{i-1} \overset{\delta_i}{\longrightarrow} D_i$) of seq.

We write $cp[P]$ representing the set which contains all the computational sequences leading program P to terminating state or divergent state. $cp[P]_{ter}$ and $cp[P]_{div}$ stand for the sets which contain all the sequences leading program P to the terminating and divergent states correspondingly. Therefore, we have $cp[P] = cp[P]_{ter} \cup cp[P]_{div}$.

From the operational semantics we know the initial state of process P can be one of the following states before it is executed.

- $< P, \sigma, \emptyset, 0 >$ (represented as $ttr = null$ in the denotational model).
- $< P, \sigma, \sigma', 1 >$ (represented as $ttr \neq null \wedge flag = 1$).
- $< P, \sigma, \sigma', 0 >$ (represented as $ttr \neq null \wedge flag = 0$).

Example 3.3: Let $P =_{df} x := 1; @(\uparrow y)$. Consider the computational sequences of process P under the state $< P, \sigma, \sigma', 1 >$ (operational semantics in the appendix):

$$
\begin{aligned}
seq1: \quad & < P, \sigma, \sigma', 1 > \\
\longrightarrow \quad & < P, \sigma, \sigma', 1, 1 > \\
\longrightarrow \quad & < @(\uparrow y), \sigma, \sigma'[1/x], 1 > \\
\longrightarrow \quad & < @(\uparrow y), \sigma, \sigma_1, 0 > \\
\overset{<\sigma,\sigma_1>}{\longrightarrow}_{\neg c} \quad & < @(\uparrow y), \sigma_1, \emptyset, 0 >
\end{aligned}
$$

$$
\begin{aligned}
seq2: \quad & < P, \sigma, \sigma', 1 > \\
\longrightarrow \quad & < P, \sigma, \sigma', 1, 1 > \\
\longrightarrow \quad & < @(\uparrow y), \sigma, \sigma'[1/x], 1 > \\
\longrightarrow \quad & < @(\uparrow y), \sigma, \sigma_1, 0 > \\
\overset{<\sigma,\sigma_1>}{\longrightarrow}_{\neg c} \quad & < @(\uparrow y), \sigma_1, \emptyset, 0 > \\
\overset{<\sigma_1,\sigma_2>}{\longrightarrow}_{\neg c} \quad & < @(\uparrow y), \sigma_2, \emptyset, 0 > \\
\overset{<\sigma_2,\sigma_3>}{\longrightarrow}_{\neg c} \quad & < @(\uparrow y), \sigma_3, \emptyset, 0 > \\
\overset{<\sigma_3,\sigma_4>}{\longrightarrow}_{c} \quad & < \epsilon, \sigma_4, \emptyset, 0 >
\end{aligned}
$$

where $c = fire(\uparrow y)$ (definition in Sect. 4.4) which means two consecutive states can trigger this guard. Also, $\sigma_1 = \sigma'[1/x]$. $\sigma'[1/x]$ is the same as σ' except mapping x to 1. Here, we find the computational sequence $seq2$ will lead the program to the terminating state (ϵ).   □

Example 3.4: Let $Q =_{df} @(\uparrow y); x := 1;$ **chaos**. Consider the computational sequences of process Q under the state $< Q, \sigma, \emptyset, 0 >$:

$$
\begin{aligned}
seq3: \quad & < Q, \sigma, \emptyset, 0 > \\
\overset{<\sigma,\sigma_1>}{\longrightarrow}_{\neg c} \quad & < Q, \sigma_1, \emptyset, 0 > \\
\overset{<\sigma_1,\sigma_2>}{\longrightarrow}_{\neg c} \quad & < Q, \sigma_2, \emptyset, 0 >
\end{aligned}
$$

$$
\begin{aligned}
seq4: \quad & < Q, \sigma, \emptyset, 0 > \\
\overset{<\sigma,\sigma_1>}{\longrightarrow}_{\neg c} \quad & < Q, \sigma_1, \emptyset, 0 > \\
\overset{<\sigma_1,\sigma_2>}{\longrightarrow}_{\neg c} \quad & < Q, \sigma_2, \emptyset, 0 > \\
\overset{<\sigma_2,\sigma_3>}{\longrightarrow}_{c} \quad & < x := 1; \textbf{chaos}, \sigma_3, \emptyset, 0 >
\end{aligned}
$$

Here $c = fire(\uparrow y)$. **chaos** can perform an infinite sequence of instantaneous transitions under any state $< \textbf{chaos}, \sigma, \sigma', 1 >$ [7]. If "$x := 1;$ **chaos**" takes control at the state $< x := 1; \textbf{chaos}, \sigma_3, \emptyset, 0 >$, it will execute an infinite sequence of instantaneous transitions. Therefore, $seq4$ is the computational sequence leading the program Q to the divergent state.   □

$cp[P]_{ter}(0)$ and $cp[P]_{div}(0)$ stand for the sets leading the program to the terminating and divergent states under $< P, \sigma, \emptyset, 0 >$ respectively. $cp[P]_{ter}(1)$ and $cp[P]_{div}(1)$ are the sets leading the program to the terminating and divergent states under $< P, \sigma, \sigma', 1 >$ correspondingly. $cp[P]_{ter}(2)$ and $cp[P]_{div}(2)$ stand for the sets leading the program to the terminating and divergent states under $< P, \sigma, \sigma', 0 >$ correspondingly. This means:

$$cp[P]_{ter} = cp[P]_{ter}(0) \cup cp[P]_{ter}(1) \cup cp[P]_{ter}(2) \quad \text{and}$$
$$cp[P]_{div} = cp[P]_{div}(0) \cup cp[P]_{div}(1) \cup cp[P]_{div}(2).$$

Definition 3.5: Let seq stands for a computational sequence of program P. Suppose $len(seq) = n$, $sem(seq)$ is the semantics of the computational sequence seq which can be defined as:
If $len(seq) = 0$ then $sem(seq) =_{df} II$.
If $len(seq) = 1$ then $sem(seq) =_{df} sem_1$.
Otherwise $sem(seq) =_{df} sem_1 ; \ldots ; sem_n$.
sem_i is the phase semantics of the i-th transition ($seq[i]$) of the computational sequence seq.   □

Example 3.6: Let $P =_{df} x := 1 ; x := 2$. There is only one computational sequence seq of P under $< P, \sigma, \sigma', 1 >$:

$$seq : < P, \sigma, \sigma', 1 > \longrightarrow < P, \sigma, \sigma', 1, 1 > \qquad \longrightarrow < x := 2, \sigma, \sigma'[1/x], 1 >$$
$$\longrightarrow < x := 2, \sigma, \sigma'[1/x], 1, 1 > \longrightarrow < \epsilon, \sigma, \sigma'[2/x], 1 >$$

The semantics of computational sequence seq is:

$$
\begin{aligned}
& sem(seq) & & \{\text{Def of 3.5}\} \\
&= sem_1 \;;\; sem_2 \;;\; sem_3 \;;\; sem_4 & & \{\text{Phase semantics, Th 2.9}\} \\
&= Inst(\overrightarrow{tr} = \overleftarrow{tr} \wedge O(notnull, ttr[2/x], 1, 1)) & & \square
\end{aligned}
$$

The denotational semantics of program P can be defined as:

Definition 3.7: (Map from operational to denotational)

$$P =_{df} P[0] \lhd ttr = null \rhd (P[1] \lhd flag = 1 \rhd P[2])$$

where

$$P[i] =_{df} \bigvee_{seq \in cp[P]_{div}(i)} (sem(seq) \;;\; \bot) \;\; \vee \;\; \bigvee_{seq \in cp[P]_{ter}(i)} (sem(seq)),$$
$$i = 0, 1, 2$$

Here $P[0]$, $P[1]$ and $P[2]$ stand for the semantics of program P under the states $< P, \sigma, \emptyset, 0 >$, $< P, \sigma, \sigma', 1 >$ and $< P, \sigma, \sigma', 0 >$ respectively. $\qquad\square$

The following definitions and theorems are useful for calculating the denotational semantics for Verilog statements.

Definition 3.8: $\quad < P, \sigma_0, \emptyset, 0 > \;\; (\xrightarrow{e}_c)^i \;\; < P, \sigma_i, \emptyset, 0 >$

means there exist i steps environment transitions,

$$< P, \sigma_0, \emptyset, 0 > \xrightarrow{<\sigma_0, \sigma_1>}_c < P, \sigma_1, \emptyset, 0 > \ldots \xrightarrow{<\sigma_{k-1}, \sigma_k>}_c < P, \sigma_k, \emptyset, 0 >$$
$$\ldots \xrightarrow{<\sigma_{i-1}, \sigma_i>}_c < P, \sigma_i, \emptyset, 0 > \quad \square$$

Definition 3.9: $\quad L1(i)$ stands for the following computational sequence:

$$< P, \sigma_0, \emptyset, 0 > \;\; (\xrightarrow{e})^i \;\; < P, \sigma_i, \emptyset, 0 > \qquad \square$$

Theorem 3.10: $\quad \bigvee_{i \geq 0} sem(L1(i)) = (ttr = null) \wedge (flag = 0) \wedge hold(0)$, where

$hold(n) =_{df} \mathbf{H}(\mathbf{true} \vdash idle \wedge ttr' = ttr \wedge flag' = flag \wedge (\delta < n \lhd wait' \rhd \delta = n))$,

$idle =_{df} \pi_3(\overrightarrow{tr} - \overleftarrow{tr}) \in 0^* \wedge incr(\pi_1(\overrightarrow{tr} - \overleftarrow{tr}))$,

$incr(s) =_{df} \forall < t_1, t_2 > \text{ in } s \bullet (t_2 - t_1) \in Nat \qquad \square$

Definition 3.11: $\quad < P, \sigma, \emptyset, 0 > \;\; (\xrightarrow{et}_c)^{j_0, \ldots, j_\delta} \;\; < P, \sigma', \emptyset, 0 >$

means the following detailed computational sequence:

$$< P, \sigma, \emptyset, 0 > (\xrightarrow{e}_c)^{j_0} \quad < P, \sigma_1, \emptyset, 0 > \xrightarrow{1} \quad < P, \sigma_1, \emptyset, 0 >$$
$$\ldots \qquad \ldots \qquad \ldots \qquad \ldots$$
$$(\xrightarrow{e}_c)^{j_{\delta-1}} < P, \sigma_n, \emptyset, 0 > \xrightarrow{1} \quad < P, \sigma_n, \emptyset, 0 >$$
$$(\xrightarrow{e}_c)^{j_\delta} \quad < P, \sigma', \emptyset, 0 > \qquad \square$$

where δ is the interval length $(\overrightarrow{time} - \overleftarrow{time})$.

Definition 3.12: $L2(c, j_0, \ldots, j_\delta)$ stands for the following computational sequence.

$$< P, \sigma, \emptyset, 0 > \quad (\xrightarrow{et}_c)^{j_0, \ldots, j_\delta} \quad < P, \sigma', \emptyset, 0 >$$

Theorem 3.13: $\bigvee sem(L2(c, j_0, \ldots, j_\delta)) =_{df} silence(c)$

where, the disjuction "$\bigvee$" is for all $j_0 \geq 0, \ldots, j_\delta \geq 0$.

$$silence(c) =_{df} \mathbf{H} \left(\mathbf{true} \vdash \left(\begin{array}{c} idle \wedge O(null, null, 0, 0) \wedge \\ \forall < \sigma_1, \sigma_2 > \ \mathbf{in} \ \pi_2(\overrightarrow{tr} - \overleftarrow{tr}) \bullet c(\sigma_1, \sigma_2) \end{array} \right) \right)$$

$silence(c)$ means during this period, the environment can do any atomic actions, but can not fire the condition $\neg c$.

4 Deriving the Semantics for Statements of Verilog

In this section we will derive the denotational semantics for the Verilog statements by strict proof. Therefore our denotational semantics is equivalent with its operational semantics.

The main purpose of the mathematical definition of Verilog operators is to deduce their interesting properties. These are most elegantly expressed as algebraic laws (equations usually). As our denotational map is based on the transition system of a program, we have two ways to prove the algebraic laws, one using the denotational semantics and the other using the transition system.

4.1 Sequential Composition

The notation $(P \, ; \, Q)$ represents the process which behaves like P before P terminates, and then behaves like Q afterwards.

Theorem 4.1: $(P \, ; \, Q) \ = \ (P) \, ; \, (Q)$

The "$;$" in the left side is the sequential composition of programs, whereas "$;$" in the right side is the semantic sequential composition of logic formulae. This theorem indicates the denotational semantics of program $P; Q$ is the sequential composition of their denotational semantics.

4.2 Skip

The role of **skip** is the same as $x := x$ (see operational semantics in the appendix).

Theorem 4.2: $\mathbf{skip} \ = \ flash \lhd (ttr \neq null \wedge flag = 0) \rhd II$
$$; \ (hold(0) \, ; \, init) \lhd ttr = null \rhd II$$

where, $init =_{df} Inst(\overrightarrow{tr} = \overleftarrow{tr} \wedge O(null, \pi_2(last(\overleftarrow{tr})), 0, 1))$

$$flash =_{df} Inst \left(\begin{array}{c} ttr' = null \wedge flag' = 0 \wedge (\overrightarrow{tr} = \overleftarrow{tr} \lhd (ttr = null \ \vee \\ \pi_2(last(\overleftarrow{tr})) = ttr) \rhd \overrightarrow{tr} = \overleftarrow{tr}\ ^\frown < (\overleftarrow{time}, ttr, 1) >) \end{array} \right)$$

4.3 Assignment

The execution of $x := e$ assigns the value of e to x. Assignment $x := e$ can be in either of the three states before its execution: $< x := e, \sigma, \emptyset, 0 >$, $< x := e, \sigma, \sigma', 1 >$ and $< x := e, \sigma, \sigma', 0 >$.

Case 1: If $ttr = null$, the corresponding computational sequence is :

$$< x := e, \sigma, \emptyset, 0 > (\xrightarrow{e})^i < x := e, \sigma_i, \emptyset, 0 > \longrightarrow < x := e, \sigma_i, \sigma_i, 1, 1 >$$
$$\longrightarrow < \epsilon, \sigma_i, \sigma_i[e/x], 1 >$$

The transitional conditions of the last two instantaneous transitions are: $\mathbf{Cond}_{1,1}$ and $\mathbf{Cond}_{2,2}$.

Let $assign(x := e) =_{df} Inst(\overrightarrow{tr} = \overleftarrow{tr} \wedge ttr' = ttr[e/x] \wedge flag' = flag)$

By proof, $Inst(\mathbf{Cond}_{1,1} ; \mathbf{Cond}_{2,2}) = init ; assign(x, e)$. Using theorem 3.10, the semantics of $x := e$ in this case is: $hold(0) ; init ; assign(x, e)$.

Case 2: If $ttr \neq null \wedge flag = 1$, the corresponding computational sequence is

$$< x := e, \sigma, \sigma', 1 > \longrightarrow < x := e, \sigma, \sigma, 1, 1 > \longrightarrow < \epsilon, \sigma, \sigma[e/x], 1 >$$

The semantics of assignment in this case can be proved as $assign(x, e)$.

Case 3: If $ttr \neq null \wedge flag = 0$, the corresponding computational sequence is:

$$< x := e, \sigma, \sigma', 0 > \xrightarrow{<\sigma, \sigma'>} < x := e, \sigma', \emptyset, 0 > \quad (\xrightarrow{e})^i < x := e, \sigma_i, \emptyset, 0 >$$
$$\longrightarrow \quad < x := e, \sigma_i, \sigma_i, 1, 1 > \longrightarrow \quad < \epsilon, \sigma_i, \sigma_i[e/x], 1 >$$

The semantics of $x := e$ under this case is: $flash ; hold(0) ; init ; assign(x, e)$.

Using the semantic map and predicate calculus, we obtain the semantics of assignment.

Theorem 4.3: $x := e = \mathbf{skip} ; assign(x := e)$

Verilog assignment statements obey the same set of algebraic laws as its counterpart in the conventional programming languages.

4.4 Event Guard

The guard event is denoted by $@(g)$. A primitive guard g can be of the following forms:
- $\uparrow v$ waits for an increase of the value of v.
- $\downarrow v$ waits for a decrease of the value of v.
- v waits for a change of v.

There are also three types of compound guards.
- g_1 *or* g_2 becomes enabled when either g_1 or g_2 is fired.
- g_1 *and* g_2 becomes enabled if both g_1 and g_2 are awaken simultaneously.
- g_1 *and* $\neg g_2$ becomes fired if g_2 remains idle and g_1 is awaken.

We introduce a predicate $fire(g)(\sigma, \sigma')$ to indicate the transition from the state σ to the state σ' can awake the guard $@(g)$.

$$fire(\uparrow v)(\sigma, \sigma') =_{df} \sigma(v) < \sigma'(v), \quad fire(\downarrow v)(\sigma, \sigma') =_{df} \sigma(v) > \sigma'(v)$$
$$fire(v)(\sigma, \sigma') =_{df} \sigma(v) \neq \sigma'(v)$$
$$fire(g_1 \ or \ g_2)(\sigma, \sigma') =_{df} fire(g_1)(\sigma, \sigma') \vee fire(g_2)(\sigma, \sigma')$$

$$fire(g_1 \ and \ g_2)(\sigma, \sigma') =_{df} fire(g_1)(\sigma, \sigma') \wedge fire(g_2)(\sigma, \sigma')$$
$$fire(g_1 \ and \ \neg g_2)(\sigma, \sigma') =_{df} fire(g_1)(\sigma, \sigma') \wedge \neg fire(g_2)(\sigma, \sigma')$$

The event guard $@(g)$ can be immediately fired after it is scheduled, it is actually triggered by the execution of its prior atomic action. According to the operational semantics of $@(g)$ (in the appendix), there are two kinds of computational sequences leading to the terminating state.

$$< @(g), \sigma, \sigma', 1 > \longrightarrow < @(g), \sigma, \sigma', 0 > \overset{<\sigma,\sigma'>}{\longrightarrow}_{fire(g)} < \epsilon, \sigma', \emptyset, 0 >$$

$$< @(g), \sigma, \sigma', 0 > \overset{<\sigma,\sigma'>}{\longrightarrow}_{fire(g)} < \epsilon, \sigma', \emptyset, 0 >$$

Another case is the guard $@(g)$ waits to be fired by the environment. There are three kinds of computational sequences leading to the terminating state.

$$< @(g), \sigma, \sigma', 1 > \longrightarrow \qquad < @(g), \sigma, \sigma', 0 > \overset{<\sigma,\sigma'>}{\longrightarrow}_{\neg c} < @(g), \sigma', \emptyset, 0 >$$
$$(\overset{et}{\longrightarrow}_{\neg c})^{j_0, \dots, j_\delta} < @(g), \sigma_n, \emptyset, 0 > \overset{<\sigma_n,\sigma_{n+1}>}{\longrightarrow}_c < \epsilon, \sigma_{n+1}, \emptyset, 0 >$$

$$< @(g), \sigma, \sigma', 0 > \overset{<\sigma,\sigma'>}{\longrightarrow}_{\neg c} < @(g), \sigma', \emptyset, 0 > (\overset{et}{\longrightarrow}_{\neg c})^{j_0, \dots, j_\delta} < @(g), \sigma_n, \emptyset, 0 >$$
$$\overset{<\sigma_n,\sigma_{n+1}>}{\longrightarrow}_c < \epsilon, \sigma_{n+1}, \emptyset, 0 >$$

$$< @(g), \sigma, \emptyset, 0 > (\overset{et}{\longrightarrow}_{\neg c})^{j_0, \dots, j_\delta} < @(g), \sigma_n, \emptyset, 0 > \overset{<\sigma_n,\sigma_{n+1}>}{\longrightarrow}_c < \epsilon, \sigma_{n+1}, \emptyset, 0 >$$

Here $c = fire(g)$. There is a corresponding phase semantics for each type of transition. Using the definition of phase semantics and Theorem 2.9, 3.13, we obtain:

Theorem 4.4: $@(g) = selftrig(g) \vee (await(g) \, ; \, trig(g))$

where,

$$selftrig(g) =_{df} \mathbf{H}(\mathbf{true} \vdash ttr \neq null \wedge fire(g)(\pi_2(last(\overleftarrow{tr})), ttr)) \wedge II \, ; \, flash$$

$$await(g) =_{df} \mathbf{H}(\, \mathbf{true} \vdash (ttr = null \vee \neg fire(g)(\pi_2(last(\overleftarrow{tr})), ttr))) \wedge II$$
$$; \, flash \, ; \, silence(\neg fire(g))$$

$$trig(g) =_{df} Inst \left(\begin{array}{c} idle \wedge len(\overrightarrow{tr} - \overleftarrow{tr}) = 1 \wedge O(null, null, 0, 0) \\ \wedge fire(g)(\pi_2(last(\overleftarrow{tr})), \pi_2(last(\overrightarrow{tr}))) \end{array} \right)$$

4.5 Other Statements

chaos represents the worst process. Its behaviour is totally unpredictable. The conditional **if** $b(v)$ **then** P **else** Q behaves the same as the "**then**" branch if b is true when activated, and the same as the "**else**" branch otherwise. The delay event $\#n$ holds the execution for n units. An assignment guard $@(x := e)$ is a special assignment representing an atomic action. It is used in supporting the parallel expansion laws.

Let $\{g_i \mid 1 \leq i \leq n\}$ be a finite family of event guards, and $\{P_i \mid 1 \leq i \leq n\}$ a family of Verilog processes. The notation $(@(g_1) \, P_1) [\![\, \dots \,]\!] (@(g_n) \, P_n)$ denotes the program which initially waits for one of the guards to be fired, and then behaves the same as the corresponding guarded process. The program $(@(x_1 := e_1) \, P_1) [\![\, \dots \,]\!] (@(x_n := e_n) \, P_n)$ performs one of its alternative , and the choice is made non-deterministically.

In accordance with the semantic map and operational semantics of these statements [7], we obtain the denotational semantics for these statements.

Theorem 4.5

(1)   **if** b **then** S_1 **else** S_2 = **skip** ; $S_1 \lhd b(ttr) \rhd S_2$

(2) $\#n$ = $flash$; $hold(n)$

(3) $@(x := e)$ = $flash$; $hold(0)$; $trig(@(x := e))$

(4) $(@(x_1 := e_1) \, P_1) \, [\!] \, \dots \, [\!] \, \{ @(x := e_n) \, P_n)$

   = $\bigvee \{ @(x_i := e_i); P_i \mid 1 \le i \le n \}$

(5)   **chaos** = $(flash \lhd (ttr \ne null \wedge flag = 0) \rhd II)$; $\bot$

(6) $(@(g_1) \, P_1) \, [\!] \, \dots \, [\!] \, (@(g_n) \, P_n)$

   = $\bigvee \{ (selftrig(g_i) \vee (await(g); trig(g_i))) \, ; \, P_i \mid 1 \le i \le n \}$

where $trig(@(x := e)) =_{df} Inst(\, \overrightarrow{tr} = \overleftarrow{tr} \wedge O(null, \pi_2(last(\overleftarrow{tr}))[e/x], 0, 0) \,)$

   g stands for the compound guard g_1 or $\dots$ or g_n.

4.6 Parallel

Although we have not derived the universal formula representing the denotational semantics for a parallel process, we can write down its transition system. Its semantics can
be calculated based on its transition steps. Algebraic laws dealing with parallel can also
be proved using the denotational map based on its specific transition systems.

Example 4.6: Let P = $(\#1; x := 2) \parallel (x := 1; \#1)$; $\#1$. Consider the denotational
semantics of P.

We can write down the computational sequences leading program P to the terminating
state under three cases ($ttr = null$, $ttr \ne null \wedge flag = 1$ and $ttr \ne null \wedge flag = 0$)
based on the parallel transition [7]. The semantics of P can be calculated based on the
semantic map and its computational sequences. Therefore, the denotational semantics
of P is

$$flash \, ; \, hold(0) \, ; \, Inst(\, S(1) \,) \, ; \, hold(1) \, ; \, Inst(\, S(2) \,) \, ; \, hold(1)$$

where, $S(u) =_{df} \overrightarrow{tr} = \overleftarrow{tr}\hat{} < (\overleftarrow{time}, \{ x \mapsto u \}, 1) > \wedge O(null, null, 0, 0)$ $\square$

Theorem 4.7 (Expansion laws)

(par-1) Let $P_i =_{df} @(\eta_i) \, Q_i$ for $i = 1, 2$. Then

$$P_1 \parallel P_2 \; = \; \left(\begin{array}{l} (@(\eta_1 \; and \; \neg\eta_2) \, (Q_1 \parallel P_2)) \\ [\!] \; (@(\eta_1 \; and \; \eta_2) \, (Q_1 \parallel Q_2)) \\ [\!] \; (@(\eta_2 \; and \; \neg\eta_1) \, (P_1 \parallel Q_2)) \end{array} \right)$$

(par-2) Let $P_i =_{df} @(x_i := e_i) \, Q_i$ for $i = 1, 2$. Then

$$P_1 \parallel P_2 \; = \; \left(\begin{array}{l} (@(x_1 := e_1) \, (Q_1 \parallel P_2)) \\ [\!] \; (@(x_2 := e_2) \, (P_1 \parallel Q_2)) \end{array} \right)$$

5   Conclusion

The main contribution of our work is to derive the denotational semantics for a subset of
Verilog from its operational semantics [7]. Thus, our denotational semantics presented
here is equivalent with its operational semantics. We provide a discrete denotational
model and design a refinement calculus for it. Our approach for the derivation is new.

We define a transitional condition and the phase semantics for each type of transition. The denotational semantics can be derived as the sequential composition of those phase semantics. Verilog's algebraic laws are also discussed, which can support program transformation and system partitioning for hardware/software co-design. Proofs are undertaken in two ways, one using the denotational semantics and the other using the operational semantics. Thus, the three semantics form a unifying model for (a subset of) Verilog.

For the future, we are continuing to explore unifying theories for Verilog. We wish to extend the scope of the derivation of denotational semantics for Verilog to further constructs in the language such as iteration. The derivation of operational semantics from denotational semantics for Verilog is another interesting topic for study.

References

1. J. P. Bowen, He Jifeng and Xu Qiwen. An Animatable Operational Semantics of the VERILOG Hardware Description Language. *Proc. ICFEM2000: 3rd IEEE International Conference on Formal Engineering Methods*, IEEE Computer Society Press, pp. 199–207, York, UK, September 2000.

2. M. J. C. Gordon. The Semantic Challenge of Verilog HDL. *Proc. Tenth Annual IEEE Symposium on Logic in Computer Science*, IEEE Computer Society Press, pp. 136–145, June 1995.

3. He Jifeng and Xu Qiwen. *An Operational Semantics of a Simulator Algorithm*. Technical Report 204, UNU/IIST, P.O. Box 3058, Macau, 2000.

4. He Jifeng and Zhu Huibiao. Formalising Verilog. *Proc. IEEE International Conference on Electronics, Circuits and Systems*, IEEE Computer Society Press, pp. 412–415, Lebanon, December 2000.

5. C. A. R. Hoare and He Jifeng. *Unifying Theories of Programming*. Prentice Hall International Series in Computer Science, 1998.

6. IEEE *Standard Hardware Description Language based on the Verilog® Hardware Description Language*. IEEE Standard 1364-1995, 1995.

7. Li Yongjian and He Jifeng. *Formalising VERILOG: Operational Semantics and Bisimulation*. Technical Report 217, UNU/IIST, P.O. Box 3058, Macau, November 2000.

8. Zhou Chaochen, C. A. R. Hoare and A. P. Ravn. A Calculus of Durations. *Information Processing Letters*, 40(5):269–276, 1991.

9. Zhu Huibiao and He Jifeng. A Semantics of Verilog using Duration Calculus. *Proc. International Conference on Software: Theory and Practice,* pp. 421–432, Beijing, China, August 2000.

10. Zhu Huibiao, Jonathan Bowen and He Jifeng. *From Operational Semantics to Denotational Semantics for Verilog*. Technical Report SBU-CISM-01-08, South Bank University, London, UK, May 2001.

Appendix

Below are the transition system definitions for the assignment and event guard constructs. Definitions for other commands can be found in [7].

1. Assignment

$\mathbf{T}_1$: $< v = e, \sigma, \emptyset, 0 > \longrightarrow < v = e, \sigma, \sigma, 1, 1 >$
        $< v = e, \sigma, \sigma', 1 > \longrightarrow < v = e, \sigma, \sigma', 1, 1 >$
$\mathbf{T}_2$: $< v = e, \sigma, \sigma', 1, 1 > \longrightarrow < \varepsilon, \sigma, \sigma'[e(\sigma')/v], 1 >$
$\mathbf{T}_5$: $< v = e, \sigma, \sigma', 0 > \overset{<\sigma,\sigma'>}{\longrightarrow} < v = e, \sigma', \emptyset, 0 >$
        $< v = e, \sigma, \emptyset, 0 > \overset{<\sigma,\sigma'>}{\longrightarrow} < v = e, \sigma', \emptyset, 0 >$

2. Event Guard

$\mathbf{T}_3$: $< @(\eta), \sigma, \sigma', 1 > \longrightarrow < @(\eta), \sigma, \sigma', 0 >$
$\mathbf{T}_5$: $< @(\eta), \sigma, \sigma', 0 > \overset{<\sigma,\sigma'>}{\longrightarrow}_{fire(\eta)} < \varepsilon, \sigma', \emptyset, 0 >$
        $< @(\eta), \sigma, \emptyset, 0 > \overset{<\sigma,\sigma'>}{\longrightarrow}_{fire(\eta)} < \varepsilon, \sigma', \emptyset, 0 >$
        $< @(\eta), \sigma, \sigma', 0 > \overset{<\sigma,\sigma'>}{\longrightarrow}_{\neg fire(\eta)} < @(\eta), \sigma', \emptyset, 0 >$
        $< @(\eta), \sigma, \emptyset, 0 > \overset{<\sigma,\sigma'>}{\longrightarrow}_{\neg fire(\eta)} < @(\eta), \sigma', \emptyset, 0 >$
$\mathbf{T}_6$: $< @(\eta), \sigma, \emptyset, 0 > \overset{1}{\longrightarrow} < @(\eta), \sigma, \emptyset, 0 >$

Efficient Verification of a Class of Linear Hybrid Automata Using Linear Programming*

Li Xuandong, Pei Yu, Zhao Jianhua, Li Yong, Zheng Tao, and Zheng Guoliang

State Key Laboratory of Novel Software Technology
Department of Computer Science and Technology
Nanjing University, Nanjing
Jiangsu, P.R. China 210093
lxd@nju.edu.cn

Abstract. In this paper, we show that for a class of linear hybrid automata called *zero loop-closed automata*, the satisfaction problem for linear duration properties can be solved efficiently by linear programming. We give an algorithm based on depth-first search method to solve the problem by traversing all the simple paths (with no repeated node occurrence) in an automaton and checking their corresponding sequences of locations for a given linear duration property.

1 Introduction

The model checking problem for real-time hybrid systems is very difficult, even for a well-formed class of hybrid systems - the class of linear hybrid automata - the problem is still undecidable in general [1-4]. So an important question for the analysis and design of hybrid systems is identification of subclasses of such systems and corresponding restricted classes of analysis problems that can be settled algorithmically [2].

In this paper, we consider the problem of checking linear hybrid automata for linear duration properties. We show that for a class of linear hybrid automata called *zero loop-closed automata*, the satisfaction problem of linear duration properties can be solved efficiently by linear programming. We give an algorithm based on depth-first search method to solve the problem by traversing all the simple paths (with no repeated node occurrence) in an automaton and checking their corresponding sequences of locations for a given linear duration property.

The paper is organized as follows. In next section, we recall the notion of linear hybrid automata, and introduce linear duration properties. In Sect. 3, we define *zero loop-closed automata*. Section 4 gives an efficient algorithm to check *zero loop-closed automata* for linear duration properties. The last section discusses the related work and contains some conclusion.

---

* This work is supported by the National Natural Science Foundation of China under Grant 60073031 and Grant 69703009, and by International Institute for Software Technology, The United Nations University (UNU/IIST).

T. Margaria and T. Melham (Eds.): CHARME 2001, LNCS 2144, pp. 465–479, 2001.

2   Linear Hybrid Automata and Linear Duration Properties

2.1   Linear Hybrid Automata

A linear hybrid automaton is a conventional automaton extended with a finite set of real-valued variables. We use a simplified version of linear hybrid automata defined in [1]. The simplification is that any linear hybrid automaton considered in this paper has just one initial location, no initial condition, and no transition to the initial location (we suppose that each variable with an initial value is reset to the initial value by the transitions from the initial location) .

Definition 1. *A linear hybrid automaton is a tuple* $\mathcal{H} = (Z, X, V, E, v_I, \alpha, \beta)$, *where*

- Z *is a finite set of system states.*
- X *is a finite set of real-numbered variables.*
- V *is a finite set of locations.*
- E *is transition relation whose elements are of the form* (v, ϕ, ψ, v') *where* v, v' *are in* V, ϕ *is a set of variable constraints which are of the form* $a \leq x \leq b$, *and* ψ *is a set of reset actions which are of the form* $y := c$ $(x \in X, y \in X, a, b, c$ *are real numbers,* a *and* b *may be* ∞; *if* a (b) *is* $-\infty$ (∞), *then* $a \leq x \leq b$ *is taken to be* $x \leq b$ $(a \leq x)$; *if* $a = b$, *then* $a \leq x \leq b$ *is taken to be* $x = a$).
- v_I *is an initial location.*
- α *is a labeling function which maps each location in* V *to a state in* Z.
- β *is a labeling function which maps each location in* V *to a set of change rates which are of the form* $\dot{x} = a$ $(x \in X$ *and* a *is a real number). For any location* v, *for any* $x \in X$, *there is one and only one* $\dot{x} = a \in \beta(v)$.

□

Let us consider an example of a water-level monitor in [3]. The water level in a tank is controlled through a monitor, which continuously senses the water level and turns a pump on and off. The water level changes as a piecewise-linear function of time. When the pump is off, the water level falls by two inches per second; when the pump is on, the water level rises by one inch per second. Suppose that initially the water level is one inch and the pump is on. There is a delay of two seconds from the time that the monitor signals to change the status of the pump to the time that the change becomes effective. The requirement of the water-level monitor is that the monitor must keep the water level in between 1 and 12 inches. A design of the monitor can be modelled by the hybrid automaton depicted in Fig. 1. The initial location of the automaton is v_0. The other four locations v_1, v_2, v_3, v_4 are assigned with the system states s_1, s_2, s_3, s_4 respectively. In the locations v_1 and v_2, the pump is on; in the locations v_3 and v_4, the pump is off. The variable y is used to model the water-level, and x is used to specify the delays: whenever the control is in location v_2 or v_3, the value of x indicates how long the signal to switch the pump off or on has been sent.

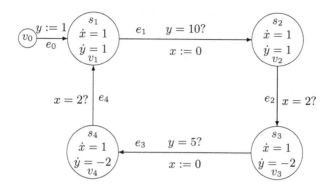

Fig. 1. A hybrid automaton modelling a water-level monitor

We use *sequences of locations* to represent the untimed behaviour of linear hybrid automata. A sequence of locations is of the form

$$v_0 \xrightarrow{(\phi_0,\psi_0)} v_1 \xrightarrow{(\phi_1,\psi_1)} \ldots \xrightarrow{(\phi_m,\psi_m)} v_{m+1} \,,$$

which indicates that the automaton start from location v_0, move to v_{i+1} from v_i with executing the reset actions in set ψ_i when the variable constraints in set ϕ_i are satisfied. For a linear hybrid automaton $\mathcal{H} = (Z, X, V, E, v_I, \alpha, \beta)$, if v_0 is v_I and $(v_i, \phi_i, \psi_i, v_{i+1}) \in E$ for each i $(0 \leq i \leq m)$, then the sequence of locations represents a untimed behaviour of $\mathcal{H}$.

The behaviour of linear hybrid automata can be represented by *timed sequences*. Any timed sequence is of the form $(s_1, t_1)\hat{\ }(s_2, t_2)\hat{\ } \ldots \hat{\ }(s_m, t_m)$ where s_i $(1 \leq i \leq m)$ is a state and t_i $(1 \leq i \leq m)$ is a nonnegative real number, which represents a behaviour of an automaton that the system starts at the state s_1, stays there for t_1 time units, then changes to s_2 and stays in s_2 for t_2 time units, and so on.

Definition 2. *For a linear hybrid automaton $\mathcal{H} = (Z, X, V, E, v_I, \alpha, \beta)$, a timed sequence $(s_1, t_1)\hat{\ }(s_2, t_2)\hat{\ } \ldots \hat{\ }(s_m, t_m)$ $(m \geq 1)$ represents a behaviour of $\mathcal{H}$ if and only if there is a untimed behaviour of the automaton*

$$v_0 \xrightarrow{(\phi_0,\psi_0)} v_1 \xrightarrow{(\phi_1,\psi_1)} \ldots \xrightarrow{(\phi_m,\psi_m)} v_{m+1}$$

such that

- *for each i $(1 \leq i \leq m)$, $\alpha(v_i) = s_i$; and*
- *$t_1, t_2, \ldots, t_m$ satisfy all the variable constraints in ϕ_i $(1 \leq i \leq m)$, i.e. for each variable constraint $a \leq x \leq b$ in ϕ_i, if there is a reset action $x := c$ in ψ_j $(0 \leq j < i)$ and $x := d$ is not in ψ_k for any k $(j < k < i)$, then*

$$a \leq c + w_{j+1}t_{j+1} + w_{j+2}t_{j+2} + \ldots + w_i t_i \leq b \,,$$

where for each l $(j < l \leq i)$, $\dot{x} = w_l \in \beta(v_l)$. □

For example, for the linear hybrid automaton depicted in Fig. 1, the timed sequence $(s_1, 9)^\frown(s_2, 2)^\frown(s_3, 3.5)^\frown(s_4, 2)$ is a behaviour.

For a linear hybrid automaton $\mathcal{H}$, for a transition $e = (v, \phi, \psi, v')$ in $\mathcal{H}$, if e is labeled with a variable constraint $a \leq x \leq b$, i.e. $a \leq x \leq b \in \phi$, then we say that x is *tested* by e; if e is labeled with a reset action $x := c$, i.e. $x := c \in \psi$, then we say that x is *reset* by e. Notice that if a transition is labeled with a variable constraint $x = c$, we can take it as the transition resets the variable x to c. For example, for the automaton depicted in Fig. 1, we can say that the transitions e_1 and e_3 reset the variable y to 10 and 5 respectively, and the transitions e_2 and e_4 reset the variable x to 2.

2.2   Linear Duration Properties

Linear duration properties are linear inequalities on integrated durations of system states. Here we use Duration Calculus (DC) [5] to describe this kind of properties. In DC, states are modelled as Boolean functions from reals (representing continuous time) to $\{0, 1\}$, where 1 denotes state presence, and 0 denotes state absence. For a state S, the integral variable $\int S$ of DC is a function from bounded and closed intervals to reals which stands for the accumulated presence time (duration) of state S over the intervals, and is defined formally by $\int S[a, b] \hat{=} \int_a^b S(t)dt$, where $[a, b]$ ($b \geq a$) is a bounded interval of time. A linear duration property in DC is of the form $\sum_{i=1}^n c_i \int S_i \leq M$, where S_is are system states and M, c_is are real numbers.

The requirement of the water-level monitor is that the monitor must keep the water level in between 1 and 12 inches, which can be expressed by linear duration properties as well. We know that when the control is in locations v_1 or v_2, the water level rises 1 inch per second, and when the control is in locations v_3 or v_4, the water level falls by 2 inch per second. Furthermore, for an interval $[0, t]$, the accumulated time that the system stays in s_1 or s_2 is $\int s_1 + \int s_2$, and the accumulated time that the system stays in s_3 or s_4 is $\int s_3 + \int s_4$. Therefore, the water level at time t, given that at the beginning the water level is one inch, is $1 + \int s_1 + \int s_2 - 2(\int s_3 + \int s_4)$. Hence, the requirement for the water-level monitor can be described by the following linear duration properties

$$1 + \int s_1 + \int s_2 - 2(\int s_3 + \int s_4) \leq 12;$$
$$1 + \int s_1 + \int s_2 - 2(\int s_3 + \int s_4) \geq 1.$$

3   Zero Loop-Closed Automata

Zero loop-closed automata form a subclass of linear hybrid automata. In the following, we define this class of linear hybrid automata.

For a linear hybrid automaton $\mathcal{H} = (Z, X, V, E, v_I, \alpha, \beta)$, a *path segment* is a sequence of locations $v_1 \xrightarrow{(\phi_1, \psi_1)} v_2 \xrightarrow{(\phi_2, \psi_2)} \ldots \xrightarrow{(\phi_{m-1}, \psi_{m-1})} v_m$ which satisfies $(v_i, \phi_i, \psi_i, v_{i+1}) \in E$ for each i ($1 \leq i \leq m-1$). A *path* is a path segment starting at v_I. A path is called *simple* if all locations in the path are distinct. For a simple path $v_1 \xrightarrow{(\phi_1,\psi_1)} v_2 \xrightarrow{(\phi_2,\psi_2)} \ldots \xrightarrow{(\phi_{m-1},\psi_{m-1})} v_m$, if there is v_i $(1 < i \leq m)$ such that $(v_m, \phi, \psi, v_i) \in E$, then the sequence

$$v_i \xrightarrow{(\phi_i,\psi_i)} v_{i+1} \xrightarrow{(\phi_{i+1},\psi_{i+1})} \ldots \xrightarrow{(\phi_{m-1},\psi_{m-1})} v_m \xrightarrow{(\phi,\psi)} v_i$$

is a *loop*, v_i is the *loop-start node* of the loop, $v_1 \xrightarrow{(\phi_1,\psi_1)} v_2 \xrightarrow{(\phi_2,\psi_2)} \ldots \xrightarrow{(\phi_{i-1},\psi_{i-1})} v_i$ is a *loop-enter path* of the loop, and (v_m, ϕ, ψ, v_i) is the *end* transition in the loop. Notice that a loop may have many different loop-enter paths. For a loop ρ, if ρ_1 is a loop-enter path of ρ, we say that ρ *can be entered through* ρ_1. For example, in the automaton depicted in Fig. 2, the sequence of locations

$$v_2 \xrightarrow{(\emptyset,\{y:=0\})} v_3 \xrightarrow{(\{x\geq 5\},\emptyset)} v_4 \xrightarrow{(\{y\geq 3\},\{x:=0\})} v_2$$

is a loop, and the sequence of locations

$$v_0 \xrightarrow{(\emptyset,\{x:=-3,y:=1,z:=0\})} v_1 \xrightarrow{(\{z\geq -5\},\{x:=0\})} v_2$$

is a loop-enter of the loop; and the sequence of locations

$$v_5 \xrightarrow{(\emptyset,\{x:=0\})} v_6 \xrightarrow{(\{1\leq y\leq 5\},\emptyset)} v_7 \xrightarrow{(\{-1\leq x\leq 2\},\{y:=1\})} v_5$$

is a loop, and the sequence of locations

$$v_0 \xrightarrow{(\emptyset,\{x:=-3,y:=1,z:=0\})} v_1 \xrightarrow{(\{z\geq -5\},\{x:=0\})} v_2 \xrightarrow{(\emptyset,\{y:=1,z:=2\})} v_5$$

is a loop-enter of the loop.

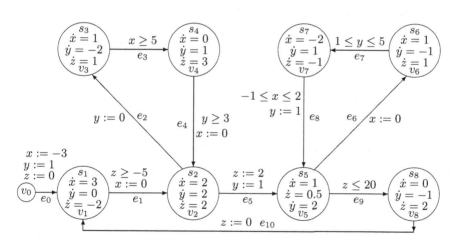

Fig. 2. A loop-closed automaton

Let $v_1 \xrightarrow{(\phi_1,\psi_1)} v_2 \xrightarrow{(\phi_2,\psi_2)} \ldots \xrightarrow{(\phi_{m-1},\psi_{m-1})} v_m$ be a path in a linear hybrid automaton $\mathcal{H}$. For a variable constraint $a \leq x \leq b$ labeled on a transition $(v_i, \phi_i, \psi_i, v_{i+1})$ $(1 < i < m)$, its *reference point* is a transition $(v_j, \phi_j, \psi_j, v_{j+1})$ $(1 \leq j < i)$ such that

- x is reset by the transition $(v_j, \phi_j, \psi_j, v_{j+1})$, and
- x is not reset by any transition $(v_k, \phi_k, \psi_k, v_{k+1})$ $(j < k < i)$,

which means that for calculating the value of x when the automaton stay in v_i along the path in order to check if the variable constraint $a \leq x \leq b$ is satisfied, we need to refer the value of x which is reset to by the transition $(v_j, \phi_j, \psi_j, v_{j+1})$. We say that the variable constraint $a \leq x \leq b$ *combines* the transitions $(v_i, \phi_i, \psi_i, v_{i+1})$ and $(v_j, \phi_j, \psi_j, v_{j+1})$.

Let $\mathcal{H}$ be a linear hybrid automaton, and ρ be a loop in $\mathcal{H}$ which is of the form $v_1 \xrightarrow{(\phi_1, \psi_1)} v_2 \xrightarrow{(\phi_2, \psi_2)} \ldots \xrightarrow{(\phi_{m-1}, \psi_{m-1})} v_m$. We say that any transition $(v_i \xrightarrow{(\phi_i, \psi_i)} v_{i+1})$ $(1 \leq i < m-1)$, which is not the end transition of ρ, *is inside ρ*. We defined that ρ is *closed* if any variable constraint does not combine transition occurrences inside and outside of the loop, i.e. the following condition holds:

- for any variable constraint $a \leq x \leq b$ labeled on a transition $(v_i, \phi_i, \psi_i, v_{i+1})$ $(1 \leq i < m)$ in ρ, its reference point is in ρ, i.e., for any simple path or loop

$$u_1 \xrightarrow{(\phi'_1, \psi'_1)} u_2 \xrightarrow{(\phi'_2, \psi'_2)} \ldots \xrightarrow{(\phi'_{n-1}, \psi'_{n-1})} u_n$$

such that $v_1 = u_n$, if there is no transition $(v_j, \phi_j, \psi_j, v_{j+1})$ $(1 \leq j < i)$ in ρ resetting x, then x is reset to c by the end transition $(v_{m-1}, \phi_{m-1}, \psi_{m-1}, v_m)$ of ρ and by the transition $(u_{n-1}, \phi'_{n-1}, \psi'_{n-1}, u_n)$, i.e. $x := c \in \psi_{m-1}$ and $x := c \in \psi'_{n-1}$; and
- for any variable constraint labeled on a transition outside ρ, its reference point is not inside ρ, i.e., for any simple path segment

$$u_1 \xrightarrow{(\phi'_1, \psi'_1)} u_2 \xrightarrow{(\phi'_2, \psi'_2)} \ldots \xrightarrow{(\phi'_{n-1}, \psi'_{n-1})} u_n$$

such that $v_1 = u_1$, there is no variable constraint $a \leq x \leq b$ labeled on the transition $(u_{n-1}, \phi'_{n-1}, \psi'_{n-1}, u_n)$ such that
- x is reset by a transition $(v_{i-1}, \phi_{i-1}, \psi_{i-1}, v_i)$ $(1 < i < m-1)$ inside ρ, and
- x is not reset by the end transition $(v_{m-1}, \phi_{m-1}, \psi_{m-1}, v_m)$ of ρ and by any transition $(u_{k-1}, \phi'_{k-1}, \psi'_{k-1}, u_k)$ $(1 < k < n)$.

For example, in the automaton depicted in Fig. 2, the loop

$$v_2 \xrightarrow{(\emptyset, \{y:=0\})} v_3 \xrightarrow{(\{x \geq 5\}, \emptyset)} v_4 \xrightarrow{(\{y \geq 3\}, \{x:=0\})} v_2$$

is a closed loop. But it is not closed if we remove the reset action $y := 0$ from the transition e_2 since now for the variable constraint $y \geq 3$ labeled on e_4, its reference point is e_0 which is outside the loop. That a loop is closed implies that the variable values inside the loop do not depend on their values outside the loop, and that the variable values outside the loop do not depend on their values inside the loop.

Definition 3. *A linear hybrid automaton $\mathcal{H}$ is loop-closed if and only if any loop in $\mathcal{H}$ is closed.* □

For example, the automaton depicted in Fig. 2 is a loop-closed automaton. Notice that for some linear hybrid automata which are not loop-closed, we can construct loop-closed automata with the same behaviour from them. For example, the automaton depicted in Fig. 1 is not loop-closed, but we can construct a loop-closed automaton with the same behaviour, which is depicted in Fig. 3.

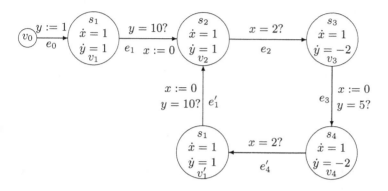

Fig. 3. A loop-closed automaton which has the same behaviour as the automaton in Fig. 1

For a loop-closed automaton $\mathcal{H}$, let ρ be a loop in $\mathcal{H}$ which is of the form $v_1 \xrightarrow{(\phi_1,\psi_1)} v_2 \xrightarrow{(\phi_2,\psi_2)} \ldots \xrightarrow{(\phi_{m-1},\psi_{m-1})} v_m$. We say that ρ is a *zero* loop if for any variable constraint $a \le x \le b$ in ϕ_i $(1 \le i < m)$, $a - d \le 0$ and $b - d \ge 0$ where d is the value of x reset by the reference point of the variable constraint $a \le x \le b$, i.e. d satisfies one of the following conditions:

- x is reset to d by a transition $(v_j, \phi_j, \psi_j, v_{j+1})$ $(1 \le j < i)$, but not reset by any transition $(v_k, \phi_k, \psi_k, v_{k+1})$ $(j < k < i)$; or
- x is reset to d by the transition $(v_{m-1}, \phi_{m-1}, \psi_{m-1}, v_m)$, but not reset by any transition $(v_k, \phi_k, \psi_k, v_{k+1})$ $(1 \le k < i)$.

A loop is called *nonzero* loop if it is not a zero loop. According to the variable constraints on the transitions of a loop, if a loop is a zero loop, then a repetition of the loop may take no time; if a loop is a nonzero loop, then a repetition of the loop must take time. For example, in the automaton depicted in Fig. 2, $v_5 \xrightarrow{(\emptyset,\{x:=0\})} v_6 \xrightarrow{(\{1 \le y \le 5\},\emptyset)} v_7 \xrightarrow{(\{-1 \le x \le 2\},\{y:=1\})} v_5$ and

$$v_1 \xrightarrow{(\{z \ge 5\},\{x:=0\})} v_2 \xrightarrow{(\emptyset,\{z:=2,y:=1\})} v_5 \xrightarrow{(\{z \le 20\},\emptyset)} v_8 \xrightarrow{(\emptyset,\{z:=0\})} v_1$$

are a zero loop, while $v_2 \xrightarrow{(\emptyset,\{y:=0\})} v_3 \xrightarrow{(\{x \ge 5\},\emptyset)} v_4 \xrightarrow{(\{y \ge 3\},\{x:=0\})} v_2$ is a nonzero loop.

For a loop-closed automaton $\mathcal{H}$, let ρ be a loop in $\mathcal{H}$ which is of the form $v_1 \xrightarrow{(\phi_1,\psi_1)} v_2 \xrightarrow{(\phi_2,\psi_2)} \ldots \xrightarrow{(\phi_{m-1},\psi_{m-1})} v_m$. We say that ρ is *free* if it is not constrained by any variable constraint outside ρ, i.e. for any simple path segment

$$u_1 \xrightarrow{(\phi_1', \psi_1')} u_2 \xrightarrow{(\phi_2', \psi_2')} \ldots \xrightarrow{(\phi_{n-1}', \psi_{n-1}')} u_n$$

such that $v_1 = u_1$, for any variable constraint $a \le x \le b$ labeled on the transition $(u_{n-1}, \phi_{n-1}', \psi_{n-1}', u_n)$, x is reset by a transition $(u_i, \phi_i', \psi_i', u_{i+1})$ $(1 \le i < n - 1)$ or by the end transition $(v_{m-1}, \phi_{m-1}, \psi_{m-1}, v_m)$ of ρ. For example, in the automaton depicted in Fig. 2, the nonzero loop

$$v_2 \xrightarrow{(\emptyset, \{y:=0\})} v_3 \xrightarrow{(\{x \ge 5\}, \emptyset)} v_4 \xrightarrow{(\{y \ge 3\}, \{x:=0\})} v_2$$

is free, but the zero loop $v_5 \xrightarrow{(\emptyset, \{x:=0\})} v_6 \xrightarrow{(\{1 \le y \le 5\}, \emptyset)} v_7 \xrightarrow{(\{-1 \le x \le 2\}, \{y:=1\})} v_5$ is not free since it is constrained by the variable constraint $z \le 20$ on the transition e_9.

Definition 4. *A zero loop-closed automaton is a loop-closed automaton in which any nonzero loop is free.* □

For example, the automata depicted in Figs. 2 and 3 are zero loop-closed automata.

Although the definition of zero loop-closed automaton is not simple, we can develop an efficient algorithm to check if a linear hybrid automaton is zero loop-closed, which is described in the appendix. Zero loop-closed automata form a decidable class of linear hybrid automata. We think there are a number of real systems that are of the loop-closed property so that some of them can be modelled by zero loop-closed automata. For example, for control systems, the loop-closed property means that every repetition of a control process starts from the same control conditions. In next section, we will show that the satisfaction problem of zero loop-closed automata for linear duration properties can be solved by linear programming.

4   Checking Zero Loop-Closed Automata for Linear Duration Properties

In this section, we solve the problem of checking zero loop-closed automata for linear duration properties.

The satisfaction problem of linear hybrid automata for linear duration properties are defined as follows. Let $\mathcal{P} = \sum_{i=1}^{n} \int c_i S_i \le M$ be a linear duration property, and $\sigma = (s_1, t_1)\hat{}(s_2, t_2)\hat{}\ldots\hat{}(s_m, t_m)$ be a timed sequence. The integrated duration of S_i over σ can be calculated as $\int S_i = \sum_{j \in \alpha_i} \delta_j$ where $\alpha_i = \{j \mid (0 \le j \le m) \wedge (s_j \Rightarrow S_i)\}$. Let $\theta(\sigma, \mathcal{P}) = \sum_{i=1}^{n} c_i (\sum_{j \in \alpha_i} \delta_j)$. A timed sequence σ satisfies a linear duration property $\mathcal{P}$ if and only if $\theta(\sigma, \mathcal{P}) \le M$. A linear hybrid automaton satisfies a linear duration property if and only if every timed sequence representing its behaviour satisfies the linear duration property.

Let $\mathcal{H} = (Z, X, V, E, v_I, \alpha, \beta)$ be a linear hybrid automaton, and ρ be a sequence of locations of the form $v_1 \xrightarrow{(\phi_1, \psi_1)} v_2 \xrightarrow{(\phi_2, \psi_2)} \ldots \xrightarrow{(\phi_m, \psi_m)} v_{m+1}$. We define

$\tau(\rho)$ to be the set of the timed sequences of the form $(s_1, t_1)\,\hat{}\,(s_2, t_2)\,\hat{}\,\ldots\,\hat{}\,(s_m, t_m)$ satisfying that

- $s_i = \alpha(v_i)$ for each i $(1 \le i \le m)$, and
- $t_1, t_2, \ldots, t_m$ satisfy all the variable constraints in all ϕ_i $(1 \le i \le m)$, i.e. for each variable constraint $a \le x \le b$ in ϕ_i,
  - if there is a reset action $x := c$ in ψ_j $(1 \le j < i)$ and $x := d$ is not in ψ_k for any k $(j < k < i)$, then

$$a \le c + w_{j+1}t_{j+1} + w_{j+2}t_{j+2} + \ldots + w_i t_i \le b;$$

  - if $x := c$ is not in any ψ_j $(1 \le j < i)$ and $x := d$ is in ψ_m, then

$$a \le d + w_1 t_1 + w_2 t_2 + \ldots + w_i t_i \le b;$$

  - if $x := c$ is not in any ψ_j $(1 \le j < i)$ and in ψ_m, then

$$a \le w_1 t_1 + w_2 t_2 + \ldots + w_i t_i \le b,$$

where for each l $(1 \le l \le i)$, $\dot{x} = w_l \in \beta(v_l)$.

We define that a sequence ρ of locations satisfies a linear duration property if every timed sequence in $\tau(\rho)$ satisfies the linear duration property. It follows that a linear hybrid automaton satisfies a linear duration property if its every path satisfies the linear duration property.

Now let us consider to solve the problem for a finite path of a linear hybrid automaton. Let $\rho = v_1 \stackrel{(\phi_1, \psi_1)}{\longrightarrow} v_2 \stackrel{(\phi_2, \psi_2)}{\longrightarrow} \ldots \stackrel{(\phi_m, \psi_m)}{\longrightarrow} v_{m+1}$ be a path in a linear hybrid automaton $\mathcal{H}$, and every timed sequence in $\tau(\rho)$ be of the form $(s_1, t_1)\,\hat{}\,(s_2, t_2)\,\hat{}\,\ldots\,\hat{}\,(s_m, t_m)$. We know that $t_1, t_2, \ldots, t_m$ should satisfy all the variable constraints in all ϕ_i $(1 \le i \le m)$, which forms a group of linear inequalities denoted by C. If the group C of linear inequalities has no solution, then $\tau(\rho) = \emptyset$, which means that there is no timed sequence representing the behaviour of $\mathcal{H}$ corresponding to ρ; otherwise the problem of checking if ρ satisfies a linear duration property $\mathcal{P} = \sum_{i=1}^n \int c_i S_i \le M$ is equivalent to the problem of finding the maximum value of the linear function

$$\sum_{i=1}^n c_i (\sum_{u \in \alpha_i} t_u) \quad \text{where} \quad \alpha_i = \{u \mid (1 \le u \le m) \wedge (s_u \Rightarrow S_i)\}$$

subject to the linear constraint C and checking whether it is not greater than M. The latter is a linear programming problem. So we reduce the problem into a linear programming problem for a finite path of a linear hybrid automaton.

We know that for a linear hybrid automaton, there could be infinite paths and the number of paths could be infinite. So we attempt to solve the problem based on a finite set of finite paths. In the following, we show how to solve the problem for zero loop-closed automata based on a finite set of finite sequences of locations.

Let $\mathcal{H}$ be a linear hybrid automaton, and ρ be a path in $\mathcal{H}$ which is of the form $v_1 \xrightarrow{(\phi_1,\psi_1)} v_2 \xrightarrow{(\phi_2,\psi_2)} \ldots \xrightarrow{(\phi_{m-1},\psi_{m-1})} v_m$. If ρ is not a simple path, then we can find v_i and v_j $(1 \le i < j \le m)$ such that $v_i = v_j$, and then we can get a path ρ_1 which is constructed from ρ by removing any v_k $(i < k \le j)$. By applying the above *elimination* step repeatedly, we can get a simple path ρ'. We say that ρ is an *extension* of ρ'. We define that any simple path is an extension of itself. It is clear that a linear hybrid automaton $\mathcal{H}$ satisfies a linear duration property $\mathcal{P}$ if and only if for any simple path ρ in $\mathcal{H}$, any extension of ρ satisfies $\mathcal{P}$. In the following, for a linear duration property $\mathcal{P}$, for a simple path ρ in a zero loop-closed automaton, we define a sequence of locations such that any extension of ρ satisfies $\mathcal{P}$ if and only if the sequence of locations satisfies $\mathcal{P}$.

For a sequence ρ of locations in a zero loop-closed automaton, for a linear duration property $\mathcal{P}$, by linear programming we can calculate the supremum of the set $\{\theta(\sigma, \mathcal{P}) \mid \sigma \in \tau(\rho)\}$. If the supremum is larger than zero, we say that ρ is *violable* for $\mathcal{P}$, otherwise we say that ρ is not violable for $\mathcal{P}$. Notice that if a sequence ρ of locations is violable for a linear duration property $\mathcal{P}$, by repeating ρ with finite many times we can construct a sequence of locations which does not satisfy $\mathcal{P}$.

Let $\mathcal{H} = (Z, X, V, E, v_I, \alpha, \beta)$ be a linear hybrid automaton, and $\mathcal{P}$ be a linear duration property which is of the form $\sum_{i=1}^{n} \int c_i S_i \le M$. For a location v in $\mathcal{H}$, if there is S_i $(1 \le i \le n)$ such that $\alpha(v) \Rightarrow S_i$ and $c_i > 0$, then we say that v is *positive* for $\mathcal{P}$. Notice that for a zero loop-closed automaton $\mathcal{H}$, for a sequence ρ of locations in $\mathcal{H}$ which is violable for a linear duration property $\mathcal{P}$, there must be a location in ρ which is positive for $\mathcal{P}$.

For two sequences of locations

$$\rho_1 = v_1 \xrightarrow{(\phi_1,\psi_1)} v_2 \xrightarrow{(\phi_2,\psi_2)} \ldots \xrightarrow{(\phi_{m-1},\psi_{m-1})} v_m \text{ and}$$

$$\rho_2 = u_1 \xrightarrow{(\phi_1',\psi_1')} u_2 \xrightarrow{(\phi_2',\psi_2')} \ldots \xrightarrow{(\phi_{n-1}',\psi_{n-1}')} u_n$$

such that $v_m = u_1$, let $\rho_1 \diamond \rho_2$ be the sequence of locations

$$v_1 \xrightarrow{(\phi_1,\psi_1)} v_2 \xrightarrow{(\phi_2,\psi_2)} \ldots \xrightarrow{(\phi_{m-1},\psi_{m-1})} v_m \xrightarrow{(\phi_1',\psi_1')} u_2 \xrightarrow{(\phi_2',\psi_2')} \ldots \xrightarrow{(\phi_{n-1}',\psi_{n-1}')} u_n.$$

For a linear hybrid automaton, we can find all loops using depth-first search method whose algorithm is described in the appendix. According to the order that the loops are found out, for any loop ρ we let $o(\rho)$ be a integer which represents the position of ρ in the order.

Let $\rho = v_1 \xrightarrow{(\phi_1,\psi_1)} v_2 \xrightarrow{(\phi_2,\psi_2)} \ldots \xrightarrow{(\phi_{m-1},\psi_{m-1})} v_m$ be a loop in a zero loop-closed automaton, ρ_1 be a loop-enter path of ρ, and for each i $(1 < i < m)$, $\rho_i = \rho_1 \xrightarrow{(\phi_1,\psi_1)} v_2 \xrightarrow{(\phi_2,\psi_2)} \ldots \xrightarrow{(\phi_{i-1},\psi_{i-1})} v_i$. For any linear duration property $\mathcal{P} = \sum_{i=1}^{n} \int c_i S_i \le M$, let $\mu(\rho_1, \rho, \mathcal{P})$ be a sequence of locations, which is defined recursively as follows:

– if ρ is a nonzero loop, then

$$\mu(\rho_1, \rho, \mathcal{P}) = \begin{cases} v_1 & \text{if } \rho' \text{ is not violable for } \mathcal{P} \\ v_1 \xrightarrow{(\emptyset,\emptyset)} v \xrightarrow{(\emptyset,\psi_{m-1})} v_m & \text{if } \rho' \text{ is violable for } \mathcal{P} \end{cases},$$

where v is a location in ρ' which is positive for $\mathcal{P}$ and

$$\rho' = v_1 \xrightarrow{(\phi_1,\psi_1)} w_2 \xrightarrow{(\phi_2,\psi_2)} w_3 \xrightarrow{(\phi_3,\psi_3)} \ldots \xrightarrow{(\phi_{m-2},\psi_{m-2})} w_{m-1} \xrightarrow{(\phi_{m-1},\psi_{m-1})} v_m$$

where for each i $(1 < i < m)$, if there is not any loop which can be entered through ρ_i, then $w_i = v_i$, otherwise

$$w_i = \mu(\rho_i,\rho_{i1},\mathcal{P}) \diamond \mu(\rho_i,\rho_{i2},\mathcal{P}) \diamond \ldots \diamond \mu(\rho_i,\rho_{in_i},\mathcal{P})$$

where $\rho_{i1},\rho_{i2},\ldots,\rho_{in_i}$ are all the loops which can be entered through ρ_i and which are such that $o(\rho_{ij}) < o(\rho_{ij+1})$ for any j $(1 \le j < n_i)$;
- if ρ is a zero loop, then

$$\mu(\rho_1,\rho,\mathcal{P})$$
$$= v_1 \xrightarrow{(\phi'_1,\psi_1)} w_2 \xrightarrow{(\phi'_2,\psi_2)} w_3 \xrightarrow{(\phi'_3,\psi_3)} \ldots \xrightarrow{(\phi'_{m-2},\psi_{m-2})} w_{m-1} \xrightarrow{(\phi'_{m-1},\psi_{m-1})} v_m$$

where
- for each i $(1 < i < m)$, if there is not any loop which can be entered through ρ_i, then $w_i = v_i$, otherwise

$$w_i = \mu(\rho_i,\rho_{i1},\mathcal{P}) \diamond \mu(\rho_i,\rho_{i2},\mathcal{P}) \diamond \ldots \diamond \mu(\rho_i,\rho_{in_i},\mathcal{P})$$

where $\rho_{i1},\rho_{i2},\ldots,\rho_{in_i}$ are all the loops which can be entered through ρ_i and which are such that $o(\rho_{ij}) < o(\rho_{ij+1})$ for any j $(1 \le j < n_i)$, and
- for each i $(1 < i < m)$, $\phi'_i = \phi_{i1} \cup \phi_{i2} \cup \phi_{i3}$,

$$\phi_{i1} = \{d \le x \mid a \le x \le b \in \phi_i \text{ and } a = d\},$$
$$\phi_{i2} = \{x \le d \mid a \le x \le b \in \phi_i \text{ and } b = d\},$$
$$\phi_{i3} = \{x = d \mid a \le x \le b \in \phi_i \text{ and } a = b = d\},$$

where d is such that $x := d \in \psi_j$ $(1 \le j < i)$ and $x := c \notin \psi_k$ for any k $(j < k < i)$ or that $x := d \in \psi_{m-1}$ $(1 \le j < i)$ and $x := c \notin \psi_k$ for any k $(1 \le k < i)$.

Let ρ be a simple path of a zero loop-closed automaton which is of the form $v_1 \xrightarrow{(\phi_1,\psi_1)} v_2 \xrightarrow{(\phi_2,\psi_2)} \ldots \xrightarrow{(\phi_{m-1},\psi_{m-1})} v_m$, and for each i $(1 < i < m)$, $\rho_i = v_1 \xrightarrow{(\phi_1,\psi_1)} v_2 \xrightarrow{(\phi_2,\psi_2)} \ldots \xrightarrow{(\phi_{i-1},\psi_{i-1})} v_i$. For any linear duration property $\mathcal{P}$, let $\omega(\rho,\mathcal{P})$ be a sequence of locations, which is defined as follows:

$$\omega(\rho,\mathcal{P}) = v_1 \xrightarrow{(\phi_1,\psi_1)} w_2 \xrightarrow{(\phi_2,\psi_2)} w_3 \xrightarrow{(\phi_3,\psi_3)} \ldots \xrightarrow{(\phi_{m-2},\psi_{m-2})} w_{m-1} \xrightarrow{(\phi_{m-1},\psi_{m-1})} v_m$$

where for each i $(1 < i < m)$, if there is not any loop which can be entered through ρ_i, then $w_i = v_i$, otherwise

$$w_i = \mu(\rho_i,\rho_{i1},\mathcal{P}) \diamond \mu(\rho_i,\rho_{i2},\mathcal{P}) \diamond \ldots \diamond \mu(\rho_i,\rho_{in_i},\mathcal{P})$$

where $\rho_{i1},\rho_{i2},\ldots,\rho_{in_i}$ are all the loops which can be entered through ρ_i and which are such that $o(\rho_{ij}) < o(\rho_{ij+1})$ for any j $(1 \le j < n_i)$.

Lemma 1. *Let $\mathcal{H}$ be a zero loop-closed automaton, and $\mathcal{P}$ be a linear duration property. For any simple path ρ in $\mathcal{H}$, any extension of ρ satisfies $\mathcal{P}$ if and only if $\omega(\rho, \mathcal{P})$ satisfies $\mathcal{P}$.* $\quad\square$

The detailed proof of Lemma 1 is omitted here because of space consideration. By Lemma 1, we can get the following theorem.

Theorem 1. *A zero loop-closed automaton $\mathcal{H}$ satisfies a linear duration property $\mathcal{P}$ if and only if for each simple path ρ of $\mathcal{H}$, $\omega(\rho, \mathcal{P})$ satisfies $\mathcal{P}$.* $\quad\square$

$currentpath := \langle v_I \rangle$;
repeat
    $node :=$ the last node of $currentpath$;
    if $node$ has no new successive node
    then delete the last node of $currentpath$
    else begin
            $node :=$ a new successive node of $node$;
            construct $\omega(\rho, \mathcal{P})$ for the current path ρ;
            check if $\omega(\rho, \mathcal{P})$ satisfies $\mathcal{P}$;
            if no, **then return false**;
            if $node$ is not in $currentpath$
            then append $node$ to $currentpath$;
    end
until $currentpath = \langle \rangle$;
return true.

Fig. 4. Algorithm checking zero loop-closed automata for linear duration properties

---

Based on Theorem 1, we can develop an efficient algorithm to check if a zero loop-closed automaton $(X, V, E, v_I, \alpha, \beta)$ satisfies a linear duration property. The algorithm is based on depth-first search method and described in Fig. 4. The main data structure in the algorithm is a list $currentpath$ of locations which is used to record the current paths. We traverse all simple paths to check if there is a simple path ρ such that $\omega(\rho, \mathcal{P})$ does not satisfy $\mathcal{P}$. In the algorithm, we need to solve linear programs. Linear programming is well studied, and can be solved with a polynomial-time algorithm in general. The number of the linear programs we need to solve equals the number of all nonzero loops and simple paths in the automaton. The number of variables in each linear program is not larger than the numbers of the locations in the longest simple paths and in all the loops.

5   Conclusion

In this paper, we have shown that for a class of linear hybrid automata called zero loop-closed automata, the satisfaction problem for linear duration properties can be solved efficiently by linear programming. In general the model checking problem is undecidable for the class of linear hybrid automata. This paper gives a new result for the decidability of the model checking problem because the class of zero loop-closed automata is not contained by the decidable classes of hybrid systems we have found in the literature so far. In [2], the decidability of a class of linear hybrid systems called *integration graphs* is reduced to the verification problem for timed automata. In integration graphs, it is not allowed to test a variable in a loop which has different change rate in different locations. So the class of integration graphs does not contain that of zero loop-closed automata. In [4], a class of hybrid automata, *initialized rectangular automata*, are proved to be decidable for linear temporal logic (LTL) requirements. A symbolic method is presented in [7] such that the tool HYTECH [8] which runs a symbolic procedure can terminate on initialized rectangular automata. Any initialized rectangular automaton requires that any variable must be reset when its change rate is changed. So the class of zero loop-closed automata is not contained by that of initialized rectangular automata. In [3], an automatic approach, which attempt to construct the reachable region by symbolic execution, has been presented. But the procedures often do not terminate.

The idea to check linear duration properties by linear programming comes from [6] in which the problem for real-time automata is solved by linear programming technique, which is well established. By developing the techniques in [6], we show that by linear programming the problem can be solved totally for a class of linear hybrid automata in [9,10]. In [9,10], we describe the decidable class of linear hybrid automata by using an extension of regular expressions with time constraints, but do not give any direct definition of the decidable hybrid automata, and the presented algorithm is of high complexity in some case because we need to unfold loops so that the number and size of the linear programs we need to solve become very large. Compared with the work in [9,10], this paper makes two new contributions: First the class of zero loop-closed automata presented this paper is not included by the decidable class of linear hybrid automata defined in [9,10], and secondly the approach in this paper is based on automata directly and leads itself to an efficient implementation.

The algorithm presented in this paper has been implemented. We think there are a number of real systems that are of the loop-closed property (for example, for control systems, the loop-closed property means that every repetition of a control process starts from the same control conditions). An important topic for future work is to do case studies in practical use.

References

1. Thomas A. Henzinger. The theory of hybrid automata. In *Proceedings of the 11th Annual IEEE Symposium on Logic in Computer Science (LICS 1996)*, pp. 278-292.
2. Y. Kesten, A. Pnueli, J. Sifakis, S. Yovine. Integration Graphs: A Class of Decidable Hybrid Systems. In *Hybrid System, LNCS 736*, pp.179-208.

3. R. Alur, C. Courcoubetis, N. Halbwachs, T.A. Henzinger, P.-H.Ho, X. Nicollin, A. Olivero, J. Sifakis, S. Yovine. The algorithmic analysis of hybrid systems. In *Theoretical Computer Science*, 138(1995), pp.3-34.
4. Thomas A. Henzinger, Peter W. Kopke, Anuj Puri, and Pravin Varaiya. What's Decidable About Hybrid Automata? In *Journal of Computer and System Sciences*, 57:94-124, 1998.
5. Zhou Chaochen, C.A.R. Hoare, A.P. Ravn. A Calculus of Durations. In *Information Processing Letter*, 40, 5, 1991, pp.269-276.
6. C. Zhou, J. Zhang, L. Yang, and X. Li. Linear Duration Invariants. In *Formal Techniques in Real-Time and Fault-Tolerant Systems, LNCS 863*, pp.88-109.
7. Thomas A. Henzinger, Rupak Majumdar. Symbolic Model Checking for Rectangular Hybrid Systems. In *Proceedings of the Sixth Workshop on Tools and Algorithms for the Construction and Analysis of Systems (TACAS 00)*, Lecture Notes in Computer Science, Springer, 2000.
8. T.A. Henzinger, P.-H. Ho, and H. Wong-Toi. HYTECH: a model checker for hybrid systems. In *Software Tools for Technology Transfer*, 1:110-122, 1997.
9. Li Xuandong, Dang Van Hung, and Zheng Tao. Checking Hybrid Automata for Linear Duration Invariants. In *Advances in Computing Science - ASIAN'97, LNCS 1345*, Springer-Verlag, 1997, pp.166-1180.
10. Li Xuandong, Zheng Tao, Hou Jianmin, Zhao Jianhua, and Zheng Guoliang. Hybrid Regular Expressions. In *Hybrid Systems: Computation and Control, LNCS 1386*, Springer-Verlag, 1998, pp.384-399.

A   Algorithm to Check if a Linear Hybrid Automaton is Zero Loop-Closed

An efficient algorithm is described in Fig. 5, which is to check if a linear hybrid automaton $(X, V, E, v_I, \alpha, \beta)$ is zero loop-closed. The algorithm is based on depth-first search method. The main data structure in the algorithm includes a list *currentpath* of locations which is used to record the current paths, and a set *loopset* of loops which records all the loops in the automaton. The algorithm consists of three steps. First, we find out all loops and check if any simple path is such that any loop satisfies that for any variable constraint in the loop, its reference point is in the loop. Then we check if any loop is such that any other loop with the same loop-start node satisfies that for any variable constraint in the loop, its reference point is in the loop. Last, for any loop, from the loop-start node we traverse all simple path segment to check if any simple path segment satisfies that for any variable constraint outside a loop, its reference point is not inside the loop; and if any simple path segment is such that any nonzero loop is free. The complexity of the algorithm is proportional to the number of the simple paths and the size of the longest simple path in an automaton.

$currentpath := \langle v_I \rangle$; $loopset := \emptyset$;
repeat
    $node :=$ the last node of $currentpath$;
    if $node$ has no new successive node
    then delete the last node of $currentpath$
    else begin
        $node :=$ a new successive node of $node$;
        if $node$ is in $currentpath$ (we discover a loop)
        then begin
            put the loop into $loopset$;
            check if the current path is such that the loop satisfies
            that for any variable constraint in the loop,
            its reference point is in the loop;
            if no (the loop is not closed), **then return false**;
        end
        else append $node$ to $currentpath$;
    end
until $currentpath = \langle \rangle$;

---

for any loop in $loopset$ **do**
    begin check if the loop is such that any other loop with the same
        loop-start node satisfies that for any variable constraint
        in the loop, its reference point is in the loop;
        if no, **then return false**;
    end;

---

for any loop in $loopset$ with a loop-start node v **do**
    begin $currentpath := \langle v \rangle$;
    repeat
        $node :=$ the last node of $currentpath$;
        if $node$ has no new successive node
        then delete the last node of $currentpath$
        else begin
            $node :=$ a new successive node of $node$;
            check if the current path satisfies that
            for any variable constraint outside a loop,
            its reference point is not inside the loop;
            if no, **then return false**;
            check if the current path is such that any nonzero loop is free;
            if no, **then return false**;
            if $node$ is not in $currentpath$
            then append $node$ to $currentpath$;
        end
    until $currentpath = \langle \rangle$;
    end;
return true.

Fig. 5. Algorithm for checking if a linear hybrid automaton is zero loop-closed

Author Index

Lecture Notes in Computer Science

For information about Vols. 1–2086
please contact your bookseller or Springer-Verlag